DATE DUE

Spatial Accuracy Assessment: Land Information Uncertainty in Natural Resources

Edited by

Kim Lowell
Annick Jaton

Ann Arbor Press
Chelsea, Michigan

Library of Congress Cataloging-in-Publication Data

Spatial accuracy assessment : land information uncertainty in natural resources / edited by Kim Lowell and Annick Jaton.
    p.   cm.
    Based on papers presented at the Third International Symposium on Spatial Accuracy Assessment in Natural Resources and Environmental Sciences in Quebec City, Canada. May 20–22, 1998.
    ISBN 1-57504-119-7
    1. Geographic information systems—Congresses. 2. Spatial analysis (Statistics)—Congresses. 3. Natural resources—Management—Statistical methods—Congresses. I. Lowell, Kim. II. Jaton, Annick. III. International Symposium on Spatial Accuracy Assessment in Natural Resources and Environmental Sciences (3rd : 1998 : Quebec City, Quebec)
G70.212.S636   1999
910′.285—dc21   99-18424
                                              CIP

ISBN 1-57504-119-7

ANN ARBOR PRESS
121 South Main Street, Chelsea, Michigan 48118
Ann Arbor Press is an imprint of Sleeping Bear Press, Inc.

PRINTED IN THE UNITED STATES OF AMERICA
10 9 8 7 6 5 4 3 2 1

Acknowledgments

Organizing a conference requires the support of a number of different groups. The editors of this volume gratefully acknowledge the support of the following individuals and organizations.

Conference Steering Committee

Michael Goodchild, NCGIA, University of California, Santa Barbara, California 93106-4060 United States
Annick Jaton, Centre de recherche en géomatique, Université Laval, Québec City, Québec G1K 7P4 Canada
Adam Lewis, Department of Tropical Environment Studies and Geography, James Cook University, Townsville 4819, Australia
Kim Lowell, Centre de recherche en géomatique, Université Laval, Québec City, Québec G1K 7P4 Canada
Todd Mowrer, U.S.D.A. Forest Service, 240 W. Prospect, Fort Collins, Colorado 80526-2098 United States

Scientific Advisory Board

Kate Beard
Joe Berry
Ray Czaplewski
Don Edwards
Mike Goodchild
Edward Kennedy
Barbara Koch
Michael Köhl
Adam Lewis
Kim Lowell

H. Gyde Lund
Ann Maclean
Todd Mowrer
Sue Nichols
G.P. Patil
Jocclyn Prince
Steve Prisley
J.P. Skovsgaard
Risto Pavinen

Sponsoring Organizations

American Society of Photogrammetry and Remote Sensing—GIS Division
American Statistical Association
Biometrics Society
Canadian Institute of Geomatics
Ecological Society of America—Section on Statistical Ecology
Geomatics Industry Association of Canada
Interface Foundation of North America
International Association of Ecology
International Union of Forestry Research Organizations (IUFRO) 4.01—Mensuration, Growth, and Yield
IUFRO 4.02—Forest Resource Inventory
IUFRO 4.11—Statistical Methods
IUFRO 4.12—Remote Sensing/GIS
National Centre for Geographic Information and Analysis
Society of American Foresters

Preface

From May 20 to May 22, 1998, some 150 people attended the Third International Symposium on Spatial Accuracy Assessment in Natural Resources and Environmental Sciences in Québec City, Canada. The first symposium in this series occurred in 1994 (in Williamsburg, Virginia) and was organized by a number of individuals from various disciplines who recognized that professionals in natural resources were beginning to be aware of the presence of, and problems caused by, spatial uncertainty in natural resources databases. The organizers of the first symposium believed that general awareness of spatial uncertainty in the natural resources community had become elevated and widespread enough in 1994 that it would be useful to provide a forum for the exchange of ideas on the subject. At the three symposiums held thus far (the second being held in 1996 in Fort Collins, Colorado), multidisciplinarity has been a key focus of the meeting, with professionals from natural resources domains such as forestry, wildlife biology, and hydrology mixing with individuals from other areas such as geography, statistics, and computer science. Moreover, the international flavor of the symposium has remained, with presenters and attendees coming from the four corners of the world. What all people attending have in common is an interest in spatial uncertainty in natural resources databases.

At the three symposiums held thus far, the editors of this volume have noticed that very few attendees claim to have "solved" a particular problem. Rather, those who make presentations do so almost cautiously, hoping that what is being presented makes at least a small contribution to the advancement of knowledge in the field and is pertinent for at least one other attendee. Similarly, conversations during coffee breaks seem to constantly find people scratching their heads at the complexity of a particular problem, with participants in the conversation grateful for *any* insight into how to resolve it. Often during these conversations someone will delightedly exclaim "I never thought of that!" when someone else has provided them a useful suggestion as to how to advance past what previously seemed to be a scientific dead end. It is a stimulating and fascinating dynamic.

This spirit of openness, questing, and, well...uncertainty about problems encountered is apparent in the papers submitted to the third symposium. The present volume is a record of all presentations made at the third symposium for which an accompanying article was written. These articles represent the state of research on, and knowledge about, spatial uncertainty in natural resource databases as of 1998. It is notable that in almost every paper, authors have taken pains to describe not just possible applications and uses of what they have accomplished, but also to clarify relevant limitations. In examining these clarifications it becomes obvious that we do not yet have a clear idea of all the sources of spatial uncertainty resource databases, nor of how to quantify such uncertainty, nor of how to represent it in a computer, nor of how to communicate it to a human user. Nonetheless, it is encouraging to see the diversity of work being undertaken in this domain, and to be able to take stock of what it is we do know about spatial uncertainty in 1998, rather than all the things that we do not know.

In closing, it has been a very positive experience to have organized the third symposium and to be able to contribute in some small way to advancing our understanding of spatial uncertainty in natural resources databases. We are grateful to all those who contributed to making this third conference a success—speakers, session moderators, article reviewers, sponsoring organizations...the list goes on and on. We would be remiss, however, if we did not specifically mention those individuals from the Centre for Research in Geomatics of Laval University who edited papers, sent out calls for papers, helped with registration, etc. Without your help, this conference would not have attained the success it did.

Merci beaucoup à toutes et à tous.

Kim Lowell and Annick Jaton

About the Editors

Kim Lowell has a Ph.D. in forest biometrics from Canterbury University in Christchurch, New Zealand. He has worked actively with spatial data and geomatic tools for over ten years and has interests specifically in the practical implications of spatial uncertainty in natural resource management. He served as the president of the Third Symposium on Spatial Data Accuracy in Natural Resource Data Bases and is among a handful of people who have now attended all three symposiums. He currently is employed as a full professor at Laval University in Québec City, Canada and is the Director of the Center for Research in Geomatics at Laval University.

Annick Jaton has an M.Sc. in remote sensing from Sherbrooke University. She has worked with geomatic technologies for some ten years and during that time has provided critical leadership and assistance in a number of provincial and national geomatic initiatives. She coordinated all aspects of the Third Symposium on Spatial Data Accuracy in Natural Resource Data Bases, oversaw the organization of the proceedings, and even developed the design of the cover of these proceedings (which is no ordinary world map, as you will see if you look closely). She currently holds the position of Research Coordinator of the Center for Research in Geomatics at Laval University.

Contributors

Aggrey Agumya
Centre for GIS and Modelling
Department of Geomatics
The University of Melbourne
Parkville, Victoria 3052
Australia

Ali A. Alesheikh
Department of Geomatics Engineering
The University of Calgary
2500 University Dr., N.W.
Calgary, AB
Canada

Rodney C. Allan
RMIT Centre for Remote Sensing and GIS
Department of Land Information
RMIT University
124 Latrobe Street
Melbourne, Victoria 3000
Australia

Giuseppe Arbia
Dipartimento di Metodi Quantitativi e Teoria
    Economica
Facolta di Economia
University "G. D'Annunzio"
Viale Pindaro, 42
65127 Pescara
Italy

Mounir Azouzi
Geodetic Engineering Laboratory
Institute of Geomatics
Swiss Federal Institute of Technology, Lausanne
CH-1015 Lausanne
Switzerland

David J.B. Baldwin
Ontario Forest Research Institute
1235 Queen Street East
Sault Ste. Marie, Ontario P6A 2E5
Canada

Lawrence E. Band
Department of Geography
University of Toronto
100 St. George St.
Toronto, Ontario M5S 3G3
Canada

L. Bastin
Department of Geography
University of Leicester
Leicester LE1 7RH
United Kingdom

Pierre Y. Bernier
Canadian Forestry Service
Laurentian Forestry Centre
1055 du P.E.P.S., P.O. Box 3800
Sainte-Foy, Québec G1V 4C7
Canada

J.A.R. Blais
Department of Geomatics Engineering
The University of Calgary
2500 University Dr., N.W.
Calgary, AB
Canada

P.E. Bonhomme
Natural Resources Canada
Canadian Forest Service
Laurentian Forestry Centre
1055 du P.E.P.S, P.O. Box 3800
Sainte-Foy, Québec G1V 4C7
Canada

Daniel G. Brown
Assistant Professor of Geography
Michigan State University
East Lansing, MI 48824

Naomi Burden
Department of Computer Science
University of Glasgow
Glasgow G12 8QQ
United Kingdom

J. Bygrave
Department of Geography
University of Leicester
Leicester LE1 7RH
United Kingdom

Frank Canters
Centre for Cartography and GIS
Department of Geography
Brussels Free University (V.U.B.)
Pleinlaan 2
B-1050 Brussels
Belgium

Vincent Chaplot
ENSAR
Soil Science Laboratory
65 rue de Saint Brieuc
35042 Rennes
France

Arthur D. Chapman
Environmental Resources Information Network
Environment Australia
P.O. Box 787
Canberra ACT 2601
Australia

M.A. Chapman
Department of Geomatics Engineering
The University of Calgary
2500 University Dr., N.W.
Calgary, AB
Canada

Nicholas R. Chrisman
Associate Professor
Department of Geography
Box 353550
University of Washington
Seattle, WA 98195-3550

Keith C. Clarke
National Center for Geographic Information &
    Analysis and Department of Geography
University of California
Santa Barbara, CA 93106

Peter Cowell
Coastal Studies Unit
School of Geosciences
University of Sydney, NSW 2006
Australia

Noel Cressie
Department of Statistics
Iowa State University
Ames, Iowa 50011-1210

F. Csillag
Department of Geography
University of Toronto at Mississauga
3359 Mississauga Road
Mississauga, ON L5L 1C6
Canada

Pierre Curmi
INRA, Soil Science and Bioclimatology Unity
Soil Science Laboratory
65 rue de Saint Brieuc
35042 Rennes
France

Kevin Curtin
University of California
Santa Barbara, CA 93106-4060

Raymond L. Czaplewski
U.S. Forest Service
Rocky Mountain Research Station
240 W. Prospect Road
Fort Collins, CO 80526

P. Defourny
Department of Environmental Sciences
Université Catholique de Louvain
Place Croix du Sud, 2 Bte 16
1348 Louvain La Neuve
Belgium

William De Genst
Centre for Cartography and GIS
Department of Geography
Brussels Free University (V.U.B.)
Pleinlaan 2
B-1050 Brussels
Belgium

T. De Groeve
Centre de Recherche en Géomatique (CRG)
Pavillon Casault 0722
Université Laval, G1K 7P4

Jane Drummond
Department of Geography and Topographic Science
University of Glasgow
Glasgow G12 8QQ
United Kingdom

Matt Duckham
Department of Geography and Topographic Science
University of Glasgow
Glasgow G12 8QQ
United Kingdom

Jiunn-Der Duh
Graduate Student, Geography
Michigan State University
East Lansing, MI 48824

Robert S. Dzur
Center for Advanced Spatial Technologies
University of Arkansas
Ozark Hall, Room 12
Fayetteville, AR 72701

Geoffrey Edwards
Chaire Industrielle en Géomatique Appliquée à la
    Foresterie
Centre de Recherche en Géomatique
Pavillon Casault, Université Laval
Sainte-Foy, Québec, G1K 7P4
Canada

Herman Eerens
Centre for Teledetection and Atmospheric Processes
Flemish Institute for Technological Research (VITO)
Boeretang 200
B 2400 Mol
Belgium

Peter L. Fellows
Department of Geography
Maxwell School of Citizenship and Public Affairs
Syracuse University
144 Eggers Hall
Syracuse, NY 13244-1090

P.F. Fisher
Department of Geography
University of Leicester
Leicester LE1 7RH
United Kingdom

Antonio Floris
Research Technician
Forest Management Departement
Forest and Range Management Research Institute
    (ISAFA)
Piazza Nicolini
6-38050 Villazzano di Trento – I
Italy

Peter Fohl
National Center for Geographic Information and
    Analysis, and
Department of Geography
University of California
Santa Barbara, CA 93106-4060

Bruce C. Forster
School of Geomatic Engineering
The University of New South Wales
Sydney 2052
Australia

Michelle Fortin
Centre de Recherche en Géomatique
Pavillon Casault, Université Laval
Sainte-Foy, Québec, G1K 7P4
Canada

Richard A. Fournier
Natural Resources Canada
Canadian Forest Service
Laurentian Forestry Centre
1055 du P.E.P.S, P.O. Box 3800
Sainte-Foy, Québec G1V 4C7
Canada

Chris Funk
University of California
Santa Barbara, CA 93106-4060

John Gabrosek
Department of Statistics
Iowa State University
Ames, Iowa 50011-1210

Jonathon Gascoigne
Benfield Greig Ltd.
Devon House
58-60 St. Katherine's Way
London E1 9LB
United Kingdom

Dean B. Gesch
Raytheon STX Corporation
EROS Data Center
U.S. Geological Survey
Sioux Falls, SD 57198

Michael F. Goodchild
National Center for Geographic Information &
    Analysis, and Department of Geography
University of California
Santa Barbara, CA 93106-4060

Pierre Goovaerts
Professor
Department of Civil and Environmental Engineering
The University of Michigan
EWRE Bldg., Room 117
Ann Arbor, MI 48109-2125

Jonathan Gottsegen
National Center for Geographic Information and
    Analysis
Department of Geography
University of California, Santa Barbara
Santa Barbara, CA 93106-4060

Philip Gray
Department of Computer Science
University of Glasgow
Glasgow G12 8QQ
United Kingdom

Daniel A. Griffith
Department of Geography
Maxwell School of Citizenship and Public Affairs
Syracuse University
144 Eggers Hall
Syracuse, NY 13244-1090

Robert P. Haining
Department of Geography
University of Sheffield
Sheffield S10 2TN
United Kingdom

R.N. Handcock
Department of Geography
University of Toronto at Mississauga
3359 Mississauga Road
Mississauga, ON L5L 1C6
Canada

G. Hecquet
Department of Environmental Sciences
Université Catholique de Louvain
Place Croix du Sud, 2 Bte 16
1348 Louvain La Neuve
Belgium

Rachel Riemann Hershey
USDA Forest Service
Northeastern Forest Experiment Station
5 Radnor Corporate Center, Suite 200
Radnor, PA 19087-4585

Gerard B.M. Heuvelink
Faculty of Environmental Sciences
University of Amsterdam
Nieuwe Prinsengracht 130
1018 VZ Amsterdam
The Netherlands

Hsin-Cheng Huang
Institute of Statistical Science
Academia Sinica
Taipei 115
Taiwan

Yanping Huang
Centre de Recherche en Géomatique
Université Laval
Sainte-Foy, Québec G1K 7P4
Canada

M. Hughes
Department of Geography
University of Leicester
Leicester LE1 7RH
United Kingdom

Gary J. Hunter
Centre for GIS and Modelling
Department of Geomatics
The University of Melbourne
Parkville, Victoria 3052
Australia

Wolfgang Jacquet
Statistics Research Unit
Department of Mathematics
Brussels Free University (V.U.B.)
Pleinlaan 2
B-1050 Brussels
Belgium

H. Karimi
North Carolina Supercomputing Center

Mathias J.P.M. Lemmens
Delft University of Technology
Faculty of Civil Engineering and Geo-Sciences
Subfaculty of Geodetic Engineering
Division of GIS-Technology
Thijsseweg 11, 2629 JA Delft
The Netherlands

Adam Lewis
Lecturer
School of Tropical and Environmental Studies and
    Geography
James Cook University
Townsville, Queensland 4811
Australia

Lihua Li
School of Geomatic Engineering
The University of New South Wales
Sydney 2052
Australia

K.E. Lowell
Centre de Recherche en Géomatique (CRG)
Pavillon Casault
Université Laval
Québec (QC) G1K 7P4
Canada

S. Mitchell
Department of Geography
University of Toronto at Mississauga
3359 Mississauga Road
Mississauga, ON L5L 1C6
Canada

Melinda Moeur
USDA Forest Service
Rocky Mountain Research Station
1221 South Main Street
Moscow, ID 83842

Daniel Montello
National Center for Geographic Information and
    Analysis
Department of Geography
University of California, Santa Barbara
Santa Barbara, CA 93106-4060

H. Todd Mowrer
USDA Forest Service
Rocky Mountain Research Station
Fort Collins, Colorado

Wayne L. Myers
Associate Professor
The Pennsylvania State University
Room 124, Land & Water Research Bldg.
University Park, PA 16802

Val Noronha
Digital Geographic Research Corporation

Ajith H. Perera
Ontario Forest Research Institute
1235 Queen Street East
Sault Ste. Marie, Ontario P6A 2E5
Canada

T. Philippart
Department of Environmental Sciences
Université Catholique de Louvain
Place Croix du Sud, 2 Bte 16
1348 Louvain La Neuve
Belgium

Jibo Qiu
Centre for GIS and Modelling
Department of Geomatics
University of Melbourne
Parkville, Victoria 3052
Australia

Jacques Régnière
Canadian Forestry Service
Laurentian Forestry Centre
1055 du P.E.P.S., P.O. Box 3800
Sainte-Foy, Québec G1V 4C7
Canada

Keith Rennolls
University of Greenwich
London SE18 6PF
United Kingdom

Gianfranco Scrinzi
Scientific Director
Forest Management Department
Forest and Range Management Research Institute
    (ISAFA)
Piazza Nicolini
6-38050 Villazzano di Trento – I
Italy

Stephen V. Stehman
College of Environmental Science and Forestry
State University of New York (SUNY)
320 Bray Hall, 1 Forestry Dr.
Syracuse, NY 13210

Ashton M. Shortridge
National Center for Geographic Information and
    Analysis, and
Department of Geography
University of California
Santa Barbara, CA 93106-4060

David Tait
Department of Geography and
    Topographic Science
University of Glasgow
Glasgow G12 8QQ
United Kingdom

B. Thierry
Centre de Recherche en Géomatique
Pavillon Casault
Université Laval
Québec (QC) G1K 7P4
Canada

Keith P.B. Thomson
Centre de Recherche en Géomatique
Pavillon Casault 0722
Université Laval
Sainte-Foy, Québec, G1K 7P4
Canada

Chhun-Huor Ung
Natural Resources Canada
Canadian Forestry Service
Laurentian Forestry Centre
1055 du P.E.P.S., P.O. Box 3800
Sainte-Foy, Québec G1V 4C7
Canada

Dr. François Vauglin
COGIT Laboratory
Institut Géographique National
2 avenue Pasteur
94160 Saint-Mandé
France

Frank Veroustraete
Centre for Teledetection and Atmospheric Processes
Flemish Institute for Technological Research (VITO)
Boeretang 200, B 2400 Mol
Belgium

Alain A. Viau
Centre de Recherche en Géomatique
Université Laval
Sainte-Foy, Québec G1K 7P4
Canada

Richard Wadsworth
Institute of Terrestrial Ecology
Monks Wood, Abbots Ripton
Huntingdon, Cambs PE17 2LS
United Kingdom

Christian Walter
ENSAR
Soil Science Laboratory
65 rue de Saint Brieuc
35042 Rennes
France

J. Wood
Department of Geography
University of Leicester
Leicester LE1 7RH
United Kingdom

Limin Yang
Raytheon STX Corporation
EROS Data Center
Sioux Falls, SD 57198

Thomas Q. Zeng
Coastal Studies Unit
School of Geosciences
University of Sydney, NSW 2006
Australia

Zhi-Liang Zhu
Raytheon STX Corporation
EROS Data Center
Sioux Falls, SD 57198

Contents

Part V
Spatial Uncertainty Methods

Part VI
Generalization and Aggregation

Part VII
Decreasing Spatial Uncertainty

Part VIII
Decomposing Digital Images to Improve Classification and Describe Uncertainty

Part IX
Characterizing and Obtaining Spatial Uncertainty Information for Specific Situations

Section A

Section B

Section C

Section D

Spatial Accuracy Assessment: Land Information Uncertainty in Natural Resources

Part I
Keynote Speeches

Each day of the three-day conference began with a keynote address from invited speakers before breaking into concurrent sessions. These keynote speakers were chosen because of their long-time work with spatial uncertainty, the innovation of their work, and also because their approaches to problems of spatial uncertainty are indicative of the diversity of work that is ongoing in the domain.

T. Mowrer, who spoke on the first day of the conference, was among the individuals who were involved in the first symposium in 1994. His keynote speech addressed the need of being able to demonstrate and communicate the importance of spatial uncertainty to "nonbelievers." That is, those who currently work in the domain understand already that spatial uncertainty can have pronounced yet hidden impacts on decisions concerning public and private lands when those decisions are made with the aid of spatial decision support tools. In his chapter, Dr. Mowrer discusses ways of communicating this.

D. Griffith, the second keynote speaker, has worked in the area of spatial statistics for a number of years. Dr. Griffith chose to address the uncertainty that can arise when two or more spatial data sources are combined. His chapter discusses the nature of error and uncertainty resulting from operations commonly performed on geographic databases—overlaying and buffering, for example—and provides a set of research questions that should help guide future research in spatial uncertainty.

N. Chrisman, the final keynote speaker, was among the first individuals to develop a model of spatial uncertainty—the epsilon band of his 1982 doctoral thesis. Dr. Chrisman's keynote address and accompanying paper touch on a number of philosophical issues. Principal among these is the question "What are geographic data?" In this chapter, the point is made that people working with geographic information systems (GIS) tend to see spatial data as a collection of objective information, whereas everyone must recognize that this information is partly a function of political and social power. He closes by emphasizing that this difference in perception of what spatial databases contain can have profound ethical consequences for researchers and users alike.

CHAPTER 1

Accuracy (Re)assurance: Selling Uncertainty Assessment to the Uncertain

H.T. Mowrer

INTRODUCTION

The process of uncertainty assessment involves developing reliable procedures to estimate the accuracy, precision, bias, variance, probability and/or measurement, sampling, calibration, and propagated error from data through resulting predictions of natural resource and environmental phenomena. We generally think of our clients as being analysts who use geographic information systems (GIS), remote sensing, and other spatial techniques in their work. However, in a larger context, our clients also include upper-level managers, politicians, special interest groups, and many others with an interest in predictions affecting public and private lands. Both within and outside of our user community, some remain skeptical of the value of our work. This is understandable: ours is an endeavor fraught with negative connotations, complex and confusing nomenclature, and incompatible and competing methodologies. The goal of this chapter is threefold. First, we should consider adopting a set of definitions that are consistent among ourselves and with other disciplines. Some simple examples are provided to help explain these to "uncertain" clients. Second, while we continue to improve the state of the art, we should also consider Monte Carlo techniques as both a standard benchmark for comparison of new methods and for comprehensive uncertainty assessment of spatial analyses, from start to finish. And third, we must work to convince all our clients of the utility of comprehensive uncertainty assessments and their importance in the overall technical, social, and political decision space. These proposals are made with the expressed intent of stirring debate leading to consensus and consistency, thus facilitating widespread recognition of the value of our work.

UNCERTAIN DEFINITIONS

Perhaps the worst nightmare of a natural resources manager is to appear "uncertain" to the public, or to admit that there is "error" in the decision process being presented. At the same time, the entire range of professionals dealing with natural resources and environmental assessments, from ecologists to environmental engineers, recognizes that natural processes are inherently variable. To help alleviate negative connotations in the minds of our clients, we need to agree on consistent definitions for the terms we use to describe our work. Proper use of GIS requires understanding and integrating disciplines such as geodesy, cartography, cadastral surveying, statistics, remote sensing, photogrammetry, relational databases, computer science, and others. These multiple disciplines have previously evolved with some degree of independence. Communicating between these disciplines in the common context of GIS has resulted in conflicts in the common definition of terms, creating "uncertainty" in the user community. We must clarify the terms and definitions for ourselves, adopting a consistent usage that does not contradict other disciplines. The terms accuracy, precision, bias, scale, and uncertainty all require clarification in this context.

Over the past 15 years, several descriptive terms have been applied to our field of endeavor. (Perhaps it is not surprising that those drawn to our field of work might be "uncertain" about what to call it.) Initially, in the analysis of computer simulation models, the term "error analysis" was adopted from the field of physics (Taylor, 1982) and "propagation of error" from statistics and surveying. Regrettably, in my early experience natural resource managers assumed that propagation of error meant making one blunder after another. An early reference to "uncertainty analysis" appeared in Downing et al. (1985). Ronen's 1988 book, *Uncertainty Analysis,* was a highly quantitative application of Monte Carlo and sensitivity techniques in nuclear engineering. Goodchild et al. (1994) differentiated between uncertainty and error, with uncertainty being a relative measure of discrepancy, while error was a value for that measure. According to *Webster's New World Dictionary* (1964), "Uncertainty ranges in implication from a mere lack of absolute sureness (*uncertainty* about a date of birth) to such vagueness as to preclude anything more than guesswork (the *uncertainty* of the future)." Regrettably, by Webster's definition, error "implies deviation from truth, accuracy, correctness, right, etc." In the strictest sense, these coincide with Goodchild's definition and the definition of accuracy discussed below; however, vernacular use has imparted negative connotations.

Accuracy

The *Dictionary of Statistical Terms* (Marriott, 1990) defines accuracy as "the closeness of computations or estimates to the exact or true values." Accuracy assessment is applied to discrete cases, such as the classification of pixels into unique classes in remote sensing images (Fenstermaker, 1994). The concept of classification accuracy with an associated error matrix can be used to determine "producer's accuracy" and "user's accuracy." Classification accuracy was initially determined by the overall accuracy from the error matrix, and later by normalized accuracy and the *KHAT* statistic (Congalton, 1991). Because these terms are well accepted in the remote sensing literature, many assume "accuracy assessment" applies to these techniques for discrete classification.

The other context for accuracy assessment, and one in which the terminology becomes more ambiguous, is for continuous variables. For continuous variables, accuracy is described statistically as mean-squared error: the sum of the variance and a squared bias term (Mood et al., 1974). Bias systematically distorts the representativeness of the measurement, as opposed to random error which balances out on the average

(Marriott, 1990). An example of variance and bias can be explained to "uncertain" clients in the context of weighing one's dog by holding it and stepping on a scale several times. One's weight adds a systematic error to the dog's weight. If one's weight is known accurately, it can be subtracted from the dog's to correct this bias. However, one may obtain (random) variation in the weights displayed by the scale each time the person weighs the dog (if the dog wiggles as badly as mine does, perhaps). This is sample variance, or lack of precision, associated with the dog's estimated weight. The ways in which the term "accuracy" are used are generally in fairly good agreement. However, it is often not clearly understood by our clients that measurements of accuracy include components of statistical precision and of bias or systematic error, and require a "true" value of the quantity intended to be observed to correctly measure this bias. In many cases involving natural resource and environmental phenomena, the "true" state of nature is not known. Accuracy in this context is also relevant to "map accuracy," which is discussed below with regard to scale.

Precision

The term "precision" is generally used in two ways. The first ("statistical precision") is based on the calculation of sample variance in classical statistics (Cochran, 1977). Marriott (1990) defines precision as the way in which repeated observations conform to themselves. Statistical precision has nothing to do with how repeated observations conform to the "true" value. Calculations of statistical precision do not require "truth" as do accuracy measurements. In the statistical context, precision refers to the dispersion of observations around their own (sample) mean value. Usage in natural resources (Avery and Burkhart, 1994), surveying (Wolf and Brinker, 1994), photogrammetry (Slama, 1980), and geopositioning (Keating, 1993) generally conform to this usage. The criteria we should use to judge whether we are reporting accuracy or statistical precision is whether we are comparing estimates to "true" values (for measures of accuracy) or solely to the mean of repeated measurements of the same phenomenon (for measures of statistical precision). In many, if not most cases, we will only be able to measure statistical precision, not accuracy. The second way precision is used, (which I will refer to as "numerical precision," but is also called "machine precision") has arisen through computer and information management sciences (Laurini and Thompson, 1992; ESRI, 1997).

Numerical precision relates to the word length of the computer hardware, and measures the degree of detail in reporting each observed value; i.e., the number of significant digits. This conforms more closely to common usage in *Webster's Dictionary* (1964), "minutely exact, as the precise amount."

In explaining accuracy, precision, and bias to "uncertain" clients, it is useful to consider the analogy of target shooting. Consider three different examples of shot groupings on a "bull's-eye" target. The first example consists of a tight grouping of shots, the center of which is well below the bull's-eye. The tight grouping indicates high statistical precision (accidental or random errors are minimized; e.g., "good technique"). However, the distance from the center of the bull's-eye to the center of this tight grouping of shots indicates poor accuracy (which includes both a small variance or statistical precision component and a large systematic error or bias component). By definition, systematic errors follow a quantifiable regime, for which corrections may be calculated. For example, the windage and elevation on a gun sight may be adjusted to compensate for systematic error. Second, consider a very loose grouping of shots that are centered on the bull's-eye, but scattered randomly all over the target. This example also demonstrates poor accuracy, but in this case it is primarily due to poor statistical precision (the loose grouping). The overall centering on the bull's-eye indicates the contribution from systematic error or bias is small. Third, consider a tight grouping of shots (similar to the first example), but exactly in the center of the bull's-eye. This result is highly accurate, which necessarily implies low bias (small systematic errors) and high statistical precision (small random errors): there is no fixed "direction" or bias in the departures from exact center and the tight grouping indicates it is highly precise in a statistical sense. The important thing to note is that we have knowledge of the "true" value (e.g., the bull's-eye center). Estimation of accuracy requires this information, estimation of statistical precision does not.

Note that the above example does not address numerical precision, which is the exactness or degree of detail with which we measure the location of each shot, or the "detail in recording the observed properties" (Laurini and Thompson, 1992). Poor numerical precision may give the impression of high statistical precision, in that all the values for the shots in the last example might all be given a value of 5 which would indicate infinitely high precision, or zero variance. However, improved numerical precision might yield values (e.g., radial distance from exact target center) of 5.01, 5.48, 5.25, etc., which indicate a measurable lack of statistical precision.

Scale

Extending this discussion into the realm of spatial analysis requires the additional considerations of resolution and scale. Again, use of the term "scale" requires careful definition, as laymen often intuitively interpret something described at a large scale as covering a vast area, while cartographers would assume just the opposite. (Some ecologists prefer the terms grain and extent; see Turner and Gardner, 1991.) National Map Accuracy Standards state that not more than 10% of points tested may be in error more than 0.8 mm for map scales larger than 1:20,000, and 0.5 mm for smaller map scales (Wolf and Brinker, 1994). Thus, in the bull's-eye analogy above, not more than 10% of the "shots" could fall outside the limiting distance (±0.8 or 0.5 mm) from the center of the bull's-eye. This, of course, implies that all the points are vanishingly small positions in space with no size or shape, while in reality, all map objects have a limiting spatial resolution. Quattrochi and Goodchild (1997) suggest that metric (map) scale (the ratio of map distance to ground distance) should be augmented, or perhaps replaced by statements of positional accuracy and content (the smallest observable object, or limiting spatial resolution). Similarly, ESRI (1997) defines "map resolution" as the "accuracy of location and shape of features that can be depicted for a given map scale." When maps are viewed digitally using GIS, the scale becomes continuously variable as one zooms in or out on features. Highly aggregated maps generally result in low numerical precision if expanded digitally to too large a metric scale. Stated differently, "the effect of generalization is to introduce uncertainty into the representation of a real phenomenon that could only be mapped perfectly at a much larger scale," (Quattrochi and Goodchild, 1997). Goodchild (1996) appropriately concludes that metric scale alone is "not sufficient as a specification parameter for digital databases."

An overall recommendation for clear and consistent professional usage would be to never use the word "precision" without the appropriate qualifier; e.g., *statistical* precision or *numerical* precision. A calculation of mean square error (*accuracy*) is appropriate when a true value is available for comparison. If not, the estimated *precision* should be indicated. This information should be provided for both locational (positional) information and for attribute values. In addition to information on *positional accuracy,* the limiting spatial resolution (*content*) of spatially represented features should be provided.

UNCERTAINTY AND RISK ASSESSMENT

In addition to providing technical expertise, we must establish the relative importance of the overall effect of technical uncertainty relative to social and political factors as they all affect the decision process. Risk and uncertainty assessments generally take all three of these factors into account. In the classical definitions of risk and uncertainty, Knight (1921) distinguishes between two different levels of knowledge under which a choice among alternatives would be made. *Uncertainty* occurs when outcomes can be enumerated, but the associated probabilities *cannot be estimated*, while *risk* incorporates explicit estimation of probabilities or "likelihoods." Thus, within this discipline, uncertainty cannot be measured or quantified, while risk can. Later authors (Faber et al., 1992) have extended Knight's definition to include *ignorance*, the inability to enumerate all possible outcomes.

Under these definitions, uncertainty occurs when outcomes can be enumerated, but associated probabilities cannot be estimated. To these practitioners, an "estimate of uncertainty" or "quantified uncertainty" might seem a contradiction in terms. So-called "uncertainty assessments" (Finkel, 1990) evaluate different levels of predictability of measures of risk (the "probability of the probability," if you will). In this context, uncertainty is an inherent part of risk, but remains value-neutral (Cleaves and Haynes, 1996). Risk, however, has both quantitative and qualitative aspects as will be discussed below.

Formally stated, *risk* is the exposure to a chance of loss. When a specific outcome cannot be predicted, but the possible outcomes can be defined and their probabilities or "likelihoods" estimated, the risk can be quantified. This characterization must include an estimate of the magnitude of the loss, the probability (or "likelihood") of occurrence, and the *exposure*. Exposure describes what would incur the loss, when it would occur, and to what extent. In a spatial context, a resulting statement of risk might read, "In the current analysis, there is a 10% probability that the recommended polygon area may not include all of the old growth forest, though the expected loss would only be 3% of the total old growth biomass." Haight (1995) provides a highly relevant and useful approach to in-

tegrating measures of probability ("uncertainty") into ecological risk assessment.

Risk and Scale

In an ecological context, risk is scientifically difficult to assess because of different stressors acting individually and in combination on species, populations, communities, and entire ecosystems (CENR, 1995). The issue of *scale* is again relevant here, in that risk assessments at specific sites within a large region provide insight on processes and point-of-impact consequences, while assessments across the entire geographic region provide the contextual information necessary to assess exogenous and cumulative endogenous impacts. Regional risk may be affected by exogenous events such as climate change and by the spatial and temporal aggregation of hundreds of individual decisions at the (endogenous) site level. The interactions across scales are generally not simple and may reflect nonlinear, multivariate, and discontinuous processes, which in extreme circumstances may result in irreversible bifurcation between ecosystem states (Moir and Mowrer, 1993). Because regional environmental problems affect broad segments of society and large geographic areas by their very nature, the ultimate solution is often political, expressed as government policy (Caldwell, 1996).

Social and Political Factors

The concept of risk also implies qualitative components involving concern (fear) that something of societal value is at risk. The public's conscious and unconscious perceptions of risk are often far from objective (Cotgrove, 1981). Cleaves (1996) cites three factors for this. First is the confusion caused by the casual use of the terms risk and uncertainty by both scientific and lay audiences. Second, many people have difficulty accepting probabilistic forecasts. They may demand certainty where none is possible or discount probabilistic evidence that does not agree with preconceived notions. Third, psychological attitudes toward risk perception affect the way in which people perceive the magnitude and possibility of loss. These include the degree of control over the outcome, its dread or vividness, its familiarity, the extent or degree of the losses, and impacts on future generations (Covello et al., 1986). These factors are often the underlying motivation for challenges to the environmental and natural resource decision process. The factual basis for a particular challenge may have scientific credibility and also provide leverage to further an underlying agenda motivated by psychological attitudes.

Caldwell (1996) therefore concludes that problems involving natural resources and the environment are often solved on two levels: scientific (or technical) and political. Human attitudes, beliefs, and behavior may thereby provide a large area beyond technical uncertainty in the ultimate solution to environmental problems. When environmental decision-making takes place through political processes, little tolerance for admitted doubt or uncertainty is allowed. Unfounded certainty about a perceived problem (particularly those shaped by societal beliefs and normative assumptions) may far outweigh technical or scientific uncertainty in the decision-making process (ibid.). When rigorous treatment of scientific uncertainty is not included explicitly in problem analyses, Lemons (1996) warns that natural resource managers often accept them as being more reliable than is warranted. Such unfounded acceptance can bias subsequent decision-making processes such as risk analyses and economic cost-benefit calculations. Conversely, Lemons also points out that the standard 95% confidence required for scientific proof is often too stringent, creating a burden of proof of environmental harm that cannot be met.

Rauscher (in press) proposes a three-dimensional decision space in which decisions are formed. He refers to the "rational/technical dimension" as the mathematical formulation process in decision-making. Quantitative uncertainty assessment allows confidence intervals to be placed on point estimates in this dimension. Rauscher's second dimension is the "political/power dimension" that would include that of Caldwell (1996) described above. The third and final dimension is the "value/ethical" or societal dimension (Cleaves and Haynes, 1996; Cotgrove 1981). The context in which a final decision is determined is somewhere in this three-dimensional decision space. *The widespread utility and acceptance of our technical approach to uncertainty assessment may hinge upon our ability to convince those engaged in the social and political dimensions of the importance of technical uncertainties that affect this overall decision space.*

UNIVERSAL UNCERTAINTY ASSESSMENT METHODOLOGY

The magnitude of uncertainty in the technical dimension, and its affect on the social and political aspects, cannot be evaluated until we can provide comprehensive uncertainty assessments across entire analyses, from start to finish. These must encompass the entire range of data and tools brought to bear on the technical problem, and (where appropriate) give a quantifiable estimate of the uncertainty in the final technical recommendation. Providing comprehensive assessments of the probability distribution of particular outcomes is necessary if we are going to aid in providing probabilities for traditional risk assessments (for example, see Haight, 1995). While many managers are uncomfortable supporting anything they don't understand, researchers continue to be rewarded for new and different approaches that merit publication in the literature. In our quest to improve the state of the art, the techniques and methods we propose are diverse, complex, and often incompatible. However, what we need as a discipline is a common system for uncertainty assessment that serves as the "lowest common denominator" that can be applied universally to all phases of an analyses and different sources of data. The Monte Carlo approach can serve in this regard as a benchmark for comparison of new techniques, and as a comprehensive tool for overall uncertainty assessment.

Openshaw (1989) recognized the Monte Carlo approach as providing a generally applicable technique for quantitative uncertainty assessment. The technique is straightforward to apply, and can usually treat the analytical tool (e.g., GIS) as a "black box." Mowrer (1991) found the result from Monte Carlo error propagation to be significantly different, and assumedly more robust, than statistical error propagation formulas rigorously applied to a temporal (aspatial) forest growth simulation. Factors cited for this difference include the numerous cross correlations that occurred within the relatively simple simulation model. In spatial analyses, not only do cross correlations exist between variables, spatial autocorrelations also exist, compounding the problem even further (Heuvelink, 1993). To avoid these problems, Mowrer (1994a, 1994b, 1997) recommended the Monte Carlo technique of sequential Gaussian simulation to quantify the uncertainty introduced through spatial interpolation of point location information. While mean vectors and covariance matrices must be stated for external input data sources, cross-correlations within an analytical process are inherently included through the Monte Carlo process. The only situation where access to the internal code of an analytical tool is required is where predictive coefficients are statistically calibrated within the software (e.g., regression or multivariate techniques). This may be overcome by externalizing the estimation process and perturbing the calibrated coefficients appropriately. This places the major computational burden on the computer, as opposed to tedious and intricate calibration of error propagation equations that must be changed every time the computer code is changed. With each successive improvement in computer processor speeds, and the consistent decrease in the cost of computer storage devices, Monte Carlo techniques make more and more sense.

These techniques work well for raster GIS, but what of vector representations? Lowell (1992, 1994) recommended fuzzy polygon boundaries through Voronoi diagram-based area stealing. The uncertainty analysis was then accomplished using a raster GIS. Lowell (1996) further concluded that it was inappropriate to model forest stands as a continuous (raster) surface, but by "a set of polygons whose boundaries show a range of possible locations." Perhaps the best approach to date has been recommended by Heuvelink (1996) through the "mixed model of spatial variation." This approach combines both a "discrete model of spatial variation" (vector GIS; e.g., Lowell's work) and "continuous models of spatial variation" (raster GIS; e.g., Mowrer's work) simultaneously. Heuvelink (1996) demonstrates that considerable differences in prediction error standard deviation occur, depending on which of these three approaches are used.

CONCLUSION

While risk assessment in the environmental sciences has been well accepted, recent work (Liu and Herrington, 1996; Pukkala and Kangas, 1996) indicates an increasing awareness of uncertainty and risk assessment in the natural resources arena. In order to ensure increasing acceptance as a necessary and valued component in this and other environmental and natural resource management strategies, we must work toward professional consensus in three main areas. We must communicate what we do in a positive and unambiguous context through conscientious use of appropriately defined terminology that does not conflict with uses in other disciplines. The terms accuracy, precision, bias, scale, and uncertainty all require careful use. Because the term "uncertainty" is often used in various contexts, we must be particularly conscious of how we use it in our discipline. We must be able to provide an understandable overall measure of uncertainty across an entire analytical process, through Monte Carlo techniques, for example. We must educate our expanded client base in

the need for uncertainty assessment in technical analyses, and its importance to the overall decision in the combined technical, social, and political dimensions.

Natural systems are inherently variable, and although managers generally prefer certainty and predictability, natural systems seldom conform to their desires. Adept natural resource managers take a precautionary approach, not to the exclusion of human use, but acknowledging and preparing for the possibility of low probability occurrences with high potential impact (surprises). We, as a discipline, must develop the quantitative basis to estimate the probability of these and other types of risks. In applying any technique for uncertainty assessment, a final caveat cannot be expressed better than by Goovaerts (1996), who states that "...uncertainty is not intrinsic to the phenomenon under study: rather, it arises from our imperfect knowledge of that phenomenon, it is data-dependent and most importantly model-dependent.... No model, hence no uncertainty measure, can ever be objective: the point is to accept the limitation and document clearly all aspects of the model."

ACKNOWLEDGMENT

The author wishes to acknowledge Dr. Ronald E. McRoberts, USDA Forest Service, for his encouragement and shared insight in addressing the statistical definitions discussed above.

REFERENCES

Avery, T.E. and H.E. Burkhart. *Forest Measurements, 4th ed.,* McGraw-Hill, New York, 1994, p. 11.

Caldwell, L.K. Science Assumptions and Misplaced Certainty in Natural Resources and Environmental Problem Solving, in *Scientific Uncertainty and Environmental Problem Solving,* Blackwell Science, Cambridge, MA, 1996, p. 394.

CENR (Center on Environment and Natural Resources). Draft Research Strategy and Implementation Plan, in *Risk Assessment Research in the Federal Government*, Subcommittee on Risk Assessment, National Science and Technology Council, 1995, p. 11.

Cleaves, D. and R. Haynes. Uncertainty, Risk, and Ecosystem Management, in *Proceedings of the Ecological Stewardship Workshop,* Available from: USDA Forest Service Ecosystem Management Staff, Washington, DC, via the INTERNET, 1996.

Cochran, W.G. *Sampling Techniques, 3rd ed.,* John Wiley & Sons, New York, 1977, p. 16.

Congalton, R.G. A Review of Assessing the Accuracy of Classifications of Remotely Sensed Data, *Remote Sensing Environ.,* 46, pp. 35–36, 1991.

Cotgrove, S. Risk, Value Conflict, and Political Legitimacy, in *Dealing with Risk,* Manchester University Press, Manchester, Great Britain, 1981.

Covello, V.T., D. von Winterfeldt, and P. Slovic. Risk Communication: A Review of the Literature, *Risk Abstracts,* 3, pp. 171–182, 1986.

Downing, D.J., R.H. Gardner, and F.O. Hoffman. An Examination of Response-Surface Methodologies for Uncertainty Analysis in Assessment Models, *Technometrics,* 27, p. 151, 1985.

ESRI. *Arc/Info ArcDoc On-Line Documentation, ver. 7.0.,* Environmental Systems Research Institute, Redlands, CA, 1997.

Faber, M., R. Manstetten, and J.L.R. Proops. Humankind and the Environment: An Anatomy of Surprise and Ignorance, *Environ. Values,* 1(3), pp. 217–242, 1992.

Fenstermaker, L.K. *Remote Sensing Thematic Accuracy Assessment: A Compendium,* American Society of Photogrammetry and Remote Sensing, Bethesda, MD, 1994, p. 413.

Finkel, A.M. *Confronting Uncertainty in Risk Management: A Guide for Decision-Makers*, Resources for the Future, Center for Risk Management, Washington, DC, 1990.

Goodchild, M.F., B. Buttenfield, and J. Wood. Introduction to Visualizing Data Quality, in *Visualization in Geographical Information Systems,* John Wiley & Sons, New York, 1994.

Goodchild, M.F. Generalization, Uncertainty, and Error Modeling, *GIS/LIS '96 Conference Proceedings,* American Society for Photogrammetry and Remote Sensing, pp. 765–774, 1996.

Goovaerts, P. *Geostatistics for Natural Resources Evaluation,* Oxford University Press, New York, 1996, p. 442.

Haight, R.G. Comparing Extinction Risk and Economic Cost in Wildlife Conservation Planning, *Ecol. Appl.,* 5(3), pp. 767–775, 1995.

Heuvelink, G.B.M. Identification of Field Attribute Error Under Different Models of Spatial Variation, *Int. J. Geogr. Inf. Syst.,* 10(8), pp. 921–935, 1996.

Heuvelink, G.B.M. *Error Propagation in Quantitative Spatial Modeling,* Nederlandse geografische studies, Universiteit Utrecht, The Netherlands, 1993, p. 30.

Keating, J.B. *The Geo-Positioning Selection Guide for Resource Management,* Bureau of Land Management, Cheyenne, WY, 1993, pp. 52, 61.

Knight, F.N. *Risk, Uncertainty, and Profit,* Houghton-Mifflin, Boston, MA, 1921, p. 133.

Laurini, R. and D. Thompson. *Fundamentals of Spatial Information Systems,* Academic Press, San Diego, CA, 1994, pp. 300–301.

Lemons, J. *Scientific Uncertainty and Environmental Problem Solving,* Blackwell Science, Cambridge, MA, 1996, p. 394.

Liu, R. and L.P. Herrington. The Expected Cost of Uncertainty in Geographic Data, *J. For.,* 94(12), pp. 27–31, 1996.

Lowell, K. On the Incorporation of Uncertainty into Spatial Data Systems, *Proceedings of the GIS/LIS '92 Conference, San Jose, California, November 1992,* American Society of Photogrammetry and Remote Sensing, Bethesda, MD, 1992, pp. 484–493.

Lowell, K. Initial Studies in Fuzzy Surface-Based Cartographic Representations of Forests, *Proceedings of the International Symposium on the Spatial Accuracy of Natural Resource Data Bases, Williamsburg, Virginia, May 1994,* American Society of Photogrammetry and Remote Sensing, Bethesda, MD, 1994, pp. 168–177.

Lowell, K. Discrete Polygons or a Continuous Surface: Which is the Appropriate Way to Model Forests Cartographically, *Spatial Accuracy Assessment in Natural Resources and Environmental Sciences: Proceedings of the Second Symposium, Fort Collins, Colorado, May 1996,* USDA Forest Service, Rocky Mountain Research Station, General Technical Report RM-GTR-277, Fort Collins, CO, 1996, pp. 235–242.

Marriott, F.H.C. *A Dictionary of Statistical Terms, 5th ed.,* Longman Scientific and Technical, Essex, England, 1990, pp. 83, 159.

Moir, W.H. and H.T. Mowrer. Unsustainability: The Shadow of Our Future, in *Sustainable Ecological Systems,* USDA Forest Service, General Technical Report RM-247, Rocky Mountain Forest and Range Experiment Station, Fort Collins, CO, 1993, p. 147.

Mood, A.M., F.A. Graybill, and D.C. Boes. *Introduction to the Theory of Statistics, 3rd ed.,* McGraw-Hill, New York, 1974, p. 293.

Mowrer, H.T. Estimating Components of Propagated Variance in Growth Simulation Model Projections, *Can. J. For. Res.,* 21, pp. 379–386, 1991.

Mowrer, H.T. Spatially Quantifying Attribute Uncertainties in Input Data for Propagation through Raster-Based GIS, in *GIS '94 Proceedings, Vancouver, BC, Canada, February 1994,* Bowne Printers, Vancouver, BC, 1994a, pp. 371–382.

Mowrer, H.T. Monte Carlo Techniques for Propagating Uncertainty through Simulation Models and Raster-based GIS, *International Symposium on the Spatial Accuracy of Natural Resource Data Bases, Williamsburg, Virginia,* American Society of Photogrammetry and Remote Sensing, Bethesda, MD, 1994b, pp. 179–188.

Mowrer, H.T. Propagating Uncertainty through Spatial Estimation Processes for Old-Growth Subalpine Forests Using Sequential Gaussian Simulation in GIS, *Ecol. Modeling,* 98, pp. 73–86, 1997.

Openshaw, S. Learning to Live with Errors in Spatial Databases, in *Accuracy of Spatial Databases,* M. Goodchild and S. Gopal, Eds., Taylor & Francis, New York, 1989.

Pukkala, T. and J. Kangas. A Method for Integrating Risk and Attitude Toward Risk into Forest Planning, *For. Sci.,* 42(2), pp. 198–205, 1996.

Quattrochi, D.A. and M.F. Goodchild. *Scale in Remote Sensing and GIS,* Lewis Publishers, Boca Raton, FL, 1997, pp. 1–6.

Rauscher, H.M. Decision-making Methods for Ecosystem Management, in *For. Ecol. Manage.,* in press.

Ronen, Y. *Uncertainty Analysis,* CRC Press, Boca Raton, FL, 1988, p. 282.

Slama, C.C. *Manual of Photogrammetry, 4th ed.,* American Society of Photogrammetry, Falls Church, VA, 1980, pp. 996–1033.

Taylor, J.R. *An Introduction to Error Analysis,* University Science Books, Sausalito, CA, 1982, p. 270.

Turner, M.G. and R.H. Gardner. *Quantitative Methods in Landscape Ecology,* Springer-Verlag, New York, 1991, p. 7.

Webster's New World Dictionary. World Publishing Company, New York, 1964.

Wolf, P.R. and R.C. Brinker. *Elementary Surveying, 9th ed.,* HarperCollins, New York, 1994, pp. 24–32.

CHAPTER 2

Uncertainty and Error Propagation in Map Analyses Involving Arithmetic and Overlay Operations: Inventory and Prospects

D.A. Griffith, R.P. Haining, and G. Arbia

MOTIVATION

In recent years spatial scientists have become actively concerned about uncertainty in geospatial information, as well as consequences of error propagated during analyses in which it is present, spawning an increasing awareness of additional complications that accompany noisy, messy, and dirty georeferenced data. Many of these complexities and difficulties, which impact upon the nature and quality of spatial decision-making, are well-known in spatial statistics, where their failure to be efficiently and effectively handled by conventional statistical theory is extensively documented (see, e.g., Cliff and Ord, 1981; Cressie, 1991; Haining, 1993). The research reported on in this chapter was motivated by a need to enhance the accuracy of final map products in the presence of uncertainty in geospatial information. But an awareness of more problematic, less exoteric complications stemming from uncertainty in geospatial information has emerged only recently in the scientific community at large. Goodchild and Gopal's (1989) innovative compendium has motivated considerable research on this topic, including that by Haining and Arbia (1992), and that by Amrhein and Griffith (1994). Numerous articles on this topic have appeared in recent years in journals such as the *International Journal of Geographical Information Systems* and *Photogrammetric Engineering & Remote Sensing,* and specialty books such as Guptill and Morrison's *Elements of Spatial Data Quality* (1995) are beginning to appear now. Clearly, increasing attention has begun to focus on this problem, making it topical and hence further motivating its investigation, as well as the research summarized in this chapter.

INTRODUCTION TO THE PROBLEM

The study of map error and its propagation raises a distinct set of problems that go beyond traditional error analysis (Taylor, 1982). This is because in the case of map data, observations consist of attributes recorded at locations, and attribute values at adjacent locations tend to be similar (are spatially correlated) because of the continuity of ground truth. Also, attribute measurement error may not be independent between adjacent locations and there may be errors in specifying the locations of attributes. Discussions of different sources and forms of map error can be found in, for instance, Goodchild and Gopal (1989), Lunetta et al. (1991), Amrhein and Griffith (1994), Thapa and Bossler (1992), Haining and Arbia (1992), Heuvelink (1993), and Veregin (1995). These publications, especially, summarize the present state of knowledge in this academic field.

Error propagation effects arise because of the types of operations performed on map data. For example, maps constructed using the arithmetic operations of addition and/or ratioing are used separately and together in deriving vegetation indexes (Curran, 1980), in seismic risk assessment (Emmi and Horton, 1995), and in techniques frequently used in remote sensing, such as principal components and linear discriminant analysis. Map overlay and buffering are used for purposes of resource exploitation, factory siting and environmental risk assessment, for instance (Berry, 1987; Veregin, 1994). Particularly when multiple operations are performed, errors are likely to compound in ways that could seriously undermine objectives of analyses. Hence, understanding how results from single and multiple map operations are affected by errors in source maps is important (Richards, 1986).

Background

Reviewing especially the aforementioned literature discloses that error properties of maps include both the magnitude of the errors and the spatial structure (or geography) of the errors. Any study of map error needs to consider how errors are distributed in geographic space. Further, there is a practical reason for studying the spatial structure of errors. Isolated errors or errors with no discernible pattern may be more likely to stand out and be detected by the trained eye, particularly if they are large. An analyst familiar with a region or the phenomena under study may be better at spotting this type of error as opposed to the error that possesses spatial continuity, particularly if it blends in with the underlying map structure. For this reason short distance and long distance error properties need to be examined in order to determine whether or not a tendency exists for error regions to form, alluding to the need for semivariogram descriptions of error.

Let error be the discrepancy between a reference or source map (ground truth) and the corresponding observed or error infected map. Propagated errors can be defined as the discrepancies that exist after performing identical operations on a set of source maps and their corresponding set of observed maps. Some of the sources of error are a product of measurement instruments, and as such can be reduced by improving these instruments, and perhaps by upgrading the technology or preprocessing the georeferenced data measurements. Other sources of error are a product of the geographic landscape on which measurements have been taken. Accordingly, there exists a need to identify those classes of geographic landscapes that tend to be more or less prone to serious error propagation. In addition, propagated error is a function both of the magnitude and the spatial distribution of the source errors; the critical question here asks whether or not errors stand out or blend into their statistical as well as their underlying true map geographic distributions. A need also exists to quantify conspicuous differences in the contribution that spatial and aspatial sources of error make to any final propagated error.

The signal-to-noise ratio comprises natural or stochastic attribute variability (the numerator) and error process variance (the denominator). The former is usually given, dependent on the particular landscape being studied. However, error process variability will be inflated by recording/measurement errors, and careful data coding and editing will help to reduce its impact. Further, error process variability may be correlated with the size of attribute values (e.g., census population undercounts) so that particular attention may need to be paid to error process variability in the case of those geographic regions where, for example, there is marked attribute variability. Findings reported in Arbia et al. (1998, 1999) suggest that this source of error interacts with location error, determining whether or not the only effective way of reducing it is by exploiting its interaction with location error; any success in reducing location error should in turn help reduce error propagation due to the signal-to-noise ratio. Accordingly, future research needs to seek a better understanding of relationships between the signal-to-noise ratio, location error, and propagated error.

While spatial correlation in errors associated with collecting census data, for example, has not been found to be a serious problem (Griffith et al., 1994), the same cannot be said of remotely sensed data where the point spread function describes how measurement error is smeared across adjacent pixels. Remotely sensed images cannot be treated as though pixels can be cut apart into separate pieces, and then simply shuffled for analysis purposes, mimicking the classical statistics textbook balls-in-an-urn experiment. Hence, both spatial and aspatial elements need to be incorporated into error models for maps and into the set of error properties; of note is that good descriptions of the spatial attributes of error often are the most elusive.

RESEARCH TO DATE: AN INVENTORY

The underlying error model assumed here is considered relevant to both remotely sensed and GIS databases. First, for analysis purposes consider data recorded for a set of regular pixels. Such regularity is not usually encountered in GIS databases, although it is the normal data structure for remotely sensed data. Second, different remotely sensed maps (bands) are mutually correlated and indexes are often constructed using bands that are known to be (positively) correlated. In the case of GIS, indexes are usually constructed using variables that are based on (largely) correlated maps.

For analytical purposes consider an error model that is a special case of the Geman and Geman (1984) "corruption" model in which a "true" or source map (**S**) comprising an n-by-m grid of n×m pixels is transformed to an error infected or corrupted map (**Z**). Let S(i,j) denote the true attribute value associated with pixel (i,j). Let Z(i,j) denote the corresponding error corrupted value, such that for all (i,j)

$$Z(i,j) = \sum_g \sum_h w_{i,j}(g,h)\, S(i+g, j+h) + u(i,j) \qquad (1)$$

where

$$\sum_g \sum_h w_{i,j}(g,h) = 1$$

and

$$w_{i,j}(g,h) \ge 0$$

The values {u(i,j)} may be either stochastic or fixed.

This model is discussed in, among other sources, Ripley (1988) and Cressie (1991), and has been recognized as a good model for remote sensing errors in the case of field restoration (Richards, 1986, p. 36). This model can be used to capture many elements of map error. Location error is represented through {$w_{i,j}(g,h)$}. Setting $w_{i,j}(0,0) = 1$ for all (i,j) implies no location error. Different forms for this matrix (sometimes called the "point spread function" or "blurring matrix") have been proposed in the remote sensing literature where the specification is defined after geometric correction of an image (Forster, 1980; Rosenfeld and Kak, 1982; Geman and Reynolds, 1992).

Systematic attribute error (or bias) is introduced by fixing nonzero values in {u(i,j)}. Stochastic attribute error is introduced by setting the {u(i,j)} to a random variable with mean μ(i,j) and dispersion matrix $\sigma_u^2 \Sigma$. The mean quantifies the bias in the attribute measurement process, σ_u^2 is a scalar measuring the size of the random error, and Σ denotes an n×m-by-n×m correlation matrix where nonzero values in the off-diagonal cells allow the representation of spatially correlated attribute measurement error. This last component is a particular feature of remotely sensed data over short distances due to light scattering (Forster, 1980). To describe this, nonzero values in the row corresponding to pixel (i,j) of Σ often are associated with only the geographically-near neighbors of pixel (i,j) (Ripley,

1988). The resulting error map, denoted {e(i,j)}, may be defined by

$$e(i,j) = \left[Z(i,j) - S(i,j)\right] = \left[w_{i,j}(0,0) - 1\right]S(i,j)$$
$$+ \sum_g \sum_h w_{i,j}(g,h)\, S(i+g, j+h) + u(i,j) \qquad (2)$$

Given Equation 2, three sources of spatial pattern can be detected in an error map: (a) spatial continuity in {S(i,j)}; (b) spatial continuity associated with μ(i,j), since this may be correlated with {S(i,j)}; and (c) spatial structure in Σ. The first two sources are associated with the mean, and the third with the variance-covariance properties of the model. Note that it is through the effects of location error that source map structure influences spatial and aspatial error properties.

Haining and Arbia (1992) report preliminary simulation results in which true and error infected maps (according to Equation 1) are constructed and then each subjected to the map operations of addition, ratioing, and overlaying. Contrasting values for the parameters, in particular {$w_{i,j}(g,h)$} and Σ, were taken from the literature about remote sensing data errors. Two contrasting map structures (**S**) were used—one pair of maps with strong spatial continuity (spatial correlation) and one pair of maps with weak spatial continuity. Results from operating on true maps are compared with results from operating on error infected maps. Aspatial statistics [RMSE and Pr(θ)—the probability of a severe error—where θ = 0.1, 0.3 and 0.5] and spatial statistics (low order spatial correlation in the errors) are reported. Recently these results were extended, explicitly quantifying the contribution of different components of error and their interactions (Arbia et al., 1998, 1999).

Now denote the values in location (i,j) for two different maps by $_1S(i,j)$ and $_2S(i,j)$, respectively. Suppose the geographic distributions of attribute values across these maps are stationary with moments: $E[_kS(i,j)] = {_k}\mu$, and $Var[_kS(i,j)] = {_k}\sigma^2$, for all (i,j) and k=1,2; and correlation function: $Corr[_kS(i,j), {_k}S(i+g,j+h)] = {_k}R(g,h)$, for all (i,j) and k=1,2. These two maps may be correlated, too: $Corr[_1S(i,j), {_2}S(i+g,j+h)] = {_{1,2}}R(g,h) = \rho\, {_k}R(g,h)$, for all (i,j), k, (g,h) and −1≤ρ≤1. Meanwhile, the two observed maps $_1$**Z** and $_2$**Z** are obtained through a corruption of $_1$**S** and $_2$**S**. The following isotropic structure is assumed for {$w_{i,j}(g,h)$}, k=1,2:

$$_kZ(i,j) = a\ _kS(i,j) + b\ _kS(i+1,j) + c\ _kS(i,j+1) + d \times\ _kS(i+1,j+1) + _ku(i,j) \tag{3}$$

The location error parameters do not depend on (i,j) and are identical on both maps. Further, assume b = c (this simplification also appears in Welsch et al., 1985, was used in Haining and Arbia, 1992, and assists in presenting results). Then, since a + b + c + d = 1,

$$b = c = a^{1/2} - a \quad \text{and} \quad d = (1 - a^{1/2})^2 \tag{4}$$

Location error increases as "a" decreases. Finally, suppose u(i,j) is a random variable with $E[_ku(i,j)] = 0$, $Var[_ku(i,j)] = _k\sigma_u^2$, and $Corr[_ku(i,j),\ _ku(i+g,j+h)] = _kR_u(g,h)$, for all (i,j) and k=1,2. The map $\mathbf{S}$ and the measurement process $\mathbf{u}$ are uncorrelated at all spatial lags, but the map $\mathbf{S}$ and the corresponding $\mathbf{Z}$ map are correlated. Therefore, it can be shown that

$$Var[_kZ(i,j)] = _k\sigma^2[(a^2 + b^2 + c^2 + d^2) + 4b(a+d) \times\ _kR(0,1) + 2(ad+bc)\ _kR(1,1) + SNR^{-1}]$$

where the final term in the right-hand expression is the inverse of the signal-to-noise ratio $(_k\sigma^2/_k\sigma_u^2 = SNR)$. Furthermore, $Cov[_kZ(i,j),\ _kS(i,j)] = [a + 2b\ _kR(0,1) + d\ _kR(1,1)]\ _k\sigma^2$.

Findings for Overlay Map Operations (Arbia et al., 1998)

Veregin (1994, 1995) provides a critical review of error propagation modeling for various overlay operations, particularly in relation to obtaining the limits to derived layer accuracy. An earlier paper by Lantner and Veregin (1992) provides further background to this problem. Other similar reviews can be found in Fisher (1991), Lunetta et al. (1991), and Unwin (1995). These papers also provide background on error sources in remote sensing and GIS. But none of these researchers employ the Geman and Geman model, or explicitly incorporate formal spatial operators into the error process and then examine spatial properties of the propagated errors. Such an approach does, however, share common ground with work by Heuvelink and Burrough (1993) and Heuvelink (1993). Also, by seeking to find a representation of the spatial form of the error process, this research shares common ground with Goodchild et al. (1992), which draws on the work in spatial statistics concerned with modeling spatial variation (see, for example, Haining, 1993).

Assume two error-free source maps upon which an n-by-m regular grid of quadrants is superimposed, say $_1S_{i,j} \in\ _1\mathbf{S}$ and $_2S_{i,j} \in\ _2\mathbf{S}$ (i=1,...,n; j=1,...,m). Of note is that considering more than two maps greatly increases the complexity of the formalism without contributing proportionally to an understanding of the mechanism of error propagation. Consider the following four basic logical operations (after Veregin, 1996; p. 603):

1. "Univariate overlay," which identifies the set of areas on a single attribute map where attribute values lie in a specified interval. This forms the building block for AND, OR, and XOR overlay operations. Boundaries of soil regions employ this operator.
2. "Overlay-AND," which identifies the set of areas where attribute conditions are met in both maps $_1\mathbf{S}$ and $_2\mathbf{S}$. The concept of "high erosion hazard" can be formalized through this operation (Heuvelink and Burrough, 1993). Any one of the two conditions, taken without reference to the other, would be an example of a "univariate overlay" operation.
3. "Overlay-OR," which identifies the set of areas where attribute conditions are met in one or both images $_1\mathbf{S}$ and $_2\mathbf{S}$. The identification of "land unavailable for development" can be formalized through the OR operation (Parrott and Stutz, 1991).
4. "Overlay-XOR," which identifies the set of areas where attribute conditions are met in only one image, either $_1\mathbf{S}$ or $_2\mathbf{S}$.

These definitions implicitly include the buffering operation. Buffering identifies areas in terms of their spatial relationship to an initial set of areas that possess certain attribute properties. Propagated errors that arise under buffering, then, are fundamentally a consequence of the errors that arise in defining the initial set of areas (which are covered by the four cases defined here) plus a second component of error that arises from the spatial operator acting on these initial maps. Error propagation properties will vary principally as a function of the size of the buffer interacting with the map edge and the number and geographical distribution (relative to one another and to the edge of the map) of the errors in the initial maps (Veregin, 1994, 1996).

Suppose that source maps $_1\mathbf{S}$ and $_2\mathbf{S}$ are observed through an interval/ratio level measurement process (i.e., $_1\mathbf{S} \in \mathfrak{R}$ and $_2\mathbf{S} \in \mathfrak{R}$) that can corrupt them with both attribute measurement error and location error (see Haining and Arbia, 1992). Next, assume that these two error components are independent. The location error component operates directly on the source map; a stochastic measurement error, say $\mathbf{u}$, may be added to this term, producing a blurring of the map in the sense of running a spatial filter or "mask" over the map (the point-spread function). The observed maps, say $_1\mathbf{Z}$ and $_2\mathbf{Z}$, can be expressed in terms of the error-free source maps as:

$$_1\mathbf{Z} = \mathbf{H} {}_1\mathbf{S} + {}_1\mathbf{u} \text{ and } {}_2\mathbf{Z} = \mathbf{H} {}_2\mathbf{S} + {}_2\mathbf{u} \qquad (5)$$

with matrix $\mathbf{H}$ representing location error, where the row sums of $\mathbf{H}$ are unity. The difference between $_{op}\mathbf{Z}$ and $_{op}\mathbf{S}$ ("op" meaning "UNI," "AND," "OR," or "XOR") measures the error propagated by the operation.

A basic tool for constructing the most widely used error propagation measures, in the case of overlay operations, is the misclassification or classification of error matrix (henceforth CEM; Veregin, 1995, p. 597). From the entries of the CEM two common measures can be defined: the percent of misclassified pixels (PMP; Haining and Arbia, 1992), which may be defined as

$$PMP = 1 - PCC = PEC + PEO \qquad (6)$$

where PCC denotes the percentage of correctly classified pixels, PEC the percentage of errors of commission, and PEO the percentage of errors of omission; and the Kappa coefficient (Stehman, 1996), which may be defined as

$$Kappa = \frac{PCC - E(PCC)}{1 - E(PCC)} \qquad (7)$$

where E($\bullet$) indicates the expected value under randomization (i.e., a random allocation of observations to the four states of a classification table).

Accuracy is evaluated as the discrepancy between actual and expected PCC. PMP and Kappa appear to be the two most widely used measures and their relative strengths are discussed by Veregin (1995). There are other measures, such as user's and producer's accuracy, which distinguish between errors of omission and commission (Congalton, 1991; Veregin, 1995), as well as the Rousseau index (Rousseau, 1980). These indices use the same information contained in the

CEM (for instance, PMP is a weighted sum of user's and producer's accuracy), and hence should behave similarly.

The elements of the CEM can be expressed as integrals of the joint distribution of $_{op}S$ and $_{op}Z$, say $\Phi(S,Z)$. When $\Phi(S,Z)$ can be specified, this approach can be used to analytically evaluate error propagation. For example, recalling that Pr$\{\bullet\}$ denotes probability,

$$\Pr\left\{ {}_{op}S_{ij} = 1 \cap {}_{op}Z_{ij} = 1 \right\} = \iint\limits_{op\Re_A} \Phi(S, Z)dSdZ \qquad (8)$$

where $_{op}\Re_A$ is the domain over which the conditions implied by operation "op" are satisfied in both $_{op}S$ and $_{op}\mathbf{Z}$. The domains of integration in the simplest case of "univariate overlay" are only double integrals, while in all other cases Equation 8 refers to quadruple integrals, which to date have defied being expressed in closed form.

The preceding measures shed light on the global amounts of error that are propagated from applying overlay operations to corrupted maps. However, a second class of measure is needed to describe the spatial distribution of errors. Spatial measures help characterize error propagation effects—are errors distributed at random across a map (perhaps standing out as local anomalies or outliers) or are they clustered (forming "plausible" error regions)? Isolated errors are more likely to be detected by the trained analyst than errors that cluster together and possess spatial continuity properties that blend in with the underlying scene.

Haining and Arbia (1992) propose the use of the BB and BW join count statistics under free sampling to characterize the spatial distribution of errors (Cliff and Ord, 1981). Both indices can be expressed as functions of the CEM entries. Let

$$_{op}E_{i,j} = \begin{cases} 0 \text{ if } \left({}_{op}S_{i,j} = 1 \text{ and } {}_{op}Z_{i,j} = 1 \right) \\ \text{or} \\ \left(0 \text{ if } {}_{op}S_{i,j} = 0 \text{ and } {}_{op}Z_{i,j} = 0 \right) \\ 1 \text{ otherwise} \end{cases} \qquad (9)$$

so that $_{op}E_{ij}$ (i=1,...n; j=1,...,m) is the error map for the given overlay operation (op). Then,

$$_{op}BB = Pr\{E_{i,j} = 1 \text{ and } E_{i+1,j} = 1\}$$

$$_{op}BW = Pr\{E_{i,j} = 1 \text{ and } E_{i+1,j=0}\}$$
$$+ Pr\{E_{i,j} = 0 \text{ and } E_{i+1,j} = 1\}$$

Combining Equations 8 and 9:

$$_{op}BB = Pr\{[(_{op}S_{i,j} = 1 \text{ and } _{op}Z_{i,j} = 0)$$

$$\text{or } (_{op}S_{i,j} = 0 \text{ and } _{op}Z_{i,j} = 1)] \text{ and}$$

$$[(_{op}S_{i+1,j} = 1 \text{ and } _{op}Z_{i+1,j} = 0) \text{ or}$$

$$(_{op}S_{i+1,j} = 0 \text{ and } _{op}Z_{i+1,j} = 1)]\}$$

A similar expression can be defined for the BW statistic.

To make illuminating progress in terms of deriving analytical expressions for the various error propagation measures, the following two definitions are used with the corruption model (1):

$$_1Z_{i,j} = a\ _1S_{i,j} + (\sqrt{a} - a)(_1S_{i+1,j} + _1S_{i,j+1})$$
$$+ (1 - \sqrt{a})^2\ _1S_{i+1,j+1} + _1u_{i,j} \qquad (10a)$$

and

$$_2Z_{i,j} = a\ _2S_{i,j} + (\sqrt{a} - a)(_2S_{i+1,j} + _2S_{i,j+1})$$
$$+ (1 - \sqrt{a})^2\ _2S_{i+1,j+1} + _2u_{i,j} \qquad (10b)$$

where "a" is the single parameter in **H** that specifies the degree of location error in a map (see Arbia et al., 1998, 1999 for details).

In Equation 10 $_1u_{i,j}$ and $_2u_{i,j}$ are the additive attribute errors. Further, assuming that both **S**s and **u**s are Gaussian results in the **Z**s being Gaussian; hence, the only specification necessary is for the means, variances and covariances of source maps and of the error process. If both maps have been corrupted by stochastic error processes (**u**) with zero-means, and these error processes are independently distributed both from the

source maps and from one another, but possibly spatially correlated, then:

$$E[_bu_{i,j}] = 0,\ b = 1,2;\ Cov[_bu_{i,j},_bu_{i+k,j+h}]$$
$$= \sigma^2 r_u(k,h),\ b = 1,2, \forall\ k,h$$

and, hence,

$$Cov[_bu_{i,j},_gS_{i+k,j+h}] = 0 \text{ and}$$
$$Cov[_bu_{i,j},_gu_{i+k,j+h}] = 0,\ b,g = 1,2;\ \forall\ k,h \qquad (11a,b)$$

For two maps:

$$E[_bS_{i,j}] = \mu,\ b = 1,2;\ Cov[_bS_{i,j},_bS_{i+k,j+h}]$$
$$= \sigma^2 r(k,h),\ b = 1,2, \forall\ k,h$$

and, hence,

$$Cov[_bS_{i,j},_gS_{i+k,j+h}] = \sigma^2 \rho r(k,h),$$
$$b,g = 1,2, \forall\ k,h \qquad (11c)$$

Accordingly, given Equation 10, all joint probability characteristics of **S** and **u** (and hence of **Z**) can be summarized by the following parameters:

the means of the source maps $_1\mu,\ _2\mu$; intersource map, quadrat-by-quadrat, bivariate correlation ρ; source map variances $_1\sigma^2,\ _2\sigma^2$; error process variances $_1\sigma_u^2,\ _2\sigma_u^2$; spatial correlation in the source maps, at various lags (k,h), $_1r,\ _2r$; error process spatial correlation $_1r_u(k,h),\ _2r_u(k,h)$; and, location error "a."

These parameters can be classified into two distinct groups. One group of parameters has to do with the process of data acquisition; these are the location error parameter "a," the ratio between source map variance and error process variance $\sigma^2/\sigma_\varepsilon^2$ (the signal-to-noise ratio, SNR), and the spatial pattern of errors $[r_u(k,h)]$. These parameters often can be specified by users in practical circumstances. For instance, in LANDSAT, TM imagery estimates are available for spatial correlation in the error process via the point-spread function (Forster, 1980) and location error (e.g.,

Welsch et al., 1985). Furthermore, in LANDSAT data, often a reasonable assumption is that the SNR does not exceed 20, with an error variance that is 5% of the signal variance corresponding to about 13 dB. In the case of census data there is unlikely to be location error (hence setting the value for a), although some data sets may possess location error introduced through the process of translating data reported for one spatial framework to another, such as census tract data mapped onto zip code areas. There is no evidence to indicate that errors arising in the process of census data collection are spatially correlated. However, in the case of nonsampled census data there is evidence of error at the level of individual reporting units. Undercounting may be correlated both with unit size and socioeconomic composition (Mulry and Spencer, 1993; Schenker, 1993). In the case of U.S. sampled census data recent work has provided estimates of the level of sampling error on a tract-by-tract basis and suggests that this will tend to be correlated with tract size and socioeconomic characteristics (Griffith et al., 1994). From this a signal-to-noise ratio could be computed for individual census tracts. These findings may well be relevant to georeferenced natural resources and/or environmental data, such as those collected for forest stands or by the United States Department of Agriculture (USDA).

In contrast, a second group of parameters refers to source map characteristics. These are the mean value (μ), attribute variability (σ^2), and spatial correlation $[r(k,h)]$ of the individual source maps plus the bivariate correlation between the different source maps (ρ). These parameters obviously vary with the particular geography being stored in a GIS database. An estimate of these could be obtained from the observed image (in the case of remote sensing) or map (in the case of census data), providing that what is observed is not thought to be too gross a distortion of ground truth.

Simulation experimental results reported by Arbia et al. (1998) may be tabulated into a table of quantitatively ranked contributors to error propagation (Table 2.1). These findings suggest that attribute error and location error play the most important role in pixel misclassification. With regard to spatial autocorrelation, simulation experimental results reported by Arbia et al. (1998) may be tabulated into a table of quantitatively ranked contributors to error propagation (Table 2.2)—Z(BB) and Z(BW) respectively are the z-scores for the join count statistics used to index spatial autocorrelation latent in dichotomous nominal data. These findings suggest that attribute autocorrelation

is the single most important contributor to autocorrelation characterizing the map pattern of misclassified pixels.

Findings for Arithmetic Map Operations: Addition (Arbia et al., 1999)

When two maps, say $_1\mathbf{S}$ and $_2\mathbf{S}$, are added, the properties of errors may be derived from $T(i,j) = [(_1Z(i,j) + _2Z(i,j)) - (_1S(i,j) + _2S(i,j))]$. The aspatial statistics RMSE (root mean square error) and $\Pr(\theta)$, and the geostatistical semivariogram can be evaluated for these two error propagation situations, illuminating impacts of uncertainty in geospatial information. RMSE may be defined algebraically as $\{E[T(i,j)^2]\}^{1/2} = \{E[_1Z(i,j) - _1S(i,j)]^2 + E[_2Z(i,j) - _2S(i,j)]^2 + 2E[(_1Z(i,j) - _1S(i,j))(_2Z(i,j) - _2S(i,j))]\}^{1/2}$. When the two maps have the same isotropic variance-covariance matrix and both are contaminated by the same errors with respect to $\{w(g,h)\}$ and u (allowing simplification of the subscripts), then, after algebraic manipulation, the following analytical solution can be written:

$$RMSE = \left\{2\sigma^2\left[(1+p)[\alpha_1 + \alpha_2 R(0,1) + \alpha_3 R(1,1)] + SNR^{-1}\right]\right\}^{1/2} \tag{12}$$

where $\alpha_1 = 2(1 - a^{1/2})^2(1 + 2a)$, $\alpha_2 = -8(a^{1/2} - a)^2$, and $\alpha_3 = 2(1 - a^{1/2})^2(2a - 1)$. Numerically evaluating Equation 12 reveals that RMSE for maps with weak spatial structure is dominated by α_1, the effect of which is progressively offset on maps with stronger spatial structure by α_2 and α_3. Considering the spatial correlation components, Equation 12 reveals that average pixel error increases as location error increases and as source map structure decreases. Interaction between these two elements arises because effects of the location error operator are strongest on maps where adjacent values are dissimilar. RMSE increases sharply as location error approaches its maximum for source maps with weak spatial structure. And, RMSE increases as intermap correlation, ρ (≥ 0), increases.

If the two source maps have different structure, and error contamination differs, the function specification yielding Equation 12 depends on nine parameters, the additional parameters corresponding to the parameters in Equation 12 doubled up for the two source maps and error process parameters. Next,

Table 2.1. Quantitatively Ranked Contributors to Error Propagation in Simulation Experimental Results.[a]

| | GIS Operations/Map Properties | | | | | | | |
| | Overlay | | AND | | OR | | XOR | |
Error Source	PMP	Kappa	PMP	Kappa	PMP	Kappa	PMP	Kappa
AA: attribute autocorrelation	3	3.5	3.5	3.5	3	3	3	3
AE: attribute error	1.5	2	2	2	1.5	2	1.5	1.5
LE: location error	1.5	1	1	1	1.5	1	1.5	1.5
Interaction terms: AAAE	*	*	3.5	3.5	*	*	*	*
AALE	4.5	*	*	*	4	4	4	4
AELE	4.5	3.5	*	*	*	*	*	*

[a] Asterisks denote less than 5% of variance accounted for; ties indicate lack of statistically significant differences in percent accounted variance, based upon a binomial distribution.

Table 2.2. Quantitatively Ranked Contributors to Error Propagation with Regard to Spatial Auto-Correlation.

| | GIS Operations/Map Properties | | | | | | | |
| | Overlay | | AND | | OR | | XOR | |
Error Source	Z(BB)	Z(BW)	Z(BB)	Z(BW)	Z(BB)	Z(BW)	Z(BB)	Z(BW)
AA: attribute autocorrelation	1	1	1	1	1	1	1	1
AE: attribute error	*	3.5	*	3.5	*	3	*	*
LE: location error	*	2	2.5	*	2	2	*	*
EA: error autocorrelation	*	*	2.5	3.5	*	*	*	2.5
Interaction terms: AAAE	*	3.5	*	*	*	*	*	*
AAEA	*	*	*	*	*	*	2	2.5

and

$$\Pr(\theta) = \Pr\left\{\frac{|T(i,j)|}{{}_1S(i,j) - {}_2S(i,j)} > \theta\right\}$$

$$= \Pr\left\{\frac{T(i,j)}{{}_1S(i,j) - {}_2S(i,j)} > \theta\right\}$$

$$+ \Pr\left\{\frac{T(i,j)}{{}_1S(i,j) - {}_2S(i,j)} < -\theta\right\}$$

$$= \Pr\{L(i,j) > 0\}$$

$$+ \Pr\{K(i,j) < 0\}$$

where $L(i,j) = [{}_1Z(i,j) + {}_2Z(i,j)] - (1 + \theta)\,[{}_1S(i,j) + {}_2S(i,j)]$, and $K(i,j) = [{}_1Z(i,j) + {}_2Z(i,j)] - (1 - \theta)[{}_1S(i,j) + {}_2S(i,j)]$. Assuming the source maps and error processes have the same properties as described above,

$$E[L(i,j)] = -2\theta\mu, \quad E[K(i,j)] = 2\theta\mu$$

$$\mathrm{Var}[K(i,j)] = E[K(i,j)^2] - (2\theta\mu)^2$$

Paralleling derivation of the RMSE Equation 12,

$$\mathrm{Var}[K(i,j)] = \sigma_K^2$$

$$= 2\sigma^2\big\{(1+\rho)[\delta_{1K} + \delta_{2K}R_s(0,1)$$

$$+ \delta_{3K}R_s(1,1)] + \mathrm{SNR}^{-1}\big\}$$

where $\delta_{1K} = (a - 1 + \theta)^2 + 2(a^{1/2} - a)^2 + (1 - a^{1/2})^4$, $\delta_{2K} = 4(a^{1/2} - a)[a + (1 - a^{1/2})^2 - 1 + \theta]$, and $\delta_{3K} = 2[2(a^{1/2} - a)^2 - (1 - \theta)(1 - a^{1/2})^2]$. The expression $\mathrm{Var}[L(i,j)]$ $(= \sigma_L^2)$ does not change when dealing with $-\theta$ rather than θ. If the random variables are assumed to be Gaussian (normal), then, since linear combinations of normal variables are still normally distributed,

$$Pr(\theta) = Pr\{SND < (-2\,\theta\,\mu\,/\,\sigma_K)\}$$
$$+ Pr\{SND > (2\,\theta\,\mu\,/\,\sigma_L)\} \tag{13}$$

where SND denotes standard normal deviate. The graphical behaviors of $\{\delta_{\bullet K}\}$ and $\{\delta_{\bullet L}\}$ are similar to one another and to the function described by Equation 12.

Analytical results for the spatial properties of error propagation are derivable from the following semi-variance function $\{\gamma(k, h)\}$:

$$\gamma(k,h) = Var[T(i,j)T(i+k,j+h)]\,/\,2$$
$$= E[T(i,j)^2] - E[T(i,j)T(i+k,j+h)]$$

The first term in this equation is the Mean Square Error. Algebraic manipulations render

$$E[T(i,j)T(i+k,j+h)]$$
$$= 2\sigma^2\{(1+\rho)[\eta_1 R(k,h) + \eta_2 R_2 + \eta_3 R_3 + \eta_4 R_4]$$
$$+ SNR^{-1}\,R_u(k,h)\}$$

where: $\eta_1 = 2(1 - a^{1/2})^2\,(1 + 2a) = \alpha_1,$
$\eta_2 = -8\,(a^{1/2} - a)^2 = \alpha_2,$
$\eta_3 = 2(a - 1)\,(1 - a^{1/2})^2,$
$\eta_4 = 2(a^{1/2} - a)^2,$
$R_2 = [R(k+1,h) + R(k,h+1) + R(k-1,h) + R(k,h-1)]/4,$
$R_3 = [R(k-1,h-1) + R(k+1,h+1)]/2,$ and
$R_4 = [R(k-1,h+1) + R(k+1,h-1)]/2$

Combining these expressions yields the following expression for $\gamma(k, h)$:

$$2\sigma^2\left\{(1+\rho)\begin{bmatrix}\eta_1(1 - R(k,h)) + \eta_2(R(0,1) - R_2) + \\ \eta_3(R(1,1) - R_3) + \end{bmatrix}\right.$$
$$\left.\eta_4(R(1,1) - R_4)\right] + SNR^{-1}(1 - R_u(k,h))\right\} \tag{14}$$

Numerical evaluation of Equation 14 reveals that, in the case of weak source map spatial correlation, pattern is present in the distribution of errors only in the case where the error process is spatially correlated. The presence of intermap correlation has no effect on the correlation proportion of the error, but error map variance increases with increasing intermap correlation.

In contrast with the overlay analysis, analytical results reported by Arbia et al. (1999), based upon Equations 12–14, reveal the following prominent importance rank orderings:

RMSE, $Pr(\theta > 0.1)$: attribute error, location error, bivariate correlation, attribute autocorrelation/location error interaction semivariance: attribute error/error autocorrelation interaction, bivariate correlation

These findings suggest that interattribute correlation plays an important role in error propagation, which has salient implications for multivariate georeferenced data analyses. In addition, location error has little impact on semivariance, while having considerable impact on more conventional summaries of error. Finally, the single most important contributor to error propagation when performing the operation of addition is attribute error.

Arithmetic Map Operations: Ratioing (Arbia et al., 1999)

In contrast to the preceding addition situation, when maps $_1S$ and $_2S$ are ratioed, the properties of errors may be derived from

$$V(i,j) = \left[\frac{_1Z(i,j)}{_2Z(i,j)} - \frac{_1S(i,j)}{_2S(i,j)}\right]$$

As with overlay operations, and distinct from the case of map addition, it is not possible to obtain analytical results for most of the error properties under ratioing. Some insights can be gained from addition, though, by noting that the operations of addition and ratioing are related through the logarithm operator; hence, the dependency of error properties on source map and error process properties in the case of ratioing should include the same set of characteristics as arose in the case of addition for any measure, although the quantitative responses will be different and there could be additional interaction effects.

Because addition and ratioing may be related through a logarithmic transformation, the general behavior of the RMSE when ratioing maps should be

Table 2.3. Quantitatively Ranked Contributors to Error Propagation from Simulation Experimental Results.[a]

Error Source	RMSE	$Pr(\theta > 0.1)$	$R_e(1,0)$
AA: attribute autocorrelation	*	3.5	2
AE: attribute error	1	1	*
LE: location error	2.5	2	*
EA: error autocorrelation	*	3.5	1
Interaction terms: AELE	2.5	*	*

[a] Again asterisks denote less than 5% of variance accounted for; ties indicate lack of statistically significant differences in percent accounted variance, based upon a binomial distribution.

very similar to the case for addition. Meanwhile, $Pr(\theta)$ for ratioing is given by

$$Pr\left\{\frac{V(i,j)}{{}_1S(i,j)/{}_2S(i,j)} > \theta\right\}$$

$$+ Pr\left\{\frac{V(i,j)}{{}_1S(i,j)/{}_2S(i,j)} < -\theta\right\}$$

$$= Pr\{Q > 1 + \theta\} + Pr\{Q < 1 - \theta\}$$

where $Q = [{}_2S(i,j)/{}_2Z(i,j)] [{}_1Z(i,j))/{}_1S(i,j)]$ is a ratio of two quadratic forms.

Simulation experimental results reported by Arbia et al. (1998a) may be tabulated into a table of quantitatively ranked contributors to error propagation (Table 2.3). These findings suggest that attribute error is the single most important factor contributing to error propagation in terms of conventional map accuracy measures, with location error also playing an important role. Meanwhile, semivariance is impacted upon most by the underlying autocorrelation structuring of both attribute and error.

An Empirical Case Study: Pediatric Lead Poisoning in Syracuse, New York

Griffith et al. (1998) present a spatial statistical analysis of one outcome of urban environmental pollution, namely children being poisoned by lead. The particular database used in this study comprises 28,521 measures taken for children residing in the city of Syracuse, New York, over the period 1992–1995. The error analysis summarized in this section is for an aggregation of data results into the 1980 census tract geography resolution (58 areal units).

Two components of attribute error are associated with these data: measures recorded in the interval 0–4 micrograms of lead per deciliter of blood—current technology cannot reliably measure serum lead concentrations in this range; and, method of obtaining a blood sample—either with a finger prick (less accurate) or venipuncture (more accurate). Using an E–M algorithm to reestimate values in the range [0,4], with an objective of making this lower tail of the frequency distribution better align with that for a normal distribution, reduces the variance in these data a mere 0.5%. Comparison of capillary versus venipuncture cases reveals a marked discrepancy between results, with the respective variances being 50.5 and 97.0. A subset of 1,274 cases in this data set consists of replicate measures involving one capillary (F) and one venipuncture (V) sample taken in the same calendar year. A regression analysis of these cases yields $\hat{L}_V = 4.730 + 0.389 L_F$, where L_j denotes the serum lead concentration measure yielded by method j, $R^2 = 0.286$ and $s^2 = 23.8$. In other words, these data contain considerable attribute and error variability.

Spatial autocorrelations latent in these data are positive and modest in magnitude. The average pediatric blood lead levels, by census tract, across the city render a Moran Coefficient of 0.38469 (and a Geary Ratio of 0.54683), indicating the presence of weak positive spatial dependency. The corresponding index values for the regression residuals generated with the above equation are 0.14225 and 0.69922; the geographic distribution of this error term displays very weak positive spatial autocorrelation.

Finally, the extent of location error contained in this data set was explored. Of the total cases for the city, 28,251 had recorded census tracts attached to them by personnel obtaining the blood samples; 672 cases had a nonexisting census tract number of 0 attached to them, while 46 had this field left blank. Address matching of these data to the 1990 U.S. Census Bureau TIGER files was completed using ARC/INFO®[1], with 27,621 of the 28,251 cases successfully matched (a 97.8% success rate). Of these, 1,553 cases have an address-matched census tract that differs from the recorded census tract (a 5.6% location error rate); 1,027 of these mismatches are affiliated with adjacent census tracts. Thus, of the location error 3.7% is comparable to that explored in the foregoing simulation analyses, whereas 1.9% arises from a somewhat different source.

Two types of operation commonly performed with such environmental pollution data are overlay and calculation of percentages of elevated lead levels, usually in a quest to identify hot spots. The preceding overlay and ratioing error propagation results suggest that the extent of attribute error here may be quite problematic; this source more than likely overshadows the low levels of location error and error or attribute spatial autocorrelation. The implication is that resources should be concentrated on obtaining better measures of lead poisoning, while GIS address matching could be an economical and effective way to manage location error.

RESEARCH PROSPECT: WHERE TO NEXT?

The general problem of how the quality/uncertainty latent in geospatial information impacts upon spatial decision-making needs to have findings regarding univariate and bivariate cases extended to the multivariate case, and from the rather well-behaved geospatial information situations reported on here to messy, dirty, noisy situations. In order to achieve these two goals, future research needs to address the nine specific questions posed in this section.

Both analytical and simulation experimental results reported in this chapter identify the importance of underlying (true) map structure in relation to error propagation, and in particular how different types of errors interact with statistical properties of an underlying (true) landscape. The relative magnitudes of error propagation sources in georeferenced data have been quantified in terms of their impacts on indices like PMP, Kappa and the join count statistics BB and BW.

Therefore, relaxing assumptions about the error process should further illuminate these elements:

1. What are the elements of uncertainty in real-world geospatial information?

While a basic set of elements of uncertainty is established here, together with quantitative rankings of importance, this set needs to be extensively analyzed in order to ascertain its generality. Assumptions about means, variances, correlations, isotropy, and frequency distributions of variables (i.e., assumptions characterized by Equations 11a, 11b, and 11c) need to be relaxed; the answer to this particular question can verify or modify preliminary rankings already established and reported here.

There is every indication that model specification for the overlay and ratioing map operations can be guided by the specification of Equations 12, 13, and 14. To date, factor interaction complexities have made these equations difficult to posit; nevertheless,

2. What are explicit models of uncertainty for the overlay and ratioing map operations?

Since modeling the error propagated through these map operations defies analytical solution, even in ideal situations, simulation experimentation furnishes a database that can be explored using regression techniques in order to obtain mathematical descriptions of error propagation. In addressing this question, a goal of future research should be to develop extensions to Equations 12, 13, and 14, further determining how uncertainty is propagated through the fusion and analysis of geospatial information.

Apparently in the bivariate case there is a general increase in error property values with increasing intermap correlation, although the nature of relationships remains unaffected, while the relative importance of terms in the analytical expressions 12, 13, and 14 remain unaffected. But intermap correlation may not have much of a pronounced impact on results as the number of maps increases, since as intermap correlations converge to one, maps become redundant, and the multivariate case moves toward the bivariate case described here. Thus,

3. What are the elements of uncertainty in geospatial information that result from linear combinations of many maps?

Often spatial scientists and practitioners are faced with multivariate georeferenced data situations. Frequently, for example, socioeconomic and remotely

---

[1] Registered Trademark of Environmental Systems Research Institute, Redlands, California.

sensed data are subjected to principal components/factor analysis in a quest for parsimony. In addressing this question, a goal of future research should be to extend Equations 12, 13, and 14.

Next, consider the case of water quality analysis (Miles et al., 1992). One useful index for studying water turbidity that is computable from remotely sensed data is the ratio

$$\frac{BAND4}{BAND6}$$

which is the type of map ratioing operation discussed in this chapter. Another index, useful for studying salinity, is

$$\frac{BAND6}{BAND4 + BAND6}$$

which involves a combination of the addition and ratioing map operations. An even more sophisticated index, also useful for studying turbidity, is

$$\frac{BAND4 - BAND5 \times BAND6}{BAND5 - BAND6 \times BAND7}$$

How geospatial data error models, such as Equations 12, 13, and 14, will alter in their specifications in order to accommodate these far more complicated cases remains unknown. Hence,

4. What are the elements of uncertainty in geospatial information that result from sophisticated ratios constructed from many maps?

One major empirical problem affiliated with this question has to do with the construction of variables with remotely sensed data, where measures often are combined arithmetically. In addressing this question, a goal of future research should be to extend the models described here that already have been derived for the bivariate case.

Traditional regression analysis, the workhorse of statistical applications work, employs an interval/ratio response variable, Y, and interval/ratio predictor variables, X. Using this technique for geospatial data analysis relates to error propagation through map addition operations. It also can embrace error propagation through the addition of map ratioing operations. As soon as indicator variables are included in a regres-

sion equation (analysis of variance or covariance as a regression problem), the map overlay operation is introduced. Consequently, commonly georeferenced data analyses may well involve all three types of map operations. As such, error propagation in this composite instance merits considerable attention, too. Therefore,

5. What are the elements of uncertainty in geospatial information that result from mixing overlay and arithmetic map operations in a spatial analysis?

Moreover, most georeference data analyses involve numerous map operations, rather than simply a single type. In addressing this question, a goal of future research should be to extend those models developed for individual map operations to the hybrid situation of a mixture of map operations.

The emphasis in this chapter is on error components arising from standard GIS database manipulations. Of particular concern may be the merging of different remotely sensed images, or the combining of images based upon different sized pixels. With reference to vector data, the failure of different surface partitionings to nest, or the changing of areal unit boundaries through time (e.g., census tracts for different decennial censuses) also can introduce the type of location error defined here. Of note is that GPS is being used in an attempt to remove some of these types of location error. But certain classes of geography, such as those involving aggregate areal units, may well be more or less prone to serious error propagation. A new and improved taxonomy of location error and its propagation effects may be called for here. Thus,

6. What kinds of error structures are introduced into data sets as a result of carrying out (a) merging remotely sensed images, (b) GIS address matching, (c) common raster-to-vector/vector-to-raster conversion, (d) use of inappropriate areal unit polygon centroids, (e) merging noncompatible surface partitionings, or (f) hardening fuzzy boundaries?

Other sources of location error, which may be more or less severe than surface misalignment, need to be explored in order to attain a generalized conclusion. In addressing this question, a goal of future research should be to investigate the spectrum of location error sources, as well as their relative importance rankings among the set of uncertainty elements.

Spatial sampling error is a component of the signal-to-noise ratio discussed here. Meanwhile, "error coincidence" (Veregin, 1995, p. 599) refers to bivariate correlation existing between the error processes at the same location on two source maps. This potentially could have serious effects in the case of census data, where identical levels of error in the form of under- or overreporting can be introduced by the same household over a set of questions. Thus, another version of the geospatial data model is called for here. The affiliated research question takes on a particular importance in the context of interpolated data where the sampling error is linked to, say, the use of kriging, especially since the resulting confidence map is used for visualizing sampling error across a surface. Therefore,

7. How does sampling error, like that associated with national census data (e.g., USDA) or special programs such as the United States Environmental Protection Agency's EMAP project, relate to the other elements of uncertainty latent in geospatial information?

One of the recognized differences between census data and many other types of georeferenced data, such as remotely sensed data, is the apparent absence of measurement error spatial correlation. In addressing this question, a goal of future research should be to extend models already formulated for this special case.

Overall, the preceding seven questions help establish a research agenda that seeks to contribute to improving the quality of "uncertainty reporting" for georeferenced databases. Given facts (which a practitioner might glean from a simulation experiment) about the probable types of error and their quantitative magnitudes in a data set, a practitioner should be cautioned as to which kinds of map analyses may give acceptable results (in terms of level of accuracy) for a map. Uncertainties associated with georeferenced data (a consequence of data errors) are magnified by performing certain map operations, suggesting that analysis needs to be underpinned by providing selected measures of this propagated error. Details about location error should accompany a spatial database, possibly taking the form of an index. Measures of attribute variability across the map and between regions of the map appear to be necessary. An appropriate spatial correlation statistic should be reported for each attribute for map segments. Properties of the method of data collection should routinely provide information on the variance of the error process associated with taking measurements and any known spatial correlation among the errors. The main impact on data interpretation concerns the extent to which analysis has been undermined. A practitioner needs to generate the distribution of indices like PMP and Kappa under assumptions that are consistent both with the empirical

properties of a geographic attribute and known error properties of the method of data collection. As noted above, the issue of map accuracy includes both the magnitude of errors and their spatial distribution, and in particular whether they stand out or blend in with the underlying true map structure. A practitioner can be advised about the propensity for error regions to form on a given map. It may not be sufficient for an error to be an outlier in the statistical distribution sense; it must also be a spatial outlier—different from its set of neighboring values—before it will be readily detectable. In other words, propagated error is a function both of the magnitude and the spatial distribution of errors in observed maps; the critical question here asks whether or not errors stand out or blend in to their statistical as well as their underlying true map geographic distributions. The spatial scientist also needs to know, at this point, how georeferenced data interpretations might be qualified in the presence of propagated error. Thus,

8. What geospatial data quality toolbox should a researcher have at her/his disposal in order to better cope with complexities associated with decision-making based upon georeferenced data containing error?

The effective communication of uncertainty is key to better understanding it, managing it, and coping with its effects on spatial decision-making. Just as results for an opinion poll need to be accompanied by a margin-of-error statement, geospatial information needs to be accompanied by instruments that enable it to become an innate part of the way geospatial information is described. In addressing this question, a goal of future research should be to formulate a toolbox that will allow users, from novice to professional, to better comprehend uncertainty and error propagation effects for any georeferenced database of interest.

A special instance of this last question meriting attention asks what kinds of spatial queries can be posed and how robust they are. A distinction must be made between classification-type queries and spatial relational queries (e.g., asking questions about errors arising in the identification of all areas that are within distance *x* units of a location with attribute value *y*). Not only may certain indices be more robust than others in the presence of errors, but certain generic classes of queries also may be more robust than others; an obvious distinction is between queries that involve pixel classification (Is this area correctly classified?) and those that involve spatial relationships (how many pixels are correctly identified as falling within distance band x of a site with a certain property). A map might have all sorts of horrendous errors of the first type while queries of the second type remain largely unaffected.

Finally, certain sources of error are a product of measurement instruments, and as such can be reduced through instrumentation advancements and/or applying instrument theory. Meanwhile, effects of most sources of error appear to result from interactions between them and location error; hence, lessening location error will tend to reduce the impact of other sources on error propagation. A good strategy where error exists may well be to employ appropriate spatial statistical techniques—for example, including a valid spatial error model when using regression. Therefore,

9. What recommendations can be made regarding (a) where resources might be invested in order to reduce error propagation, and (b) approaches to error propagation management in spatial databases?

Since location error repeatedly emerged as an important component in results summarized in this chapter, and since it is a source of error that can be controlled, every attempt should be made to do so. Producers of remotely sensed data already devote considerable resources to rectifying these data, aligning maps with ground truth markers. But investing resources in order to reduce other sources of error, and to devise routines for managing these other sources of error remains undefined. In addressing this question, a goal of future research should be to achieve a broad-based understanding of uncertainty, encouraging improvements in the way in which geospatial information is manufactured.

CONCLUSIONS

Selected analytical and simulation experimental findings for overlay and simple arithmetic map operations are inventoried here. These results reveal the relative importance of attribute and location error, and attribute and error spatial autocorrelation in error propagated with simple map operations. A better understanding of the relative importance of these factors can be achieved through future work that addresses the nine questions posed as a prospective research agenda. A summary of a detailed analysis of an empirical case study demonstrates how the type of results reported in this chapter can be used to decide upon where resources should be marshaled in order to better contain, minimize, and/or manage sources of error in georeferenced data sets, with the ultimate ob-

jective of controlling error propagation and hence strengthening spatial decision-making support for the natural resources and environmental sciences.

ACKNOWLEDGMENTS

Research for parts of this chapter was made possible by a grant from the NATO Scientific Affairs Division, Reference 900999.

REFERENCES

Amrhein, C. and D. Griffith. Errors in Spatial Databases: A Summary of Results From Several Research Projects, in *Proceedings, International Symposium on the Spatial Accuracy of Natural Resources Databases*, R. Congalton, Ed., American Society for Photogrammetry and Remote Sensing, Bethesda, MD, 1994, pp. 214–226.

Arbia, G., R. Haining, and D. Griffith. Error Propagation Modelling in GIS Overlay Operations, *Int. J. Geogr. Inf. Syst.*, 12, pp. 145–167, 1998.

Arbia, G., R. Haining, and D. Griffith. Modelling Error Propagation in Maps: Arithmetic Operations. *Cartography and Geographic Information Systems*, forthcoming, 1999.

Berry, J. Fundamental Operations in Computer Assisted Map Analysis. *Int. J. Geogr. Inf. Syst.*, 1, pp. 119–136, 1987.

Cliff, A. and J. Ord. *Spatial Processes*, Pion, London, 1981.

Congalton, R. A Review of Assessing the Accuracy of Classifications of Remotely Sensed Data, *Remote Sensing Environ.*, 37, pp. 35–46, 1991.

Cressie, N. *Statistics for Spatial Data.* John Wiley & Sons, New York, 1991.

Curran, P. Multispectral Remote Sensing of Vegetation Amount, *Prog. Phys. Geogr.*, 4, pp. 315–341, 1980.

Emmi, P. and C. Horton. A Monte Carlo Simulation of Error Propagation in a GIS Based Assessment of Seismic Risk, *Int. J. Geogr. Inf. Syst.*, 9, pp. 447–461, 1995.

Fisher, P. Modelling Soil Map-Unit Inclusion by Monte Carlo Simulation, *Int. J. Geogr. Inf. Syst.*, 5, pp. 193–208, 1991.

Forster, B. Urban Residential Ground Cover Using LANDSAT Digital Data, *Photogrammetric Eng. Remote Sensing*, 46, pp. 1725–1734, 1980.

Geman, S. and D. Geman. Stochastic Relaxation, Gibbs Distributions and the Bayesian Restoration of Images, *IEEE Trans. in Pattern Anal. Mach. Intelligence*, 6, pp. 721–742, 1984.

Geman, D. and G. Reynolds. Constrained Restoration and the Recovery of Discontinuities, *IEEE Trans. Pattern Anal. Mach. Intelligence*, 14, pp. 367–383, 1992.

Goodchild, M. and S. Gopal. *Accuracy of Spatial Databases,* Taylor and Francis, London, 1989.

Goodchild, M., S. Guoging, and Y. Shiren. Development and Test of an Error Mode for Categorical Data, *Int. J. Geogr. Inf. Syst.,* 6, pp. 87–104, 1992.

Griffith, D., R. Haining, and G. Arbia. Heterogeneity of Attribute Sampling Error in Spatial Data Sets, *Geogr. Anal.,* 26, pp. 300–320, 1994.

Griffith, D., P. Doyle, D. Wheeler, and D. Johnson. A GIS and Spatial Statistical Analysis of Urban Childhood Lead Pollution Exposure: A Case Study of Syracuse, NY, Annals, Association of American Geographers, 88, pp 640–665, 1998.

Guptill, S. and J. Morrison, Eds. *Elements of Spatial Data Quality,* Pergamon, Tarrytown, NY, 1995.

Haining, R. *Spatial Data Analysis in the Social and Environmental Sciences,* Cambridge University Press, 1993.

Haining, R. and G. Arbia. Error Propagation Through Map Operations, *Technometrics,* 35, pp. 293–305, 1992.

Heuvelink, G. *Error Propagation in Quantitative Spatial Modelling,* Cip-Gegevens Koninklijke Bibliotheek, Den Haag, 1993.

Heuvelink, G. and P. Burrough. Error Propagation in Cartographic Modelling Using Boolean Logic and Continuous Classification, *Int. J. Geogr. Inf. Syst.,* 7, pp. 231–246, 1993.

Lantner, D. and H. Veregin. A Research Paradigm for Propagating Error in Layer-Based GIS, *Photogrammetric Eng. Remote Sensing,* 58, pp. 825–883, 1992.

Lunetta, R., R. Congalton, L. Fenstermaker, J. Jensen, K. McGwire, and L. Tinney. Remote Sensing and Geographic Information System Data Integration: Error Sources and Research Issues, *Photogrammetric Eng. Remote Sensing,* 57, pp. 677–687, 1991.

Miles, M., D. Stow, and J. Jones. Incorporating the Expansion Method Into Remote Sensing-Based Water Quality Analysis, in *Applications of the Expansion Method,* J. Jones and E. Casetti, Eds., Routledge, London, 1992, pp. 279–296.

Mulry, M. and B. Spencer. Accuracy of the 1990 Census and Undercount Adjustments, *J. Am. Stat. Assoc.,* 88, pp. 1080–1091, 1993.

Parrott, R. and F. Stutz. Urban GIS Applications, in *Geographical Information Systems. Principles and Applications,* D. Maguire, M. Goodchild, and D. Rhind, Eds., Longman, Harlow, 1991.

Richards, J. *Remote Sensing Digital Image Analysis,* Springer-Verlag, Berlin, 1986.

Ripley, B. *Statistical Inference for Spatial Processes,* Cambridge University Press, 1988.

Rosenfeld, A. and A. Kak. *Digital Picture Processing,* Vols. 1 and 2, Academic Press, New York, 1982.

Rousseau, D. A New Skill Score for the Evaluation of Yes/No Forecasts, *WMO Symposium on Probabilistic and Statistical Methods in Weather Forecasting,* Nice, 8–12 September, 1980, pp. 167–174.

Schenker, N. Undercount in the 1990 Census, *J. Am. Stat. Assoc.,* 88, pp. 1044–1046, 1993.

Stehman, S. Estimating the Kappa Coefficient and Its Variance Under Stratified Random Sampling, *Photogrammetric Eng. Remote Sensing,* 62, pp. 401–407, 1996.

Taylor, J. *An Introduction to Error Analysis,* University Science Books, Mill Valley, CA, 1982.

Thapa, K. and J. Bossler. Accuracy of Spatial Data Used in Geographic Information Systems, *Photogrammetric Eng. Remote Sensing,* 58, pp. 835–841, 1992.

Unwin, D. Geographical Information Systems and the Problem of "Error and Uncertainty," *Prog. Human Geogr.,* 19, pp. 549–558, 1995.

Veregin, H. Integration of Simulation Modelling and Error Propagation for the Buffer Operation in GIS, *Photogrammetric Eng. Remote Sensing,* 60, pp. 427–435, 1994.

Veregin, H. Developing and Testing of an Error Propagation Model for GIS Overlay Operations, *Int. J. Geogr. Inf. Syst.,* 9, pp. 595–619, 1995.

Veregin, H. Error Propagation through the Buffer Operation for Probability Surfaces, *Photogrammetric Eng. Remote Sensing,* 62, pp. 419–428, 1996.

Welsch, R., T. Jordan, and M. Ehlers. Comparative Evaluations of the Geodetic Accuracy and Cartographic Potential of LANDSAT-4 and LANDSAT-5 Thematic Mapper Image Data, *Photogrammetric Eng. Remote Sensing,* 51, pp. 1799–1812, 1985.

CHAPTER 3

Speaking Truth to Power: An Agenda for Change

N.R. Chrisman

DATA QUALITY CONCERNS IN GIS

Fifteen years ago, I traveled to Hull, Québec to deliver a talk entitled "The Role of Quality Information in the Long-Term Functioning of a GIS" (Chrisman, 1984). The issue of accuracy and data quality was not considered the most important concern at Auto-Carto 6. My session was at the very end of the day, when most attendees were on their way to more relaxed pursuits. Many of the well-established experts on GIS were highly skeptical of my claim that data quality information would grow to become a major proportion of a total system. I do not need to detail the extent of the changes over the past 15 years, but data quality is certainly much more prominent. The term metadata has created new kinds of information resources, as clearinghouses and digital libraries become a part of the information landscape. So it is no longer possible to claim that data quality issues have been ignored.

This third conference devoted to studies of accuracy of spatial data demonstrates the expanding interest. The research community has recognized the need and has vastly expanded the knowledge of many aspects of accuracy assessment and data quality more generally. Though it took many years to come to fruition, the concept of "fitness for use" has become an element of most data standards efforts worldwide (National Institute of Standards and Technology, 1992, for example; see also: Moellering, 1991; Guptill and Morrison, 1995). This signals a remarkable change from the era of technological paternalism when agencies promulgated their own criteria for accuracy assessment, and simply informed a user whether the map sheet "conforms to National Map Accuracy Standards."

Before we give ourselves too much credit, these events are by no means unusual. The overall trends during the past 15 years have seen enormous alterations. The world economy has continued to globalize and has continued to place great reliance on information to reduce all forms of wasted effort. This trend has favored "flexible production" in the place of the old "Fordist" model of uniform products (Scott and Storper, 1986). The emphasis on fitness for use can be interpreted as the cartographic community's reaction to these larger trends (Chrisman, 1991). Since 1983, the trend in computer capacities has increased at ever expanding rates. Yet, the nature of our GIS software has not evolved as rapidly. Data quality, while it has become recognized in the research arena, has not become a driving force in the software packages available to most users. Most of the software still emulates the procedures of the predigital era. There are many challenges that have not been adequately addressed. This chapter will present one detailed example of a procedure that would have direct impact on virtually all GIS users. Before beginning the example, it is important to establish some revisions in the underlying reasons for studying data quality and map accuracy.

SPEAKING TRUTH TO POWER

The main title of this chapter comes from the Quaker community. It encapsulates a spirit of resistance to established power structures, and the belief that persuasion will defeat prejudice. The connection to the study of map accuracy might seem immediate. Accuracy makes a claim to compare a measurement to the "truth," thus giving a strength in any argument. While this interpretation of accuracy is quite common, and may characterize the majority of those attending this conference, I have come to doubt this simple interpretation. I have found a different way of interpreting the work of accuracy assessment that leads me to a more indirect reading of the Quaker maxim.

The "scientific method," at least since the time of Francis Bacon, has placed great emphasis on direct observations of nature. Truth is supposed to lie ready for discovery by the properly prepared scientist. The story of "objective discovery" implies that the researcher is not really responsible for their discovery; the truth was implicit in nature, simply waiting to be revealed. By contrast, untrue scientific statements are entirely due to human errors by the researcher. Desire to conform to social expectations, pressure to support political structures, economic weakness can be marshaled to explain away the mistakes of a scientist influenced by social distractions. As Bruno Latour (1993, p. 92) characterizes the common interpretation of science:

> Errors, beliefs could be explained socially, but truth remained self-explanatory. It was certainly possible to analyze a belief in flying saucers, but not the knowledge of black holes.

It is very hard to avoid the hero worship of the great explorers of science whose efforts revealed such a succession of grand new truths. Like a history written after the fact to justify the actions of the victors, these heroic accounts do not stand up to careful scrutiny. Boyle's experiments on air pressure, to pick one well-documented account (Shapin and Schaffer, 1985), did not seem at the time to have such pure objective truth due to the problems with the nature of his mechanical equipment—his air pump leaked and his observations flew in the face of long-established explanatory logic (dating back to Aristotle). Now we may think that Boyle's critics (such as Hobbes) were influenced by their context, their scholastic training, their reliance on authority, but so was Boyle who adopted a jury-like form of "witnessing" for his demonstrations. Boyle's scientific method was as strongly rooted in the image of the gentleman as it was in technical details of his pump. As a result of such studies, the sociologists of science have come to understand that error and truth do not come from two different kinds of work

(Bloor, 1976). Latour (1993) goes further in demanding a strict symmetry in seeing the constant interaction between nature and society. Truth is not directly or necessarily located in "nature," and error is not simply the result of human failings.

What do such academic arguments matter to practitioners of GIS? Any project manager knows that error is inevitable. No database can hope to represent every object in nature without any flaw. No one expects a Nobel prize for their efforts in operating a GIS; the work is closer to routine utility maintenance than high science. Yet, in the accuracy assessment sphere, we often set up standards of truth without thinking them through sufficiently. The standards and procedures of our measurements do not come from a pristine nature, but are deeply social in their origins and in their operations. As one who has sat through 10 years of standards committee meetings, I can attest to the multiple levels of political, cultural, psychological interactions that produce the agreements that make science (or GIS data transfers) possible. Yet, the social component does not wall itself off from the rigor of observing the world, and trying to confront it as it is, not as we would like it to be. The critical element of data quality assessment is the willingness to confront all the array of powerful human institutions with the truth. It is in this act of persuasion, itself intensely social, that the Quaker maxim takes on its application to data quality assessment. All the social forces in the world become unconvincing when they suggest some phenomenon that simply does not exist. Persuasion, and the assembly of allies who will agree with your statements, is the real core of the scientific method (Latour, 1987).

AN AGENDA FOR CHANGE

One of the most pervasive forces in technology is the tyranny of the way things have always been done. The research community can produce great advances, but they do not influence the practice of GIS until the research concepts change the routine ways of doing things. This chapter cannot recount all the possible topics, but it will focus on one problem that highlights missed opportunities.

Coordinate Transformations

If the central commonality of GIS is the integration of different sources, almost every application requires some form of geometric registration. To connect the measurements obtained on a digitizer, the device units are transformed into map coordinates by solving the correspondence for a set of known points. To transform a remotely sensed image to other layers, a set of known features is located on the image. Perhaps the greatest quantity of research on this process is in the field of photogrammetry. This procedure is critical to the whole GIS enterprise, yet the tools presented to users are considered so simple that they are rarely called into question. Like many elements of technology, once the decisions are made, it becomes a "black box" whose internal structure is not worth considering (Latour, 1987).

To pry open the black box, some set of points are used to estimate the parameters for some mathematical formulation that will translate other points, not just those whose values are known. The decisions embedded in the software concern the nature of the mathematical formulas, for instance, the use of linear equations as in the affine, the projective, or piecewise equations (White and Griffin, 1985). The estimation method typically uses least squares, a decision that embeds certain assumptions. In addition, the number of points used for registration strongly influences the result.

The first issue is the number of points used for the transformation. This may seem like the least controversial from the research perspective. Considering that the user is fitting a linear equation with six unknowns (in the case of an affine), having three coordinate pairs provides no redundancy, hence no check for errors. How many points are "enough"? One might debate whether 12 or 20 would begin to be enough, but four (4) would not pass muster as a statistical procedure. Yet, the market leader in GIS software suggests four points (ESRI, 1991, p. 5–13): "Select 4 widely spaced points common to maps A and B to be used as tics for A." ESRI software will handle many more points, but four is considered adequate. Some other systems only allow four, no more (Planet One Corp., 1997a). In the translation from the research community to the world of practice, important messages are not being communicated. In the case of Planet One, it seems to be a drive for a simple, minimal interface that led to a fixed number of control points. Also, they are simply following the lead of their Business Partner, ESRI.

Another part of the process involves the mathematics of the coordinate transformation. The affine, the overwhelming favorite in GIS software, is a rather odd choice. If the sources are in the same projection, and simply need to be subject to the rotation, translation, and scaling described in most textbooks, then the simi-

larity transformation should be adequate. This assumes that the scaling on the input axes should be preserved. The affine relaxes the assumption about a common scale and permits some degree of distortion (varying scales from one side to another). In many cases, sources run through these transformations are not in the same projection. In the USA, this is common when topographic quadrangles (compiled on state plane projections or the antiquated polyconic) are merged with UTM sources such as digital elevation matrices. The affine might paper over some of the local differences between two projections, but it is not the appropriate method to apply. Users are unlikely to be able to decipher the report of the transformation parameters to be able to see if the differences between the X and Y scales are reasonable, since the parameters reported include the rotation component in the same numbers. Here the software designers are serving their own needs to carry out routine mathematical calculations, but they are not reporting them back to the user in such a way to warn them of possible problems with data quality. It is most amusing to see the dialog box (Planet One Corp., 1997b) provided in one ArcView tool that calculates an affine transformation although the software provides no way to actually rotate a background image. The user is told to go rotate by many degrees in some other package. Some software packages also provide Helmert's projective transformation. This is particularly appropriate for unrectified air photographs and any source where the scale would vary outward from a center. Of course, this requires more parameters, and four points do not provide any redundancy. To solve for a center and the scaling away from it, the points should also be distributed more densely than the four-corner approach that seems to be common in GIS documentation. The four corner points are all equidistant from the center of a rectangle, so they provide no information on change in scale from the center. Though the option to use the projective is there, most users have very little reason to deviate from the default.

The deepest dark corner of the black box is the method of estimation. Software packages will use least squares solutions to generate the parameters that fit the observed points. Least squares is a common fixture of statistical methods and other applied mathematics. The time-tested technique does indeed produce the most efficient use of the observed information, under a set of assumptions that need to be recalled and more clearly advertised to users. The conditions for least squares are simple, and quite strict. And they are fundamentally social at their core, they create a division of labor between the estimation technique and the user. They also distinguish "good" (random) error from "bad" (human-created) error. These divisions are counterproductive.

Least squares provides a Best Linear Unbiased Estimate (BLUE) of the parameters under the condition that the errors in the points come from a common normal distribution. The shorthand is *iid*—independently and identically distributed normal variates. Academic treatment of error and adjustments (Mikhail and Ackermann, 1976, for example) divides error into three basic categories: systematic, random, and blunders. Each of the coordinate transformation methods will remove various kinds of systematic effects. The least squares method is designed to provide the best estimate of the parameters, given the random error in the points. Blunders are meant to be removed prior to invoking the procedure. Here is a grand chasm between the world of research and the world of practice. With only four points, the user has next to no information to alert to the existence of blunders. In some software, the text message about the fit of the parameters flashes by on a screen that is covered over by the graphics display. Here the software conspires to hide the numerical details from the user, thus creating a dysfunctional situation.

Blunders are a residual category, created by the mathematical model. Since "random" error is defined as having iid Normal distribution, anything else is simply not the responsibility of the statistical model. Blunders are blamed on the user; they are tied to human failings such as reversed digits, selecting the wrong object on the photo, and whatever else might go wrong. The logic here is airtight; least squares is the most efficient procedure given random error, blunders do not behave according to random distributions, hence all blunders must be removed prior to using least squares. There is a substantial literature on removing blunders from photogrammetric and surveying adjustments (Kavouras, 1982; Kubik et al., 1988, for example). Detecting blunders certainly requires more points than the bare minimum, so the standards of practice in GIS do not support this necessary step.

Perhaps in an earlier period such a two-phase procedure made sense. On a manual calculator, least squares can be done in a single pass, but more fancy iterative techniques would impose a serious time cost. The explosion of computer capacity has changed all the ratios of effort, but the software industry has not fully exploited these changes. In other numerical disciplines such as statistics, there has been a continual development of outlier detection methods, of exploratory procedures more generally, and of robust estimation procedures (Draper and John, 1981; Hoaglin et al., 1985; Hampel et al., 1986, for example). The research community in the mapping sciences has taken notice of these developments (Kubik et al., 1987, for example). One particular estimation procedure, least median squares (Rousseeuw, 1984; Rousseeuw and Leroy, 1987), has been applied for estimation of mapping transformations (Shyue, 1989). Rather than separating the steps, the points can be contaminated with up to 50% blunders, and the estimation will ignore the blunders and fit the parameters to the rest of the data. Unlike some of the complex weighting techniques, this requires virtually no user intervention or tricky tuning parameters. In the spirit of bootstrap methods, the least median procedure (Rousseeuw and Leroy, 1987) is obtained by sampling the possible combinations of data points. The regression is estimated for each combination, and the one with the best fit (least median squared) is selected. For large numbers of points, the number of possible combinations rises very rapidly, but for the typical numbers of points used for map registration (10–20), a few thousand samples are bound to separate out the outliers. Of course, least medians do not wring all the value out of every point, but they remove the division of labor. This estimation technique is built to compensate for the possibility of blunders.

Least median squares and other robust techniques are a technology with clear advantages over the earlier techniques. The academic research sector has done its job in exploring these techniques and reporting them in academic outlets. None of this changes the tools that users see until the software vendors adopt these innovations. There is nothing inevitable about this next step. Good ideas do not move inexorably into the large software packages. As long as the documentation presented to the user is so slim, the users barely recognize how the process is done by their current software. It is unlikely that consumer demand will work. The academic sector tends to continue chasing off toward every better improvement, without ensuring that their innovations become converted into practice. And, to recognize the harried circumstances of the software team, no software manager wants to waste effort on some feature that no one will recognize. There are plenty of other priorities for each actor in the network. Consequentially, the story of coordinate transformations is the story of many other situations where the software remains stuck on the tried and true, not the most recent research results.

CONCLUSION

The example of coordinate transformation procedures points out that data quality issues cannot be described by setting up some external "truth" as the arbiter. The concept of best fit has developed from a given historical context when certain kinds of computing were more possible than others. These approaches, once adopted, do not simply shrivel up when the original rationale changes. Each group involved, users, programmers, salespersons, and researchers shares the responsibility for the current failures of communication. The power of the way things have always been is what must be resisted. New ideas must be demonstrated so that all can benefit. Too frequently the research community thinks their job is over when the publication has been accepted. There is a responsibility to convert important ideas into changes in practice that benefit a larger community.

REFERENCES

Bloor, D. *Knowledge and Social Imagery*, Routledge & Kegan Paul, London, 1976.

Chrisman, N.R. The Role of Quality Information in the Long-Term Functioning of a Geographic Information System, *Cartographica,* 21(3&4), pp. 79–87, 1984.

Chrisman, N.R. Building a Geography of Cartography: Cartographic Institutions in Cultural Context, *Proceedings International Cartographic Association 15th Conference,* Bournemouth, UK, 1, pp. 83–92, 1991.

Draper, N.R. and J.A. John. Influential Observations and Outliers in Regression, *Technometrics,* 23, pp. 21–26, 1981.

ESRI, *Map Projections and Coordinate Management: Concepts and Procedures*, Environmental Systems Research Institute, Redlands CA, 1991.

Guptill, S.K. and J. Morrison, Eds. *Elements of Spatial Data Quality,* published on behalf of the International Cartographic Association, Elsevier, Oxford, 1995.

Hampel, F.R., E.M. Ronchetti, P.J. Rousseeuw, and W.A. Stahel. *Robust Statistics: The Approach Based on Influence Functions,* John Wiley & Sons, New York, 1986.

Hoaglin, D.C., F. Mosteller, and J.W. Tukey. *Exploring Data Tables, Trends, and Shapes,* John Wiley & Sons, New York, 1985.

Kavouras, M. *On the Detection of Outliers and the Determination of Reliability in Geodetic Networks,* Technical Report 87, Department of Surveying Engineering, University of New Brunswick, 1982.

Kubik, K., K. Lyons, and D. Merchant. Photogrammetric Work without Blunders, *Photogrammetric Eng. Remote Sensing,* 54, pp. 167–169, 1988.

Kubik, K., D. Merchant, and T. Schenk. Robust Estimation in Photogrammetry. *Photogrammetric Eng. Remote Sensing,* 53, pp. 167–169, 1987.

Latour, B. *Science in Action.* Harvard University Press, Cambridge, MA, 1987.

Latour, B. *We Never Were Modern.* Harvard University Press, Cambridge, MA, 1993.

Mikhail, E.M. and F. Ackermann. *Observations and Least Squares.* IEP-Dunn-Donnely, New York, 1976.

Moellering, H., Ed. *Spatial Database Transfer Standards: Current International Status.* Elsevier, New York, 1991.

National Institute of Standards and Technology. *Spatial Data Transfer Standard.* National Institute of Standards and Technology, Department of Commerce, Washington, DC, 1992.

Planet One Corp., Image Registration and Reference Point Tools, http://www.planetonegis.com/pages/image/text_image13.htm, 1997a.

Planet One Corp, Recommending a Rotation, http://www.planetonegis.com/pages/image/text_ image 22.htm, 1997b.

Rousseeuw, P.J. Least Median Squares Regression, *J. Am. Stat. Assoc.,* 79, pp. 871–880, 1984.

Rousseeuw, P.J. and A.M. Leroy. *Robust Regression and Outlier Detection*, John Wiley & Sons, New York, 1987.

Scott, A.J. and M. Storper, Eds. *Production, Work, Territory: The Geographical Anatomy of Industrial Capitalism,* Allen & Unwin, London, 1986.

Shapin, S. and S. Schaffer. *Leviathan and the Air Pump: Hobbes, Boyle, and the Experimental Life*, Princeton University Press, Princeton NJ, 1985.

Shyue, S.W. High Breakdown Point Robust Estimation for Outlier Detection in Photogrammetry, unpublished Ph.D. dissertation, University of Washington, 1989.

White, M. and P. Griffin. Piecewise Linear Rubber Sheet Map Transformation, *The American Cartographer,* 12, pp. 123–131, 1985.

Part II
Sensitivity of Decision-Making to Spatial Uncertainty

As more and more individuals become involved in the study of spatial uncertainty, one of the principal questions remains "How important is spatial uncertainty relative to decisions that affect society?" The answer to this question depends on the use to which spatial data are being put and who is answering the question. Scientists who like to work with exact quantities may consider any uncertainty to be too much. To users of spatial data, however, the latent uncertainty inherent in spatial databases may be important only if it would cause a given decision to be changed. The chapters in this part address both of these groups.

The first chapter, by Agumya and Hunter, describes a general framework for deciding whether or not spatial data are suitable for any given use. The second two chapters, by Stehman, and Gascoigne and Wadsworth, address the issue of spatial data uncertainty for decision-making relative to specific domains: wildlife habitat and air pollution. The final four chapters by Defourny et al., Lewis, Zeng and Cowell, and Chaplot et al. discuss the sensitivity of certain common data sources to various analytical techniques, to different magnitudes and distributions of errors, and to errors inherent in different data sources.

CHAPTER 4

Assessing "Fitness for Use" of Geographic Information: What Risk Are We Prepared to Accept in Our Decisions?

A. Agumya and G.J. Hunter

INTRODUCTION

Geographic information is today increasingly used in decision-making, thanks to technological developments that have made its collection, handling, and analysis in digital form more accessible to an ever-growing community of users. However, the utility of this information is contingent upon its appropriate use; that is, through avoidance of misuse or erroneous use. In order to ensure appropriate application of geographic information, it is essential for users to assess its "fitness for use" prior to using it. For this purpose metadata that reports uncertainty in the information is essential, and while this is now a common supplement to geographic information, the problem of how to ascertain fitness for use still remains.

This problem is particularly critical in the case of nonexpert users—who constitute a considerable and expanding proportion of the user-base—and for decisions that rely very much on geographic information and/or for which the consequences of inappropriate use are severe. The significance of assessing fitness for use is largely attributed to recognition of the potential for adverse consequences arising from using data of inappropriate quality. Other important reasons include averting wastage, and opportunities that are lost when users pay more to access and process data of unnecessarily high quality.

Fitness for use is essentially evaluated by establishing whether the consequences of error and uncertainty in the information are acceptable to the user. One approach is currently being developed by Agumya and Hunter (1997a) whereby such consequences are expressed in terms of risk. According to their method, which is valid only for structured decision problems, assessing fitness for use reduces to comparing risk in the decision (arising from uncertainty in the information) with the risk acceptable to the decision-maker. Information is considered to be fit for use when the risk due to its uncertainty does not exceed the acceptable risk. Unfortunately, the comparison of the two risks is the easiest part of this approach—the biggest challenge lies in estimating the two risks.

While a framework for estimating risk due to uncertainty in geographic information has already been presented in Agumya and Hunter (1997a, 1997b), this chapter is primarily concerned with estimating acceptable risk. The chapter is divided into two main sections—the first discusses the standards-based procedure traditionally employed to assess fitness for use and compares it with its risk-based counterpart, while the second section examines the issue of acceptable risk in the latter procedure and suggests how it can be estimated.

PROCEDURES TO ASSESS FITNESS FOR USE

Assessing fitness for use can be likened to quality control—that is, determining the acceptability of a product on the basis of a given specification (Balce, 1987). The three basic components of quality control are: specification, product, and procedure. In the case of fitness for use, the product is the information, while the procedure is the method of assessing whether or not the information is fit for use, subject to meeting certain specifications. The procedure determines the form in which the specifications are defined while the reliability of the quality control depends on the procedure and specifications used. This chapter is primarily concerned with setting specifications; however, because of the link between procedure and specifications it has been deemed appropriate to devote the first part of it to procedures, and in particular to compare the risk-based procedure with the more commonly used standards-based procedure. In order to facilitate the comparison, the desirable characteristics of a fitness for use procedure are presented first.

Desirable Characteristics of a Procedure for Assessing Fitness for Use

- Simple and Cheap: the procedure should be easy (even for nonexpert users) and inexpensive to apply. A common treatment of uncertainty has been simply to ignore it (Coward and Heywood, 1991; Brunsdon and Openshaw, 1993), partly because solutions such as uncertainty handling functions in commercial GIS packages are not yet widely available, and partly because it is perceived that treatment of uncertainty comes at an unjustifiable price.
- Easily Understood: the procedure should be founded on concepts such that users can easily associate information uncertainty with the impact on their decisions.
- Use of Available Metadata: the procedure should not demand more than is provided by the metadata compiled for the data set, which includes: lineage, positional and attribute accuracy, logical consistency, and completeness. The requirement for more metadata will undoubtedly make the procedure more expensive.
- Informative: rather than merely stating whether the information is fit for use, the procedure should additionally inform the user about the extent to which this is the case. The extent of fitness for use is necessary to judge how much uncertainty reduction is required or how much residual uncertainty will need to be absorbed by the decision-maker. It should be recognized that fitness for use is strongly linked to these two concepts, since it is insufficient to accept information as suitable yet remain unaware of the consequences

of any residual uncertainty. Knowledge of such consequences is essential for controlling their effects.
- Measurable Reliability: finally, the procedure should provide a measure of the reliability of its results. This is essential for informing users about the confidence with which they may accept the information as fit for use.

Procedures for Assessing Fitness for Use

Fitness for use of geographic information can be assessed either by the standards-based procedure which is the most commonly used approach, or the risk-based procedure which is recognized for its promise but still remains under development.

The Standards-Based Procedure

With this technique, fitness for use is assessed by directly comparing the quality elements of information against a set of standards that represent the corresponding acceptable quality components (see Figure 4.1). To facilitate direct comparison, the standards are defined using the same elements as those used for describing data quality. These may include: scale (of the source document); Root Mean Square Error (RMSE); resolution; Percentage of Correctly Classified pixels (PCC); currency; and percentage completeness. For example, consider the prediction of crop yield based on crop conditions interpreted from remotely sensed images. The fitness for use of any image intended to be used in this application would depend on whether that image satisfies all predetermined standards for, say, image resolution, PCC, currency, and completeness.

However, it is our belief that standards for information quality should be ultimately determined by the acceptable margin of uncertainty in the final decision—a margin that should ideally be independent of the information. Nevertheless, in many instances the characteristics of information (which are mostly of an economic nature) will dictate this margin, such as:

- when alternatives to the information being tested for its suitability are neither available, affordable, nor cost-justifiable; and /or
- when the cost or delay involved in reducing uncertainty to an acceptable level in the information to be used is simply not feasible.

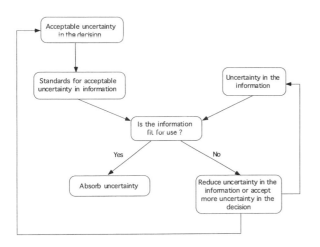

Figure 4.1. The standards-based approach.

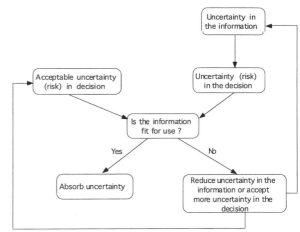

Figure 4.2. The risk-based approach.

At this point, it is worth noting that these problems are not just restricted to the standards-based approach but also extend to risk-based procedures. Because it is the level of acceptable uncertainty in decisions that determines any quality standards, its estimation should take precedence over setting the standards. In the example above, the standards might well be derived from the acceptable financial loss attributable to error in crop prediction—which implies that reliable estimation of this loss is essential to the development of meaningful and effective standards.

The Risk-Based Procedure

In this procedure, the impact of information uncertainty is expressed as the amount of risk in the final decision (see Figure 4.2), and assessment of fitness for use now involves comparing this residual risk with the risk acceptable to the decision-maker. Risk and uncertainty are closely related, and Goodchild (1992, p. 7)—commenting on the relevance of risk analysis to the appropriate use of geographic information—observes that "risk analysis is an important link in the chain which stretches from error models and database concerns through to decision making." Indeed, because risk is an established concept for estimating and dealing with uncertainty in decisions, its relevance as a basis for dealing with uncertainty in geographic information cannot be doubted. However, until now risk has not been embraced as a basis for assessing fitness for use. Nonetheless, as awareness of the presence and magnitude of uncertainty in geographic information grows, it is expected that attention will increasingly

be directed to the impact of this uncertainty upon decisions, and subsequently to the merits of risk analysis for handling this uncertainty.

Comparison of the Two Procedures

The standards-based procedure is comparatively simple as long as the standards are known, but this is often a major hurdle. It involves estimating the acceptable uncertainty in the decision, which is then converted into the various information quality elements (Figure 4.1). The conversion, which is a reverse propagation of uncertainty from decision to information, may involve estimating several unknowns from a single value—the theoretical solution of which is an infinite number of possible combinations of the unknowns. The conversion becomes more complicated when the decision involves integration of various data sets, each having different metadata. Moreover, the forward propagation of uncertainty from information to decision is very often not well understood, which means that the reverse propagation is even more difficult to carry out. When reverse propagation is not practicable, a possible option is iterative forward propagation starting with approximate values for the set of standards, and then refining them progressively as the propagation converges to the acceptable uncertainty in the decision.

Once a set of standards has been specified for a given application, the information being assessed must satisfy *all* of the standards before it can be considered suitable. This disregard for possible compensatory characteristics among information quality elements is a major weakness of this procedure, which can be traced back to the absence of a composite descriptor or index for information quality. Other drawbacks of this procedure include: its inability to indicate the extent to which information is fit for use and the reliability of this assessment; and its grounding on a basis that does not readily associate uncertainty with its potential impact on decisions. In addition, owing to their persistent nature, standards are rarely reviewed with the regularity that reflects changes in the environments of their applications, and are therefore prone to being out of date. On the other hand, the main strengths of this procedure are the simplicity of its application after the objective derivation of the standards has been resolved, and availability of the metadata it requires. The latter benefit is not surprising given that the current content of metadata was specified to suit this procedure.

Turning to the alternative procedure, the main tasks of the risk-based approach are propagating uncertainty in the information into risk in the decision, and determining the acceptable risk. Because it involves a systematic identification of what could go wrong due to uncertainty in the information, the former task affords the user a more intimate understanding of the limitations of both the information and the subsequent decisions. Thus, the user proceeds with a decision well aware of the risks due to the information employed. Moreover, because it is an established mechanism for dealing with uncertainty, risk management provides various tested options for dealing with residual risk, such as risk retention, self insurance, and risk transfer. In addition, the growing awareness of the effects of uncertainty by users of geographic information is expected to attract their attention to mechanisms of controlling this impact. Hence, the risk-based procedure fits well as a means of satisfying user expectations of a technique for assessing fitness for use.

Another benefit is that risk analysis should result in more economically efficient solutions, since it strives to strike a rational balance between risk reduction and its associated cost (McDonald, 1995). Also, by combining the various elements that describe information uncertainty into one quantity—risk—this procedure allows for compensation among these elements. Determining acceptable risk, though still a complex process with this approach, is comparatively more comprehensive, especially when using formal analysis (discussed later in this chapter).

Putting these arguments for the risk-based procedure aside, there are some drawbacks. They include

the requirement for extra data (in addition to metadata as currently reported) for the uncertainty-to-risk conversion, and difficulty in evaluating the reliability of calculated risk. In addition, the effort and skills required to carry out a comprehensive risk analysis render this procedure expensive, time-consuming and less practicable for the more routine and simpler decision tasks. Finally, although the representation of risk as a single (often pecuniary) unit is a strength of this procedure, in so doing its attempt to combine incommensurable entities on the same scale such as the value of lives, the environment, and real property, is also seen as a weakness.

Despite its drawbacks, the risk-based procedure is important, especially for large applications where the consequences of information uncertainty can be severe, and where the benefits of its economic efficiency are large enough to exceed its costs. Accordingly, it is not surprising that organizations which deal with such applications, for example the U.S. Environmental Protection Agency (EPA) (Rowe, 1977) and the Australian National Commission of Large Dams (ANCOLD) (McDonald, 1995), have initiated the transition from standards-based to the risk-based procedure.

ACCEPTABLE RISK

In the risk-based approach, fitness for use is conditional upon the estimated risk in a decision (due to uncertainty in the information) being acceptable—that is, not greater than a specified risk threshold. This threshold defines the acceptable risk, and the main aim of this chapter is to show how acceptable risk can be determined—a task that is recognized to be "extremely complex" (Melchers, 1993) and perhaps the most vexed of all risk management steps. The enormity of this task is evident from recurrent problems entailed in setting standards for the design, operation, and regulation of technologies that employ risk management to address health and safety concerns (for example, the siting of chemical production plants and nuclear power reactors).

An examination of the literature suggests that in such technologies the concept of acceptable risk has been more widely discussed than in disciplines where risk management is used to transfer risks or manage financial loss. Presumably, this is because the former often deal with the contentious issue of value of life, and are required to accommodate the usually conflicting views of communities exposed to risk. In addition, most of their hazard scenarios are of low probability and severe consequences, which means that their risks are sensitive to uncertainties in probability values and in general cannot be reliably estimated. These problems extend to risks due to uncertainty in geographic information, albeit with diminished significance. Hence, the concept of acceptable risk in these technologies, and the various methods used to judge this risk should be instructive to estimating the acceptable risk required to assess fitness for use. Accordingly, the first part of this section will present a general overview of acceptable risk, drawing on the treatment of this concept of these technologies, while the second part will discuss how the problem of acceptable risk may be handled in the risk-based procedure.

What is Acceptable Risk?

Various interpretations of acceptable risk have been suggested, for example according to Fischhoff et al. (1981) it is the risk associated with the most acceptable option in a decision problem. Rowe (1977), on the other hand, describes it as the risk that we are willing to accept in return for a benefit, while Lowrance (1976) depicts it in terms of safety; that is, something is considered safe if its risks are acceptable. The Conservation Foundation (1985) classifies acceptable risks into three categories, namely:

- zero risk: whereby no risk is tolerated. This has been dismissed as an irrational, and in many cases, impossible goal;
- technology-based risk: whereby the focus is on the best practicable technology for reducing risk, rather than on the risk per se.
- reasonableness of risk balanced with benefits: whereby a balance is sought between the risk and the economic costs of reducing or controlling it. This is the most preferred conception of acceptable risk. However, resolving when the balance is "reasonable" remains a challenge.

It is evident from all the above definitions—with the exception being that of Lowrance (1976)—that economic considerations are paramount in determining acceptable risk. Indeed, zero risk is commonly dismissed as irrational because it disregards economics. As for the word "acceptable," it elicits the questions, "Acceptable to whom, in whose view, and under what circumstances?" These questions point to an impor-

tant characteristic of acceptable risk; namely, that it is highly subjective and influenced by the continually changing values of parties to the risk and the circumstances under which it is determined.

The use of the word "acceptable" has also been criticized because it does not reflect the reluctance that those exposed usually show toward risk. In its place the word "tolerable" has been suggested (Pidgeon et al., 1992) and is often used; however, according to the British Health and Safety Executive (HSE, 1988, p. 1), "'Tolerability' does not mean 'acceptability.' It refers to the willingness to live with a risk to secure certain benefits in the confidence that it is being properly controlled." This distinction and its significance are demonstrated in the ALARP (As Low As Reasonably Practicable) principle (HSE, 1988). The principle posits that there is a maximum limit beyond which risk is not acceptable and a minimum limit below which risk is acceptable and can be ignored. The interval between these limits is called the tolerability or ALARP region, whereby the risk is tolerated only if its reduction is impracticable or its cost is disproportionate to the resulting marginal benefits. With respect to acceptable risk in the context of assessing the fitness for use of information, we observe that the significance of a risk band such as the ALARP region does not principally lie in the semantics of "tolerability" and "acceptability." Instead, it is due to the attention it draws toward a somewhat fuzzy (as opposed to a crisp) threshold for acceptable risk, since estimates of decision risk are inexact. The subjectivity and context dependence of acceptable risk can be traced to many factors, including characteristics of and what is known about the risk, which in turn shape the perceptions and attitudes of parties to the risk. Lowrance (1976) lists such characteristics as whether:

- the risk is assumed voluntarily,
- its effect is immediate,
- alternatives are available,
- risk is known with certainty,
- exposure is an essential or luxury,
- risk is encountered occupationally,
- the risk is due to a common hazard,
- the risk affects average people, and
- the consequences are reversible.

An important additional characteristic is the severity of consequences, and for the exposed party the ability to pay for risk reduction has a considerable influence on what is considered to be an acceptable risk.

Methods of Estimating Acceptable Risk

Fischhoff et al. (1981) have extensively investigated systematic approaches for estimating acceptable risk and they suggest three generic approaches, *viz.*:

- professional judgment, as embodied in individual professional skills or institutionally agreed standards such as engineering codes of practice;
- the so-called "boot strapping" approaches, which include the method of revealed preferences, and base acceptability upon extrapolation from statistics summarizing behavior toward existing hazards. The basic assumptions in these approaches, respectively, are that the current risk-benefit trade-offs of society are satisfactory and that new hazards should not impose a greater risk than those presently tolerated by society; and
- formal analysis such as cost-benefit or decision analysis. This approach is hailed for its rigor and comprehensiveness.

These three approaches are not that conceptually distinct. For example, formal analyses require a large element of professional judgment, and professionals often base their judgments on formal analysis. Alternatively, Rowe (1977) has proposed a method whereby acceptable risk is estimated by observing (as in "boot strapping") the risk that is generally acceptable by society over a wide range of applications (risk reference), and then modifying the reference to reflect risk attitude to a particular risk-exposing application.

Estimating the Acceptable Risk Required to Assess Fitness for Use

In this section, only the three methods proposed by Fischhoff et al. (1981) are discussed. For all three methods, conversion of information uncertainty to risk in the decision is an important process.

"Boot Strapping"

Individual or organized groups of geographic information users may, through experience or observation, establish that information with particular quality characteristics has historically been considered satisfactory for their applications. In other words, the risk attributable to uncertainty in that information has his-

torically been acceptable. Based on the "boot strapping" assumption that past trade-offs of risks and benefits are satisfactory even in present times, this risk can be presumed to be currently acceptable. Clearly, the validity of this assumption in turn determines the validity of the acceptability. For example, in instances where previously the choice of information has been limited, but present developments promise to increase that choice and offer better information, it can be shown that the acceptable risk obtained by applying the "boot strapping" assumption will most likely be unreliable and misleading. Yet, this is a typical scenario for many types of geographic information. Hence, for such information this approach is not suitable. It is also noted here that by relying on experience and judgment of users, and by considering the risk-benefit trade-offs, a "boot strapping" approach manifests characteristics of professional judgment and formal analysis, respectively. This illustrates how nebulous the differences between the various approaches are.

Professional Judgment

In this approach, the personal experience, intuition, and collective expertise of knowledgeable users (experts) is harnessed to judge the acceptable level of information uncertainty for a particular application. Accordingly, for this application the risk associated with the information uncertainty constitutes its acceptable risk. Professional judgment assumes that the experience and knowledge of experts is sufficient to enable judgment of acceptable information uncertainty. As with "boot strapping," this approach relies on experience; however, this experience goes beyond observing historical risk-benefit trade-offs and, likewise, the results of the approach are not restricted to these trade-offs. Rather, experience with and knowledge of the application are used to judge what level of information uncertainty should be acceptable, even for new applications which do not yet have an established record of such trade-offs. The application of this approach relies on the availability of experts with the necessary skills. The experts may, where appropriate, use "boot strapping" and formal analysis to arrive at their judgment. Hence, because it is adaptable, professional judgment is an appropriate approach, but only so long as judgments are based on valid assumptions. Nevertheless, the acceptable uncertainty derived in this way is undoubtedly influenced by the perceptions and biases of experts.

Formal Analysis

We shall restrict the discussion of formal analysis to cost-benefit analysis. In the context of estimating acceptable risk, benefit is the reduction in risk, while cost is the expense required to achieve that benefit. The analysis involves quantifying the benefits and corresponding costs, and then determining a desirable balance between them which is largely a value judgment. To illustrate how cost-benefit analysis may be applied for determining acceptable risk, consider a user who has access to four alternative information sets of different quality (DS1-4), as shown in Table 4.1. Suppose that this user has also established the cost of accessing—for example, collecting, purchasing, leasing, or licensing—each of these information sets, and from a study of their respective metadata, has calculated the risk traceable to uncertainty in the information. Assuming that the information sets listed in Table 4.1 are in order of increasing quality, and that the adage "you only get what you pay for" (Hunter, 1996) is true in this case, then the cost of information should increase from top to bottom, while the corresponding risks would be expected to decrease. Figure 4.3 illustrates a typical risk-cost graph from the values in Table 4.1.

An analysis of Figure 4.3 can be used in the following way to judge acceptable risk. Starting with the cheapest information (DS1) and assuming its cost is affordable, the following are examined: the risk (R1); the benefit gained by using the next information set in the hierarchy (R1 minus R2) and its corresponding cost (C2 minus C1); and the ability to pay this extra cost. If the extra cost cannot be afforded, regardless of the size of benefit gained by using better information, then DS1 is the sole choice and its risk (R1) will be the acceptable risk. If the extra cost can be afforded, the decision to proceed to DS2 and repeat the analysis depends on whether the benefit gained sufficiently exceeds the cost. Obviously, "sufficiently" is subjective and depends on the risk attitude of the user as well as on the characteristics and magnitude of the risk. On the other hand, if the best information (DS4) can be afforded and the user is risk averse, the investigation may start with this information and progressively assess whether the extra exposure to risk caused by using the next best information in the hierarchy can be justified by the corresponding cost savings. Again, this justification is influenced by value judgments of the user.

The data sets in Table 4.1 should not necessarily be restricted to available information. Extra entries in the table could be generated by simulating information sets

Table 4.1. Costs and Corresponding Risks of Alternative Data Sets.

Data Set	Cost	Risk
DS1	C1	R1
DS2	C2	R2
DS3	C3	R3
DS4	C4	R4

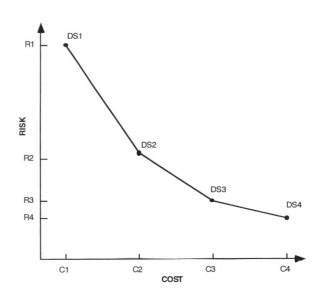

Figure 4.3. A typical risk versus cost graph.

of typical quality characteristics obtained either by improving the existing information sets or by collecting fresh data, and then estimating their costs and corresponding risks. For instance, it is possible that uncertainty reduction in DS1 may produce information with a better benefit-cost trade-off than that afforded by DS2. An example of research in this direction is by Liu (1994), who used Monte Carlo simulation to investigate the effect of random errors in DEMs on the cost of harvesting forest compartments. He simulated three error levels and estimated the "expected cost of uncertainty," which is equivalent to risk, for each error level. Given this information, the only extra information required to plot Figure 4.3 is the cost of the DEM for each error level. However, while we are interested in simulations that affect uncertainty reduction, Liu's simulations perturb a DEM to generate realizations with greater elevation uncertainty.

From the previous discussion it is clear that acceptable risk determined by formal analysis is contingent upon the quality of available information, the cost of improving this quality (either directly or by choosing better alternatives), the ability of users to pay for this improvement and the users' risk attitude. It follows that unless these factors cancel themselves out, then changes within them should cause changes in acceptable risk. Accordingly, acceptable risk is nonpersistent. It is also evident that cost-benefit analysis not only aids in determining acceptable risk but also in selecting, from an uncertainty viewpoint, the most suitable information among competing alternatives. In this way, it helps to answer both of the questions: "Is this information fit for my application?" and "Of the available information that can be afforded, which is the most suitable?" We submit that the latter question is of greater significance and should be the one that users pose before they purchase information in the first place.

CONCLUSIONS

The primary reason behind the growing concern about uncertainty in geographic information is to avoid its erroneous use and misuse, by ensuring that the information used is always fit for its purpose and is not of unnecessarily high quality—especially if that superfluous quality comes at a high cost. Traditionally, the assessment of fitness for use has mainly been carried out by comparing the metadata for the information against a set of standards for the task at hand. The authors believe this standards-based approach has several drawbacks and should either be improved or replaced with alternative approaches. In this chapter we have highlighted these disadvantages and proposed that the risk-based approach is a more suitable option for specific cases; namely, very large applications dealing with structured problems where the consequences of erroneous use of information can be severe. To corroborate this, examples have been cited of applications with these characteristics in which it was necessary to initiate the transition from a standards-based to risk-based approach to assessing the fitness for use of information. However, in order to increase the utility of the risk-based approach, its drawbacks must be addressed. For example, current metadata content needs to be extended to adequately cater for estimating risk, while techniques for calculating the reliability of risk estimates still require refinement.

ACKNOWLEDGMENT

The authors acknowledge funding support received from the University of Melbourne and the Australian Research Council Large Grant No. A49601183, "Modelling Uncertainty in Spatial Databases."

REFERENCES

Agumya, A. and G.J. Hunter. Determining Fitness for Use of Geographic Information, *ITC Journal,* 1997-2, pp. 109–113, 1997a.

Agumya, A. and G.J. Hunter. Estimating Risk in GIS-Supported Decisions, *Proceedings of the 1997 Urban and Regional Information Systems Association (URISA) Conference, Toronto,* 20–24 July, 9 pp. (on CD only), 1997b.

Balce, A.E. Quality Control of Height Accuracy of Digital Elevation Models, *ITC Journal,* 1987-2, pp. 327–332, 1987.

Brunsdon, C. and S. Openshaw. Simulating the Effects of Error in Geographical Information Systems, in *Geographic Information Handling: Research and Applications,* P.N. Mather, Ed. John Wiley & Sons, Chichester, England, 1993, pp. 47–61.

Conservation Foundation. *Risk Assessment and Risk Control,* Conservation Foundation, Washington DC, 1985.

Coward, P. and I. Honeywood. Aspects of Uncertainty in Spatial Decision Making, *Proceedings of the 1991 European Conference on Geographical Information Systems (EGIS 91),* Vol. 1, EGIS Foundation, Utrecht, 1991, pp. 233–242.

Fischhoff, B., S. Lichtenstein, P. Slovic, S.L. Derby, and R.L. Keeney. *Acceptable Risk,* Cambridge University Press, Cambridge, 1981.

Goodchild, M.F. *Closing Report on National Centre for Geographic Information and Analysis (NCGIA) Research Initiative I: Accuracy of Spatial Databases,* NCGIA, University of California, Santa Barbara, 1992.

Health and Safety Executive. *The Tolerability of Risk from Nuclear Power Stations,* Her Majesty's Stationery Office (HMSO), London, 1988.

Hunter, G.J. Management Issues in GIS: Accuracy and Data Quality, *Proceedings of the Conference on "Managing Geographic Information Systems for Success,"* University of Melbourne, 1996, pp. 95–101.

Liu, R. The Effects of Spatial Data Errors on the GIS-Based Forest Management Decisions. Unpublished Ph.D. thesis, State University of New York, 1994.

Lowrance, W.W. *Of Acceptable Risk: Science and the Determination of Safety,* William Kaufmann, Los Altos, CA, 1976.

McDonald, L.A. ANCOLD Risk Assessment Guidelines, in *Acceptable Risks for Major Infrastructure: Proceedings of the Seminar on Acceptable Risks for Extreme Events in the Planning and Design of Major Infrastructure,* Sydney, 26–27 April, 1994, P. Heinrichs and R. Fell, Eds., A.A. Balkema, Rotterdam, 1995, pp. 105–121.

Melchers, R.E. Society, Tolerable Risk and the ALARP Principle, in *Proceedings of the Conference on Probabilistic Risk and Hazard Assessment*, Newcastle, Australia, 22–23 September 1993, R.E. Melchers and M.G. Stewart, Eds., A.A. Balkema, Rotterdam, 1993, pp. 243–252.

Pidgeon, N., C. Hood, D. Jones, B. Turner, and R. Gibson. Risk Perception, in *Risk Analysis, Perception and Management,* The Royal Society, London, 1992.

Rowe, W.D. *An Anatomy of Risk*, John Wiley & Sons, New York, 1977, p. 488.

CHAPTER 5

Alternative Measures for Comparing Thematic Map Accuracy

S.V. Stehman

INTRODUCTION

Comparisons of map accuracy are employed to determine which of several classifiers may be best for a given mapping project, to determine which of two land-cover classification schemes results in higher accuracy, or to assess if accuracy differs among different spatial regions. Maps may be compared on the basis of several accuracy measures. Because these measures reflect different components of accuracy, determining which of two maps is better should consider the information provided by the various measures. For example, the map comparison could be based on user's accuracies and producer's accuracies for the different land-cover classes. However, sometimes map comparisons employ a single accuracy measure, either for convenience, or because a very large number of comparisons need to be conducted in a somewhat automated fashion. For those cases in which the comparison is based on a single measure, that measure should be chosen to reflect the objectives motivating construction and use of the map; that is, the metric used to compare the two maps should somehow represent the value of the map for a particular use.

The results of an accuracy assessment are often summarized via an error matrix (Table 5.1). Assume that each pixel (or polygon) displayed on the map is labeled as one land-cover class, and a reference ("true") land-cover classification is also obtained for that polygon or pixel. The error matrix then summarizes the comparisons between the map and reference classifications for the collection of pixels or polygons. When evaluating accuracy measures, it is useful to think in terms of the population error matrix that would result if we were to obtain a census of reference data (i.e., complete coverage of reference data). Taking this population perspective allows us to focus on the accuracy measures rather than on the sample-based estimates of these measures.

If the error matrix represents q land-cover classes, some of the commonly used accuracy measures are the following:

(i) Overall proportion of area correctly classified,

$$P_c = \sum_{i=1}^{q} p_{ii} \tag{1}$$

(ii) Kappa coefficient of agreement,

$$\kappa = \frac{P_c - \sum_{i=1}^{q} p_{i+}p_{+i}}{1 - \sum_{i=1}^{q} p_{i+}p_{+i}} \tag{2}$$

where

$$p_{i+} = \sum_{k=1}^{q} p_{ik}$$

and

$$p_{+i} = \sum_{k=1}^{q} p_{ki}$$

Table 5.1. Error Matrix Notation for a Population.[a]

		Reference				
		1	2	$\cdots$	q	
	1	p_{11}	p_{12}	$\cdots$	p_{1q}	p_{1+}
Map	2	p_{21}	p_{22}	$\cdots$	p_{2q}	p_{2+}
	$\vdots$	$\vdots$	$\vdots$	$\cdots$	$\vdots$	$\vdots$
	q	p_{q1}	p_{q2}	$\cdots$	p_{qq}	p_{q+}
		p_{+1}	p_{+2}	$\cdots$	p_{+q}	

[a] p_{ij} is the proportion of area in mapped land-cover class i and reference land-cover class j.

[b] $p_{i+} = \sum_{j=1}^{q} p_{ij}, \ p_{+j} = \sum_{i=1}^{q} p_{ij}.$

(iii) Kappa, with random chance agreement as suggested by Foody (1992),

$$\kappa^* = \frac{P_c - 1/q}{1 - 1/q} \qquad (3)$$

(iv) User's accuracy for cover-type i, the conditional probability that an area classified as land-cover type i by the map is classified as category i by the reference data,

$$P_{Ui} = p_{ii} / p_{i+} \qquad (4)$$

(v) Producer's accuracy for cover-type j, the conditional probability that an area classified as land-cover type j by the reference data is classified as category j by the map,

$$P_{Aj} = p_{jj} / p_{+j} \qquad (5)$$

(vi) Tau coefficient of agreement (Ma and Redmond, 1995),

$$\tau = \frac{P_c - \sum_{i=1}^{q} \beta_i p_{+i}}{1 - \sum_{i=1}^{q} \beta_i p_{+i}} \qquad (6)$$

where β_i is a user-specified, a priori probability of membership in map class i.

Congalton (1991), Janssen and van der Wel (1994), and Stehman (1997) review properties of these measures. P_c, κ, κ^*, and τ are all single summary mea-

sures, whereas user's and producer's accuracies both result in q measures of accuracy, one for each land-cover class. Averaging either user's accuracies or producer's accuracies creates a single summary measure that could also be used to compare maps. The issue is then which of these summary measures is best for comparing two maps, or are there still other measures derivable from this basic set?

Determining an appropriate comparison measure depends on the mapping objectives, so users may employ a different comparison measure depending on their intended use of the map. The measure chosen also depends on the type of comparison involved. Four general types of map comparisons may be defined depending on whether the comparison is over the same or different regions, and whether the classification scheme is the same or different for both maps:

Case 1: the two maps represent the same geographic region and are based on the same land-cover classification scheme.
Case 2: the two maps represent the same geographic region but the land-cover classification scheme is different.
Case 3: the two maps represent different regions and the land-cover classification scheme is the same.
Case 4: the two maps represent different regions and the land-cover classification scheme is different.

Two confounding features potentially arise when attempting to compare maps. If the maps represent different regions, one region may be easier to classify than the other. As a simple illustrative example, suppose the objective is to compare accuracy of an area in which edge-matching techniques have been used to a region in which edge-matching was unnecessary. The two areas are classified using the same land-cover classification scheme, and suppose the five land-cover categories include water and four forest types. Further, assume that water is very easy to distinguish from the four forest types so water is classified with high accuracy. If one region (say the edge-matched region) is predominantly water and the other region is predominantly forest, the former region is likely to be more accurate because it is easier to classify. Thus the accuracy comparison of edge-matched versus nonedge-matched regions is confounded by the land-cover differences between the two regions. The regions being compared may be two administrative regions within the mapping project (e.g., states). If the accuracy of the two maps is compared, the comparison is confounded by differences in the land-cover proportions in the two regions.

Confounding also occurs if the land-cover classification scheme is different for the two maps being compared. For example, suppose two maps representing the same region are constructed using different classification schemes. Further, suppose the two schemes are hierarchical in the sense that one scheme consists of five classes, four of which are different forest types, the other class being nonforest, and the other scheme is a collapsed version of the first, using only two classes, forest and nonforest. The map based on only two classes will almost surely be more accurate than the map derived from the more complex, five-class scheme, so a direct comparison of accuracy seems inappropriate. Rather, the comparison somehow should account not only for accuracy, but for the differences in the information available from the different classification schemes. That is, how valuable is the land-cover information, relative to mapping objectives, obtained from the map based on the more complex classification compared to the map derived from the simpler two-class scheme? In terms of uses of the map, what is lost by being able to identify an area only as "Forest" in the two-class scheme compared to being able to identify the area as, say, "Oak-Hickory Forest" in the five-class scheme?

The four cases reflect an increasing degree of confounding, with Case 4 comparisons confounded by both differences in classification scheme and in the land-cover itself, whereas Case 1 comparisons have neither of these confounding factors. Cases 2 and 3 each represent one source of confounding. When these confounding factors are present, the map accuracy comparison should account for this. For example, if the confounding arises from different regions being compared, the comparison could be based on an accuracy measure which incorporates an adjustment to a common land-cover regional representation. This common representation might be a hypothesized set of land-cover area proportions in which all land-cover classes common to both regions are equally represented. The nature of such adjustments will be elaborated in the following section. But even if the comparison fits the less confounded Case 1 classification, map comparisons should still take into account the mapping objectives, which may place greater importance on some land-cover classes relative to others. That is, when the mapping objectives are taken into consideration, all correct classifications may not be equally important, and all misclassifications may not be equally problematic.

ACCURACY COMPARISONS BASED ON MAP VALUE

Instead of comparing maps solely on the basis of map accuracy, the objectives motivating map comparisons may be addressed more directly by basing the comparison on measures of map value which explicitly incorporate differences in importance of various correct classifications and/or misclassifications. "Importance" is still determined by the intended uses of the map. These map value measures may be derived from user's or producer's accuracies, with weights assigned to each land-cover class according to the importance of that class. Map value measures derived from user's or producer's accuracies implicitly regard all misclassifications within a land-cover class as equally important. An alternative approach (described in the next section) allows for assigning different importance to various misclassification errors. The choice of whether to base the measure on user's or producer's accuracy depends on which is more relevant to mapping objectives.

Map value based on user's accuracy will be described, but a similar derivation could be carried out for map value based on producer's accuracy. Using the notation described earlier, let P_{Ui} represent the user's accuracy of class i, p_{i+} represent the proportion of the map in class i, and w_i be a weight assigned to class i depending on the importance or "value" of class i to mapping objectives. Then a measure of map value summarizing the information in the q user's accuracies and also incorporating the relative importance of the different classes (as defined by the w_i's) is

$$V_U = \sum_{i=1}^{q} w_i p_{i+} P_{Ui} \qquad (7)$$

Because $P_{Ui} = p_{ii}/p_{i+}$, an equivalent formulation of Equation 7 is

$$V_U = \sum_{i=1}^{q} w_i p_{ii} \qquad (8)$$

This general form of V_U weights by both importance of the classes (w_i) and by the proportion of mapped area in each class (p_{i+}). The weights w_i are user-specified, so in a sense arbitrary, and these weights need not sum to 1.

For the special case in which all land-cover classes are considered equally important ($w_i = w$ for i = 1,..., q),

$$V_U = w \sum_{i=1}^{q} p_{i+} P_{Ui} = wP_c \qquad (9)$$

Thus the map value measure weighted by the mapped land-cover proportions (p_{i+}) and regarding all land-cover classes as equally important is proportional to P_c. The selected weight, w, becomes important when comparing two maps.

If area weighting by p_{i+} is not employed, p_{i+} must be replaced by another set of proportions, say p^*_{i+}, such that

$$\sum_{i=1}^{q} p^*_{i+} = 1$$

This scales V_U similarly to the measure obtained using the area weighting supplied by p_{i+} (7). Setting p^*_{i+} = 1/q (i = 1,..., q) results in

$$V_U = \sum_{i=1}^{q} w_i P_{Ui} / q \qquad (10)$$

and the special case form when all classes are regarded as equally important ($w_i = w$, i = 1,..., q) is

$$V_U = \frac{w}{q} \sum_{i=1}^{q} P_{Ui} = w\overline{P}_U \qquad (11)$$

where $\overline{P}_U$ is the average user's accuracy. Replacing the area proportions p_{i+} by 1/q results in a map value measure for the hypothetical case in which the land-cover map has equal area proportions for all land-cover types. Alternatively, any arbitrary set of proportions, p^*_{i+}, satisfying the constraint

$$\sum_{i=1}^{q} p^*_{i+} = 1$$

could be selected. To interpret V_U, we must recognize that it is based on the assumption that the mapped land-cover proportions produced were the specified set of proportions, either p_{i+} or p^*_{i+}. The use of an arbitrary set of proportions, p^*_{i+}, becomes relevant when the map comparison is confounded by regional differences. If map value measures are adjusted to a common set of proportions, p^*_{i+}, then the regional confounding is eliminated.

Two commonly used accuracy measures are recognizable as special case forms of V_U. If $w_i = 1$ for all q classes and area weighting (by p_{i+}) is used, $V_U = P_c$ (i.e., set $w_i = 1$ in Equation 9). If $w_i = 1$ for all classes and $p^*_{i+} = 1/q$,

$$V_U = \overline{P}_U = \sum_{i=1}^{q} P_{Ui} / q$$

average user's accuracy. Recognizing P_c and $\overline{P}_U$ as measures of map value provides insight into the implicit weighting scheme used by each. For both P_c and $\overline{P}_U$, all land-cover classes are regarded as equally important (or of equal value), so the difference between P_c and $\overline{P}_U$ depends on whether the comparison employs area weighting by p_{i+}, the mapped land-cover proportions, or uses instead an equal area representation, 1/q, for each land-cover class.

Some recommendations can be provided for the four cases of map comparisons described. To distinguish features of the two maps, a superscript prime will be used to indicate a characteristic of the second map. The recommendations will be phrased in terms of the versions of map value based on user's accuracy, but similar recommendations apply to map value based on producer's accuracy.

Case 1

The p_{i+} area-weighted form of V_U is recommended. Because both maps represent the same geographic region, no confounding attributable to differences in land-cover area proportions exists between the two maps. Further, if the land-cover classes differ in value relative to the user's objectives, correctly classifying a valuable land-cover class, say class k, should contribute more to total map value as the area proportion correctly classified in that cover type (p_{kk}) increases. Because the land-cover classes are the same for both maps, $w_i = w'_i$ (i.e., the weights are equal for the two maps because they represent the same land-cover class in the same region). The difference in map value is

$$V_U - V'_U = \sum_{i=1}^{q} w_i (p_{ii} - p'_{ii}) \qquad (12)$$

If all land-cover classes are considered equally important ($w_i = w'_i = w$), then

$$V_U - V'_U = w \sum_{i=1}^{q} (p_{ii} - p'_{ii}) = w(P_c - P'_c) \quad (13)$$

and the map comparison thus depends on the difference between the overall proportion of area classified correctly by each map. The choice of the weight w affects the comparison only in scaling the magnitude of the difference between the overall proportion correctly classified by each map.

Case 2

The area weighted form of V_U is again recommended because the two maps are still attempting to represent the same region and hence the same land cover. The difference in map value is

$$V_U - V'_U = \sum_{i=1}^{q} w_i p_{ii} - \sum_{k=1}^{q'} w'_k p'_{kk} \quad (14)$$

If *within* each map the land-cover classes are considered equally important so that $w_i = w$ and $w'_k = w'$, the difference in map value simplifies to

$$V_U - V'_U = wP_c - w'P'_c \quad (15)$$

If the first map represents a more detailed land-cover scheme (i.e., $q > q'$), it is likely that $w > w'$ because the more detailed classification scheme is more valuable to mapping objectives. Consequently, the more complex map classification scheme will have higher value ($V_U > V'_U$) if both maps result in the same overall accuracy ($P_c = P'_c$).

Case 3

If the two maps represent different regions, some adjustment of the comparison to a common set of land-cover proportions seems called for. Otherwise, differences in the land-cover area proportions between the two regions confound the comparison. Adjusting the comparison to equal land-cover proportions, 1/q, creates an equal area weighting. Because both maps have the same land-cover classification scheme, $w_i = w'_i$, and the difference in value is

$$V_{II} - V'_{II} = \frac{1}{q} \sum_{i=1}^{q} w_i (P_{Ui} - P'_{Ui}) \quad (16)$$

If all land-cover classes are deemed equally important ($w_i = w'_i = w$),

$$V_U - V'_U = \frac{w}{q} \sum_{i=1}^{q} (P_{Ui} - P'_{Ui}) = w(\overline{P}_U - \overline{P}'_U) \quad (17)$$

so the comparison reduces to a difference based on average user's accuracy. This is an intuitively appealing result in that if we want to eliminate the confounding effect of the different area proportions in the two regions, we compare the two maps on the basis of average user's accuracy.

Case 4

This is the least commonly occurring case and the most confounded. Because of the confounding attributable to differences in the land-cover area proportions in the two regions, adjusting the map value measures for the two regions to some common set of area proportions is again necessary. But the comparison is further confounded because of the different land-cover classification schemes, which may be based on different numbers of classes ($q \neq q'$). If the land-cover proportions within both maps are assumed equal, then the map value comparison is

$$V_U - V'_U = \frac{1}{q} \sum_{i=1}^{q} w_i p_{ii} - \frac{1}{q'} \sum_{k=1}^{q'} w'_k p'_{kk} \quad (18)$$

If for each map equal weights are assigned to all land-cover classes so that $w_i = w$ and $w'_k = w'$,

$$V_U - V'_U = w\overline{P}_U - w'\overline{P}'_U \quad (19)$$

This equation illustrates that the map value comparison depends on both average user's accuracy and the value of the land-cover classification scheme as reflected by the choices of w and w'.

MAP VALUE BASED ON ALL CELLS OF THE ERROR MATRIX

Map value could also be defined using all cells of the error matrix. These measures would require specifying weights to represent the value of both correct classifications and misclassifications, the correct classifications representing a positive contribution to map value, and the misclassifications generally representing a negative contribution to map value. If w_{ij} is the weight assigned to cell (i,j) of the error matrix, then map value may be defined as

$$V = \sum_{i=1}^{q} \sum_{j=1}^{q} w_{ij} p_{ij}$$

$$= \sum_{i=1}^{q} w_{ii} p_{ii} + \sum_{\substack{i=1 \\ j \neq i}}^{q} \sum_{j=1}^{q} w_{ij} p_{ij} \qquad (20)$$

The weights (w_{ij}) need not sum to 1, because the map comparisons depend on the absolute utility or value of the maps, not their relative utility.

Two special case results show the relationship of V to other map value measures, and illustrate the implicit weighting inherent in these measures. If $w_{ij} = 0$ for all off-diagonal cells,

$$V = \sum_{i=1}^{q} w_{ii} p_{ii}$$

which is V_U. So if no penalty is assessed to the misclassification errors, V is equivalent to V_U. If in addition $w_{ii} = 1$, the area weighted map value is $V = P_c$ (the area weights are now p_{ij}).

Returning to the general form (20), for the Case 1 class of map comparisons (with $w_{ij} = w'_{ij}$, because both maps have the same classification scheme), the difference in map value is

$$V - V' = \sum_{i=1}^{q} \sum_{j=1}^{q} w_{ij} \left(p_{ij} - p'_{ij} \right)$$

$$= \sum_{i=1}^{q} w_{ii} \left(p_{ii} - p'_{ii} \right) + \sum_{\substack{i=1 \\ j \neq i}}^{q} \sum_{j=1}^{q} w_{ij} \left(p_{ij} - p'_{ij} \right) \qquad (21)$$

If all correct classifications are considered equally important ($w_{ii} = w'_{ii} = w$) and all misclassifications are considered equally serious ($w_{ij} = -w^*$), the difference in map value is

$$V - V' = w \sum_{i=1}^{q} \left(p_{ii} - p'_{ii} \right) - w^* \sum_{i=1}^{q} \sum_{j=1}^{q} \left(p_{ij} - p'_{ij} \right) \qquad (22)$$

$$= (w - w^*)(P_c - P'_c) \qquad (23)$$

because

$$\sum_{\substack{i=1 \\ j \neq i}}^{q} \sum_{j=1}^{q} p_{ij} = (1 - P_c)$$

By setting $w - w^* = 1$, a map comparison based on $P_c - P'_c$ may be interpreted as a difference in map value in which equal weights are specified for the diagonal cells (i.e., all correct classifications are equally valuable), and all misclassifications are regarded as equally problematic.

ESTIMATING MAP VALUE

The map value measures have so far been described in terms of parameters (i.e., characteristics of a census of reference data). In practice, map value must be estimated from the reference sample. The formulas for estimating map value will depend on the sampling design used to collect the reference data. The simplest estimation approach is to obtain $\hat{p}_{ij}$, the estimate of p_{ij} for each cell of the error matrix, and then to substitute $\hat{p}_{ij}$ for p_{ij} in the map value formulas. That is, compute the estimated error matrix ($\hat{p}_{ij}$), and then use these estimated proportions to estimate user's and producer's accuracies, or any other needed summary statistic for the sample error matrix. Stehman (1995) describes the basic general estimation approach.

SUMMARY

Several different measures of map value have been described for use in map comparisons. These measures permit selecting weights that reflect the value of different components of map accuracy to project objectives. For example, if some land-cover classes are more critical to the intended uses of the map, then those classes can be weighted more heavily in the map value measure. Viewing map comparisons in the framework of map value also provides insight into the implicit weighting scheme employed in some common measures used for comparisons, such as differences in overall accuracy (P_c) or differences in average user's accuracy ($\overline{P}_U$) or average producer's accuracy ($\overline{P}_A$). At first, the user-specified nature of the weights (w_i or w_{ij}) may engender resistance to this map value approach because of concern over the seeming arbitrariness of the selected weights. However, recognizing that the commonly used comparison measures also employ some weighting scheme suggests that it is advantageous to select the explicit weighting scheme used, and to construct the map value measure on that basis. Current practice is to choose the comparison measure and to let the weighting be defined implicitly. The variety of options available for defining map value should provide individual users with the information needed to compare maps on the basis of a summary measure that best reflects their intended use of the maps.

REFERENCES

Congalton, R.G. A Review of Assessing the Accuracy of Classifications of Remotely Sensed Data. *Remote Sensing Environ.*, 37, pp. 35–46, 1991.

Foody, G.M. On the Compensation for Chance Agreement in Image Classification Accuracy Assessment, *Photogrammetric Eng. Remote Sensing,* 58, pp. 1459–1460, 1992.

Janssen, L.L.F. and F.J.M. van der Wel. Accuracy Assessment of Satellite Derived Land-Cover Data: A Review. *Photogrammetric Eng. Remote Sensing,* 60, pp. 419–426, 1994.

Ma, Z. and R.L. Redmond. Tau Coefficients for Accuracy Assessment of Classification of Remote Sensing Data. *Photogrammetric Eng. Remote Sensing,* 61, pp. 435–439, 1995.

Stehman, S.V. Thematic Map Accuracy Assessment from the Perspective of Finite Population Sampling. *Int. J. Remote Sensing,* 16, pp. 589–593, 1995.

Stehman, S.V. Selecting and Interpreting Measures of Thematic Classification Accuracy. *Remote Sensing Environ.,* 62, pp. 77–89, 1997.

CHAPTER 6

Mapping Misgivings: Monte Carlo Modeling of Uncertainty and the Provision of Spatial Information for International Policy

J. Gascoigne and R. Wadsworth

INTRODUCTION

This chapter describes an investigation into the uncertainty of "critical load" exceedances of acidity for the soils of Great Britain using a Monte Carlo (MC) simulation approach. In so doing, the work aims to provide a practical illustration of Mowrer's (1997) concluding comment that:

[T]he results of Monte Carlo uncertainty assessment techniques provide the opportunity for improved decision making through knowledge of alternative levels of uncertainty in the results of the spatial analyses.

Communication of both the derivation and result of uncertainty estimates for improved decision-making is hopefully greatly enabled by the chosen methodology. This approach should be relevant to any spatial assessment which makes use of classed data to describe environmental phenomena. Thus the approach represents a generic Monte Carlo technique that can identify areas associated with various levels of probability of satisfying predefined criteria and which itself qualifies as one of Openshaw's "GISable" spatial analyses (Openshaw, 1993).

Critical Loads

The critical loads approach to emission controls of gaseous pollutants has come to play an ever-increasing role in European and International policy-making (see Bull, 1992), leading to revisions of UN/ECE protocols for sulfur and nitrogen. A critical load (CL) has generally been defined as:

...a quantitative estimate of an exposure to one or more pollutants below which significant harmful effects on specified elements of the environment do not occur according to present knowledge

Nilsson and Grennfelt, 1988

For soils, a workshop at Skokloster in Sweden concluded that the rate of chemical weathering of minerals was the single most important factor in determining CL. Soil materials were divided into five classes on the basis of their dominant weatherable minerals. CL ranges were then assigned to these classes (Table 6.1), according to the amount of acidity neutralizable by the weathering of these materials. The CL can also be adjusted within these ranges by including factors such as precipitation, vegetation, and topography. Class 1, with the lowest critical load value, is thus the most sensitive to the effects of acidifying pollution and class 5 is the least sensitive.

Figure 6.1 shows the empirical CL map (see Hornung et al., 1995 for a detailed description), where darker shading denotes higher sensitivity. The exceedance is the excess deposition over the CL. Deposition of acid compounds, especially sulfur, is estimated at 20 km resolution using the University of Hull Acid Rain Model (HARM). Deposition is influenced by

- clustered sources around London and on the Yorkshire/Nottingham coalfield

Table 6.1. Mineralogical Classification of Soil Materials and Critical Loads for Soils.[a]

Class	Minerals Controlling Weathering	Critical Load (keq H^+ ha^{-1} $year^{-1}$)	Assigned Critical Load Value
1	Quartz K-feldspar	< 0.2	0.10
2	Muscovite Plagioclase Biotite (<5%)	0.2 – 0.5	0.35
3	Biotite Amphibole (<5%)	0.5 – 1.0	0.75
4	Pyroxene Epidote Olivine (<5%)	1.0 – 2.0	1.50
5	Carbonates	> 2.0	3.0

[a] After Nilsson and Grennfelt, 1988.

Figure 6.1. Empirical map of critical loads (CL) of acidity for soils in Great Britain.

Figure 6.2. Estimated areas at risk from acid deposition: exceeded low deposition scenario.

- European emissions, approximately 50% of UK deposition
- the enhancement of deposition in orographic rainfall areas

Figure 6.2 shows a typical map of exceeded 1 km squares for Great Britain; the coarser scale of deposition, however, is responsible for the blocky nature of exceeded areas in Scotland. In general, the exceeded areas exhibit strong coincidence with upland areas with shallow soils and high rainfall.

Note should briefly be made here that current UK critical loads research (Hall et al., 1997) for the calculation and mapping of critical thresholds in Europe (Posch et al., 1997) for the UN/ECE Convention on Long-Range Transboundary Air Pollution utilizes a

simple mass balance model to calculate CL based on ecosystem type.

Uncertainty

Heraclitus's observation of never being able to step into the same river twice and his view that we exist in a universe of constant flux of all things has practical as well as philosophical implications for describing the world around us. Terms such as accuracy, precision, and uncertainty have often been defined and distinguished in geographical literature since the ascendancy of geographical information systems (GIS) as a modus operandi for the keen (or a modus vivendi for the reluctant). Freksa and Barkowsky (1986) make a common and useful distinction:

In representing knowledge about the geographic world, we must distinguish three kinds of uncertainty:

1. we may not know the precise location of crisply classified geographic entities and thus may be uncertain of their location, or
2. we know the precise locations of the geographic entities including the (possibly gradual) transitions between them, but we are uncertain how to classify them, or
3. we may have a combination of the two possibilities.

Uncertainty is increasingly more narrowly defined in light of artificial intelligence research. For example, when discussing soil surveys, Lagachiere et al. (1996) consider that "uncertainty corresponds to an ill-known position of the boundary, fuzziness indicates ill-defined soil attributes close to the boundary."

Monte Carlo Simulation

Mowrer (1997) provides a description of the MC approach taken in this paper which is worth repeating:

> Monte Carlo simulation is an alternative uncertainty assessment technique that involves re-running an analysis many times. Each time the analysis is re-run, the variables subject to uncertainty (stochastic variables) are perturbed, or altered somewhat, according to some underlying assumption (usually a probability distribution function). Each alternative, equally probable result is termed a realization of the Monte Carlo process. Repeatedly perturbing the values, then running the analysis for each set of perturbed values, produces a large set of equally likely alternative values for the outcome of the analysis. Internal correlations and variance (uncertainties) inherent in the analysis are realized across the multiple output layers through these multiple repetitions. In raster-based spatial analyses the distribution of these alternative outcomes across identical raster cell locations in multiple realization layers provides a measure of the uncertainty propagated through the analysis at each spatial location.

Early examples of Monte Carlo application to geographical concerns can be found, such as Cliff and Ord, 1973. The main drawback of MC is its "brute force" computational intensivity, but as Openshaw (1998) reiterates, computers should be doing the work for us, with techniques that are robust, noise-resistant, and distribution-free, so negating the need for fragile statistical assumptions about the underlying spatial data. Furthermore,

> GIS databases contain errors and uncertainties of various kinds and it is important that they do not mislead the innocent. The basic null hypothesis in GIS is not of randomness but of database error.
>
> Openshaw, 1993

More recent work highlights a wide variety of MC applications for assessing spatial pattern that includes rare disease clustering (Openshaw et al., 1987; Besag and Newell, 1991), urban land use (Gascoigne et al., 1995), old-growth subalpine forests (Mowrer, 1997), and seismic risk (Emmi and Horton, 1995).

GIS is predominantly an applied activity which can add considerable value to governmental and commercial activities, for few activities are without spatial consequence. The "Wow!" factor of introducing colorful cartographic output to boardroom and committee meetings is waning as spatial mapping, if not analysis, become incorporated into everyday business pursuits. With the increased sophistication of decision-making with respect to things geographical, and more importantly, with the establishment of comparable spatial results over a period of time for particular activities, the need to derive, communicate, and also be able to readily understand and communicate this understanding, of measures of uncertainty in output becomes crucial. MC approaches are both pertinent and powerful and the authors agree with Besag and Diggle (1977) that:

> Monte Carlo tests are relatively straightforward to explain in the course of consulting with non-statisticians.

METHODOLOGY AND ALGORITHMS

Three algorithms were used to simulate three different types of uncertainty in the use of classed data to estimate the environmental vulnerability of soils to acidification. These three sources of uncertainty were:

1. within-class uncertainty
2. between-class uncertainty
3. class-boundary uncertainty

While Mowrer's (1997) paper shows how three input variables, each with one "dimension" of uncertainty can be combined, this chapter takes one input (CL class) with three dimensions of uncertainty.

The algorithms were operationalized using the Arc Macro Language (AML) of proprietary Arc/Info 7.0 GIS software on a Unix platform. (Copies of the programs are available from the authors.) The objective of each algorithm is described below, along with assumptions that are made and pseudocode which explicates its implementation. Alternative assumptions are also provided which have not yet been tested but which could all throw more light upon inherent uncertainties.

1. Within-Class Uncertainty

Objective: to assess the uncertainty introduced by using the midclass value rather than the true (but unknown) values for critical load.

Assumptions:
- the distribution of values within a class is uniform; i.e., all values within a class are equally probable
- no spatial correlation exists within a class; i.e., high or low values are not likely to be locally clustered (but see alternative assumptions below)

Implementation (Pseudocode)

map_u = original CL map (Figure 6.1) reclassed so that each cell holds the upper limit of the class; i.e., cells with class 1 = 0.2, class 2 = 0.5, etc.

map_r = original class map reclassed so that each cell hold the range for the appropriate class; i.e., cells with class 1 = 0.2, class 2 = 0.3, etc.

For each trial:
    rand   = a grid of random numbers (from a uniform distribution range 0–1)
    new_map = map_u - map_r * rand
    estimate exceedance
next trial

Alternative Assumptions
◊ if the classed data represented an uninterrupted variable, the range of values could be adjusted to reflect the values in neighboring cells. For example, where a class 1 cell is adjacent to a class 2 cell, the assigned value could be biased toward the high end of the range while the value in the class 2 cell could be biased toward the low end of the range
◊ introduce spatial correlation in the production of the random grid by perhaps using a nested approach
◊ instead of changing each cell independently, all cells in a homogeneous patch could be assigned the same value

2. Between-Class Uncertainty

Objective: there is some uncertainty in the allocation of particular cells to particular classes. This algorithm investigates the effect of a given amount of misallocation of class.

Assumptions:
- Misallocations are uniform (as many cells are misclassified as being of a more sensitive class as those of a less sensitive class)
- All classes have an equal number of misclassified cells
- No spatial autocorrelation for misallocations (i.e., misallocations are randomly distributed in space)

Implementation (Pseudocode)

cert   = a "certainty" value (between 0 and 1)
min_a = area of the smallest class
For each class calculate a probability of a misclassification in both directions:
        up = 1.0 – (min_a * cert) / area_of_class
        dn = (min_a * (1.0 – cert)) / area_of_class
        For each trial:
            rand = random grid (values between 0 and 1)
            if (rand > up), reassign cell to the next higher class
            else if (rand < dn), reassign cell to the next lowest class
            estimate exceedance
        next trial
next class

Alternative Assumptions
◊ instead of a fixed number of cells changing class, use a fixed proportion
◊ certainty of allocation to a class could vary with class

3. Class-Boundary Uncertainty

Objective: because CL classes are based on soil type and soil boundaries are inferred rather than observed, the boundaries of patches of a particular class are uncertain.

Assumptions:
- error in boundaries is unbiased (equally likely to move in any direction)
- true location of the boundary is plus or minus one cell width from the current position

Implementation (Pseudocode)

Adjacent cells of the same class are grouped into homogeneous patches, each patch having a unique identifier.

Specify the "certainty" that a boundary is accurately located.

```
For each trial:
    For each patch:
        generate a random number (between 0 and
            1)
        if random number > certainty, expand that
            patch
    next patch
    allocate new boundary cells to correct class
    estimate exceedance
next trial
```

Alternative Assumptions

◊ alter the probability of a critical load patch boundary being correct in relation to the size of the patch or its neighbors

◊ alter the probability of a patch being selected on the basis of its critical load class type

◊ select a fixed number of patches of each critical load class type

◊ select a fixed total area of patch types

◊ ignore patches below a given threshold size

RESULTS

Three deposition scenarios were examined:

- high   best estimate of deposition without protocol control
- low   best estimate of deposition with protocol controls enforced
- uniform   spatial variation in deposition removed

Figures 6.3a to 6.3c display the cumulative standard deviations for the 250 trials of each of the three deposition scenarios. Each graph, through its stabilization, shows that enough trials were undertaken. However, the scales of the coordinate (Y) axes vary considerably between these graphs.

- *between-class uncertainty* shows low variation. Table 6.2 shows that variation in exceeded area is less than about 300 km^2
- *within-class uncertainty* shows roughly twice as much variation as between-class. Table 6.2 shows that variation in exceeded area is around 600 km^2
- *class-boundary uncertainty* shows that CL exceedance is strongly driven by where class boundaries are located. Table 6.2 shows that variation in exceeded area is between 10,000 and 20,000 km^2

CONCLUSION

With respect to the CL results, spatial uncertainty is shown to be much more important than attribute uncertainty. However, most CL research emphasis centers upon making refinements to ever more complex models which attempt to pin down the critical load values more exactly. Simple mass balance developments, which modify the CL "attribute" are incorporating *more* spatial uncertainty through the inclusion of land cover data. Recent work by Lagachiere et al. (1996) perhaps signals a way forward. Attendant simplifications in soils mapping are examined and soil "boundaries" are replaced by "fuzzy" transition zones (in standard Arc/Info GIS) to indicate the possibility of a site being in or outside a soil polygon.

Table 6.2 shows the relative importance of the different types of uncertainty. An alternative view is to consider the spatial aspects of certainty. Figure 6.4 shows the areas in Wales which are always exceeded with the three deposition scenarios; these are the areas that we can be certain are at risk. Obviously, more sophisticated (complex) maps can be produced delineating isolines of certainty based on these MC trials.

Having some estimate for uncertainty must be better than having no estimate at all. Best of all is having an estimate which can be intuitively appreciated by the decision-maker. Moreover, fear of ever calling into question the procedures of geographic analyses, lest the "GIS bubble" should burst, must be dispelled in

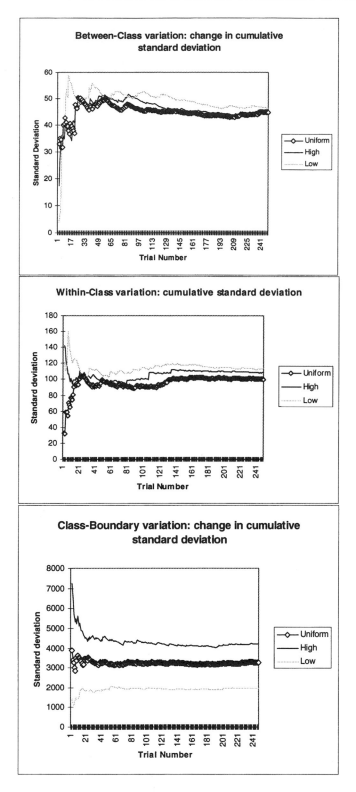

Figure 6.3. (a) Between-Class Uncertainty: variation in cumulative standard deviation; (b) Within-Class Uncertainty: variation in cumulative standard deviation; (c) Class-Boundary Uncertainty: variation in cumulative standard deviation.

Table 6.2. Monte Carlo Exceedance Results.

Deposition	No. of 1 km² Between-Class	Cells Within-Class	Exceeded Class-Boundary
High	251	689	19839
Low	271	533	11688
Uniform	253	539	14321

Figure 6.4. Maps of certainty: areas always exceeded under three deposition scenarios: (a) uniform, (b) high, (c) low.

the interests of longer-term credibility, as suggested by Mowrer (1998).

REFERENCES

Besag, J. and J. Newell. The Detection of Clusters in Rare Diseases. *J. R. Stat. Soc.,* 154, Part 1, pp. 143-155, 1991.

Besag, J. and P. Diggle. Simple Monte Carlo Tests for Spatial Pattern. *Appl. Stat.,* 26(3), pp. 327-333, 1977.

Bull, K.R. An Introduction to Critical Loads. *Environ. Pollut.,* 77, pp. 173–176, 1992.

Cliff A.D. and J.K. Ord. *Spatial Autocorrelation,* Pion, London, 1973.

Emmi, P. and C. Horton. A Monte Carlo Simulation of Error Propagation in a GIS-Based Assessment of Seismic Risk. *Int. J. Geogr. Inf. Syst.,* 9(4), pp. 447–461, 1995.

Freska, C. and T. Barkowsky. On the Relations Between Spatial Concepts and Geographic Objects, in *Geographic Objects with Indeterminate Boundaries,* P. Burrough and P. Frank, Eds., Taylor and Francis, London, 1996.

Gascoigne, J., S. Alvanides, and R. Cockburn. *Urban Land Use Patterning: Implementing the Besag and Newell Approach in Arc/Info.* Project presented to the School of Geography, University of Leeds as coursework for the Master's Degree in Geographical Information Systems, 1995.

Hall, J., M. Hornung, P. Freer-Smith, P. Loveland, I. Bradley, S. Langan, J. Gascoigne, H. Dyke, and K.R. Bull. *Current Status of UK Critical Loads Data: December 1996,* Report to the Department of the Environment, 1996.

Hornung, M., K. Bull, M. Cresser, J. Hall, S. Langan, P. Loveland and C. Smith. An Empirical Map of the Acidity for Soils in Great Britain. *Environ. Pollut.,* 90(3), pp. 301–310, 1995.

Lagacherie, P., P. Andrieux, and R. Bouzigues. Fuzziness and Uncertainty of Soil Boundaries: From Reality to Coding in GIS, in *Geographic Objects with Indeterminate Boundaries,* P. Burrough and P. Frank, Eds., Taylor and Francis, London, 1996.

Mowrer, H.T. Propagating Uncertainty through Spatial Estimation Processes for Old-Growth Subalpine Forests Using Sequential Gaussian Simulation in GIS. *Ecol. Modelling,* 98, pp. 73–86, 1997.

Mowrer, H.T. Selling Uncertainty to the Uncertain, in *Spatial Accuracy Assessment: Land Information Uncertainty in Natural Resources,* Ann Arbor Press, Chelsea, MI, 1999, (forthcoming).

Nilsson, J. *Critical Loads for Sulphur and Nitrogen.* Nordic Council of Ministers, Copenhagen, Denmark, 1986.

Nilsson, J. and P. Grennfelt, Eds. *Critical Loads for Sulphur and Nitrogen (Report 1988:15),* Nordic Council of Ministers, Copenhagen, Denmark, 1988.

Openshaw, S., M. Charlton, C. Wymer, and A. Craft. A Mark 1 Geographical Analysis Machine for the Automated Analysis of Point Data Sets. *Int. J. Geogr. Inf. Syst.,* 1, pp. 335–358, 1987.

Openshaw, S. What is GISable Spatial Analysis? in *Proceedings of Workshop on New Tools for Spatial Analy-*

sis, Lisbon, 18–20 November, Eurostat, Luxembourg, 1993, pp. 36–48.

Openshaw, S. Personal communication, 1998.

Posch, M., J-P. Hettelingh, P.A.M. de Smet, and R.J. Downing. *Calculation and Mapping of Critical Thresholds in Europe: Status Report 1997,* RIVM, Netherlands, 1997.

CHAPTER 7

Digital Terrain Modeling: Accuracy Assessment and Hydrological Simulation Sensitivity

P. Defourny, G. Hecquet, and T. Philippart

INTRODUCTION

Terrain modeling is a key component of any digital representation of landscape. A wide spectrum of mathematical solutions has been developed to model the topography for several decades (Millet and Laflamme, 1958; Sharpnack and Akin, 1969; Peucker and Fowler, 1978) and was more recently implemented by the software industry. While this modeling challenge is a very old issue, new contributions are still regularly published (Hodgson, 1995; Carrara et al., 1997). The state of the art can be hardly documented quantitatively because of the wide range of geomorphological patterns, the diversity of data acquisition procedures, and the various computer implementations of the different theoretical solutions (Tahiri, 1994). In addition, a main constraint for cross-comparison study is the poor availability and the poor accuracy of reference data sets.

On the other hand, user accuracy requirements vary significantly according to the application and are usually not defined in spite of the common use of digital terrain model (DTM) in many geographic information system (GIS) applications. The literature (Lee, 1991; Wood and Fisher, 1994; Felicisimo, 1994) mainly addressed the performance of the interpolation methods for the elevation estimation. Many fewer authors (Skidmore, 1989; Carter, 1992; Bolstad and Stowe, 1994) investigated the accuracy of the slope gradient and aspect computation, while these are often the only relevant variables for GIS applications.

A comprehensive error assessment throughout the digital terrain processing chain from the data capture to the estimation of three main variables of interest; i.e., the elevation, the slope gradient, and the orientation, can hardly be found. Furthermore, very little concern is paid to the actual performances of such "well-established" and popular DTM methods, whatever the complexity of the final application is.

OBJECTIVES

The main study objective is to assess the actual performances of various algorithms dedicated to terrain modeling in a processing chain perspective. First, the experiment documents the accuracy of commonly used methods for (i) the elevation interpolation, (ii) the estimation of the slope gradient, and (iii) the estimation of the slope orientation. More specifically, the study also aims to investigate the sensitivity of the output accuracy to the data input quality according to the estimation methods. Therefore, this study will quantitatively assess the influence of (iv) the input data type on the performance of the slope estimation methods, and (v) the DTM quality on the performance of the slope gradient and aspect estimation methods. Finally, the sensitivity of hydrological simulations to the DTM's accuracy is analyzed in order to discuss these results in a user requirement perspective.

EXPERIMENTAL DATA SET

The availability of a large, realistic and accurate digital data set to serve as reliable reference information is a main issue. While many published results rely on simulated data sets, a data set describing an actual terrain has been preferred to stick as close as possible to existing topographical features. This covers a representative study area of 4,000 hectares including valleys, slight slopes, and plateau in the central part of Belgium. The elevation ranges between 59 m to 147 m.

Two independent sources of information are available for the elevation over this area. The Institut Géographique National of Belgium completed a 1:10,000 map with contour interval of 2.5 meters. The Projet Informatique de Cartographie Continue of the Ministère Wallon de l'Equipement et des Transports made available a unique elevation data set corresponding to a 50×50 m grid sampling covering the whole area, complemented by characteristic points sampled along the rivers every 1 to 3 meters. This was produced in the framework of the 1:1,000 map production from detailed and supervised digital photogrammetric restitution from 1:4,000 aerial photographs. While the technical specifications require an accuracy of 12 cm in X,Y coordinates and 15 cm in Z for the validation points, the absolute error mean for the whole set is around 25 cm for the 3-D coordinates. The consistency of both data sets was checked by analytical quality control.

In addition, a field measurement campaign of slope gradient and aspect has been carried out using a differential GPS Trimble Pro XR receptor, compass, and clinometer.

METHODOLOGY

The comparison of the various methods relies on a standard experimental protocol using a dozen complementary parameters for the accuracy measurement.

Experimental Protocol

The photogrammetric data set; i.e., 18,735 measured points, was split by systematic sampling into two sets: the input data and the reference data. They both correspond to a 70.7×70.7 m grid complemented by the river features for the input data. The raster output resolution was set to 5 m.

Elevation

From the point data, 14 interpolations belonging to 3 different methods were run using 4 different softwares. Height versions of the linear interpolation

based on the average weighted by the inverse distance were defined by 2 different search radius and the 4 following weighting functions as proposed by the Imagine software developed by ERDAS:

$$W = (1-Q)/Q \quad (1) \qquad W = (1-Q)^2 \quad (5)$$
$$W = 1-Q \qquad (4) \qquad W = 1-Q^2 \quad (9)$$

with Q = (distance to point i/search radius)

An alternative implementation of this linear method uses a fixed number of points and was selected from the Easy/Pace software developed by PCI. Three interpolations by kriging were run using an exponential function as proposed by default, the linear and the spherical function based on the computed variogram. Finally, two Triangular Irregular Network (TIN) approaches proposed by StarCarto and Arc/Info were also tested including the often required vector–raster conversion step. The extensive mathematical expression of these algorithms can sometimes be found in the software documentation or in the literature (Carrara, 1988).

Most of these interpolation methods, except the kriging, have been applied from the contour lines to document the impact of these data capture techniques. In addition, 4 different solutions taking automatically into account contextual information about the concerned morphological feature were selected from three different softwares; i.e., Easy/Pace, Idrisi developed by Clark University, and the homemade GERU-UCL.

In order to assess the respective method accuracy, the predicted values are compared with the reference values for the corresponding X,Y coordinates; i.e., 7604 points.

Slope Gradient and Aspect

Three methods to derive the slope gradient and aspect from a raster file were selected: the method of Ritter (1987), the 3rd order finite difference without weighting, and with weighting by the closest diagonal points (Horn, 1981). These methods were applied from three different digital elevation models (DEM) selected according their respective quality ranging from poor to best accuracy. The slope estimation directly derived from the TIN model was also tested.

The reference data consist of the set of 100 georeferenced representative points measured from the field.

Accuracy Assessment

The output performance is analyzed using 4 different approaches. The comparison of the estimated and the reference distributions of the variable of interest is based on their respective mean and standard deviation. The statistical analysis of the error distribution relies on the maximum negative error (Emax −), the maximum positive error (Emax +), the averaged absolute error or absolute mean error (AME), the root mean square error (RMSE), and the standard deviation of the error.

Thirdly, the comparison between the estimated and the reference distributions of the variable of interest is also described by a linear regression analysis and eventually the correlation coefficient of Pearson. The regression line intercept and the slope values allow to discriminate the random error and the systematic error. Similarly, the root mean square error is decomposed into 2 components; i.e., the random one (RMSEr) and the systematic one (RMSEs), as follows:

$$RMSE^2 = RMSEr^2 + RMSEs^2$$

with

$$RMSEs = \sqrt{\frac{\sum_{i=1}^{n}(\hat{x}_i - X_i)^2}{n}}$$

$$RMSEr = \sqrt{\frac{\sum_{i=1}^{n}(\hat{x}_i - x_i)^2}{n}}$$

Finally, the visual observation of output maps and graphical histograms provides other relevant information on the methods performance.

Hydrological Simulation Sensitivity

The processing chain perspective starting from the elevation data capture to the digital terrain application has been completed using a regionalized hydrological model. This is a mechanistic model namely Modèle Hydrologique Maillé, developed in-house to estimate the flow profile at the outlet from a rainfall event. A small watershed of 9.2 km² was selected within the study area. Based on the methods performances analy-sis, 3 DTMs of different accuracy have been generated. These DTMs range from a most accurate representation to a very smoothed one.

The flow profile predicted using the accurate DTM will serve as the reference in order to test only the terrain modeling accuracy impact and not the hydrological model. Four different recorded rainfall events are used to illustrate the hydrological sensitivity to the terrain models based on their predicted flow profiles.

RESULTS ANALYSIS

Elevation

The whole set of parameters converges to discriminate three levels of performance for the elevation interpolation from point data (Table 7.1). First, the accuracy of the interpolation by kriging exceeds any other methods, with a RMS of 0.71 and a maximum error of about 7 m. The TIN model and the interpolation using distance weighting with a fixed number of points also provide good results. It is surprising that the TIN model does not appear as the most efficient option, as often considered. Similarly, it is unexpected to find a raster interpolation performing almost as well as the TIN model (RMSE = 0.88 versus RMSE = 0.85). The methods included in the third group show high absolute errors and very high systematic error as shown by the intercept and the RMSEs.

The conversion from TIN structure to raster format must be taken into account because of the perspective of a processing chain. This conversion induces a slight additional error that is best minimized by a nonlinear function except for few a points where large errors are produced.

The overall elevation mean obtained by kriging is singularly higher than the reference mean. This slight overestimation illustrates the kriging capabilities to model the tops and the depressions, unlike the other methods. The selection of the appropriate semi-variogram function seems not critical, as most of them perform similarly.

As observed in the point interpolation test, the weighting function 1 using a small search radius shows best results with regard to the plain raster interpolation method (Table 7.2). The adaptative methods taken into account, the morphological features provide much better results, however. Surprisingly, some of them perform as well as the two tested TIN versions.

Finally, the comparison between the two sources of data highlights the critical influence of the input

Table 7.1. Accuracy Assessment of the Elevation Interpolation Methods from the Point Data Set. (The interpolation names refer to the selected software implementation and the associated options; e.g., ERDAS 1/200 means the ERDAS implementation using the inverse distance function number 1 with a search radius of 200 m.)[a]

Reference	Mean	Std. Dev.	Error Distribution				Regression Analysis			
			Emax −	Emax +	AME	RMSE	Slope	Intercept	RMSEr	RMSEs
Reference	96.71	15.76								
Weighted average according to an inverse distance function										
ERDAS 1/200	96.64	15.58	−9.76	9.42	0.70	1.00	0.987	1.20	0.98	0.22
ERDAS 4/200	96.63	15.44	−10.78	10.56	1.10	1.50	0.976	2.29	1.45	0.39
ERDAS 5/200	96.64	15.50	−9.93	10.20	0.93	1.29	0.981	1.82	1.25	0.31
ERDAS 9/200	96.63	15.41	−11.38	10.76	1.19	1.61	0.973	2.53	1.55	0.43
ERDAS 1/800	96.59	15.07	−11.61	9.47	1.54	1.95	0.950	4.77	1.78	1.80
ERDAS 4/800	96.59	13.77	−17.51	21.33	4.23	5.14	0.821	16.46	4.38	2.70
ERDAS 5/800	96.58	14.09	−16.47	19.46	3.69	4.52	0.860	13.58	3.93	2.23
ERDAS 9/800	96.60	13.62	−18.03	22.30	4.52	5.48	0.811	17.97	4.62	2.95
EASI/PACE	96.69	15.65	−9.22	8.97	0.59	0.88	0.990	0.95	0.87	0.16
TIN interpolation										
StarCarto	96.72	15.66	−8.59	13.46	0.54	0.85	0.993	0.68	0.84	0.11
Arc/Info linear	96.72	15.66	−8.59	13.46	0.54	0.85	0.993	0.68	0.84	0.11
Kriging (Arc/Info)										
Exponential	96.84	15.67	−7.27	7.20	0.46	0.71	0.995	0.45	0.71	0.072
Spheric	96.84	15.67	−7.27	7.20	0.46	0.71	0.995	0.45	0.71	0.072
Linear	96.84	15.67	−7.27	7.20	0.46	0.71	0.995	0.45	0.71	0.072
TIN interpolation and grid conversion (Arc/Info)										
Linear	96.81	15.61	−18.41	8.93	0.56	0.92	0.991	0.86	0.91	0.142
Quintic	96.84	15.70	−15.07	35.12	0.49	0.89	0.997	0.29	0.89	0.047

[a] All values are in meters, except slope which is adimensional.

Table 7.2. Accuracy Assessment of the Elevation Interpolation Methods from Digitized Contour Lines.[a]

Reference	Mean 96.71	Std. Dev. 15.76	Error Distribution				Regression Analysis			
			Emax −	Emax +	AME	RMSE	Slope	Intercept	RMSEr	RMSEs
Weighted average according to an inverse distance function										
ERDAS 1/250	97.17	15.54	−8.27	12.14	1.35	1.83	0.980	2.38	1.75	0.56
ERDAS 4/250	97.08	15.27	−9.70	12.12	1.95	2.53	0.957	4.54	2.41	0.77
ERDAS 5/250	97.12	15.36	−9.29	11.96	1.74	2.28	0.965	3.83	2.19	0.69
ERDAS 9/250	97.06	15.23	−9.96	12.25	2.06	2.67	0.953	4.89	2.54	0.82
ERDAS 1/600	97.05	15.24	−8.90	11.79	1.80	2.39	0.956	4.59	2.26	0.7
ERDAS 4/600	96.85	14.40	−14.53	16.11	3.72	4.62	0.877	12.02	4.19	1.94
ERDAS 5/600	96.88	14.63	−13.48	15.54	3.34	4.13	0.897	10.19	3.79	1.64
ERDAS 9/600	96.84	14.36	−15.15	16.50	3.99	4.88	0.867	12.97	4.40	2.10
TIN interpolation										
STAR	97.26	15.74	−7.44	15.74	1.07	1.42	0.997	0.96	1.32	0.53
Arc/Info linear	97.29	15.68	−7.44	13.47	1.10	1.51	0.991	1.40	1.39	0.58
Weighted average according to an inverse distance function and morphological features										
EASI/P. ("DIAG")	97.28	15.50	−7.56	14.03	1.38	1.93	0.977	2.84	1.81	0.68
EASI/P. ("CONIG")	97.23	15.66	−7.56	14.08	1.09	1.50	0.990	1.58	1.37	0.61
IDRISI	97.30	15.83	−7.77	18.38	1.05	1.54	1.000	0.55	1.42	0.59
GERU-UCL	97.38	15.74	−9.70	13.40	1.28	1.75	1.001	0.68	1.56	0.79

[a] All values are in meters, except slope which is adimensional.

data type. The interpolation from point data using almost any method provides better results than any methods from the contour lines as far as the RMSE is concerned.

Slope Gradient

Based on the above results, three DEMs have been generated from the whole point data set. The DEM1 was generated using the best interpolation method; i.e., kriging. The DEM2 was produced by a middle-class interpolation method; i.e., weighted average by inverse distance (ERDAS 1/800). Erratic values were also randomly distributed. The third model corresponds to a very smoothed model (ERDAS 9/800).

A systematic underestimation of the slope gradient is clearly illustrated by the gradient mean and the range of the error distribution not centered around zero (Table 7.3). This expected smoothing effect is due to the automatic sampling of the topography. The erratic values of the DEM2 mask this effect in the statistical results.

The best slope gradient estimation are provided by the 3rd order finite difference computed from 8 points, with a slight advantage for the version reducing the influence of the diagonal points. The performance of the TIN model can be ranked between these two methods. The Ritter's algorithm produces larger errors mainly due to systematic error (RMSEs = 4.3).

The influence of the DEM is very critical for the reliability of the slope gradient estimation. The influence of the slope gradient algorithm becomes negligible when the DEM is not of good quality.

Slope Aspect

The RMSE is similar for the three grid-based algorithms (RMSE = 29), while the TIN-derived estimation provides a RMSE of 33. The maximum errors can be as high as 180° in the worst case. However, the statistical parameters must be carefully discussed because the error distribution is no longer Gaussian.

The slope aspect error was discussed in relation to the slope gradient: the lower the gradient, the higher the aspect error. The distribution of the aspect values estimated for the whole study area allows for investigating further the performance of the various methods (Figure 7.1). The method of Horn (1981) provides a regular aspect distribution dominated by the natural landscape orientation; i.e., 130° and 330°. The TIN-derived aspect shows the same trend with higher variability due to the coarser resolution associated with

the TIN structure in regard to the 5-m grid cell. The second 3rd order method shows slightly preferential directions along the 16 main directions used by the algorithm. The distribution produced by the Ritter's method is clearly dominated by only 8 directions.

These results show again the critical influence of the DEM quality used as input for the slope aspect estimation. The RMSE doubles from the DEM1 to DEM2 and DEM3. The overall aspect distribution obtained from DEM2 and DEM3 is biased, even for the most accurate algorithm. From a poor quality DEM, the error distribution is similar, whatever the estimation method.

Sensitivity of Hydrological Simulation

The above results allow one to define three different processing chains for DTM generation corresponding to various quality levels: (i) DTM1: kriging for the elevation interpolation then the 3rd order finite difference of Horn for the slope variables; (ii) DTM2: average weighted by the inverse distance with search radius of 200 m then the Ritter's method; (iii) DTM3: average weighted by the inverse distance with search radius of 800 m then the 3rd order method without weighting.

Figure 7.2 illustrates the influence of the DTM quality on various actual rainfall events. The simulated flow is systematically underestimated for the DTM2 and DTM3, while DTM1 follows much closer the observed flow. The flow peak underestimation varies from 14 to 23% for the DTM2 and from 28 to 38% for the DTM3.

The estimated cumulative flow shows also a significant error ranging from 6 to 22%. On the other hand, almost no time delay is observed between the flow peak measurement at the outlet and its prediction. However, a delay of 5 hours is observed for the event 2 using the DTM3.

CONCLUSIONS

This experimental study based on a unique actual data set has provided very interesting findings. These results are directly related to the topographical features observed in the study area dominated by a gently undulating relief. The conclusions should therefore be extended with caution.

(i)   The input data type contributes the most to the accuracy of the end product. In par-

Table 7.3. Accuracy Assessment of Slope Gradient Estimation Derived from Three Different DEMs.[a]

Reference	Mean	Std. Dev.	Error Distribution					Regression Analysis				Pearson coeff r
	5.791	3.230	Emin	Emax	AME	RMSE	Std. Dev.	Slope	Intercept	RMSEr	RMSEs	
3rd order finite difference *with* weighting by the closest diagonal points (Arc/Info)												
DEM1	4.637	2.225	−9.189	1.921	1.337	2.044	1.696	0.600	1.166	1.091	1.729	0.870
DEM2	6.227	4.547	−15.595	17.761	3.755	5.248	5.255	0.168	5.257	4.494	2.711	0.119
DEM3	1.453	0.558	−17.364	0.163	4.341	5.320	3.094	0.056	1.128	0.526	5.294	0.324
3rd order finite difference *without* weighting by the closest diagonal points (Erdas)												
DEM1	4.243	2.176	−9.500	1.500	1.670	2.345	1.721	0.585	0.823	1.078	2.083	0.867
DEM2	5.714	4.484	−16.500	17.00	3.762	5.179	5.203	0.166	4.752	4.431	2.681	0.120
DEM3	0.990	0.628	−17.500	0.000	4.800	5.713	3.112	0.055	0.674	0.600	5.681	0.281
Ritter's algorithm (PCI)												
DEM1	2.393	0.733	−15.396	0.082	3.399	4.284	2.622	0.196	1.256	0.366	4.268	0.865
DEM2	1.670	1.004	−18.300	1.480	4.167	5.230	3.236	0.047	1.400	0.988	5.135	0.150
DEM3	1.342	0.185	−17.947	0.239	4.454	5.457	3.175	0.018	1.235	0.175	5.454	0.303
Arc/Info												
TIN	4.852	2.385	−7.783	3.583	1.486	2.146	1.939	0.592	1.422	1.416	1.612	0.802

[a] All values are in percent, except slope and Pearson coefficient that are adimensional.

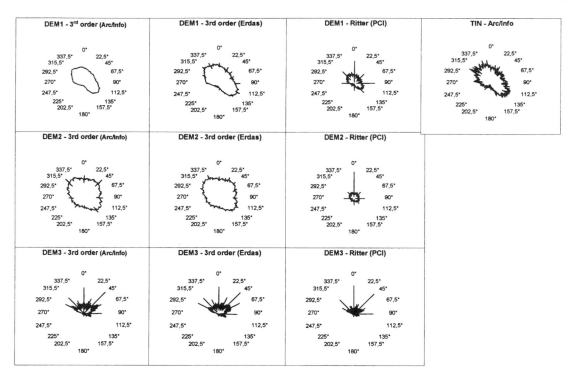

Figure 7.1. Radar histogram of the different slope aspect distributions for the whole study area as derived from three different DEMs using three different methods and from the TIN method.

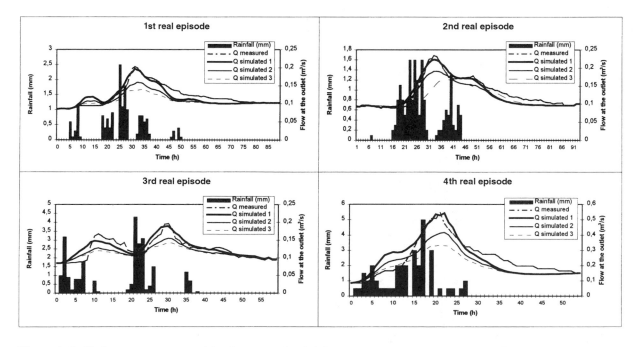

Figure 7.2. Hydrograms computed for four actual rainfall events using three different DTMs.

ticular, it is highly recommended to use point data rather than contour lines which are a symbolic abstraction of the elevation data.

(ii) While any commercial software provides at least one method for digital terrain generation, these methods and their respective computer implementations are not equivalent. It is noticeable that the same commercial softwares have implemented the most efficient method for one step of the DTM production chain, but also the worst for the other steps of the chain.

(iii) For the elevation interpolation, the kriging approach was found to be the most accurate using point data as input (absolute mean error = 0.46 m). The TIN model usually considered as the best was second, and not significantly better than the best algorithm based on the average weighted by the inverse distance. The best performance from contour line data corresponds to an absolute mean error of 1.07 m.

(iv) The slope gradient is necessarily underestimated because of the input data sampling. The estimation error of the slope aspect generally varies with the slope gradient. From a good DEM the 3rd order finite difference of Horn provides the smallest errors for the slope gradient (absolute mean error around 1.5%) and for the slope aspect (absolute mean error around 20°). While the slope gradient accuracy can be considered as well estimated, the aspect estimation error remains rather high. This experiment also demonstrates the very bad performance of some implemented methods.

(v) There is no difference between slope estimation methods when only poor quality DEM is available. However, the estimation error obtained from poor DEM is very significant for the slope gradient (absolute mean error around 5%) and for the slope aspect (absolute mean error around 52°).

From the user point of view, the main finding of this study is the high sensitivity of an application such as the hydrological modeling to the terrain representation. The accuracy requirements for terrain modeling varies significantly according to the application field. However, this study should enhance the user's concern about the DTM quality and its impact on the final results.

ACKNOWLEDGMENTS

The authors are very grateful to Prof. A. Collignon and the Ministère wallon de l'Equipement ct des Transports for his supportive interest and for the data acquired in the framework of the Projet Informatique de Cartographie Continue.

REFERENCES

Bolstad, P.V. and T. Stowe. An Evaluation of DEM Accuracy: Elevation, Slope and Aspect, *Photogrammetric Eng. Remote Sensing,* 60(11), pp. 1327–1332, 1994.

Carrara, A. Drainage and Divide Networks Derived from High-Fidelity Digital Terrain Models, in *Quantitative Analysis of Mineral and Energy Resources*, C.F. Chung, Ed., Reidel Publishing Company, 1988, pp. 581–597.

Carrara, A., G. Bitelli, and R. Carla. Comparison of Techniques for Generating Digital Terrain Models from Contour Lines, *Int. J. GIS,* 11(5), pp. 451–473, 1997.

Carter, J.R. The Effect of Data Precision on the Calculation of Slope and Aspect Using Gridded DEMs, *Cartographica,* 29(1), pp. 22–34, 1992.

Chen, Z-T. and J. Guavara. A Systematic Selection of Very Important Points (VIP) from DTM for Constructing TIN, *Proceedings of Autocarto 8,* Falls Church, VA, 1987, pp. 50–56.

Felicisimo, A.M. Parametric Statistical Method for Error Detection in Digital Elevation Models, *ISPRS J. Photogrammetry Remote Sensing,* 49(9), pp. 29–33, 1994.

Goodchild, M.F. and S. Gopal. *Accuracy of Spatial Databases,* Taylor and Francis, London, 1989.

Heuvelink, G.B.M., P.A. Burrough, and N. Stein. A Propagation of Errors in Spatial Modelling with GIS, *Int. J. GIS,* 3(4), pp. 303–322, 1989.

Hodgson, M.E. What Cell Size Does the Computed Slope/Aspect Angle Represent? *Photogrammetric Eng. Remote Sensing,* 61(5), pp. 513–517, 1995.

Horn, B.K.P. Hill Shading and the Reflectance Map, *Proc. IEEE,* 69, 1981.

Lee, J. Comparison of Existing Methods for Building TIN Models of Terrain from Grid DEM, *Int. J. GIS,* 5(3), pp. 267–285, 1991.

Lee, J., P.K. Snyder, and P.F. Fisher. Modeling the Effect of Data Errors on Features Extraction from Digital Elevation Model, *Photogrammetric Eng. Remote Sensing,* 58(10), pp. 1461–1467, 1992.

Li, Z. A Comparative Study of the Accuracy of Digital Terrain Models (DTMs) Based on Various Data

Models, *ISPRS J. Photogrammetry Remote Sensing,* 49(1), pp. 2–11, 1994.

Makarovic, C.B. Structures for Geo-Information and Their Application in Selective Sampling for Digital Terrain Models, *ITC J.,* 4, pp. 285–295, 1984.

Millet, C.L. and R.A. Laflamme. The Digital Terrain Model—Theory and Application, *Photogrammetric Eng.,* 24(3), pp. 433–442, 1958.

Openshaw, S. Accuracy and Bias Issues in Surface Representation, in *Accuracy of Spatial Databases,* M.F. Goodchild and S. Gopal, Eds., Taylor and Francis, London, 1989, pp. 263–276.

Peucker, T.K. and R.J. Fowler. The Triangulated Irregular Network, *Proc. of the ASP DTM Symposium,* Virginia, 1978, pp. 516–540.

Ritter, P. A Vector-Based Slope and Aspect Generation Algorithm, *Photogrammetric Eng. Remote Sensing,* 53(8), pp. 1109–1111, 1987.

Sharpnack, D.A. and G. Akin. An Algorithm for Computing Slope and Aspect from Elevations, *Photogrammetric Eng. Remote Sensing,* 35, p. 247, 1969.

Skidmore, A.K. A Comparison of Techniques for Calculating Gradient and Aspect from a Gridded DEM, *Int. J. GIS,* 3(4), pp. 323–334, 1989.

Tahiri, D. Les Modèles Numériques de Terrain: état de l'art, *Bulletin trimestriel de la Société Belge de Photogrammétrie-Télédétection et Cartographie,* 195-196, pp. 25–40, 1994.

Theobald, D.M. Accuracy and Bias Issues in Surface Representation, in *Accuracy of Spatial Databases,* M.F. Goodchild and S. Gopal, Eds., Taylor and Francis, London, 1989, pp. 99-106.

Torlegard, K., A. Ostman, and R. Lindgren. A Comparative Test of Photogrammetrically Sampled Digital Elevation Models, *Photogrammetria,* 41, pp. 1–16, 1986.

Weibel, R. and M. Heller. Digital Terrain Modelling, in *Geographical Information Systems Volume 1: Principles,* D.J. Maguire, M.F. Goodchild, and D.W. Rhind, Eds., Longman Scientist & Technical, G-B, 1991, pp. 269–297.

Wood, J.D. and P.F. Fisher. Assessing Interpolation Accuracy in Elevation Model, *IEEE Computer Graphics Applications,* 16, pp. 48–56, 1994.

CHAPTER 8

Depth Model Accuracy: A Case Study in the Great Barrier Reef Lagoon

A. Lewis

INTRODUCTION

The Great Barrier Reef extends 2,000 km along the northeast coast of Queensland, Australia, bounded by the shore and the edge of the continental shelf (Figure 8.1). Coral reefs are found scattered within the lagoon and fringing its outer edges, especially where the edge of the continental shelf is steep. The lagoon itself varies in depth, but is generally about 35 m deep, and smooth-bottomed between reefs.

Economic activities in the Great Barrier Reef are estimated to involve in excess of a billion dollars (Australian) per annum (McPhail, 1997). The area is Marine Park, World Heritage, and habitat to a range of charismatic flora and fauna including sea grasses, corals, and dugongs. It supports a $AUS250 million commercial fishing industry, a vital tourist industry, and incorporates several important ports and shipping routes (McPhail, 1997). Knowledge of the species, ecosystems, and processes over the Great Barrier Reef has grown rapidly in the past 10–15 years (Reichelt, 1997); however, effective management into the future will rely on the ability to address resource and conservation issues at a regional scale. In this endeavor, the availability of appropriately scaled and accurate depth models is as important to marine systems as terrain models are on land.

A digital depth model (DDM500) covering the entire Great Barrier Reef was interpolated from 178,000 irregularly located depth soundings, captured over many years by the Royal Australian Navy Hydrographic Survey Office. These soundings, which reference depth to mean sea level (MSL), are highly variable in density (ranging from an estimated one sounding per 1000 km^2 to 28 points per km^2, with an overall mean of 0.74 points per km^2), and inevitably DDM500 will be more accurate in some places than others. Typically, soundings are denser in deeper waters away from both reefs and shore lines—places conducive to ship-based hydrographic survey.

Interpolation between soundings to a 500 m lattice was addressed with the ANUDEM software (Hutchinson, 1988, 1989) as implemented in ARCINFO[1] 7.0.4 TOPOGRID. The interpolation was also controlled by two other data sources: the coastline (high-water mark) as mapped by the Great Barrier Reef Marine Park Authority (GBRMPA), which after reference to tide tables was assigned a depth of +0.8 m above MSL, and the mapped outlines of reefs where there were no soundings in the vicinity of the reef, provided that the reef was more than 12.5 km from land. About 1,280 of 3,900 reefs fall into this category. After reference to other data these mapped reef outlines were assigned a depth of −10.4 m. Coast and reef data sets have a cartographic scale between 1:100,000 and 1:250,000. TOPOGRID parameters were set to meet a root mean square error of 0.2 m, using 25 iterations and without "drainage enforcement."

DDM500 is expected to be used for quantification and visualization of a number of properties of the Great Barrier Reef at the regional scale. So far, it has formed

---

[1] ARCINFO is a registered trademark of Environmental Systems Research Institute, Inc.

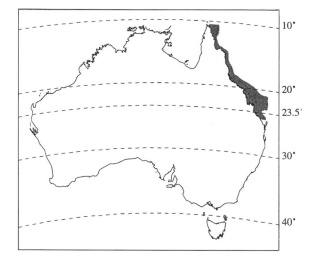

Figure 8.1. Australia's Great Barrier Reef. The reef extends to the edge of the continental shelf, and therefore varies in width.

the basis for visualization of macroscale properties of the Great Barrier Reef and estimation of the mean water volume of the system as an input to carbon budget modeling. More stringent demands in the future may include the estimation of bottom topography, including slope/break of slope points, input to light attenuation models used to model the distributions of sea grasses, and input to models of cyclone-induced wave energy over reef areas. Given the breadth of potential applications, thorough examination of the accuracy of DDM500, and how this translates into fitness for use, is essential.

Fortuitously, a number of high precision depth surveys of parts of the Great Barrier Reef have been undertaken by the Australian Surveying and Land Information Group (AUSLIG), Australia's National mapping organization. These surveys have provided digital depth data, referred to here as ground truth points (GTPs), for approximately 50 reefs with a vertical accuracy generally accepted to be within 0.2 m (Priest, 1997). The majority of these surveys cover reef areas; however, for the areas around Bowling Green Bay there are 131,000 GTPs covering 1597 km^2, giving an accurate reference for near-shore and lagoon areas. As none of these data have been used in the current version of DDM500, they provide a unique validation source for the model.

REEF AREAS VERSUS LAGOON AREAS

The Great Barrier Reef consists of a shallow (mean depth ~34 m) lagoon extending from the coast to the edge of the continental shelf, interspersed with coral reefs (Figure 8.2). The area represented by mapped reefs is less than 8% of the total; however, they are generally taken to represent the key areas of primary production, biological diversity, and physical diversity in the Great Barrier Reef. The lagoon area is also of great importance; for instance, the large, fragile sea grass beds within the lagoon form the critical food source for the dugong (*Dugong dugon*).

Requirements for DDM500 can therefore be divided into questions which pertain to the lagoon at large, and questions which relate to the reef areas. In addition, depth model data are generally superior in the nonreef areas, while topographic complexity is greatest in the reef areas. These are compelling arguments to examine the accuracy of DDM500 separately for reef and nonreef areas. In this chapter I consider reef areas to be places within 1,000 m of reef outlines as mapped by the GBRMPA.

DEPTH MODEL ACCURACY IN REEF AREAS

DDM500 was compared with ground truth points over 42 reefs by overlay of the GTPs with DDM500 and other data surfaces. These sample reefs represent approximately 1% of all reefs. The choice of sample reefs was imposed by availability of data rather than sample design considerations. However, given the a priori expectation that the accuracy of DDM500 will, to some extent, be a function of the spatial density of depth soundings, the sample reefs are suitably diverse, although concentrated spatially in the southern parts of the Great Barrier Reef.

Prior to overlays, a localized best fit of the GTPs with DDM500 was sought using an iterative algorithm which adjusted data set locations, seeking to minimize the variance between DDM500 and the GTPs. The results of this process, which was necessary to overcome inevitable misregistrations between the 1:250,000 map base used for DDM500 and the 1:10,000 scale appropriate to the survey data, were inspected visually. Where the algorithm failed to converge, no transformation was applied.

The total number of GTPs per reef varied from 39 to 132,000. Each surveyed reef was treated as a single data point for this analysis, so to ensure approximately

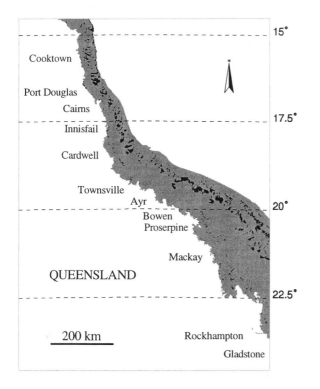

Figure 8.2. The Great Barrier Reef consists of a large lagoon interspersed with coral reefs. The area of mapped reefs is 8% of the total. Reefs and lagoons have very different structures, and the errors in DDM500 are expected to differ between them.

even weight per reef, 2,000 data points were randomly selected from all of the points available for each reef. Only three reefs had fewer than 200 GTPs.

The GTPs at each sample reef were overlaid with DDM500, using a bilinear interpolation between lattice points, to give estimates of depth from DDM500 at each GTP. The mean, standard deviation and the RMSE of the difference between the surveyed depths and the modelled depths were then calculated for the reef. For each reef, the "density" of depth soundings over the reef was calculated as the mean distance between each GTP and the closest depth sounding used in the interpolation. This measure has units of meters, but its inverse square is a true measure of point density.

Exploratory data analysis revealed that the reefs fell into two distinct groups—those which had a mean distance to the nearest sounding of less than 1,000 m, and those which did not. In the former group the error of DDM500 was predictable as a function of the mean depth predicted by DDM500; that is, the RMSE of DDM500 increased as the mean depth predicted by

DDM500 became greater (Figure 8.3). One outlier was excluded, Myrmidon Reef, which is characterized by a rapid drop-off on three sides to very deep water (150 m depth), causing overcompensation in the interpolation. However, Myrmidon Reef is generally regarded as unique. For reefs which had a mean distance to the nearest sounding of more than 1,000 m the RMSE was unpredictable, ranging from very low to very high values (Figure 8.4).

The accuracy of DDM500 over reef areas was not correlated to the density of soundings. The bias of DDM500; that is, the mean error, and the standard deviation of the error (a measure of precision), were also unrelated to the density of the soundings.

These results can be interpreted with the aid of Figure 8.5. Although not strongly correlated with the density of depth soundings on the reef, some reefs are represented by DDM500 far better than others. Figure 8.5a illustrates that Wilson Reef is well represented by DDM500, but Bait Reef is poorly represented (Figure 8.5b). Where the reef is not well represented, DDM500 tends to underestimate the overall reef structure (Figure 8.5b), giving a greater mean depth for the reef area, but also giving a greater discrepancy between DDM500 and the reef sample. Thus an increase in mean depth is strongly correlated with an increase in RMSE. (However, this interpretation is an oversimplification since the mean error is not correlated with the RMSE for these reefs.) For reef areas where the density of soundings is extremely low; that is, the mean distance to the nearest sounding is over 1,000 m, the reef outline will be used in the interpolation. In these cases the reef outline is assigned a depth of −10.4 meters, but the reef is barely detected at all, per se, and the RMSE varies widely depending on the specific details of the situation. In the extreme case shown in Figure 8.5c the absence of any depth soundings has allowed DDM500 to assume a virtually constant value of −10.4 m, corresponding to the depth allocated to reef outlines in the absence of all other data.

These results demonstrate clearly that depth over reefs can be predicted using DDM500 with limited absolute accuracy, even in places where the sounding density is relatively high. However, the RMSE associated with the depth model can be predicted from the depth model itself, provided there is sufficient density of soundings. Given the relatively abrupt topography of the reef structures, even quite "good" representations are accompanied by a high RMSE. Thus Wilson reef (Figure 8.5a) is generally well represented despite a RMSE of 6.3 meters.

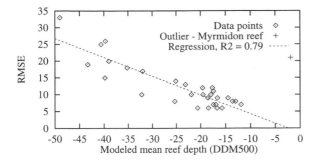

Figure 8.3. Regression of RMSE against mean depth for reefs which have a mean distance to the nearest depth sounding of less than 1000 m. $R^2 = 0.79$.

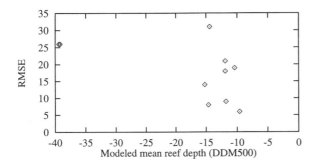

Figure 8.4. Reefs with a mean distance to the nearest depth sounding of more than 1000 m. The RMSE is not predictable for such low sounding density.

DEPTH MODEL ACCURACY IN LAGOON AREAS

The majority of the Great Barrier Reef is lagoon rather than reef, and modeling of these areas is presumed to be less demanding due to less complex topography. For lagoon areas, the accuracy of DDM500 was assessed by comparison of DDM500 with a high resolution depth model, DDM30, interpolated from 131,586 survey points within the vicinity of Cape Bowling Green (Figure 8.6). The total area of the comparison was 1597 km^2, and the density of survey points used to interpolate DDM30 was thus 82 points per km^2, two orders of magnitude greater than the mean density of soundings used in DDM500. DDM30 was used as a ground truth for DDM500 to avoid the bias in results which would occur if the GTPs themselves were used, due to increased concentration of GTPs in the center (and most complex part) of the study area. The accuracy of DDM30 itself was assessed by withhold-

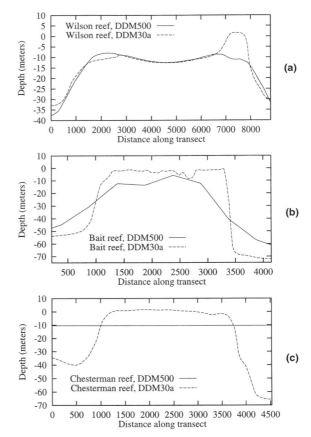

Figure 8.5. (a) A profile section through Wilson Reef showing generally good correspondence between the reef represented by DDM500 and DDM30, a more accurate depth model interpolated from the ground truth points available over the reef. (b) Bait Reef, DDM500 detects the reef structure but only approximates the form. (c) Chesterman Reef, in the absence of any depth soundings DDM500 fails to represent the reef, and assumes the value of −10.4 m associated with mapped reef outlines.

ing a random set of 10,000 points during interpolation. These points demonstrated a bias in DDM30 of 0.012 m, and a RMSE of 0.23 m, with no evidence of spatial structure in the errors. This level of error is negligible in the context of DDM500.

The depths observed in DDM30 ranged from −44.51 to 24.40 meters above MSL. An error surface (for DDM500) was derived by subtracting DDM500 from DDM30, after allowing for the 0.012 m bias in the latter. This surface demonstrated a mean error (i.e., a bias) in DDM500 of 1.25 m over the study area (a tendency for DDM500 to underestimate depth), and

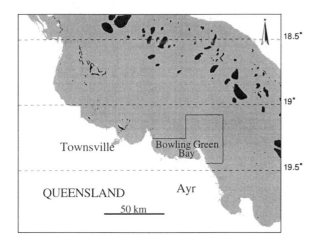

Figure 8.6. The Bowling Green Bay area. Accurate, high density, depth soundings within this area provide a basis for comparison with DDM500.

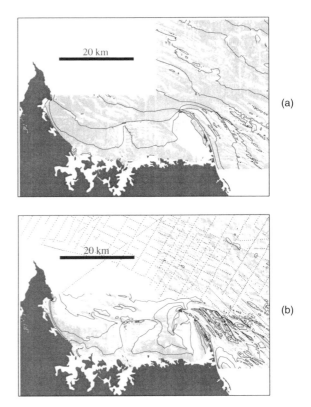

Figure 8.7. (a) DDM30, an accurate (RMSE = 0.2 m) depth model of the Cape Bowling Green study area. Shading is used to illustrate the complex small-scale variations in sea floor topography. Isolines are isobaths with an interval of 5 m. (b) The absolute error surface resulting from differencing DDM500 and DDM30. Isolines represent 2 m intervals. Dots correspond to the soundings used to interpolate DDM500. Errors are highest in places which are distant from the depth soundings, especially where there is complex sea bottom topography. Where depth soundings are dense, the error is less than 2 m.

an overall RMSE of 2.11 m. The spatial distribution of the magnitude of errors within the study area shows a clear pattern with high error where the density of depth soundings (including the shoreline) is low, more so where there is complex sea bottom topography (Figure 8.7). Unlike reef areas, it appears that over lagoon areas the accuracy of DDM500 is in part controlled by the density of soundings. The magnitude of errors over the lagoon is also substantially less than over the reef areas examined.

For each grid cell/lattice point in the study area the distance to the nearest depth sounding or data point (for example, the coastline) was calculated. This distance (*sample distance*) was converted to a localized measure of data point density (points per km^2) using $\rho = 1000000/(\pi * sample \ distance^2)$. A random sample of 997 points was generated and overlaid with the error surface, the sample density surface and the depth surface, using a bilinear interpolation between lattice points.

I found a highly significant relationship (linear correlation) between the log of the sample point density, $\log10(\rho)$, and the absolute error between DDM500 and DDM30, |*error*|, ($R^2 = 0.43$, Figure 8.8). Accuracy also improved with depth; however, this relationship is clearly an artifact of the correlation between depth and the density of soundings.

A linear regression between $\log10(\rho)$ and |*error*| was used to predict the magnitude of the errors in DDM500 over the remainder of the study area (Figure 8.9). Confidence limits on this prediction surface were

calculated nonparametrically by comparison of the actual error observed (according to DDM30) with the error predicted from the regression. This comparison gave a less pessimistic result than the confidence limits based on the standard error of the regression, due to skew in the distribution of residuals (Figure 8.10). The limitation of this approach is that the test data, while not used in the development of the model, are nonetheless from the Cape Bowling Green study area rather than the remainder of the Great Barrier Reef region to which the model is applied. Table 8.1 is a reference table for use with the error model. It gives an indication of the confidence limits associated with

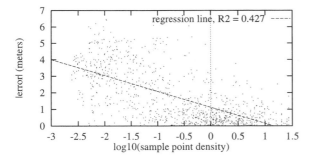

Figure 8.8. Log-linear regression of sounding point density against the absolute error of DDM500.

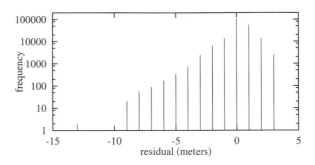

Figure 8.10. Histogram of residuals from the regression of |error| against log10 (sounding *density*).

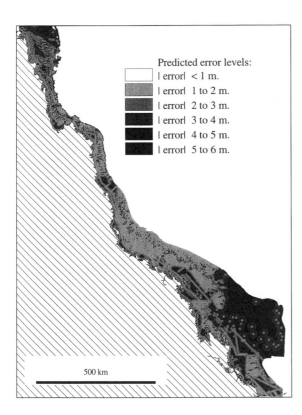

Figure 8.9. Predicted absolute error of DDM500 over the extent of the GBR.

Table 8.1. Error Model Confidence Limits.

Probability That the Discrepancy Will Not Be Greater Than That Listed	Discrepancy Between the Predicted Absolute Error and the Actual Absolute Error[a]
95%	2.4
90%	1.9
80%	0.9
70%	0.5
60%	0.1
55%	0.0
50%	−0.2
40%	−0.4
30%	−0.6
20%	−0.9
10%	−1.3
5%	−1.8

[a] Positive values indicate that the error is underestimated.

the error predictions based on the histogram of the model residuals (Figure 8.10).

CONCLUSION

This analysis of the errors within DDM500, a regional scale depth model of the Great Barrier Reef, has enabled the expected magnitude of the errors to be predicted using a simple transformation of the densities of soundings. This gives vital information to users, and insight to the general limitations of DDM500 and priorities for improvement.

The accuracy of DDM500 can be predicted reasonably well in lagoon areas, but less well in reef areas due to their topographic complexity and the tendency for soundings to avoid reefs. The fitness of DDM500 for a given use remains as always a matter of judgment, which can only be aided by appropriately designed sensitivity analysis directed specifically toward the proposed application. The results of this study are the first step to enable such investigations.

ACKNOWLEDGMENTS

This research was supported by James Cook University and the Australian Cooperative Research Centres Program through the Cooperative Research Centre for Ecologically Sustainable Development of the Great Barrier Reef, including the Australian Institute of Marine Science and the Great Barrier Reef Marine Park Authority. DDM500 was developed by Rick Smith and Terry Done of the Australian Institute of Marine Science and the author.

REFERENCES

Hutchinson, M.F. Calculation of Hydrologically Sound Digital Elevation Models, in *Proceedings of the Third International Symposium on Spatial Data Handling, Sydney, Australia,* 1988.

Hutchinson, M.F. A New Procedure for Gridding Elevation and Stream Line Data with Automatic Removal of Spurious Pits, *J. Hydrology,* 106, pp. 211–232, 1989.

McPhail, I. Partnerships and Collaboration: Management of the Great Barrier Reef World Heritage Area, Past, Present and Future, in *The Great Barrier Reef, Science, Use and Management*, Great Barrier Reef Marine Park Authority, 1997.

Priest, R. Australian Land Information Group, Personal communication, 1997.

Reichelt, R.E. Advance in Scientific Research on the Great Barrier Reef Since 1983, in *The Great Barrier Reef, Science, Use and Management*, Great Barrier Reef Marine Park Authority, 1997.

CHAPTER 9

Assessing Uncertainty in Modeling Coastal Recession Due to Sea Level Rise

T.Q. Zeng and P. Cowell

INTRODUCTION

Predicting coastal recession due to sea-level rise resulting from global warming is of vital concern to society given that 80% of Australian population (Galloway et al., 1984), and many cities around the world, arc concentrated in the coastal fringe. Mean sea-level rise, along with the increased intensity and frequency of storm surge and change in wave-climate, is commonly thought to be the most significant impact of climate change on coastal geomorphology (Terwindt and Batjes, 1991; Cowell and Thom, 1994). However, available estimations of sea-level rise under a changing climate are rather preliminary (Houghton et al., 1992), because of the uncertainties both in projection of climate change itself (Henderson-Sellers, 1993) and in the assessment of the factors that influence sea level. Furthermore, coastal processes are very dynamic, highly complex, and inherently chaotic in nature (Wright and Thom, 1977; Stive et al., 1991; De Vriend et al., 1993; Cowell and Thom, 1994). Furthermore, data that contribute to coastal modeling are usually difficult to obtain, and there are always questions of accuracy and reliability about the available data.

Cowell et al. (1996) outlined a probabilistic method for mapping estimates of coastal recession due to climate-change impacts. The method used geographic information system (GIS) procedures that delineate three zones in which the probability of coastal erosion increases from $P(x_i)=0$ to $P(x_i)=1$ along shore-normal profiles in a seaward direction across the subaerial portion of a coastal sand barrier (Figure 9.1). Where P (x_i)=1, impacts are certain because they occur already under present conditions, due to the effects of storms, or will occur based on existing trends in shoreline recession. These effects also must be added to predicted climate impacts in the domain $0<P(x_i)<1$. Probabilities also vary as $P(y_j)$ along the coast in relation to a range of environmental factors, as well as with variations in the cross-sectional sand volume.

The erosion-impact probabilities, P_{ij}, for each grid cell in the digital elevation model can be obtained from a stochastic generator based on a simplified relationship for coastal recession (R) as a function of sea-level rise, the onshore sand-body volume (i.e., dune height) and offshore seabed slope, and net losses or gains in the littoral sediment-transport budget. Each of these factors is subject to varying degrees of uncertainty. The objective of this chapter therefore is to assess the effects of these uncertainties on the reliability of overall predictions of coastal recession (R). The assessment is based on a differential methodology which is described and demonstrated by a case study of Narrabeen Beach, 12 kilometers north of Sydney Harbour, SE Australia.

UNCERTAINTIES IN COASTAL MODELING

Sources of uncertainty in coastal modeling, in general, can be classified into two types: Data Uncertainty and Model Uncertainty (Giarratano and Riley, 1989; Grzymala-Busse, 1991). Specifically, uncertainties in coastal modeling can be identified as shown in Figure 9.2.

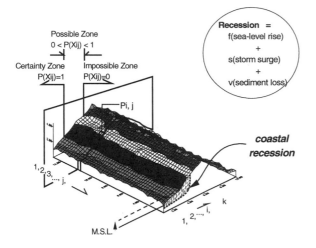

Figure 9.1. Probabilities (P_{ij}) of any cell being impacted by coastal erosion vary alongshore, in part as a function of cross-shore volume of the sand barrier. Recession probabilities must be generated stochastically for each cross-shore profile, j (from Cowell et al., 1996).

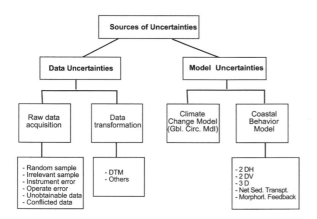

Figure 9.2. Uncertainties in Coastal Modeling (after Zeng et al., 1997).

Data Uncertainty

Coastal modeling relies upon availability of wave, wind, sediment-transport and topographical data. Data uncertainty is caused by one or a combination of the following problems:

- the stochastic and nonlinear nature of coastal processes which are generally subject to strong nonhomogeneity (Cowell and Thom, 1994);

- measurement imprecision of error, both due to instrument and operator error and limits to resolution;
- contradictory data from different sources;
- lexical imprecision;
- inaccessible long-term data that are impossible to obtain during the span of the project or at a geological scale; and
- transformation approximations induced during data interpretation and interpolation (e.g., in gridding the Digital Terrain Model).

The stochastic fluctuations in coastal processes occur over a wide range of scales (Cowell and Thom, 1994). Because of this, and the strong nonhomogeneity, sampled data are often poorly representative due to practical limitations of spatial and temporal coverage.

Model Uncertainty

Uncertainty in coastal modeling has been recognized by many researchers (Terwindt and Battjes, 1991; De Vriend et al., 1993; Cowell and Thom, 1994; Cowell et al., 1996; Zeng et al., 1997). Presently, coastal processes are still far from fully understood. In particular, the relative contributions to sediment transport from each of the many constituent hydrodynamic processes remain difficult to determine. However, modeling coastal change requires a method for including effects of the following numerous factors:

- random variations (waves and storms);
- system variations (sea-level rise) (Wright and Thom, 1977; Gordon, 1987; Healy, 1996);
- residual effects of small-scale processes to large scales (Stive et al., 1991; De Vriend et al., 1993; Cowell and Thom, 1994); and
- human interference significantly aggregates it.

From a theoretical point of view, the coast is considered as nonlinear dynamic system (NDS) which is highly complex with a multitude of factors interacting internally and externally across a hierarchy of different scales in time and space. The coastal NDS includes n interacting components (or dimensions in Phase Space) x_i, i = 1, 2,..., n. These x_i might include various factors or variables describing geology, hydraulics, morphology, and climate factors, and so on. The time behavior of any component is characterized by an ordinary differential equation describing it as a function of the other components and of parameters c_i which describe process rates. Thus,

$$\frac{dx_1}{dt} = f_1(x_1, x_2, \ldots, x_n; c_1, c_2, \ldots, c_m)$$

$$\frac{dx_2}{dt} = f_2(x_1, x_2, \ldots, x_n; c_1, c_2, \ldots, c_m)$$

$$\ldots \tag{1}$$

$$\frac{dx_n}{dt} = f_n(x_1, x_2, \ldots, x_n; c_1, c_2, \ldots, c_m)$$

In principle, the response of the system to changes in parameters c_i can be mapped in an n-dimensional state space. However, in practice, values of the parameters c_i are not precisely known and consequently, some imprecision attaches to the estimation of output (dx_n/dt). Usually, the response in a q-dimensional ($q < n$) phase space is examined since the full system is unknown or intractable. The coastal system can be evaluated for reduced dimensions, such as two-dimensional horizontal (2DH) models (Dean, 1983; De Vriend and Stive, 1987; Hsu et al., 1987) and two-dimensional vertical (2DV) models (Bruun, 1962; Schwartz, 1965; Dean, 1977, 1990, 1991; Van de Graaff and Maarten, 1990; Lee, 1994; Chcong and Rai, 1983; Thom and Hall, 1991; and Cowell et al., 1992, 1995).

METHODOLOGY

Formal techniques for dealing with uncertainty are now available to coastal impact modeling based on pioneering research, both theoretical and methodological, that is now used routinely in the field of Artificial Intelligence (Simon, 1996; Kahneman et al., 1982; Leitch et al., 1990; Morgan and Henrion, 1990; Kasabov, 1996; Zadeh, 1965, 1983), expert systems (Buchanan and Shortliffe, 1984; Giarratano and Riley, 1989; Casttile, 1991; Weichselberger, 1990), and computer simulation (Banks and Carson, 1984; Zeigler, 1984; Law and Kelton, 1991; Martinez, 1993; Shen and Leitch, 1993).

The approaches for handling uncertainty can be summarized in five types: *Probability Methods* (Good, 1983; Kaplan, 1981; Keefer and Bodily, 1983; Rubinstein, 1982; Beckman and McKay, 1987; Yang and Kushner, 1991); *Certainty Factors*; *Fuzzy Logic* (Zadeh, 1965), *Neural Network* (Kasabov, 1996), and the *Differential Approach* (Pritchard and Adelman,

1991). Intensive reviews of uncertainty handling techniques can be found in, for example, Cox (1977), Cox and Baybutt (1981), and Bhathagar and Kanal (1986).

The three main areas of uncertainty in predicting coastal recession due to climate-change impacts (Cowell et al., 1996; Zeng et al., 1997) include: (i) prediction of sea-level rise and storms (Edelman, 1970; Weggel, 1979; Vellinga, 1986); (ii) input parameters for existing coastal-process models; and (iii) preparation of the digital terrain model (Dietrich, 1991; Li, 1993). In the context of GIS modeling as proposed by Cowell et al. (1996) these uncertainties are managed by a random simulation in which samples are taken from a simplified relation for coastal recession, R:

$$\frac{R}{L} = kS \frac{(\Delta zL - \Delta V)}{D^{0.5} h_*^{1.5}} \tag{2}$$

where h_* is the water depth at the seaward limit of the shoreface, L is the width of the shoreface, Δz is the sea-level rise, ΔV is the sediment input or loss, D is the dune height, S is the coastal gradient and k is a constant (Cowell et al., 1996). The definitions of the parameters are given in Figure 9.3.

This coastal recession model (CRM) was derived from a total of 202 simulations using a more elaborate computer model, known as the *Shoreface Translation Model* (STM), that was applied to 28 different initial profile configurations and various littoral sediment budgets and rising sea levels. The STM is capable of incorporating each of these effects, based on conventional principles of mass balance along the coastal profile (Cowell et al., 1992, 1995). The *differential approach* is used to assess uncertainty of the simplified 2DV model (Equation 2). Application of the *differential methodology* and use of random simulation is schematized in Figure 9.4.

Handling Uncertainty in Sea-Level Rise

Uncertainty in sea-level rise is estimated by simulations with an exponential model (Figure 9.5):

$$z = ke^{\lambda t} \tag{3}$$

where z is the sea level, k is a scale factor, λ is a correlation factor, and t is the time for which the prediction is made. The cumulated distribution function (CDF) is determined by integrating the equation to obtain:

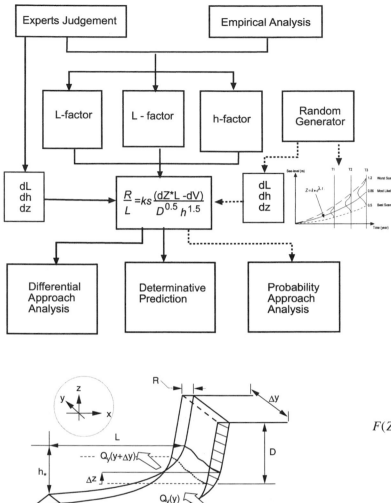

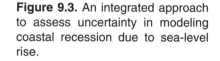

Figure 9.3. An integrated approach to assess uncertainty in modeling coastal recession due to sea-level rise.

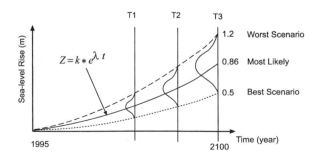

Figure 9.4. Definitions of parameters of Coastal Recession Model (from Cowell et al., 1996).

$$F(Z) = \begin{cases} 0, & t < 0 \\ \int_0^t \lambda e^{\lambda t} dt = 1 - e^{\lambda t}, & t \geq 0 \end{cases} \qquad (4)$$

Methodology of Assessing Uncertainty in the Coastal-Recession Model (CRM)

The CRM (Equation 2) was developed to provide a more flexible treatment of the Bruun Rule (Bruun, 1962; SCOR, 1991) by including the effects of net alongshore sediment transport, dune height, and the slope of the shoreface (Cowell et al., 1996). Like most of the 2DV coastal models, the CRM requires exact values for input parameters in order to obtain the output recession rate R. However, given that each of the inputs is subject to considerable uncertainty, a question may arise: how the approximation of input parameters affects the modeling outcome. To answer this question, the *differential approach* (Pritchard and Adelman, 1991) is adopted. The effect of uncertainty in input parameters on output is separately investigated in terms of one, two, and three variables.

In general, the *differential approach* for assessing uncertainty is expressed as:

Figure 9.5. The integrated approach for modeling coastal recession.

$$\frac{dR}{dx} = \frac{d[f(x)]}{dx} \qquad (5)$$

$$dR = f(x)dx \qquad (6)$$

where x is a combined representation of z, h_*, and L, which can be specified as single, double, and three parameters, as is given below:

Single-Variable Uncertainty

The absolute uncertainty of predicted recession as a function of these variables is shown in Table 9.1. The simulated results (Figure 9.6) show that the h_*-factor is more significant than the other factors. This result is consistent with the dependence of h_* on the wave climate, which is generally acknowledged as being the most dominant factor in coastal processes.

Two-Variable Uncertainty

Taking h_* and L as input variables, the projection of their uncertainty onto the predicted recession rate is calculated as follows:
Let

$$u = \Delta zL - \Delta r \qquad (7)$$

and

$$v = D^{0.5}h_*^{1.5} \qquad (8)$$

Then, replacing with Equation 7 and 8, Equation 2 can be rewritten as:

$$R = ks\frac{u}{r} \qquad (9)$$

$$dR = ks\frac{vdu - udr}{r^2} \qquad (10)$$

Inserting Equations 7 and 8 into Equation 10 gives:

$$dR =$$
$$ks\frac{(D^{0.5} * h^{1.5})\Delta zdL - (\Delta zL - \Delta V) * d(D^{0.5} * h^{1.5})}{(D^{0.5} * h^{1.5})^2}$$
$$(11)$$

Table 9.1. Single Variable and Output Uncertainty with Differential Method.

Variable	Output Absolute Uncertainty
z	$dR = \dfrac{ks}{D^{0.5} * h^{1.5}} * L * dz$
H	$dR = -1.5 * \dfrac{ks(dzL - \Delta V)}{D^{0.5} * h^{1.5}} * dh$
L	$dR = \dfrac{ks}{D^{0.5} * h^{1.5}} * z * dz$

$$dR = ks\frac{h^{1.5}\Delta zdL - (\Delta zL - \Delta V) * 1.5h^{0.5}dh}{D^{0.5} * h^3} \qquad (12)$$

$$dR = \frac{ks}{D^{0.5}}\left[\left(\frac{\Delta z}{h^{1.5}}\right)dL - 1.5\left(\frac{(\Delta zL - \Delta V)}{h^{2.5}}\right)dh\right] \qquad (13)$$

Now the uncertainty of the output is described for variables z and h_*, as

$$dR = \frac{ks}{D^{0.5}}\left[\left(\frac{L}{h^{1.5}}\right)dz - 1.5\left(\frac{(\Delta zL - \Delta V)}{h^{2.5}}\right)dh\right] \qquad (14)$$

And for z and L as

$$dR = \frac{ks}{D^{0.5}h^{1.5}}d(zL - \Delta V) \qquad (15)$$

$$dR = \frac{ks}{D^{0.5}h^{1.5}}(zdL + Ldz) \qquad (16)$$

for which the results are plotted in Figure 9.7.

Three-Variable Uncertainty

Three-variable differential is little more complicated and is done in a similar way to that of two-variable, but it is separated into two proportions (Equation 18).

$$dR = ks\frac{vdu - udv}{v^2} \qquad (17)$$

$$dR = ks\frac{du}{v} - \frac{udv}{v^2} \qquad (18)$$

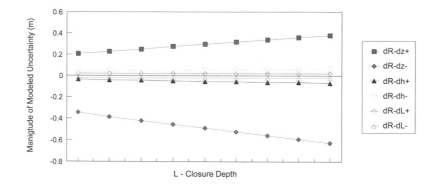

Figure 9.6. Projection of single variable uncertainty.

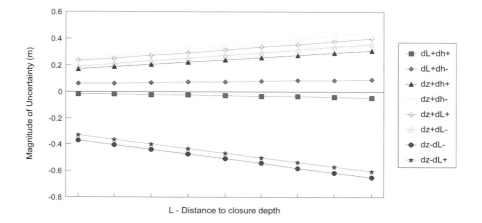

Figure 9.7. Projection of two-variable uncertainties.

$$dR =$$
$$ks\left[\left(\frac{d(\Delta zL - \Delta V)}{D^{0.5} * h^{1.5}}\right) - \left(\frac{(\Delta zL - \Delta V)d(D^{0.5} * h^{1.5})}{(D^{0.5} * h^{1.5})^2}\right)\right] \quad (19)$$

$$dR =$$
$$ks\left[\left(\frac{(\Delta zdL + Ldz)}{D^{0.5} * h^{1.5}}\right) - \left(1.5 * \frac{(\Delta zL - \Delta V)D^{0.5}h^{0.5}dh}{(D^{0.5} * h^{1.5})^2}\right)\right] \quad (20)$$

$$dR =$$
$$\frac{ks}{D^{0.5} * h^{1.5}}\left(\Delta zdL + Ldz - 1.5 * \frac{\Delta zL}{h}dh + 1.5 * \frac{\Delta V}{h}dh\right) \quad (21)$$

The results are shown in Figure 9.8.

The whole beach is divided into small sections and the calculations of erosion volume due to sea-level rise are carried for each section (Figure 9.9) using Grid Modeling techniques, and the result of the differential calculation is used as an estimated add-on sediment volume for assessing coastal recession. The final result is shown in Figure 9.10.

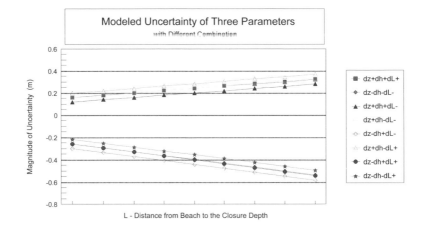

Figure 9.8. Projection of three-variable uncertainties.

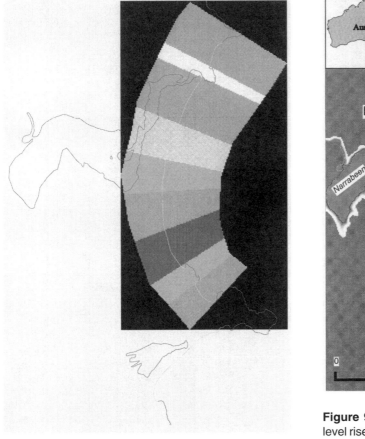

Figure 9.9. Beach sections in grid modeling.

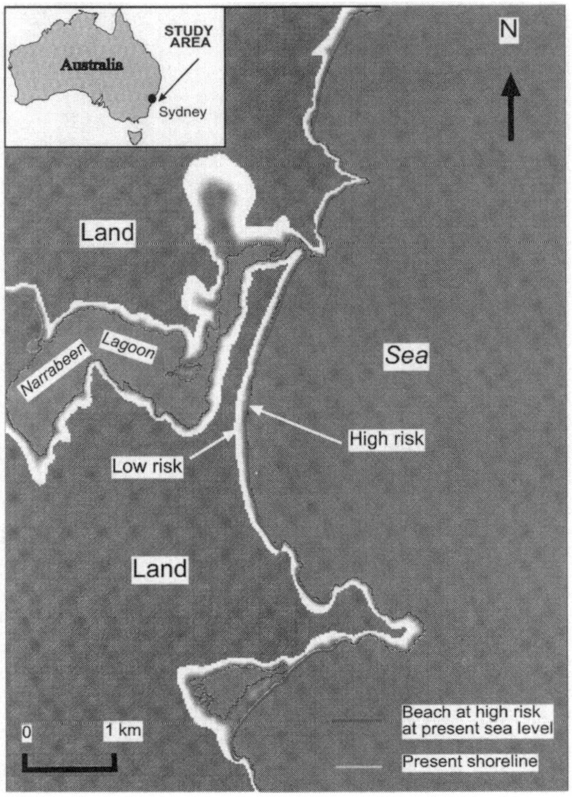

Figure 9.10. Modeled coastal recession due to sea-level rise using integrated approach, Narrabeen Beach, New South Wales, Australia.

DISCUSSION AND CONCLUSIONS

It can be concluded that the projection of the multi-variable uncertainty onto the output uncertainty equals the algebraic sum of the output uncertainties of respective single variables and can be expressed as

$$U(v_1, v_2, \ldots v_n) = \sum u(v_i)$$

This means that the modeled output uncertainty of multivariables can either aggregate or reduce. The reduction is most likely due to canceling effects among the input uncertainties, which is contrary to the conventional expectation that greater uncertainty in input parameters results in more uncertainty in modeled output. Although the modeled uncertainty of multivariables is aggregated in the worst scenario, in general they can be reduced by the "canceling effect."

This study also confirmed that the magnitude of output uncertainty is dependent on the initial state of the coastal settings, as is the same for other Nonlinear Dynamic System (Figures 9.6, 9.7, and 9.8). Built into generic tools, this integrated model can be applied to other fields involved in risk/hazard assessment. It can be incorporated with random simulation and fuzzy logic to develop a more sophisticated tool for assessing uncertainties in coastal modeling.

ACKNOWLEDGMENT

The authors would like to thank to Ms. Lucie Vallée of Centre de Recherche en Géomatique, Laval University for her assistance in correcting typographical errors in the manuscript. This research is supported by the Australian National Greenhouse Advisory Committee and is a contribution to the European Commission MAS III project, PACE.

REFERENCE

Banks, J. and J.S. Carson. *Discrete-Event System Simulation,* Prentice-Hall, Englewood Cliffs, NJ, 1984, p. 514.

Beckman, R.J. and M.D. McKay. Monte Carlo Estimation Under Different Distributions Using the Same Simulation, *Technometrics,* 29, pp. 153–160, 1987.

Bruun P. Sea Level Rise as a Cause of Shore Erosion, *J. Waterways Harbour Div.,* 88, pp. 117–130, 1962.

Buchanan, B.G. and E.H. Shortiffe. *Rule-Based Expert System: The MYCIN Experiments of the Stanford Heuristic Programming Project,* Addison Reading, Addison, MA, 1984, p. 748.

Casttile, E. *Expert Systems: Uncertainty and Learning.* Computational Mechanics, Southampton, Elsevier, London, 1991, p. 331.

Cheong, H.F. and S.P. Rai. A Study of Equilibrium Beach Profile, *Symposium of Sixth Australian Coastal and Ocean Engineering,* 13–15 July 1983, Gold Coast, 1983, pp. 240–245.

Cowell, P.J., P.S. Roy, and R.A. Jones. Shoreface Translation Model, Computer Simulation of Coastal-Sand-Body Response to Sea Level Rise, *Amsterdam, Math. Comput. Simulation,* 33, pp. 603–608, 1992.

Cowell, P.J. and B.G. Thom. Morphodynamics of Coastal Evolution, in *Coastal Evolution,* R.W.G. Carter and C.D. Woodroffe, Eds., Cambridge University Press, 1994, pp. 33–86.

Cowell, P.J., P.S. Roy, and R.A. Jones. Simulation of Large-Scale Coastal Change Using a Morphological Behaviour Model, Amsterdam, *Marine Geology,* 126, pp. 45–61, 1995.

Cowell, P.J, P.S. Roy, T.Q. Zeng, and B.G. Thom. Practical Relationship for Predicting Coastal Geomorphology Impact of Climate Change, *Ocean and Atmosphere Pacific International Conferences,* Adelaide, October 1995, 1996.

Dean, R.G. *Equilibrium Beach Profiles.* U.S. Altantic and Gulf Coast Ocean Eng. Report No. 12, University of Delaware, 1977.

Dean, R.G. Principals of Beach Nourishment, in *CRC Handbook of Coastal Processes and Erosion,* P.D. Komar, Ed., CRC Press Inc., Boca Raton, FL, 1983, pp. 217–232.

Dean, R.G. Beach Response to Sea Level Change, in *Ocean Engineering Science,* B. Le Mehaute and D.M. Danes, Eds., John Wiley & Sons, New York, 1990, pp. 869–887.

Dean, R.G. Equilibrium Beach Profiles, Characteristics and Applications, *J. Coastal Res.,* 7, pp. 53–84, 1991.

Dean, R.G., T.R. Healy, and A. Dommerhlt. A Blind-Folded Test of Equilibrium Beach Profile Concepts with New Zealand Data, *Mar. Geol.,* 109, pp. 253–266, 1993.

De Vriend, H.J., M. Capobianco, T. Chesher, H.E. De Swart, B. Latteux, and M.J.F. Stive. Approaches to Long Term Modelling of Coastal Morphology, A Review, *J. Coastal Eng.,* 21, pp. 225–269, 1993.

De Vriend, H.J. and M.J.F. Stive. Quasi-3D Modelling of Nearshore Currents, *Coastal Eng.,* 11(5/6), pp. 565–601, 1987.

Dietrich, C.F. *Uncertainty, Calibration, and Probability: The Statistics of Scientifically and Measurement,* Bristol, Philadelphia, 1991, p. 535.

Edelman, T. Dune Erosion During Storm Conditions. *Proc. 12th Int. Con. Coastal Engineering,* American Society of Civil Engineering, NY, 1970, pp. 1305–1307.

Galloway, G.R., R.C. Story, and G.A. Yapp. *Coastal Land of Australia,* National Resources Series No. 1, Institute of Australia Geographers, Melbourne, 1984.

Giarratano, J.C. and G.D. Riley. *Expert Systems: Principles and Programming,* PWS-KENT Publishing Company, Boston, MA, 1989, p. 632.

Good, I.J. *Good Thinking: The Foundations of Probability and Its Applications,* University of Minnesota Press, Minneapolis, 1983, p. 333.

Gordon, A.D. A Tentative but Tantalizing Link Between Sea-Level Rise and Coastal Recession in New South Wales, Australia, *8th Australian Conference on Coastal Engineering,* 1987, pp. 121–134.

Grzymala-Busse, J.W. *Managing Uncertainty in Expert Systems,* Ch. 6.1, Kluwer Academic Publishers, Boston, 1991, pp. 128–158.

Healy, T. Sea Level Rise and Impacts on Nearshore Sedimentation, An Overview, *Geol. Rundsch.,* Springer-Verlag, 85(3), pp. 546–553, 1996.

Henderson-Sellers, A. An Antipodean Climate of Uncertainty? *Climate Change,* 25, pp. 203–224, 1993.

Houghton, J.T., B.A. Callander, and S.Q. Varney, Eds. *Climate Change 1992, The Supplementary Report to the IPCC Scientific Assessment,* Cambridge University Press, 1992, p. 200.

Hsu, J.R.C., R. Silnester, and Y.M. Xia. New Characteristics of Equilibrium Shaped Bays, *8th Australiasian Conference on Coastal and Ocean Engineering,* Launceston, 30 Nov.–4 Dec. 1987, pp. 140–144.

Kahneman, D.P., P. Solvic, and A. Tversky, Eds. *Judgement Under Uncertainty: Heuristics and Biases,* Cambridge University Press, 1982, p. 555.

Kaplan, S. On the Method of Discrete Probability Distributions—Application Seismic Risk Assessment, *Risk Anal.,* 1, pp. 189–198, 1981.

Kasabov, N.K. *Foundations of Neural Networks, Fuzzy Systems, and Knowledge Engineering,* MIT Press, Cambridge, MA, 1996, p. 550.

Keefer, D.L. and S.E. Bodily. Three-Point Approximations for Continuous Random Variables, *Manage. Sci.,* 29, pp. 595–609, 1983.

Law, A.M. and W.D. Kelton. *Simulation Modeling and Analysis,* 2nd ed., McGraw-Hill, New York, 1991, p. 400.

Lee, P.Z.F. The Submarine Equilibrium Profile, A Physical Model, *J. Coastal Res.,* 10(1), pp. 1–17, 1994.

Leitch, R., M.E. Wiegand, and H.C. Quek. Coping with Complexity in Physical System Modeling, *Artif. Intelligence Commun.,* 3(2), pp. 48–57, 1990.

Li, Z. Theoretical Models of the Accuracy of Digital Terrain Models, An Evaluation and Some Observations, *Photogrammetric Rec.,* 14(82), pp. 651–660, 1993.

Martinez, P.A. *Simulating Nearshore Environments,* Pergamon Press, New York, 1993, p. 265.

Morgan, M.G. and M. Henrion. *Uncertainty: A Guide to Dealing with Uncertainty in Quantitative Risk and Police Analysis,* Cambridge University Press, 1990, p. 332.

Pritchard, J.I. and H.M. Adelman. Differential Equation Based Method for Accurate Model Approximations, *AIAA J.,* 29, pp. 484–486, 1991.

Rubinstein, R.Y. *Simulation and the Monte Carlo Method,* John Wiley & Sons, New York, 1982.

Schwartz, M.L. Laboratory Study of Sea-Level Rise as a Cause of Shore Erosion, *J. Geol.,* 73, pp. 528–534, 1965.

SCOR (Scientific Committee on Ocean Research) Working Group. The Response of Beaches to Sea-Level Changes, A Review of Predictive Models, *J. Coastal Res.,* 7, pp. 895–921, 1991.

Shen, Q. and R. Leitch. Fuzzy Qualitative Simulation, *IEEE Transactions on Systems, Man and Cybernetics,* 23(4), pp. 1038–1061, 1993.

Simon, H.A. *The Sciences of Artificial,* 3rd ed., MIT Press, Cambridge, MA, 1996, p. 231.

Stive, M.J.F., D.J.A. Roelvink, and H.J. de Vriend. Large-Scale Coastal Evolution Concept, in *Proceedings 22nd International Conference on Coastal Engineering,* American Society of Civil Engineers, New York, 1991, pp. 1962–1974.

Terwindt, J.H.J and J.A. Battjes. Research on Large-Scale Coastal Behaviour, *Proceedings 22nd International Conference on Coastal Engineering,* American Society of Civil Engineers, New York, 1991, pp. 1975–1983.

Thom, B.G. and W. Hall. Behaviour of Beach Profiles During Accretion and Erosion Dominated Periods, *Earth Surface Processes and Landform,* 16, pp. 113–127, 1991.

Van de Graaff, J. and J.K. Maarten. Dune and Beach Erosion and Nourishment, in *Coastal Protection,* Pilarczyk, Ed., Balkema, Rotterdam, 1990, pp. 99–120.

Vellinga, P. *Beach and Dune Erosion During Storm Surge.* Ph.D. Thesis, Delft University of Technology, also Publication No. 131, Delft Hydraulics, Delft, The Netherlands, 1986.

Weggel, R.J. *A Method for Estimating Long-Term Erosion Rates from a Long-Term Rise in Water Level,* Coastal Engineering Technical Aid 97-2, U.S. Army Corp. Coastal Eng., 1979.

Weichselberger, K. *A Methodology for Uncertainty in Knowledge-Based System,* Springer-Verlag, Berlin, New York, 1990, p. 132.

Wright, L.D. and B.G. Thom. Coastal Depositional Landforms, A Morphodynatic Approach, *Prog. Phys. Geogr.,* 1, pp. 412–459, 1977.

Yang, J. and H.J. Kushner. A Monte Carlo Method for Sensitivity Analysis and Parametric Optimisation of Nonlinear Stochastic Systems, *SIAM J. Control Optimisation,* 29, pp. 1216–1249, 1991.

Zadeh, L.A. Fuzzy Sets, *Inf. Control,* 8, pp. 338–353, 1965.

Zadeh, L.A. The Role of Fuzzy Logic in the Management of Uncertainty in Expert Systems, Memorandum No.

UCB/ERL M83/41, University of California, Berkeley, 1983.

Zeigler, B. *Multifaceted Modelling and Discrete Event Simulation,* Academic Press, 1984, p. 372.

Zeng, T.Q., P.J. Cowell, W. Hennecke, and B.G. Thom. Modeling Coastal Impact of Climate Change Under Uncertainties, *Trop. Geomorphology,* 18(2), pp. 36–53, 1997.

CHAPTER 10

Sensitivity of a Quantitative Soil-Landscape Model to the Precision of the Topographical Input Parameters

V. Chaplot, C. Walter, and P. Curmi

INTRODUCTION

The characterization of the soils of valley bottom wetlands is fundamental for land-use planning and ecological modeling. Standard soil mapping encounters in these areas both methodological and economical constraints, which makes it necessary to use complementary mapping techniques. Among these, models using terrain attributes derived from digital elevation models (DEM) are developed to investigate large areas.

Several approaches are proposed to construct models that relate the variation of soil properties to the variation of terrain attributes. A first approach is based on the spatial processes involving topography and being able to influence soil conditions such as its moisture regime. Topographic indexes derived from Darcy's law, originally developed for hydrological modeling (Beven and Kirkby, 1979) have been used to model the spatial distribution of hydromorphic soils (Merot et al., 1995). Another approach more in use is based on establishing statistical relationships between terrain and soil attributes in reference sites. These relationships are determined using multiple regression analyses (Moore et al., 1993; Thompson et al., 1997), methods of expert systems (Lagacherie et al., 1995), or interpolation techniques such as co-kriging (Bourennane et al., 1996).

In all these approaches, the topographical information constitutes the input data of the models. The impact of their precision on the prediction quality has seldom been studied. This information, available as DEM, is often much less precise than the topographical measurements that were used to create the models in the reference sites.

The aim of this study was to test the sensitivity of a soil hydromorphy prediction model to the precision of the topographic information and the interest of including auxiliary soil data. This was done in an area of some hectares where the soil distribution was precisely depicted. In this area, several DEMs were generated, with decreasing resolution (from 10 to 50 m), optionally including detailed topographic measurements or corrections by the consideration of the exact channel network location. In a first step, the intrinsic quality of these DEMs was analyzed by comparison to detailed topographic measurements. In a second step, different prediction techniques for soil hydromorphy were implemented using either regression with the topographic information only, or co-kriging utilizing a more or less important amount of soil data.

MATERIALS AND METHODS

Study Site

The study site, an agricultural field of two hectares, is located in the Armorican Massif, 20 km west from Avranches (Manche, Western France) (Figure 10.1). It is situated in the upper part of a granitic catchment and makes up the major part of a 250-m long hill slope with a difference in height of 10 m. The slopes are gentle (4.5% on average) and homogeneous. The soil cover has developed into a loamy material overlying a granitic saprolite. It comprises well drained and poorly drained soils associated with Luvisols, Albic

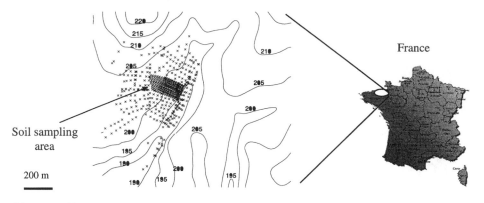

Figure 10.1. Contour map of the studied site (located in the Roche catchment: Manche department, France) indicating the 643 topographic measurement locations.

Luvisols, and Gleyic Luvisol (WRB, 1994) whose distribution is related to groundwater level controlled by topography (Curmi and Widiatmaka, 1998). Some spots of peat are observed close to the stream banks.

The Hydromorphy Index

Inspired by the work of Thompson and Bell (1996), we have chosen for the model predictions a continuous hydromorphy index that reflects the diversity of the soil hydromorphy of the site. This index (HI) takes into account the moist color of the surface horizon matrix and the proportion of hydromorphic horizons in the soil profile. At each point x, this index is given by the formula:

$$HI(x) = \frac{P(x)}{V_1(x) \cdot C_1(x)} \qquad (1)$$

where:

- $P(x)$ in %, is the cumulative thickness of the hydromorphic horizons divided by the total thickness of the soil; a hydromorphic horizon is defined as an organomineral horizon with a grey matrix or a mineral horizon with light or rusty mottles. The total thickness of the soil is the sum of the thickness of the loamy horizons plus the thickness of the loose granitic saprolite that can be hand-drilled,
- $V_1(x)$ is the Munsell value of the surface horizon observed in x,
- $C_1(x)$ is the Munsell chroma of the surface horizon observed in x.

HI ranges, in theory, from 0 (no hydromorphy) to 100 (entirely hydromorphic soil with a very dark surface color : $C_1.V_1=1$).

This index was calculated for 182 soil auger holes from 0.8 to 2.8 m depth following a 10-m square grid.

Topographic Analysis

Nine DEM, 10, 20, 30, and 50 m in resolution, were created using the "topogrid" function of ARC/INFO 7.1[®1] for an 80 ha catchment including the study site. The basic information was provided by the contour lines of the 1/25.000 topographic map, at 5 m vertical interval (IGN, card 1315 E, 3rd edition, 1990). For the creation of six of these DEMs, a correction using the real location of the channel network was realized according to Hutchinson's algorithm (Hutchinson et al., 1993). For three DEMs with the finest resolutions (10 and 20 m), 643 theodolite elevation measurements have been added, to have a more precise image of the study site's relief. The location of each measurement point in the Lambert 2 system was obtained using a GPS. Table 10.1 summarizes the topographic data used for each DEM. The 50 m DEM from IGN, available for the whole country, was also used in this study.

From these DEMs, several terrain attributes were calculated: the difference in elevation between the

---

[1]   Arc/Info from ESRI, Redlands, CA 92273.

Table 10.1. The Topographical Information Taken into Account in the DEM's (10, 20, 30, and 50 meters) Construction: Contour Line (C), Channel Network Location (R), and Additional Theodolite Measurements (T).

DEM Cell Size	DEM Code	Contour Lines	Channel Network Location	Elevation Measurements
10	10_{CTR}	*	*	*
20	20_{CTR}	*	*	*
	20_C	*		
	20_{CR}	*	*	
	20_{CT}	*		*
30	30_C	*		
	30_{CR}	*	*	
50	50_C	*		
	50_{CR}	*	*	
	IGN50	*		

point and the stream bank (DENI); the "downslope gradient" (PAV), equal to DENI divided by the distance to the stream bank; the specific catchment area (AMU), e.g., square meters per meter (Moore and Grayson, 1991). DENI and PAV were estimated using the "Grid" module of Arc/Info, while AMU was calculated using the MNTsurf software (Squividant, 1994).

DEM Intrinsic Quality Analysis

The quality of the terrain attribute estimation was tested by comparison with the 643 elevation measurements for the altitude, the DENI and PAV parameters or with the values obtained from the 10 m DEM for the AMU parameter.

Prediction of the Hydromorphy Index Using the Terrain Attributes

The hydromorphy index was considered for 140 auger holes. Forward stepwise linear regression was used to explain this index by the terrain attributes derived from the elevation measurements (altitude, DENI, PAV) or from the 10_{CTR} DEM (AMU). The significance level for the inclusion of a variable during the stepwise regression procedure was chosen equal to 0.01. The best regression model obtained was the following:

$$HI_{est} = 3.1 \cdot Log_n(AMU / PAV) - 3.4 \cdot DENI$$
$$\left(r^2 = 0.80\right) \tag{2}$$

Using this model, the quality of the prediction was tested considering, on one side, the resolution and the construction mode of the different DEMs, and on the other side, a more or less important amount of soil data (10 to 60 points). To estimate HI at each point of the study area, different prediction methods were compared:

- The simple application of the regression model based on the terrain attributes derived from DEMs: the estimation of HI at a given point utilizes the terrain attributes estimated at this point by the DEMs.
- An ordinary co-kriging considering not only the topography but also punctual observations. The main variable is the observed hydromorphy index (HI_{obs}) and the auxiliary variable is the hydromorphy index estimated by the regression model (HI_{est}). Two options were considered. In the first one, HI_{obs} is known at 10 points regularly spread on the study site (co-kriging 10). In the second one, HI_{obs} is known at 60 points regularly spread (co-kriging 60). In both cases, HI_{est} is evaluated by (1) at 140 points of a 5-m square grid. Finally, the HI estimate at a given point was obtained by taking into account 10 or 60 neighbors where HI was observed, and 140 neighbors where HI was estimated from terrain attributes.
- For comparison, an ordinary kriging was also realized: in this case, only the soil observations from 10 or 60 points are considered.

A validation set of 42 observations was used to test the different prediction techniques.

RESULTS

The Intrinsic Quality of the DEMs

The elevations estimated from the different DEMs were first compared to the 643 elevation measurements (Figure 10.2). Except for two 50-m DEMs, the estimation errors were centered on values close to 0 (median comprised between –0.8 and 0.6 m). This shows that the DEMs with fine resolution, 10 to 30 m, estimate the global slope relief in an unbiased way. Nevertheless, the error scatter increases with the grid size: for the DEM $_{10CTR}$, 90% of the estimation errors were between –1 and 1 m; for the DEM $_{30C}$, they were between –2.4 and 1.8 m; for DEM $_{50CR}$, they were between –4.3 and 1.9 m.

For 182 points, the terrain attributes derived from the DEMs (altitude, PAV, DENI, AMU) could be compared with topographic measurements (Table 10.2). The error statistics confirmed that the estimation quality is much better with the 10- to 30-m DEMs than for the 50-m DEMs. The 50 m DEM IGN turned out to be better than our DEMs of the same grid size.

Among the DEMs from 10 to 30 m, the differences of estimation quality were not clear regarding the effect of both the grid size and the construction mode (considering stream bank location or measurement points). In particular, taking into account the location of the stream has no systematic effect on the estima-

tion error: it can improve the estimation for one parameter (e.g., the elevation between DEM_{20} and DEM_{20CR}), and in the same time deteriorate the estimation for another one (e.g., the downslope gradient for these same DEMs). However, the integration of the exact location of the stream is worthwhile because it allows one to control possible drifts during the construction of the DEMs or the drainage networks.

For the next steps of the study, four DEMs with resolutions of 10, 20, 30, and 50 m were considered optimal ($_{10CTR}$, $_{20CR}$, $_{30CR}$, $_{50CR}$) and were selected to apply the prediction models.

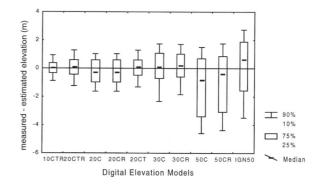

Figure 10.2. Boxplot statistics of the elevation differences between theodolite measurements and estimations with different DEMs at 643 soil sampling locations.

Table 10.2. Statistics at 182 Validation Sites of the Prediction Errors of the Elevation and Derived Terrain Attributes, for the Different DEMs.

	Altitude (m)		Elevation Above Channel Network (m) (DENI)		Downslope Gradient (%) (PAV)		Specific Catchment Area (log (m² m⁻¹)) (AMU)	
	z^*-z	$\|z^*-z\|$	z^*-z	$\|z^*-z\|$	z^*-z	$\|z^*-z\|$	z^*-z	$\|z^*-z\|$
10_{CTR}	0.2	0.4	−0.2	0.4	0.5	0.8		
20_{CTR}	−0.1	0.4	−0.7	0.8	−0.5	1.0	−0.1	0.5
20_C	0.8	0.8	0.2	0.4	−0.4	0.8	0.2	0.4
20_{CR}	0.5	0.6	0.1	0.4	−0.9	1.6	0.3	0.5
20_{CT}	−0.1	0.4	−0.6	0.7	−2.5	2.5	0.3	0.5
30_C	0.2	0.4	−0.6	0.6	−1.5	1.7	0.4	0.5
30_{CR}	0.1	0.4	−0.5	0.6	−2.6	2.8	0.4	0.5
50_C	1.6	1.8	−1.9	2.4	−2.3	2.6	0.4	0.8
50_{CR}	1.1	1.5	−1.9	2.3	−1.9	2.4	0.4	0.8
IGN50	1.1	1.5	−1.6	1.8	−0.5	1.6	0.1	0.6

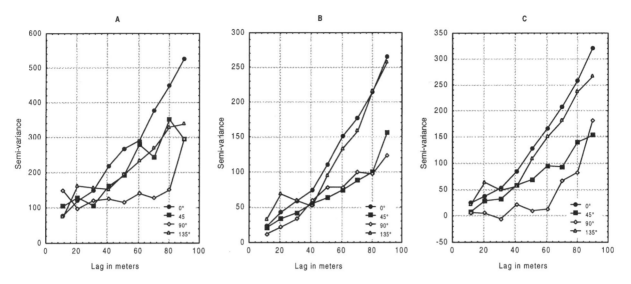

Figure 10.3. Experimental variograms and covariogram of the HI index: directional variograms of the observed index (A), the estimated index using regression (B), and the covariogram (C).

The Spatial Structure of the Hydromorphy Index (HI)

The direct and cross variograms have been computed for the observed (HI_{obs}) and estimated (HI_{est}) indexes at the 182 auger hole points (Figure 10.3). The directional variograms for HI_{obs} (Figure 10.3A) estimated up to 90 m, are linear and show a marked anisotropy in two perpendicular directions: the 0° direction, parallel to the stream bank, generates a slightly increasing variogram; the 90° direction is of highest variability; the 45° and 135° directions show medium variability. The directional variograms for HI_{est} (Figure 10.3B) are globally similar to those of HI_{obs}: increased and marked anisotropy between the directions 0° and 90°. However, in detail, some differences are observed: (i) the nugget effect is less pronounced; (ii) for a given distance, the semivariance is much lower; (iii) the directional variograms contrasted the 0° and 135° directions, of high variability, with the 45° and 90° directions, of low variability.

So, as confirmed by the covariogram (Figure 10.3C), the spatial structure of HI_{est} has features similar to those of HI_{obs}; i.e., an important variability of hydromorphy at right angles to the stream bank, and a low variability parallel to it. Nonetheless, this spatial structure is simplified: the variability is globally lower, the short distance relationships are more regular, and the anisotropy is less complex.

Validation Results

Table 10.3 allows one to compare the different estimation techniques used. The error statistics were first compared in the most favorable case of HI estimation; i.e., the 10 m DEM. The results revealed that the estimation qualities of regression and of co-kriging differ little, the co-kriging being slightly better. The estimation errors of the ordinary kriging, for which only the soil data are utilized, are larger than those of the two other techniques. With 60 points, the estimation errors are relatively low because of a high sampling density, while with 10 points, high errors are observed.

When the resolution of the DEM decreases, the differences of estimation between regression and co-kriging become more pronounced. At 20-m grid size, the performances of the techniques are still similar, even if co-kriging is slightly better. At 30-m, and then at 50-m grid size, the prediction with the regression technique drifts away significantly, since the deviation at the elevation measurement points are higher than 6 and 17 in mean, respectively. The quality of the prediction is then significantly improved by using a co-kriging with 10 points or even better with 60 points.

CONCLUSION

This study has examined the case of a soil property strongly correlated to topographic factors. Our aim was

Table 10.3. Statistics of the HI Prediction Errors at 42 Validation Sites for the Different Techniques.

DEM Cell Size (m)	Prediction Technique	$\overline{z^*-z}$	$\overline{|z^*-z|}$	Mean Rank
	Ordinary kriging 10	1.9	12.7	
	Ordinary kriging 60	1.7	5.0	
	Regression	1.4	5.4	2.5
10	Co-kriging 10	−0.5	4.2	1.8
	Co-kriging 60	−0.5	2.7	1.8
	Regression	1.8	5.3	2.5
20	Co-kriging 10	0.2	4.5	1.7
	Co-kriging 60	−0.2	3.0	1.7
	Regression	6.7	7.6	2.6
30	Co-kriging 10	−0.1	5.5	1.6
	Co-kriging 60	−0.7	2.6	1.7
	Regression	17.6	22.5	2.7
50	Co-kriging 10	5.4	12.6	1.7
	Co-kriging 60	−0.6	2.6	1.6

to test the effect of the precision of the topographic information on prediction models for soil hydromorphy using the multiple regression and co-kriging methods. The results showed that the regression method was simple and efficient, provided the topographic information was precise. A good precision of the terrain attributes was obtained, on one side, with detailed DEM (grid size less than or equal to 20 m), and on the other side by using additional terrain information (elevation measurements and exact location of the stream bank) to construct the DEM. In this case, the models can easily be generalized for large areas. If these additional data are not available, mixed methods such as co-kriging allow one to limit the deviations of the terrain information and of the models. To limit the implementation costs of the model, the required amount of soil information must be more thoroughly analyzed. Indeed, in comparison to standard surveys, the number of auger holes used in this study (10 and 60) correspond to high sampling densities.

REFERENCES

Beven K.J. and M.J. Kirkby. A Physically Based Variable Contributing Area Model of Basin Hydrology, *Hydrol. Sci. Bull.,* 24, pp. 43–69, 1979.

Bourenanne, H., D. King, P. Chery, and A. Bruand. Improving the Kriging of a Soil Variable Using Slope Gradient as External Drift, *Eur. J. Soil Sci.,* 47, pp. 473–483, 1996.

Curmi, P. and S. Widiatmaka. Soil Distribution Model in the Loamy Cover of the Armorican Massif (France): Role and Origin of Hydromorphy, *16ème Congrès Mondial de Science du Sol, Montpellier,* 20–26 août 1998.

Hutchinson, M.F. Development of a Continent-Wide DES with Applications to Terrain and Climate Analysis, in *Environmental Modeling with GIS,* M.F. Goodchild, et al., Eds., Oxford University Press, New York, 1993, pp. 392–399.

Lagacherie, P., J.P. Legros, and P.A. Burrough. A Soil Survey Procedure Using the Knowledge of Soil Pattern Established on a Previously Mapped Reference Area, *Geoderma,* 65, pp. 283–301, 1995.

Merot, P., B. Ezzahar, C. Walter, and P. Aurousseau. Mapping Waterlogging of Soils Using Digital Terrain Models, *Hydr. Process.,* 9, pp. 27–34, 1995.

Moore, I.D. and R.B. Grayson. Terrain-Based Catchment Partitioning and Runoff Prediction Using Vector Elevation Data, *Water Resour. Res.,* 27, pp. 1177–1191, 1991.

Moore, I.D., P.E. Gessler, G.A. Nielsen, and G.A. Peterson. Soil Attribute Prediction Using Terrain Analysis, *Soil Sci. Soc. Am. J.,* 57, pp. 443–452, 1993.

Squividant, E. MNTsurf, logiciel de traitement des Modèles Numériques de Terrain, Document interne, ENSAR, 1994, p. 40.

Thompson, J.A. and J.C. Bell. Color Index for Identifying Hydric Conditions for Seasonally Saturated Mollisols in Minnesota, *Soil Sci. Soc. Am. J.,* 60, pp. 1979–1988, 1996.

Thompson, J.A. J.C. Bell, and C.A. Butler. Quantitative Soil-Landscape Modeling for Estimating the Areal

Extent of Hydromorphic Soils, *Soil Sci. Soc. Am. J.,* 61, pp. 971–980, 1997.

WRB. *World Reference Base for Soil Resources,* International Society of Soil Science; International Soil Reference and Information Centre; Food and Agricultural Organization of the United Nations, O.C. Spaargaren, Ed., Rome, 1994, pp. 1–161.

Part III
Methods of Characterizing Uncertainty

An often neglected aspect of the study of spatial uncertainty is how to actually characterize uncertainty known to be present in various data. Individuals may have a conceptual idea of what uncertainty is, and may also develop ideas of what to do with it once it has been identified and quantified, but at times there is a problem in determining how to identify, characterize, measure it. This problem touches on a number of important issues, and this chapter reflects the diversity of topics that must be addressed in order to characterize and quantify spatial uncertainty.

The first chapter in this Part, by Rennolls, provides an indication of why it is not always easy to characterize spatial uncertainty: the complexity of statistical sampling increases when it is spatial data that are being collected to describe a population. The remaining chapters in Part III describe a variety of techniques for characterizing uncertainty that may be of use in a variety of situation. Among these are methods for categorical/thematic maps as described in two chapters by Allan and De Genst et al., simulation approaches as represented in chapters by Goodchild et al. and Funk et al., and a new approach named the "Reliability Concept" that is described in a chapter by Azouzi.

CHAPTER **11**

The Optimal Sample Allocation Problem for Mapped Outputs

K. Rennolls

INTRODUCTION

The wealth of information available from multi-spectral satellite imagery, and the way in which multiple sources of spatial information can be integrated in a GIS has led to the optimistic expectation that these two technologies will be the main foundations upon which future land-use inventory and monitoring are based. This is certainly the case for the European Union sponsored projects for forest inventory and monitoring, on a European and world scale; i.e., the MARS, TREES, FIRS FIMP/IFSSP projects.

However, there are a number of snags in the current use of RS/GIS technology to obtain the statistical and mapped estimates of those variables/parameters which are traditionally required for land-use and natural-resource management.

The most fundamental question is: does the available RS/GIS data "contain sufficient information of relevance" on "ground-truth" for it to be useful in the estimation of population variables/parameters, either on a regional, or mapped basis? Usually an image of a scene obtained by RS will contain data(-values) in its pixels which are correlated with the corresponding values in the appropriate ground-based pixels. We would expect that this RS, when used as a covariate in a model-based regression estimator, could only improve the estimates obtained. However, the following well-known situation, from classical sampling theory, indicates that there are approaches within which covariate information can be worse than useless.

Suppose a simple random sample is taken from a population of size N, in order to estimate the popula-tion total of the response variable "y." Suppose also an auxiliary variable, "x," is available for all elements (units) of the population. Let $\rho = \text{corr}(x,y)$. Then:

Cochran (1977), Theorem 6.2, page 157, states :

The ratio estimate $\hat{Y}_R$ has smaller variance than the expansion estimator, $Y = N \bar{y}$, (which makes no use of the auxiliary information), if

$$\rho > \frac{\text{coefficient of variation of } x}{2(\text{coefficient of variation of } y)}$$

Hence, if the coefficient of variation of the target response variable and the auxiliary variable are about the same, then the correlation between x and y has to be at least 0.5 in order for the auxiliary variable to have any value for use in a ratio estimator.

Though there are no exact theorems concerning the use of complex regression estimators from complex sample designs, the fact remains that auxiliary information which has low correlation with our target (Land-Use/Natural-Resource) variable, may be of little or no use in estimating our target parameter/variable. We need to make maximal use of the auxiliary information provided by Remote Sensing in the estimation of Land-Use/Natural-Resource Statistics and Maps. The only ways to do this are by: (i) effective modeling of the "x-y relationship," in terms of maximal multiple R^2 values, and (ii) by "efficient sample design."

We will not consider (i) in this chapter since the regression models developed will depend on the target variable concerned and the covariates that are available. See Mayaux and Lambin (1995,1997) for accounts of models which use two-stage linear, multiple-linear, nonlinear, and discontinuous, for the use of course-scale RS auxiliary information to estimate the "true" area of forests.

So, how should we sample efficiently in order to ensure that the RS information has maximal utility in the estimation of natural-resource statistics and maps?

For the estimation of regional statistics, the standard methods of sample survey are available, (stratification, multistage, multiphase, and variable probability designs), usually in combination with a multiphase regression estimator. Heuristic knowledge of the properties of such designs is well known, and has been widely applied in the MARS, TREES, and FIRS projects (FIRS, 1995), though there is scope for wider application of optimal data collection methods for model-building and model-based estimation. See Schreuder and Czaplewski (1993) for a good general review of survey design methodology in the context of a forest health inventories.

However, *when the required output is a map*, or a set of maps, the problem of the best way to sample is complicated by the fact that there are multiple error sources in the map estimation process, and due to the complexity of filters and spatial prediction models that might be used in processing of the raw RS data. However, the processes of optimal survey design must follow parallel steps to those followed in classical survey design. In the next section we review the main steps in classical sample size determination, with a view to developing a parallel set of steps when the output of the process is a *map*, rather than an estimate of the population mean.

THE CLASSICAL SAMPLE SIZE DETERMINATION PROBLEM

This may be summarized by the following sequence:

1. Define the *population* and its *elements*, and the *target population parameter* to be estimated, usually the population mean of some characteristic.
2. Define *the sampling units* and the *sampling frame(s)*.
3. Design a *sampling strategy*, without specifying the sample sizes **n** over the design.
4. Choose the *estimators* that will be used on the collected data.
5. Specify the formulas for the *variance of the estimator;* this will be a function of **n**.

Note that in doing this we have taken our measure of error of an estimator to be the mean squared error of the estimator. Also *"accuracy"* is the "inverse" of the defined error.

6. Specify a *cost model* for the data collection, as a function of **n**.

Finally *the optimal sample allocation problem* may be that stated in one of the following forms:

7i. Minimize the cost, with respect to **n**, ensuring that a *prespecified accuracy* of the estimate of the target population parameter is obtained, or,
7ii. Maximize the accuracy of the estimate of the target population parameter, with respect to **n**, subject to a *prespecified cost constraint*.

Solving the optimal sample allocation problem in practical situations is clearly very difficult since it depends on clear answers to several prior questions before the optimization task can be considered (see Traub et al. [1994] for a relevant discussion). In a number of simple cases, there are analytical solutions to the problem; for stratified sampling, and for two-stage cluster sampling, when the simple expansion estimator is used. However, for more complicated designs, and when more complex estimators are used, there are no analytical results available. In such cases, if all the earlier steps in the above process can be modeled, or approximated, then the optimal allocation problem must be solved using numerical means. This was done by Rennolls (1989) for a simple linear regression estimator on a stratified double sample, in order to estimate the total regional forest area in the 1979–82 British Forest Inventory (Rennolls and Gertner, 1993). Rennolls and Sampson (1993) have also considered the development of general purpose software which would allow such optimal sample allocation to be done for more complex designs and estimators.

FOREST MAPPING FROM REMOTELY SENSED DATA

Sample Survey Estimation, or Calibration?

It has been implicitly assumed in the above introduction that the target response variable has been actually measured (on the ground, or with high resolution imagery) on a number of sample units. The RS data is taken to be covariate, or auxiliary information. Hence such ground-truth data can be used in supervised classification of the RS data, and can also be used in the construction of regression estimators/predictors which can be applied across the whole of the surveyed scene.

An alternative view to this survey estimation approach is that of "calibration," where the RS data is taken as the primary data source, and the ground truth data is merely considered as necessary for calibrative adjustments and validation, this being the usual approach of specialists in RS. However, the former regression viewpoint approach is best suited to considering the optimum sample allocation problem. There will be little or no difference in the models constructed in the two approaches.

The Population and Its Elements: Target Variables

The population may be considered to be the mosaic, with gaps of forest blocks/compartments, however defined (elements), each having a forest-type classification attached.

In order of decreasing information, the target output variables are: a map which reflects closely the population; the frequency distribution of areas of the blocks, in different classifications and possibly in different regions; estimates of the mean of the block areas (and their total number) in different classifications and regions.

Sampling Units and Sampling Frames

Two alternatives are usually considered:

1. Mapped polygonal representations of the forest blocks, resulting from a RS classification exercise, and GIS mapping. Map area of each block would be available as auxiliary information for subsequent estimation.
2. A grid of area-based rectangular (or otherwise) units. The intersection of the mapped polygonal block representations with this unit is usually the measured response variable. The weighted segment estimation method (Gallego, 1995) is a way of selecting a variable probability sample from the sample frame of blocks.

The Sampling Design

For sampling frame (1), stratified random (by region, and class, and mapped area) and variable probability sampling are available. If the envisaged estimation method is model-based, then it is possible to abandon the random sampling paradigm and adopt a unit selection method which will allow model validation and maximize the precision of model parameter estimation.

For the area-based units, a multistage nesting of units is usual, with the lowest stage units selected either randomly (to be in accord with the "theory" of random sampling), or systematically (to be in accord with the optimal data requirements for spatial prediction modelling), or something in between.

The Map as Target Output(s)

A "corrected" map is the form of output with maximum information, from which other, possibly statistical, information may be derived.

Map "Estimators"

Final maps are obtained from RS source data through a number of stages:

1. Geometric corrections.
2. Classification methods with or without supervision on a per-pixel basis. Filters and spatial modeling methods can be used to improve the classification, and consequent estimation of the map.
3. The possible application of an area correction model, such as those of Mayaux and Lambin (1995,1997).

It is not clear how such a corrected map would be obtained in operational terms, using (3) from the initial auxiliary map obtained after (1) and (2). Some type of block shrinkage/expansion from the block centroids, with block boundaries being correspondingly adjusted, might be possible. However, different block classifications will have different reliabilities and these would probably influence the nature of areal representation error. It would probably, therefore, be necessary to apply different adjustment algorithms to the different block classes and sizes.

Map Accuracy

Again we look upon "accuracy" as the "inverse" of the "error" of a map. Maps are constructed from noisy data in a number of stages, with the errors incurred at each stage being related to what has happened at other stages of the process. Hence there is no simple single measure of the error, and hence of the "accuracy" of a map. Rather, map error/accuracy is multivalued, with the various components of error/accuracy being interrelated in complex ways. A complete characterization and estimation of such a map error/accuracy vector has not been achieved for maps constructed from RS data. A first attempt at such a characterization is given below.

Pixel-Level Error

After a pixel-based classification in which the maximum likelihood class is chosen, misclassification probabilities are available from the classification probability vector. The "magnitude" of a misclassification of a certain type can be quantified, and hence a weighted measure of misclassification error calculated. The sum of these overall pixels would give a global pixel-based measure of misclassification error, E_1, say. However, the use of filters, or spatial model adaptors, may reduce this, to E'_1 say.

Block-Level Error

At the level of a "forest block," having a given classification, and a given polygonal boundary (following GIS manipulation), and internal area, there are sources of error associated with each of these features.

Block-Classification Error

First, a weighted misclassification measure of error per block could be developed, but the weights (probabilities), would have to be determined empirically, E_2, say.

Boundary-Omission Error

Second, the major form of error of polygonal representation error will result from a failure to detect a boundary between adjacent blocks, resulting in the loss of a unit from the sampling frame, and the block that is included having a large areal error. Again, a measure of this form of areal (and possibly frame) misrepresentation error could be developed, but its determination would also have to be empirically based. Let the summed error measure over all blocks be E_3.

Boundary-Positioning Error

Third, the determined polygonal boundary of a block will have a positional error, which might possibly be represented by (the square root of ?) the difference between the areas of the union and intersection of the "true" bounded block and the polygonal representation. Call the summed error measure over all blocks E_4.

Block-Area Error

Finally, there will be an error in the area of the block representation. Let the summed form be E_5.

A Global Measure of Map Error

A global measure of map error may be obtained as a weighted sum of the above error measures, (E, say), where the weights could be chosen based on the perceived importance of each of the types of error. Though the sources of error mentioned above are not independent, and this global measure of error will involve some "duplication," we have to be heuristic in a situation of this complexity.

E will depend on how many and where the "ground truth" data samples are collected; i.e., on **n**. This dependency needs to be captured in some model. Allowing complete freedom to the magnitudes of the sample sizes in different parts of the design, and freedom to the location of sampled points would make this task intractable. Hence, to progress it would be necessary to impose minimal constraints, to impose a standard spatial sampling regime (systematic, or random, or some mix), so that empirical (pilot) studies could be used to calibrate this relationship.

The Cost Model

The cost model would need to take into account a measure of relative costs to ground-truth the various sampling units, in the different strata of the design, and the dependency on sample sizes stated explicitly. The cost model is probably not a critical component of the whole system, so it may be acceptable to use a simple approximate model, in which the various "costs" are assigned subjectively. (Note: costs do not have to be stated in terms of currency.)

THE OPTIMAL SAMPLE DETERMINATION PROBLEM FOR MAPPING

Given that the Accuracy and Cost models are specified, discrete stepwise search methods can be used, starting with minimal sampling constraints and then adding samples sequentially so as to maximize the gain in accuracy per unit cost for each additional sample. See Rennolls (1989) for an example of the application of this approach.

There is no guarantee that the sample allocation obtained will be globally optimal over the space of all possible sample allocations. However, the result will be a close approximation to the optimum, and this could be regarded as an advance in an area where such methods have not been used previously.

DISCUSSION

Having devised a near-optimal sample allocation on the basis of the RS classified map and various empirical studies, the data can be collected and the finally adjusted map estimated, and its global accuracy evaluated.

How, then, should regional statistics be estimated for forest block characteristics; e.g., total volume, by class, age, etc.? A summation of the block information from the corrected map would seem a natural estimator. But how can the accuracy of such estimates be determined? Not by reference to the map accuracy criteria introduced, since the required accuracy measure of a statistical estimate is its standard error. The only way to obtain an estimate of this standard error would be by the use of jackknife methods which involve leaving out certain parts of the information that had been used to obtain the finally corrected map and calculating the resulting variability in the partial estimates. The way in which information would need to be left out would be problematic, since there have been so many processes and steps involved in the construction of the final estimated map. It may be simpler to calculate such statistics and their standard errors by direct reference to the source ground-truth data, the auxiliary variables, and the use of standard statistical estimation methodology, not making use of the mapped output.

Estimating change of forest statistics and maps from monitoring information is rather more difficult than one-time inventory, since the target variable; i.e., change, is small in magnitude and the errors of mapping and sampling are undiminished. Estimation of biodiversity and its change would present even more difficulties, not least because of the scale noninvariance of most measures of diversity. However, the methodology should not be essentially different from that outlined above.

REFERENCES

Cochran, W.G. *Sampling Techniques,* 3rd ed., John Wiley & Sons, New York, 1977.

FIRS. Regionalization and Stratification of European Forest Ecosystems, *FIRS (Forest Information from Remote Sensing),* JRC, EU. S.P.I.95.44, 1995.

Gallego, F.J. *Sampling Frames of Square Segments,* Report EUR 16317 EN, JRC, EU. ISSN 1018-5593, 1995.

Mayaux, P. and E.F. Lambin. Estimation of Tropical Forest Area from Coarse Spatial Resolution Data: A Two-Step Correction Function for Proportional Errors Due to Spatial Aggregation, *Remote Sens. Environ.,* 53, pp. 1–15, 1995.

Mayaux, P. and E.F. Lambin. Tropical Forest Area Measured from Global Land-Cover Classifications: Inverse Calibration Models Based on Spatial Textures, *Remote Sens. Environ.,* 59, pp. 29–43, 1997.

Rennolls, K. *Design of the Census of Woodlands and Trees 1979–82,* Forestry Commission Occasional Paper 18. ISBN 0 85538 220 1, 1989.

Rennolls, K. and Sampson. A General Purpose Survey Design and Analysis Program, in *The Optimal Design of Forest Experiments and Forest Surveys, Proceedings of a IUFRO S4.11 Conference,* K. Rennolls and G.Z. Gertner, Eds., CASSM Publishing, CMS, University of Greenwich, 1993, pp. 221–232.

Rennolls, K. and G.Z. Gertner, Eds. The Optimal Design of Forest Experiments and Forest Surveys, *Proceedings of a IUFRO S4.11 Conference,* University of Greenwich, Sept. 1991. ISBN 1 897610 00 9. CASSM Publishing, CMS, University of Greenwich, 1993.

Schreuder, H.T. and R.L. Czaplewski. Long-Term Strategy for the Statistical Design of a Forest Health Monitoring System, in The Optimal Design of Forest Experiments and Forest Surveys, *Proceedings of a IUFRO S4.11 Conference,* K. Rennolls and G.Z. Gertner, Eds., CASSM Publishing, CMS, University of Greenwich, 1993.

Traub, B., P. Kleinn, and D. Pelz. *Statistical Aspects of the TREES Stratification and Calibration Methodology,* Consultancy Report to the TREES Project, 1994.

CHAPTER 12

Characterizing Local Spatial Uncertainty in the Optimization of Thematic Class Areas

R.C. Allan

INTRODUCTION

The utility of remotely sensed data and, more specifically, satellite imagery as input into a GIS is largely dependent on the accuracy of each layer in the GIS. If unclassified satellite imagery is used as a primary data source, then real world heterogeneity is, in part, displayed by this source. A problem arises, however, when this display is transferred into a GIS for analysis to meet thematic map requirements of the data user. In this context, the representation of the real world is transformed into discrete classes with no evidence of the underlying spatial variation within each class that exists on the ground. This may not be critical, depending on the intended use of the GIS *product*. Nonetheless, given only the thematic map, the data user is unable to objectively decide whether the *product* is appropriate, i.e., sufficiently accurate, for its intended use. Therefore, the utility of integrating satellite imagery with data layers produced from other sources; e.g., aerial photography and thematic maps, will be contingent upon employing appropriate methods to ascertain the various levels of spatial uncertainty over an entire classified image.

Consequently, the rationale for this study is to examine local spatial uncertainty present in a rectified and subjectively classified image through empirical examination of polygon attribute values.

BACKGROUND

Traditionally, the accuracy of thematic maps from remotely sensed data has been assessed by using the confusion (misclassification) matrix. Ground points are compared with the reference data (typically a map) and the row and column marginals are summed to give an overall estimate of classification accuracy (Ginevan, 1979; Aronoff, 1982). The row and column marginals of the matrix are used to determine the producer's and user's accuracy (Aronoff, 1982; Story and Congalton, 1986). Further work in testing map accuracy investigated appropriate sampling schemes (Congalton, 1988; Stehman, 1992); the minimum number of sampling points required (Genderen and Lock, 1977; Hay, 1979) and the efficiency of sampling designs (Moisen et al., 1994). These measures give an indication of the accuracy of the map posteriori as the boundaries between classes have been explicitly delineated. Although the variation within the map classes may be determined from the marginals, the location of these variations (differences) is not shown. Therefore, it would be useful when undertaking further analysis within a GIS to know where these differences occur, to incorporate them in the modeling process. Moreover, Lowell (1994) suggests that, as the spatial distribution of uncertainty may not be random, results may be highly variable at some locations.

Uncertainty propagation models using categorical maps have included Boolean logic (Heuvelink and Burrough, 1993), overlay operations (Newcomer and Szajgin, 1984; Walsh et al., 1987; Chrisman, 1989; Veregin, 1989, 1995); simulation (Openshaw, 1989; Fisher, 1991; Goodchild et al., 1992; Brunsdon and Openshaw, 1993; Veregin, 1994) and fuzzy set theory (Altman, 1994; Lowell, 1994; Fisher, 1994; Burrough, 1995; Wang and Hall, 1996; Davis and Keller, 1997).

Many of these stochastic and deterministic models depend on the determination of a priori uncertainty estimates of individual source layers which may be difficult to verify. Furthermore, current cartographic models are restrictive in the representation of *reality;* e.g., lines and polygons designate definite boundaries between features that may not exist.

Relatively few empirical tests have been undertaken to detect and measure the occurrence and magnitude of uncertainty for remotely sensed data. The contention herein is that uncertainty is far more pervasive than what current models elucidate and an empirical study to both quantify and manage spatial uncertainty is justified. Moreover, the uncertainty which results from the relative skill and/or bias of a human image interpreter remains a difficult component of spatial uncertainty to quantify. The existence of a more accurate data source for comparison with an interpreted image may be one solution, but rarely in practice does this source exist or is available for the specified number of classes. Consequently, there is a limited understanding of whether any or all of the classes delineated on an image by one human interpretation has any resemblance to *reality.*

The proposed methodology attempts to elicit these uncertainty measures by using different human interpreters to independently classify and rectify the same satellite image. The extraction and processing of uncertainty attributes for each polygon resulting from the synthesis of the individually interpreted images provides for an improved understanding of where uncertainty occurs, its magnitude, and how this uncertainty can be managed (reduced or absorbed) and reported (or displayed) to users.

METHODOLOGY

A study site was chosen to exemplify the characteristics needed to examine spatial uncertainty resulting from an independently and subjectively classified satellite image. The site has various land covers which exhibit both well-defined and less distinct boundaries between such land covers. While some land-cover classes are spectrally homogeneous, other classes are particularly fragmented and consequently difficult to interpret and classify. Furthermore, existing Landsat Thematic Mapper (TM) satellite imagery was available, and access to the site was sufficient to enable a number of ground control points to be established for image georeferencing.

To date, there appears to be little information concerning the consistency among human interpreters in developing a classified image of the same area using the same classes. The level of consistency among different interpretations is largely dependent on the spectral and spatial characteristics of the satellite image as well as the interpreters themselves. In particular, if the image displays significant within-class variability (heterogeneity); i.e., large disparities in the data file values for each nominated class, then the image analyst's skill and bias in interpreting and classifying the image is critical in obtaining appropriate levels of consistency. Moreover, if statistical measures; e.g., confidence levels, variances, and correlations, are to be ascribed to these classifications when input into a GIS, an understanding of both the occurrence and magnitude of spatial uncertainty which results from this interpretive classification process would be useful.

Consequently, the methodology adopted herein to detect and measure uncertainty is based on the synthesis of multiple interpreter classifications of the same Landsat TM image for the study area. Each of the four nominated image interpreters independently selected appropriate training sites by visual interpretation and performed a supervised classification. If the four classified images are then overlaid to form one composite image, the class boundaries are unlikely to exactly coincide everywhere on the composite image due to subjectivity of independent human interpretations of the same image. The result of this overlay process (composite image) is analogous to the occurrence of sliver polygons when lines, purporting to be the same class boundaries on two data layers, do not exactly coincide when overlaid in a vector-based GIS. The identification of areas (polygons) which indicate both *agreement* and *lack of agreement* between interpreters on the composite image are examined. For these areas, various attributes (semantic, geometric and spatial) can be included in the database for analyses (see Figure 12.1). The creation of polygons and development of these attributes from the composite image will be discussed in more detail subsequently.

Data Processing

One of the principal objectives of this study is to ascertain, examine, and manage the level of uncertainty on the composite image or map product. In general practice, the classification and rectification of an image is usually undertaken by a single operator. Note that the resulting image may contain a variety of types

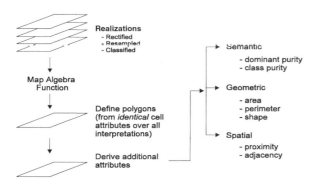

Figure 12.1. Schematic view of the methodology and database development.

of uncertainty which, in part, are due to the interpreter's knowledge, skill and bias. Extensive ground truthing may detect some of the uncertainty in the image but in many cases, ground truthing is neither viable nor feasible when dealing with indistinct boundaries between, for example, soil and vegetation classes. Consequently, to explore differences in human interpretations of a classified image, it was necessary to engage more than one remote sensing image analyst to undertake this component of the study. As previously mentioned, four analysts were chosen and each one was supplied with the same satellite image of the study area and briefed on the overall study objectives and site characteristics. Though interpreters were required to classify the image into seven classes, no attempt was made to nominate specific training sites. Each interpreter had to independently derive their own spectral signatures for each class from a number of individually selected training sites.

The procedure adopted by each analyst was initially to locate the 12 ground control points (derived from the GPS field survey) on the image. Subsequently, a first-order (linear) transformation was employed to georeference the image to AMG coordinates. All image interpreters obtained an overall root mean squared error (RMSE) for image rectification of less than 0.8 pixel. Although image georeferencing was not critical in the context of this study, it provided for a common referencing (coordinate) system if other data sources with different spectral and spatial characteristics; e.g., airborne scanner, radar, and aerial photography, were to be included in the analyses as would likely occur operationally.

At this stage, one georeferenced image existed for each of the four interpreters. The next task for each interpreter was to classify the image into seven nominated classes. Using visual interpretation, each interpreter was able to define patterns for each feature type (nominated class) over the image. Training site polygons were then drawn to enclose the area of interest. The number of training sites defined for each nominated class was allowed to be more than one, depending on the spectral complexity of the class. Each interpreter used more than one training site for each class. Once all the training sites were defined, the next phase was to create and evaluate the signature for each class. This was executed through the signature editor included in the image processing software. The signatures were created using parametric statistics. In order to evaluate these signatures, different software tools are available; e.g., alarms and contingency matrices (Erdas, 1991). The alarm tool was used by all interpreters to highlight the pixels that belong to a class according to the parallelepiped decision rule. The final adoption of a signature for a particular class is based on personal judgment, pattern recognition ability, and the available software tool. This procedure was executed for each class. Once each interpreter was satisfied with signatures for each class, a supervised classification procedure was performed. The maximum likelihood decision rule was adopted for the supervised classification where each pixel was compared to each signature in accordance with this rule. The pixels were then assigned to the class corresponding to that signature. An independently classified and rectified image existed for each of the four interpreters. Each of these classified images is considered to be a *realization;* i.e., an independently produced thematic map which has undergone the processes of rectification and classification.

Data Manipulation

This section explains how the realizations were combined to enable a per pixel (cell) extraction of attribute (class) values across all realizations. The creation of polygons from similarity in these cell attribute values in the composite image enables the subsequent development of semantic, geometric, and spatial uncertainty measures to be embedded as attributes in the database, as will be explained subsequently.

Combining Multiple Realizations

Realizations were combined using the map algebra functions within the Grid™ module of ArcInfo™. This synthesis enabled *tracking* of the land-cover class attribute of each grid cell (pixel) for each image. As the images are coregistered to the AMG, the geographic location of a pixel on one image has exactly the same position as that on another image (realization). However, the land-cover class values for coincident pixels on separate images may not be the same (refer to Figure 12.2).

More formally, in an image comprising a rectangular array of N cells with rows i and columns j, each cell C_{ij} has associated with it a vector of attributes $\{a_1, a_2, ..., a_n\}$ which represents the class assignment for each realization (Figure 12.2). In this study, each cell has four attributes (one for each interpreter) and each attribute is one of the seven possible classes. This attribute vector enables the determination of cell *purity* values over the entire image. The term *purity* has been adopted as the cell attributes have been empirically derived rather than generated by a stochastic process which would produce probability values. Furthermore, a cell having exactly the same attributes over all interpretations is considered *pure* although it may also be interpreted as a *certain* and/or *homogeneous* cell. For a cell not having identical attributes over all interpretations, the *purity* may be indicative of an uncertain classification, of mixed pixels of heterogeneous classes or fuzzy classification (Goodchild et al., 1992). For example, in those instances where cell *purity* is low—i.e., different classes were identified on each of the four interpretations—boundary definition for that part of the image may be both interpretively difficult and often inappropriate due to a high proportion of mixed pixels. This uncertainty can be shown visually to the data user using the attributes embedded in the database.

Although the concept of cell *purity* appears in Goodchild et al. (1992), those authors generated impurities (inclusions) by simulation which incorporates an underlying statistical process and assumptions about the behavior of the uncertainty. Whether the occurrence and level of inclusions was likely to occur in reality was not tested. Conversely, in this study, cell purity is determined from the cell attributes which result from the synthesis of four realizations.

Polygon Definition

As seen from the previous discussion, each grid cell in the composite image has four attributes (a_1, a_2, a_3, a_4) which represent the class identified on each of the four interpretations; each of these attributes is one of

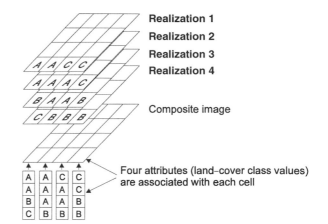

Figure 12.2. Each cell has associated with it a vector of attributes. This vector represents class assignment for each realization.

seven classes. These four attributes for a single cell are herein referred to as *interpreter* attributes. The formation of each polygon is accomplished by aggregating contiguous cells which have identical *interpreter* attributes, i.e., the attribute vector (a_1, a_2, a_3, a_4) for one cell is identical to that of an adjacent cell. Furthermore, a cell with *interpreter* attributes AABC is merged with a contiguous cell (interpreter attributes ABCA) to form one polygon regardless of which interpreter assigned the cell to a given class. Figure 12.3 illustrates the overlay of two interpretations and how each polygon is formed from the aspatial attributes in the composite image. For example, in Figure 12.3, polygon AA is derived from merging six contiguous cells, each of which has identical attributes (AA). Similarly, contiguous cells containing attribute AB create polygon AB. Note that it is not necessary to have $a_1...a_4$ equal to the same class to define a polygon. In effect, this means that all polygons are defined by the same procedure and will be treated the same. What are often considered to be *undesirable* sliver polygons are herein merely polygons that do not have identical attributes over all interpretations. However, these polygons are considered as valid as those polygons that do have the same attributes over all interpretations. Thus the GIS database now contains a set of polygons and each polygon contains a *cell count* (polygon) attribute which indicates the number of aggregated contiguous cells with identical *interpreter* attributes. The inclusion of this *cell count* attribute in the database enables extraction of additional attributes for each polygon which will be required for analyses.

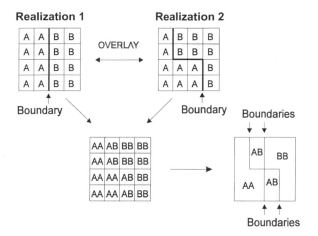

Figure 12.3. The overlay of two realizations alters the spatial and/or aspatial attributes in the composite image. Boundaries are implicitly defined from identical contiguous cell attributes.

Extraction of Attributes

Given the presence of both *interpreter* and *cell count* attributes for each polygon in the database, other descriptive (aspatial) and spatial attributes can be derived. The attributes to be added to the database for each polygon and explained in the following sections are (i) semantic—*dominant* and *class* purity; (ii) geometric—area, perimeter and *shape,* and (iii) spatial—*adjacency* (topologic) and *proximity* (metric).

Semantic

In the context of this study, the development of a *semantic* attribute for individual polygons is most easily described using a hypothetical example. Suppose that four interpreters independently classify an image into five classes A, B, C, D, and E. When the four independent realizations are combined into a composite image and polygons formed using the method described, the database contains *each* interpreter's realization and one is able to determine an additional attribute defined as *dominant purity* of each polygon. If two interpreters have classified the polygon as A, another as B, and another as D, then a *dominant purity* value of 50% is assigned regardless of whether polygons contain a particular class attribute or not. In other words, the *dominant purity* value for a polygon of type AABD is exactly the same (i.e., 50%) as a polygon with attributes BCCD. While generic *dominant purity* attribute values can be useful for analyzing polygons

over an entire image, the use of this attribute alone may be ambiguous and potentially misleading if specific classes are not considered. To overcome this limitation, a further attribute is added to the database. In the example cited; i.e., a polygon of type AABD, the polygon has *class purity* for A equal to 50%, a class purity for B of 25%, a class purity for D of 25%, with class purity for C and E set equal to zero. Searching the database for occurrences of, for example, class B at a specific *class purity* level would then enable the extraction of all possible occurrences at that level (or above or below) regardless of whether the class is dominant or not.

Geometric

The geometry of a polygon refers to its area, perimeter, and shape. These parameters, used to describe polygons regardless of purities, have been addressed by many authors, particularly in the field of remote sensing. Although area estimates for polygons are one of the most common products of a GIS (Goodchild et al., 1992), shape is also a fundamental property of geographic phenomena (Austin, 1984). Polygon shapes can be characterized by the compactness ratio or radial line index (Unwin, 1981) or by form, circularity, and radius ratios (Kidner, 1996).

The shape descriptor used in this study is the compactness ratio S_2 (Unwin, 1981) with the advantage that it is a dimensionless number that is invariant with polygon size. This ratio has values in the interval [0, 1], with a circular polygon having a compactness ratio equal to one. In a raster-based GIS, a single square cell has a compactness ratio of 0.89 which is the maximum value attainable using a cell-based GIS. A *line* (characterized as one cell wide) has a minimum ratio of 0.03 if the *line* traverses the length of the entire image (756 rows) employed in this study. The compactness ratio is used only as a relative measure of polygon shape for comparison among polygons in this study. These geometric attributes (area, perimeter, and shape) are all calculated from the *cell count* attribute and entered as additional attributes in the database for each polygon.

Spatial

The relationships among polygons in terms of distance (metric) and adjacency (topologic) are referred to as *spatial* attributes for inclusion in the database. Distances may be calculated based on Euclidean geometry or by using the *Manhattan Metric* approach (Unwin, 1981). The latter approach may be appropriate if proximity measures are required between polygons (e.g., buildings) which are located on a street network (grid pattern). However, land cover/use phenomena do not usually conform to such a grid pattern and hence Euclidean distances are adopted among polygons in this study. Distances were determined from the closest point and the furthest point of every polygon having a class purity less than 100 ($q < 100$) to the closest point of the nearest *pure* polygon with class purity equal to 100 ($q = 100$). Note that a *pure* polygon class must correspond to each of the classes that appear in the polygon ($q < 100$) vector for a distance to be calculated. For example, distances are calculated from a polygon with vector AAAB to the closest *pure* polygon A and to the closest *pure* polygon B. Also, the mean of the closest and furthest distances of each polygon ($q < 100$) to the nearest *pure* polygon is also calculated and entered as an additional attribute in the database. Furthermore, polygons ($q < 100$) that are adjacent to *pure* polygons are of interest and entered as an attribute in the database.

DISCUSSION

The next phase in the procedure will be to analyze these derived semantic, geometric, and spatial polygon attributes to attempt to elicit a spatial uncertainty structure for specified classes. Measurements ascribed to these attributes (semantic, geometric, and spatial) in the GIS database can be used to quantify the level of spatial uncertainty for particular classes. Each of the attribute measures will be analyzed separately to more explicitly define what is meant by, e.g., small, narrow, predominant, and close polygons, when dealing with local spatial uncertainty. Thus, the analysis of each polygon attribute measure undertaken will form the basis for determining the most likely class for each polygon. In order to accomplish this task, the interrelationships among polygon attribute measures must be carefully examined. For example, if a large, narrow polygon (geometric) with class *purity* of 50 (semantic) is close (or adjacent) to a *pure* polygon (class *purity* of 100) (spatial) of the same class, what is the likelihood of a polygon being that class?

CONCLUSION

The normal procedure in using satellite imagery for input as a data layer within a GIS is for a single inter-

preter to undertake the classification and rectification of an image. The resulting thematic map contains various types of uncertainty, some of which are unable to be examined from a single interpretation. Therefore, the principal objective of the proposed methodology was to be able to elicit uncertainty measures based on the synthesis of a number of independent human interpretations of the same image. To achieve this objective, and also to develop a database to support subsequent uncertainty analyses, a staged process was formulated.

Analysis of semantic and geometric attribute measures for each polygon and spatial relationships among polygons over the entire image can be undertaken to ascertain the likelihood of polygons belonging to a particular class. The likelihood (or level) of association of a polygon to a particular class can be determined using empirical inference rules.

ACKNOWLEDGMENTS

The author is most grateful to Prof. Kim Lowell for his support, encouragement, and thoughtful comments during the course of this work and for his significant contribution toward my broader understanding of spatial uncertainty.

REFERENCES

Altman, D. Fuzzy Set Theoretic Approaches for Handling Imprecision in Spatial Analysis. *Int. J. Geogr. Inf. Syst.,* 8(3), pp. 271–289, 1994.

Aronoff, S. Classification Accuracy: A User Approach. *Photogrammetric Eng. Remote Sensing,* 48(8), pp. 1299–1307, 1982.

Austin, R.F. Measuring and Comparing Two-Dimensional Shapes, in *Spatial Statistics and Models,* G.L. Gaille and C.J. Willmott, Eds., Reidel, Dordrecht, 1984, pp. 293–312.

Brunsdon, C. and S. Openshaw. Simulating the Effects of Error in GIS, in *Geographical Information Handling— Research and Applications,* P.M. Maher, Ed., Wiley, Chichester, 1993, pp. 47–61.

Burrough, P.A. Natural Objects with Indeterminate Boundaries, in *Geographic Objects with Indeterminate Boundaries,* P.A. Burrough and A.U. Frank, Eds., GISDATA 2, Taylor & Francis, London, 1995, pp. 3–28.

Chrisman, N.R. Modeling Error in Overlaid Categorical Maps, *Accuracy of Spatial Databases,* M.F. Goodchild and S. Gopal, Eds., Taylor & Francis, 1989, pp. 21–34.

Congalton, R.G. A Comparison of Sampling Schemes Used in Generating Error Matrices for Assessing the Accuracy of Maps Generated from Remotely Sensed Data.

Photogrammetric Eng. Remote Sensing, 54(5), pp. 593–600, 1988.

Davis, T.J. and C.P. Keller. Modelling Uncertainty in Natural Resource Analysis Using Fuzzy Sets and Monte Carlo Simulation: Slope, Stability, Prediction. *Int. J. Geogr. Inf. Sci.,* 11(5), pp. 409–434, 1997.

Erdas. *Erdas Field Guide,* Erdas Inc., Georgia, 1991.

Fisher, P.F. Modelling Soil Map-Unit Inclusions by Monte Carlo Simulation. *Int. J. Geogr. Inf. Syst.,* 5(2), pp. 193–208, 1991.

Fisher, P.F. Probable and Fuzzy Models of the Viewshed Operation, in *Innovations in GIS 1,* M.F. Worboys, Ed., Taylor & Francis, London, 1994, pp. 161–175.

Ginevan, M.E. Testing Land-Use Map Accuracy: Another Look. *Photogrammetric Eng. Remote Sensing,* 45(10), pp. 1371–1377, 1979.

Goodchild, M.F., S. Guoping, and Y. Shiren. Development and Test of an Error Model for Categorical Data. *Int. J. Geogr. Inf. Syst.,* 6(2), pp. 87–104, 1992.

Hay, A.M. Sampling Designs to Test Land-Use Map Accuracy. *Photogrammetric Eng. Remote Sensing,* 45(4), pp. 529–533, 1979.

Heuvelink, G.B.M. and P.A. Burrough. Error Propagation in Cartographic Modelling Using Boolean Logic and Continuous Classification. *Int. J. Geogr. Inf. Syst.,* 7(3), pp. 231–246, 1993.

Kidner, D.B. Geometric Signatures for Determining Polygon Equivalence During Multi-Scale GIS Update, in *Proceedings of Second Joint European Conference & Exhibition on Geographical Information,* M. Rumor, R. McMillan, and H.F.L. Ottens, Eds., Barcelona, Spain, Vol. 1, 1996, pp. 238–247.

Lowell, K.E. Probabilistic Temporal GIS Modelling Involving More than Two Map Classes. *Int. J. Geogr. Inf. Syst.,* 8(1), pp. 73–93, 1994.

Moisen, G.G., T.C. Edwards, and D.R. Cutler. Spatial Sampling to Assess Classification Accuracy of Remotely Sensed Data, in *Environmental Information Management and Analysis: Ecosystem to Global Scales,* W.K. Michener, J.W. Brunt, and S.G. Stafford, Eds., Taylor & Francis, London, 1994, pp. 159–176.

Newcomer, J.A. and J. Szajgin. Accumulation of Thematic Map Errors in Digital Overlay Analysis. *Am. Cartographer,* 11(1), pp. 58–62, 1984.

Openshaw, S. Learning to Live with Errors in Spatial Databases. *Accuracy of Spatial Databases,* M.F. Goodchild and S. Gopal, Eds., Taylor & Francis, 1989, pp. 263–276.

Stehman, S.V. Comparison of Systematic and Random Sampling for Estimating the Accuracy of Maps Generated from Remotely Sensed Data. *Photogrammetric Eng. Remote Sensing,* 58(9), pp. 1343–1350, 1992.

Story, M. and R.G. Congalton. Accuracy Assessment: A User's Perspective. *Photogrammetric Eng. Remote Sensing,* 52(3), pp. 397–399, 1986.

Unwin, D.J. *Introductory Spatial Analysis,* Methuen, London, 1981.

van Genderen, J.L. and B.F. Lock. Testing Land-Use Map Accuracy. *Photogrammetric Eng. Remote Sensing,* 43(9), pp. 1135–1137, 1977.

Veregin, H. Error Modeling for the Map Overlay Operation, in *Accuracy of Spatial Databases,* M.F. Goodchild and S. Gopal, Eds., Taylor & Francis, 1989, pp. 3–18.

Veregin, H. Integration of Simulation Modelling and Error Propagation for the Buffer Operation in GIS, *Photogrammetric Eng. Remote Sensing,* 60(4), pp. 427–435, 1994.

Veregin, H. Developing and Testing of an Error Propagation Model for GIS Overlay Operations. *Int. J. Geogr. Inf. Syst.,* 9(6), pp. 595–619, 1995.

Walsh, S.J., D.R. Lightfoot, and D.R. Butler. Recognition and Assessment of Error in Geographic Information Systems. *Photogrammetric Eng. Remote Sensing,* 53(10), pp. 1423–1430, 1987.

Wang, F. and G.B. Hall. Fuzzy Representation of Geographical Boundaries in GIS. *Int. J. Geogr. Inf. Syst.,* 10(5), pp. 573–590, 1996.

CHAPTER 13

Describing Uncertainty in Categorical Maps Using Correlated Categorical Data

W. De Genst, F. Canters, and W. Jacquet

INTRODUCTION

Although a large number of the available types of data sources used for environmental modeling are categorical data sources, categorical maps often fail to accurately describe the complex and continuous nature of the phenomena they represent (Goodchild et al., 1992). Vegetation cover type, for example, is more likely to vary continuously over space rather than change abruptly from one homogeneous area consisting of a specific vegetation cover type to the next. Although theoretically the homogeneous mapping units can be taken small enough to account for most of the variation in the mapped phenomenon, it is very unlikely that categorical maps will be compiled this way. Rather, in the mapping process a number of mapping units smaller than a minimum allowable mapping size will be omitted in order to enhance the readability of the map, thus introducing thematic error into the categorical map (Robinson et al., 1984; Openshaw, 1989).

Data producers have long neglected to provide information on the type and the magnitude of thematic uncertainty in categorical maps (Goodchild et al., 1994). In those cases where thematic uncertainty is documented, often only very general information is provided in the form of a global accuracy index (Lanter and Veregin, 1992). Such a general description of the overall accuracy of a data layer is often insufficient to adequately assess the sensitivity of spatial analyses to thematic uncertainty (Fisher, 1991). A number of techniques have therefore been proposed to derive more detailed information on the type and magnitude of errors in a categorical map. Perhaps one of the better known techniques is based on constructing a confusion error matrix by comparing the recorded class in a random set of locations in the map to the class identified from field inventories (Aspinall and Pearson, 1994). This procedure may, however, be very costly and very difficult to carry out, especially for less experienced users. An alternative, then, may consist of using more detailed maps for parts of a study area (Canters, 1995). Because the degree of generalization is often much lower in more detailed maps, these maps can show inclusions that were omitted from the less detailed original map. By overlaying the detailed categorical maps with the less detailed original map, an error index describing the type and magnitude as well as an error index describing the degree of spatial correlation of errors can be derived.

When it is not possible to conduct a field survey and more detailed maps of parts of the area are not available, an interesting alternative may consist of using detailed categorical maps of related spatial data to derive the probability that a location in the map was classified correctly. Depending on the strength of the relationship between the occurrence of classes in both maps, the spatial distribution of the related variable within a mapping unit of the original map may indicate the presence of inclusions of a certain type (Fisher, 1989). One of the main problems connected to the use of this technique is that it will generate inclusions of types that do not necessarily occur within a mapping unit. Information about the type of possible inclusions that can occur within a mapping unit is necessary to resolve this problem. In many categorical data sources

113

this information is not readily available. In some cases, however, when a categorical map is the result of the generalization of more detailed maps or of detailed field inventories, rather than omitting small mapping units, these units are merged together to form larger composite units (Fisher, 1989). The attribute value of a composite unit will then mention all possible types of inclusions that can occur within a polygon, but, in the main, without mention of the frequency of occurrence or the location of the inclusions within the composite unit. For maps that were compiled in this way, the underlying variation in related variables may then help to determine the frequency of occurrence and the location of inclusions within each composite unit. In this chapter we will propose a strategy to determine the frequency of occurrence of inclusions within the mapping units of a categorical map of the biotic environment of Belgium (the Biological Valuation Map), by means of the degree of correlation with a detailed categorical soil map.

DATA SOURCES AND CHOICE OF STUDY AREA

The Biological Valuation Map (BWK) is a standardized survey and evaluation of the biotic environment of Belgium (Van Straeten et al., 1993; De Blust et al., 1985). The legend units consist of approximately 130 different types which refer to ecotopes, structural elements in the landscape, and specific types of land use. About half of the number of units refer to an order, alliance, or association sensu Braun-Blanquet (De Blust et al., 1994). The BWK was compiled from detailed field surveys carried out between 1978 and 1984. Although the presence of all legend units was recorded in a very detailed way during the field surveys, these were often not represented separately in the BWK; rather, many ecotopes were merged together to form composite mapping units called complexes. All types of ecotopes occurring within a complex were recorded as a sequence of ecotopes in the attribute value of the complex. The relative frequency of the included ecotopes determines the rank of the ecotope within the sequence. The attribute value, however, does not mention the exact proportion of each ecotope, nor does it indicate how each ecotope is distributed over the complex. This clearly limits the applicability of the data source for use in environmental impact assessment and other types of spatial analysis.

A variable that is often referred to as being closely related with the distribution of ecotopes is soil type

(Paelinckx et al., 1994). In Belgium, soil maps are one of the environmental variables that are described at the highest level of detail. Because soil maps are spatially more detailed than the BWK, this variable may help to determine the magnitude and most probable distribution of inclusions within each mapping unit. To do this, a measure to describe the strength of the relationship between the occurrence of an ecotope and the occurrence of a soil type must be chosen that can be used to determine the probability of occurrence of ecotopes within a complex mapping unit.

In agreement with the Institute of Nature Conservation, a study area was chosen in which a number of ecologically valuable complexes of ecotopes occur, for which the composition of the complexes is assumed to be rather accurately described by the BWK, and for which both a digital BWK and a digital soil map are available. The study area is located in the south of the province of Antwerp near the city of Mechelen and is 8 by 10 km large (Figure 13.1). Because calculating the strength of the spatial relationship between classes in overlapping maps is much easier to carry out in a raster environment, both the BWK and the soil map of the study area were converted to raster maps with a resolution of 10 by 10 m.

DETERMINING THE PROBABILITY OF OCCURRENCE OF ECOTOPES

Calculating the amount of spatial coincidence between classes in different raster maps can be done by means of a simple map algebra overlay operation if the mapping units in the maps involved belong to a single class. Because many mapping units in the BWK are attributed to a complex of ecotopes, an alternative approach must be used to calculate the amount of spatial coincidence between classes. One possible approach may be to consider only the dominant ecotope in each mapping unit (Van Ghelue et al., 1993). Two very important problems result from this approach, however. Because many rare ecotopes often have a very small areal extent, these ecotopes are almost exclusively mapped as inclusions within a complex. By discarding all information about the composition of each complex, these rare ecotopes may be completely ignored in the analysis. Second, because the occurrence of some ecotopes will be overestimated, and the presence of other ecotopes will be underestimated, the amount of spatial coincidence with soil type classes may be completely distorted.

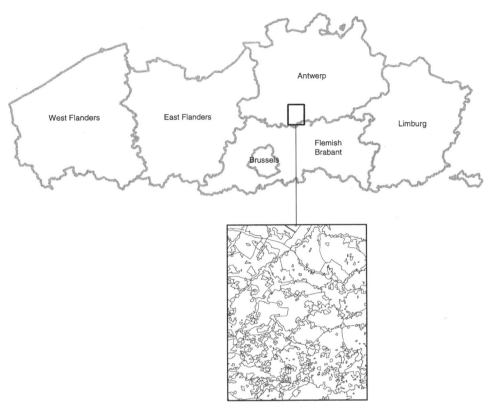

Figure 13.1. Location of the study area.

A correct estimate of the area each ecotope occupies within a complex is necessary to accurately describe the relationship between the occurrence of ecotopes and soil type. Unfortunately, this is precisely the information we are trying to obtain. As we mentioned earlier, the ecotopes grouped together to form complexes are ranked according to their dominance within the complex they belong to. De Blust et al. (1994) demonstrated that a global weighing key, based on best professional judgment, can be used to estimate the area each ecotope occupies within a considerably large region. The weight attributed to each ecotope in a complex is determined from the rank an ecotope has within a complex and the total number of ecotopes within this complex.

Assuming that the distribution of the ecotopes within a complex is homogeneous, the area of every pixel within a complex may be assumed to be partitioned over the component ecotopes, according to the weighing scheme that applies for the complex in which a pixel is located. The area each ecotope occupies within the study area, and therefore also the amount

of spatial coincidence between ecotopes and soil type can now be calculated more realistically.

Conditional Probabilities

The probability of an ecotope occurring, given the soil type, is one way of expressing the strength of the spatial relationship between ecotopes and soil type. The conditional probability of an ecotope en occurring, given a soil type B_m, can be calculated as:

$$p(e_n|B_m) = \frac{\sum\limits_{i \in (e_n \cap B_m)} w_i^{e_n} a_i}{\sum\limits_{i \in B_m} a_i} \qquad (1)$$

where $w_i^{e_n}$ is the weight of ecotope e_n in pixel i, which is dependent on the weighing scheme used, and a_i is the area of one pixel.

The estimated conditional probability will be most accurate for those ecotopes that occur mainly as separate mapping units, while the accuracy will be lowest for ecotopes that occur mainly as inclusions within complexes.

From the estimated conditional probabilities, the probability of occurrence of ecotope inclusions within a complex can be estimated, based on the soil type and the types of ecotopes that can occur in a pixel. In a pixel x belonging to a complex with n different ecotopes, the probability that this pixel belongs to an ecotope e_k can be calculated as:

$$p(e_k|\mathbf{x}) = \frac{p(e_k|B_\mathbf{x})}{\sum\limits_{i=1}^{n} p(e_i|B_\mathbf{x})} \qquad (2)$$

where $p(e_k|B_\mathbf{x})$ is the conditional probability of ecotope e_k for soil type $B_\mathbf{x}$ occurring in pixel $\mathbf{x}$.

Limits to the Use of Conditional Probabilities

One of the main problems related to the use of Equation 2 is that the resulting probabilities may contradict the prior knowledge about the dominance of the ecotopes within a complex. The dominant ecotope in a complex, as recorded in the attribute value, does not always have the highest probability of occurrence, as we would expect. This type of contradiction occurs when the dominant ecotope in a complex does not occur as frequently as some of the included ecotopes throughout the study area. We can illustrate this with the simple example shown in Table 13.1. Consider a complex of two ecotopes, E1+E2, where E1 is dominant over E2 in the complex, and consider, for the sake of illustration, that five soil types occur in the study area, namely S1, S2, S3, S4, and S5. The soil types underlying the complex E1+E2 are S1, S2, and S5. Because the frequency of occurrence of ecotope E1 is much lower than the frequency of occurrence of ecotope E2 throughout the study area, the conditional probability of E1 is much lower than the conditional probability of E2 for all soil types. Table 13.2 shows that, if we calculate the probability of occurrence of all ecotopes in the complex using Equation 2, ecotope E2 will clearly dominate the complex, which is in contradiction with how the complex was mapped. If we look at the conditional probabilities of E1 and E2 for all soil types separately, we see that E1 does seem to show a clear preference for soil types S1, S2, and S5 (the soil types underlying the complex

Table 13.1. Example Conditional Probabilities of Ecotopes E1 and E2 for a Number of Soil Types.

Soil Type	Ecotope	
	E1	E2
S1	0.1	0.2
S2	0.02	0.4
S3	0.001	0.4
S4	0.002	0.5
S5	0.1	0.3

Table 13.2. Probability of Occurrence of E1 and E2 in a Complex E1+E2 (Based on Conditional Probability).

Soil Type	Ecotope	
	E1	E2
S1	0.333	0.667
S2	0.048	0.952
S5	0.250	0.750

E1+E2), while E2 does not seem to show a strong preference for these soil types at all, and even seems to prefer other soil types more. This demonstrates that the predicted contradiction in dominance is not the result of a higher preference of the included ecotope E2 for the soil types underlying the complex, as compared to the dominant ecotope E1. Clearly, conditional probability is too strongly dependent on the frequency of occurrence of the ecotopes throughout the study area, and therefore does not allow comparing the affinity of different ecotopes for a soil type. In order to determine the probability of occurrence of an ecotope in a complex more accurately, an alternative measure expressing the strength of the relationship between ecotopes and soil type must be used. Furthermore, the measure expressing the affinity of an ecotope for a soil type should be used only to locally correct the prior estimate expressed by the weight $w_i^{e_n}$, to ensure that predictions contradictory to the recorded order of dominance in a complex occur only when the affinity of an included ecotope is much higher than the affinity of the dominant ecotope.

A Measure for Affinity

The affinity of an ecotope for a soil type should be expressed by a measure that is high when the conditional probability of an ecotope for a soil type is higher than for most other soil types, and should be low in the inverse case. Furthermore, equally high or low conditional probabilities for all soil types should result in a "neutral" affinity measure. One way in which to realize this is by defining an "average" soil type for every ecotope for which the ecotope has a neutral affinity. This "average" soil type can be defined as the soil type for which the conditional probability is equal to the average of the conditional probabilities of the ecotope for all soil types. The affinity of an ecotope for a soil type can then be expressed in function of how much the conditional probability differs from the average conditional probability. To do this, we have defined an affinity function, which converts conditional probabilities into affinity values that range between 0 and 2, and for which the affinity value becomes equal to 1 when the conditional probability is equal to the average value. The following parabolic function was chosen to define the affinity function:

$$\alpha_{B_m}^{e_n} = \frac{a_1 p(e_n|B_m)}{a_2 + p(e_n|B_m)} \tag{3}$$

where $\alpha_{B_m}^{e_n}$ is the affinity value of ecotope e_n for soil type B_m, $p(e_n|B_m)$ is the conditional probability of ecotope e_n for soil type B_m, and a_1 and a_2 are parameters that are constant for each ecotope. The value of a_1 and a_2 should be chosen in such a way that if the conditional probability is equal to 0, the affinity value is equal to 0, if the conditional probability is equal to the average over all soil types, the affinity value is equal to 1, and if the conditional probability is equal to 1, the affinity value is equal to 2. This implies that:

$$a_1 = \frac{2(1-\bar{p})}{1-2\bar{p}} \tag{4}$$

and that:

$$a_2 = \frac{\bar{p}}{1-2\bar{p}} \tag{5}$$

where $\bar{p}$ is the average of the conditional probabilities of ecotope e_n over all soil types.

Figure 13.2 shows the shape of the affinity function for a number of different values for $\bar{p}$. Conditional probabilities between 0 and 1 are represented on the X-axis and the corresponding affinity values between 0 and 2 are represented on the Y-axis.

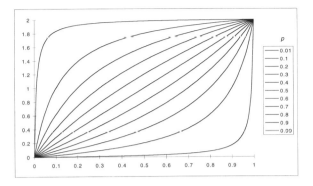

Figure 13.2. The shape of the affinity function for a number of different values for $\bar{p}$.

An affinity-corrected measure for the presence of an ecotope in a pixel can be obtained by multiplying the weight of the ecotope in a pixel by the affinity of the ecotope for the underlying soil type. Standardizing this measure for the presence of all possible ecotopes in a pixel results in an affinity-corrected probability for the occurrence of an ecotope in a pixel:

$$p_\alpha(e_k|\mathbf{x}) = \frac{\alpha_{B_\mathbf{x}}^{e_k} w_\mathbf{x}^{e_k}}{\sum_{i=1}^{n} \alpha_{B_\mathbf{x}}^{e_i} w_\mathbf{x}^{e_i}} \tag{6}$$

where $p_a(e_k|\mathbf{x})$ is the affinity-corrected probability of ecotope e_k occurring in pixel $\mathbf{x}$, $w_\mathbf{x}^{e_k}$ is the weight of ecotope e_k in pixel $\mathbf{x}$, and $\alpha_{B_\mathbf{x}}^{e_k}$ is the affinity value of ecotope e_k for the soil type $B_\mathbf{x}$ underlying pixel $\mathbf{x}$.

Let us now reconsider the conditional probabilities of the ecotopes E1 and E2 for the five soil types given in Table 13.1. Table 13.3 shows that if we calculate the affinity of the ecotopes E1 and E2 for soil types S1 through S5, using Equations 3 through 5, E1 has a higher affinity than E2 does for two out of three soil types underlying complex E1+E2. This indicates that the contradiction in dominance predicted by Equation 2 is not the result of a higher preference of the included ecotope E2 for the soil types underlying the complex, as compared to the dominant ecotope E1. Table 13.4 shows that, if we calculate the affinity corrected probability of occurrence of E1 and E2 in the complex E1+E2, using Equation 6, and assuming a weight of 0.7 for E1 and of 0.3 for E2, the probability of occurrence of E1 within the complex is higher than the probability of occurrence of E2, which is in accordance with the recorded order of dominance in the complex.

Table 13.3. Affinity Values of Ecotopes E1 and E2 for a Number of Soil Types.

Soil Type	Ecotope	
	E1	E2
S1	1.408	0.615
S2	0.608	1.085
S3	0.042	1.085
S4	0.082	1.280
S5	1.408	0.865

Table 13.4. Affinity-Corrected Probability of Occurrence of E1 and E2 in a Complex E1+E2.

Soil Type	Ecotope	
	E1	E2
S1	0.842	0.158
S2	0.567	0.433
S5	0.792	0.208

Case Study: Estimating the Area of an Ecotope in the Biological Valuation Map

The BWK is one of the most important data sources for applications in nature conservation, regional planning, and environmental impact assessment in Flanders (Paelinckx et al., 1991; Van Straeten et al., 1993; De Blust et al., 1994). For a number of spatial analyses that are of great importance to all these applications, the suitability of the BWK is limited by the use of composite mapping units. The area an ecotope occupies within a region, for example, a factor that can help to determine the rarity of an ecotope, cannot be directly derived from the BWK. In this case study we will estimate the area of ecotope Pa (very old coniferous woods with no undergrowth) in the study area defined higher, based on the technique using conditional probabilities as well as on the technique using affinity-corrected probabilities.

The ecotope Pa occurs within a number of complexes throughout the study area, and sporadically as a separate mapping unit. The complexes within the study area containing Pa are Pa+Pmb, Pmb+Pa, Pa+Pop, and Pmb+Pa+Qb. Table 13.5 shows the probability of occurrence of Pa and Pmb and their predicted area within the complex Pmb+Pa, based on the strategy using conditional probabilities. Because throughout the study area, the frequency of occurrence of Pmb is large in comparison with the frequency of occurrence of Pa, Pmb is predicted to occupy a much larger part of the area of the complex than Pa, which is in contradiction with the recorded order of dominance of the ecotopes. When estimating the total area of an ecotope in a region, these contradictions will lead to an underestimation of the area of rare ecotopes, such as Pa, and an overestimation of the area occupied by the more common ecotopes, such as Pmb. Because these rare ecotopes are often of great importance to nature conservation, a correct estimate of the area of

precisely these ecotopes is requisite. Table 13.6 shows the probability of occurrence of Pa and Pmb and their predicted area within the complex Pmb+Pa, based on the strategy using affinity-corrected probabilities. The predicted area of Pa within the complex is now larger than the predicted area of Pmb, which corresponds with the recorded order of dominance of the ecotopes within the complex.

We can estimate the total area of ecotope Pa within the study area from the predicted area Pa occupies within all complexes containing Pa and the area of Pa not contained in complexes. Table 13.7 shows the predicted area of Pa within all complexes containing Pa and the corresponding estimate of the total area of Pa within the study area, based on the predictions using conditional probabilities as well as on the predictions using affinity-corrected probabilities. Because Pmb is present in most complexes containing Pa, and Pmb is much more common than Pa is throughout the study area, the strategy based on conditional probabilities predicts a total area of Pa that is much lower than the area estimated from the predictions based on affinity-corrected probabilities. Presumably the strategy based on conditional probability underestimates the area of Pa, and the estimate based on affinity-corrected probabilities approximates the true area of the ecotope more closely. Of course, field verification is necessary in order to evaluate the correctness of this assertion.

CONCLUSIONS

The Biological Valuation Map of Belgium (BWK) is a categorical data coverage presenting a standardized survey of the biotic environment in Belgium. In the BWK, ecotopes were often merged together to form composite mapping units called complexes. The exact ecotope to which a location within a complex belongs is therefore uncertain. An error index describing the type and magnitude of thematic uncertainty in the BWK

Table 13.5. Probability of Occurrence and Area of Pa and Pmb in a Complex Pa+Pmb (Conditional Probability).

Soil Type[a]	Probability of Occurrence		Area (in m²)	
	Pa[b]	Pmb[b]	Pa	Pmb
OT	0.053	0.947	200	4000
Sbm	0.031	0.969	200	5500
Scm	0.088	0.912	1700	17800
Sdc	0.572	0.428	1000	700
Sdp	0.588	0.412	7400	5200
Sdm	0.171	0.829	2000	9500

[a] Soil type: Texture: OT: disturbed soils; S: loamy sand.
        Moisture: b: dry; c: moderately dry; d: moderately moist.
        Profile: m: deep humus A horizon; c: texture B horizon;
        p: no profile developed.
[b] Ecotopes: Pa: very old coniferous woods without undergrowth;
        Pmb: old coniferous woods with undergrowth of shrubs.

Table 6. Affinity-Corrected Probability of Occurrence, and Area of Pa and Pmb in a Complex Pa+Pmb.

Soil Type[a]	Probability of Occurrence		Area (in m²)	
	Pa[b]	Pmb[b]	Pa	Pmb
OT	0.638	0.362	2700	1500
Sbm	0.598	0.402	3400	2300
Scm	0.694	0.306	13500	6000
Sdc	0.954	0.046	1600	100
Sdp	0.809	0.191	10200	2400
Sdm	0.808	0.192	9300	2200

[a] Soil type: Texture: OT: disturbed soils; S: loamy sand.
        Moisture: b: dry; c: moderately dry; d: moderately moist.
        Profile: m: deep humus A horizon; c: texture B horizon;
        p: no profile developed.
[b] Ecotopes: Pa: very old coniferous woods without undergrowth;
        Pmb: old coniferous woods with undergrowth of shrubs.

Table 13.7. Area of Pa in the Study Area.

Complex[a]	Area Pa (in m²)	
	Conditional	Affinity
Pa+Pmb	12500	40700
Pmb+Pa	9200	25600
Pms+Pa+Qb	11900	9600
Pa+Pop	46600	50300
Pa	172700	172700
Total Area	**252800**	**298900**

[a] Pa: very old coniferous woods without undergrowth;
    Pmb: old coniferous woods with undergrowth of
    shrubs; Pms: old coniferous woods with heath; Pop:
    poplar; Qb: Querco-Betuletum.

would make it possible to assess the probability that a location within a complex belongs to a particular ecotope. Because the BWK is not accompanied by such an error index, a strategy to derive a description of thematic uncertainty from a secondary data source was developed. Because the occurrence of ecotopes is assumed to be closely related to the underlying soil type, the strength of the relation between ecotopes and soil types was considered to be appropriate to describe the probability of occurrence of each ecotope within a complex.

The conditional probability of an ecotope occurring, given the soil type, is one way of expressing the strength of the spatial relationship between ecotopes and soil type. From the conditional probabilities, the probability of occurrence of every ecotope within a complex can be derived, based on the soil types underlying the complex. As we demonstrated, the conditional probability of an ecotope for a soil type is too strongly dependent on the frequency of occurrence of an ecotope throughout, and therefore does not allow mutually comparing the preference of different ecotopes within a complex for a soil type.

An alternative measure expressing the affinity of an ecotope for a soil type was introduced to bypass this problem. The affinity value of an ecotope is used to locally correct a prior estimate of the probability of occurrence of an ecotope within a complex. Calculating the probability of occurrence of every ecotope within a complex, based on the affinity-corrected probability, respects the recorded dominance of the ecotopes within a complex. Field verification is necessary to check the validity of the estimates.

ACKNOWLEDGMENTS

The research presented in this chapter was funded by the Flemish Institute of Scientific and Technological Research for the Industry (IWT). The authors would also like to thank Desiré Paelinckx of the Institute of Nature Conservation and Johan De Smet of the Research Station of Agricultural Engineering for providing the digital data coverages used in this research.

REFERENCES

Aspinall, R.J. and D.M. Pearson. A Method for Describing Data Quality for Categorical Maps in GIS, in *EGIS/MARI '94: Proceedings of the Fifth European Conference and Exhibition on Geographical Information Systems, Paris, 1994,* EGIS Foundation, Utrecht, The Netherlands, 1994, pp. 444–453.

Canters, F. Error Modelling in Geographical Information Systems: English Summary, Research Report T/III/03/004, Belgian Office for Scientific, Technical and Cultural Affairs, Brussels, 1995.

De Blust, G., A. Froment, E. Kuijken, L. Nef, and R. Verheyen. *Biologische Waarderingskaart van België: Algemene Verklarende Tekst,* Ministry of Public Health, Brussels, Belgium, 1985, p. 98 (in Dutch).

De Blust, G., D. Paelinckx, and E. Kuijken. Up-to-Date Information on Nature Quality for Environmental Management in Flanders, in *Ecosystem Classification for Environmental Management,* F. Klijn, Ed., Kluwer, Dordrecht, The Netherlands, 1994.

Fisher, P.F. Knowledge-Based Approaches to Determining and Correcting Areas of Unreliability in Geographic Databases, in *Accuracy of Spatial Databases,* M.F. Goodchild and S. Gopal, Eds., Taylor & Francis, London, England, 1989.

Fisher, P.F. Modelling Soil Map-Unit Inclusions by Monte Carlo Simulation. *Int. J. Geogr. Inf. Syst.,* 5, pp. 193–208, 1991.

Goodchild, M.F., S. Guoqing, and Y. Shiren. Development and Test of an Error Model for Categorical Data. *Int. J. Geogr. Inf. Syst.,* 6, pp. 87–104, 1992.

Goodchild, M.F., B. Buttenfield, and J. Wood. Introduction to Visualizing Data Validity, in *Visualization in Geographical Information Systems,* H.M. Hearnshaw and D.J. Unwin, Eds., Wiley & Sons, Chichester, England, 1994.

Lanter, D.P. and H. Veregin. A Research Paradigm for Propagating Error in Layer-Based GIS. *Photogrammetric Eng. Remote Sensing,* 58, pp. 825–833, 1992.

Openshaw, S. Learning to Live with Errors in Spatial Databases, in *Accuracy of Spatial Databases,* M.F. Goodchild and S. Gopal, Eds., Taylor & Francis, London, England, 1989.

Paelinckx, D., H. Heyrman, M. Van Hove, R.F. Verheyen, and E. Kuijken. The GIS Data Base Biological Evaluation Map for Flanders: Construction and Applications, in *EGIS '91: Proceedings of the Second European Conference and Exhibition on Geographical Information Systems, Brussels 1991,* EGIS Foundation, Utrecht, 1991, pp. 826–832.

Paelinckx, D., T. Van Tilborgh, D. Van Straeten, and G. De Blust. Towards an Integration of the Biological Valuation Map and Soil Maps, in *EGIS/MARI '94: Proceedings of the Fifth European Conference and Exhibition on Geographical Information Systems, Paris, 1994,* EGIS Foundation, Utrecht, The Netherlands, 1994, pp. 2062–2063.

Robinson, A.H., R.D. Sale, J.L. Morrison, and P.C. Muehrcke. *Elements of Cartography,* 5th ed., John Wiley & Sons, New York, 1984.

Van Ghelue, P., K. Decleer, G. De Blust, D. Paelinckx, and E. Kuijken. Aanzet tot een Regionaal Landschapsecologisch Model (RELEM) voor het Gebruik in de Landinrichting, Research Report A93.91, Institute of Nature Conservation, Brussels, Belgium (in Dutch).

Van Straeten, D., D. Paelinckx, and T. Van Tilborgh. GIS en Natuuronderzoek in Vlaanderen, in *GIS en* *Leefmilieu: Notulen van de Studiedag GIS en Leefmilieu, Brasschaat,* Flemish Association for Geographic Information Systems, Brussels, 1993, pp. 75–83 (in Dutch).

CHAPTER 14

Encapsulating Simulation Models with Geospatial Data Sets

M.F. Goodchild, A.M. Shortridge, and P. Fohl

INTRODUCTION

Differences exist between real world phenomena and their digital portrayal in geospatial data sets. There is general agreement that these differences must be described and reported along with the data, so that users can make informed decisions on the fitness of the data for specific applications. There is "a strong need...to obtain detailed understanding of how errors propagate through the large number of possible combinations of model types, data types, data sources, and kinds of error, and to make this available to users in an easily accessible form" (Burrough et al., 1996). Some have argued that indeed, in the absence of metadata accuracy reports, spatial data are virtually useless (Smith et al., 1996).

Accuracy reporting typically consists of summary statistics derived from ground measurements upon a subsample of the data. For land cover maps derived from remotely sensed images, this might be the percent correctly classified for each category (Lunetta et al., 1991). For a digital elevation model, the statistic might be a root mean square error (RMSE) for a set of locations at which the true elevation is known (Shearer, 1990). These sorts of global measures of uncertainty are inadequate by themselves for analysis of uncertainty, since they provide no information about spatial structure. Indeed, map and data accuracy standards in general are not sufficient to characterize the spatial structure of uncertainty (Goodchild, 1995; Unwin, 1995).

The current paradigm holds that data producers are responsible for providing such (often inadequate) summary statistics with their data, and that data us-

ers are responsible for translating these statistics into meaningful estimates for the suitability of these data for their applications. Just what users are expected to make of these summary reports is unclear; how, for example, does a forester use RMSE to decide whether a particular elevation data set is suitable for fire tower site selection? Openshaw (1989) described general simulation approaches to modeling uncertainty in spatial data for geography, and the past decade has seen considerable progress. This research supports the notion that the general simulation and error propagation method is a complete characterization of uncertainty in spatial data and its effects on analysis. However, these methods remain both theoretically and technically challenging to implement for most spatial data users.

This chapter describes a new paradigm for both data producers and data users. Under this paradigm, data producers replace current accuracy information with an "uncertainty button" in metadata. The button ties an appropriate simulation method to the data quality report. In essence, the button becomes the accuracy metadata; the method replaces the measure. Data users adopt a new view of spatial data; instead of employing the original dataset for an application, they will use one or more realizations to produce a distribution of potential outcomes. By studying this distribution, users gain an understanding of how uncertainty in the data affects their application. Enhanced spatial operations in GIS will facilitate this approach to data handling, and provide more sensitive methods for understanding what is known about the real-world phenomenon modeled by the data.

The following section of the chapter reviews the current metadata accuracy reporting method, and describes the simulation/propagation approach for characterizing uncertainty. We suggest that this be substituted for traditional metadata reporting in the form of an uncertainty "button." The third section introduces various data examples to illustrate the proposed approach. The chapter concludes with a discussion of prospects and challenges for this framework.

METADATA CHARACTERIZATION OF SPATIAL DATA UNCERTAINTY

Much spatial data production, particularly that of federal government agencies like the USGS, is now impacted by a range of metadata specifications. The objective of the development of these specifications is to enhance the sharing of spatial information, to encourage consistency in data generation and use, and to reduce redundancy in data compilation (SDTS, 1996). Government agencies engaged in spatial data production subject their data to accuracy assessments, typically disqualifying any that fail to meet quality specifications and reporting summary information from the assessments in metadata reports for data users. These reports summarize the quality of the data as it relates to some predefined specification. As an example, consider a USGS level 2 digital elevation model (DEM). An approved DEM file must have an RMSE of less than one-half of the source contour interval, with no error exceeding one contour interval (USGS, 1995). With regard to these reports, then, producers are primarily concerned that their data meets a somewhat abstract measure of accuracy.

In contrast, data users are normally not interested in the accuracy of the data set itself, but rather in the spatial phenomenon that the data set imperfectly represents. They need to know how imperfect this representation is, as it relates to their applications. Consider a forester who wishes to use a USGS DEM to help identify promising sites for a new fire tower. The forester has calculated the size of the viewshed for a set of locations, and is interested in determining how closely the calculated viewshed matches the actual viewshed at these sites. That the RMSE for the quadrangle does not exceed 7 meters is not a detail which the forester can easily use to determine the quality of the viewshed calculations.

Indeed, analytically deriving the uncertainty of spatial attributes is frequently difficult or impossible.

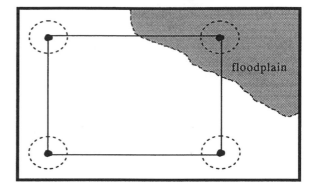

Figure 14.1. Land parcel; positional uncertainty of corner points indicated by circles.

Consider the relatively simple case presented in Figure 14.1. The area of the rectangular land parcel is defined by four corner points. According to the survey, these points are arrayed in a square one hundred meters on a side. However, the surveyed points are subject to positional uncertainty; this uncertainty is characterized by a Gaussian distribution with a mean of zero and a standard deviation of 10 meters, as depicted by the dashed circles. The application question is, what is the standard error associated with the area of the land parcel, given the positional uncertainty information?

In fact, this can only be calculated directly from the available information with some difficulty (Griffith, 1989). However, the standard error may be estimated more simply through a Monte Carlo simulation procedure, which would proceed as follows (and as illustrated in Figure 14.2). Positional error is simulated for each corner using a distribution meeting the criteria specified above. The resulting quadrilateral is a potential realization of the actual parcel. The area of this quadrilateral is calculated and stored. Then, positional error is simulated again, and the area is again noted. This process is repeated a large number of times. For each realization, uncertainty in position of the corners is propagated to variation in parcel area. By analyzing the resulting distribution of area measurements, one can estimate the standard error and characterize the variation in area due to the positional uncertainty of the corner points. Figure 14.3 portrays a histogram of areal estimates derived from 100,000 simulations. The simulation method is general, in the sense that uncertainty can be propagated to answer other questions as well. For the parcel example, the following questions might be of interest and could be answered: what is

Figure 14.2. Error propagation approach.

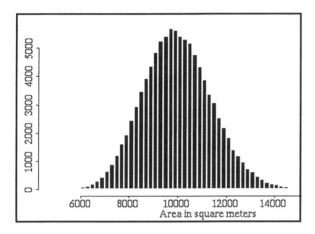

Figure 14.3. Distribution of parcel areas from 100,000 simulations. Mean is 10,000.4, standard deviation is 1427.

the chance that the parcel area is smaller than 9,000 square meters? How likely is it that more than 10% of the parcel is actually in the floodplain? A working prototype for this example is available on-line at:

http://www.ncgia.ucsb.edu/~ashton/demos/ propagate.html

These examples—the dilemma of the forester, as well as the parcel area puzzle—illustrate that traditional metadata summaries generally fall short of providing adequate measures of spatial data uncertainty to the user. The general Monte Carlo propagation approach demonstrated above, coupled with data-specific uncertainty simulation algorithms, appears to be the most adequate way of expressing what is known about some spatial phenomenon by combining the data collected about the phenomenon with relevant data quality information (Heuvelink et al., 1989; Fisher, 1991; Lee et al., 1992; Englund, 1993; Ehlschlaeger et al., 1997). For characterizing the uncertainty due to imperfect spatial data in many applications, the user requires a

set of equally probable simulations of the spatial phenomenon rather than an incomplete set of summary statistics and a data set known to be in error. In a sense, the simulations themselves become the uncertainty metadata, since the user can see the variation between them, as well as the distribution of application results across the realizations.

We propose that the responsibility for providing these simulations—as with metadata in general—rests with the data provider, not the user, since the provider has much more information concerning the quality of the data and is more equipped to perform an accuracy assessment sensitive to measuring spatial patterns of error. Additionally, simulation theory and techniques are challenging topics for most spatial data users, whose areas of expertise lie more typically with the phenomenon the data represent. The data producer can bridge this knowledge gap for the user community by encapsulating an appropriate simulation method within the metadata accompanying the spatial data set. At the U.S. federal level, at least, the mandate for this exists; data quality specification documents emphasize the responsibility of USGS data producers to "report what data quality information is known," so that users can make informed decisions about the applicability of the data for their applications (USGS, 1996).

What would such a metadata record look like? Figure 14.4 shows what might appear on the computer monitor when a user is electronically browsing a spatial data library. An "uncertainty button," following the GIS error handlers of Openshaw (1989), replaces the usual statistic or table. A short simulation algorithm replaces a line or two of text, or a number, in the record. When the user presses this uncertainty button, the specified number of simulations are generated using the producer-specified uncertainty model and simulation algorithm. These simulations are then processed by the user's GIS and a distribution of results is returned.

GIS operation functionality must be enhanced to effectively incorporate this information about uncertainty from the many data realizations. It is obvious that the main difference in computation is that the same operation must be performed n times, where n is the number of realizations. A somewhat more difficult step is deciding what the operation should return. Table 14.1 presents three very typical results from a GIS operation in the left-hand column. The central column suggests what the results might be from a compound operation, performed upon a set of realizations. The final column provides an example of each type of op-

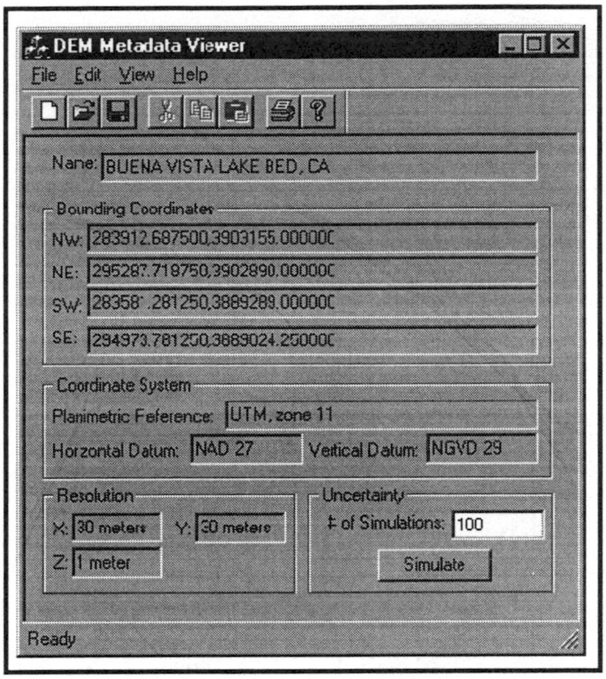

Figure 14.4. Example metadata browser window. Clicking the "Simulate" button generates DEM realizations for uncertainty propagation.

Table 14.1. Output from Different GIS Operations upon Uncertain Spatial Data.

Traditional Result	Result Incorporating Uncertainty Information	Example
Number	Mean, Standard Deviation	Query: Polygon Area
Object	Probability Field	Calculate: Buffer
Surface	Animation Frames	Generate: Cost Surface

eration. The extension of GIS functionality necessary for the implementation of uncertainty propagation is beyond the scope of the present work, but it is certainly realizable.

ILLUSTRATIVE EXAMPLES OF THE APPROACH

The previous section introduced a concept for incorporating uncertainty simulation directly into the metadata associated with an individual spatial data file. Three examples are presented here to indicate how this general method could be implemented. Point, line, and surface area data models are represented in these ex-

amples, to demonstrate the breadth of spatial data types that are amenable to this approach.

Scattered Point Data

The first example is a map of tree locations in a forest. Studies of seed dispersal for this species of tree indicate that the maximum range of dispersal from any individual is 100 meters. The data consist simply of the coordinate locations for each tree. Spatial uncertainty in this data is limited to the positional uncertainty of these coordinates. For simplicity, we assume that the organization producing these data has determined the positional uncertainty to be isotropic, with a Gaussian distribution centered on each observed location and a specified standard deviation. Uncertainty is independent for each point. An application question for these data is, given the uncertainty in the tree locations, how likely is it that all tree locations are within 100 meters of one another?

Coastline Data

The second data set is a vector coastline for a section of the central California coast near Point Conception. The data set itself is a "mean line," an average of several coinciding coastline data sets. Of potential importance is the notion that the mean line is not itself a potential coastline, due to the smoothing effect of the averaging (Goodchild et al., 1995). Many applications may require coastlines of statistically realistic texture rather than the smoother mean. The generation of such coastlines from the mean line requires simulation of removed variation, and is complicated by the prospect that this variation is spatially autocorrelated. The simulation model proposed in Goodchild et al. (1995) uses a distance decay exponent and a range parameter to characterize spatial autocorrelation of the variation about the coastline. Their model is considerably more complex to implement than the preceding point data model. However, the byte size of the algorithm code itself is not large, and could easily be transmitted with the data. An application question of interest is, how long is this stretch of coastline? A mean length (which is not equivalent to the length of the mean coastline), and standard deviation is returned.

DEM Data

The third example uses digital elevation data. The data set is a subset of the USGS one-degree DEM, Los Angeles-west. Studies comparing these data to higher resolution and accuracy collocated 7.5' USGS DEMs have developed measures of mean and variance of the difference, and of the spatial structure of this difference (Ehlschlaeger et al., 1997). An uncertainty model has been developed that uses this information to produce realizations of the difference surface. Each difference surface realization is then added to the one-degree DEM, creating a statistically probable simulation of the "actual" 7.5' DEM-quality surface. If 7.5' DEM data are adequate for a particular application, then this model creates a set of potential realizations of adequate surface representations. This is

an unconditional simulation, meaning that no locations on the DEM are necessarily spared from perturbation. An application question for these data is, what is the expected cost of a least-cost path traversing the terrain, where cost is a function of path length, path steepness, and elevation range of the path?

In each of the three cases, the uncertainty model/ simulation algorithm is encapsulated in the metadata in the form of a simulation button. Users pressing this button generate a series of simulated data sets. Through the error propagation approach, the questions proposed in these illustrations, and many others, can be answered. These answers come with confidence intervals or other measures of reliability, providing a more realistic depiction of the effects of data uncertainty on the application question.

DISCUSSION

Taken together, the three data examples are representative of much of spatial data. We chose two different object data models and a field data model. The simulation models chosen are also representative. While the first, operating on the point data, was spatially independent for each location, the remaining two simulation models directly accounted for spatial dependence in the error field. The approach advocated in this chapter is very general and extendable to any spatial data set that can be stored in a computer and can be assessed for its fidelity to the phenomenon it represents.

Several critical issues present themselves. The first is the choice of spatial uncertainty model. A growing body of research on spatial uncertainty modeling indicates the diversity of approaches, methods, and results. In the face of such diversity, how is a data producer to choose the most acceptable model? On the other hand, how is the resource manager, the ecologist, or the environmental engineer to choose? These users undoubtedly lack expert knowledge about both the data collection methods employed by the data producer and the spatial simulation model theory and implementation in vogue with spatial information scientists. By working with uncertainty modelers, data producers are in the best position to decide upon the most effective simulation approaches for specific spatial data sets. Data users can have increased confidence both in the uncertainty simulation models and in the data itself. Research on simulation model efficacy must be done to enable data producers to make informed decisions about which models to use, and to indicate needed changes to accuracy assessments to accommodate model requirements.

A second research topic concerns the distribution of computer processing for simulations. Data will be stored and queried in digital libraries. However, when the user wishes to check the uncertainty of the data, and "clicks the button," what should actually happen? One possibility is that the library maintains a large number of stored realizations for each data file. This seems unwieldy, particularly in light of the continuing rapid increase of processor speed. Instead, realizations could be generated on the fly. Where should the generation occur—at the library site or on the user's machine? From a computational perspective, it might make sense for the processing to occur on the user's machine. In this case, users would download the data file, bundled with an executable simulation routine, and generate simulations locally.

Geographic information systems algorithms require some modification under this paradigm, since they must work on multiple realizations and return meaningful, clear results. Table 14.1 identifies some relatively straightforward outputs of traditional operators and their "uncertainty-enhanced" counterparts. Research topics remain; for example, how will this method fare in compound spatial analysis, in which a large number of input data layers are combined using numerous spatial operations? How can the output of one GIS function easily be used as the input to another? How can the contribution of uncertainty from different spatial sources be easily quantified and expressed to the user? Significant representation issues arise as well. How can information about uncertainty best be communicated and understood? Which, if any, spatial models are especially resistant to effective characterization and communication of uncertainty?

Traditional metadata accuracy reports must change. Those who use spatial data increasingly demand to know how reliable their GIS results are, and standard accuracy statistics are inadequate to supply answers. Simulation-based uncertainty models have been developed for spatial data, but they remain difficult to understand and utilize for most end users. We have argued that the producer, not the user, should be responsible for providing adequate measures of spatial data uncertainty; by adequate, we mean encapsulating the simulation algorithm with the data set. This approach was demonstrated on three representative illustrations. While many challenges remain, we believe that this chapter has introduced and demonstrated a viable, general solution for adequately reporting spatial data uncertainty.

REFERENCES

Burrough, P.A., R. van Rijn, and M. Rikken. Spatial Data Quality and Error Analysis Issues: GIS Functions and Environmental Modeling, in *GIS and Environmental Modeling: Progress and Research Issues,* M. Goodchild et al., Eds., GIS World Books, Fort Collins, CO, 1996, pp. 29–34.

Ehlschlaeger, C.R., A.M. Shortridge, and M.F. Goodchild. Visualizing Spatial Data Uncertainty Using Animation. *Comput. Geosci.,* 23(4), pp. 387–395, 1997.

Englund, E.J. Spatial Simulation: Environmental Applications, in *Environmental Modeling with GIS*, M.F. Goodchild, B.O. Parks, and L.T. Steyaert, Eds., Oxford Press, New York, 1993, pp. 432–437.

Fisher, P.F. First Experiments in Viewshed Uncertainty: The Accuracy of the Viewshed Area. *Photogrammetric Eng. Remote Sensing,* 57(10), pp. 1321–1327, 1991.

Goodchild, M.F. Attribute Accuracy, in *Elements of Spatial Data Quality,* S.C. Guptill, and J.L. Morrison, Eds., Elsevier, London, 1995, pp. 59–79.

Goodchild, M.F., T.J. Cova, and C.R. Ehlschlaeger. Mean Objects: Extending the Concept of Central Tendency to Complex Spatial Objects in GIS, in *Proceedings, GIS/LIS '95,* ASPRS/ACSM, Nashville, TN, 1995, pp. 354–364.

Griffith, D.A. Distance Calculations and Errors in Geographic Databases, in *Accuracy in Spatial Databases,* M.F. Goodchild and S. Gopal, Eds., Taylor & Francis, London, 1989, pp. 81–90.

Heuvelink, G.B., P.A. Burrough, and A. Stein. Propagation of Errors in Spatial Modelling with GIS. *Int. J. Geogr. Inf. Syst.,* 3(4), pp. 303–322, 1989.

Lee, J., P.K. Snyder, and P.F. Fisher. Modeling the Effect of Data Errors on Feature Extraction from Digital Elevation Models. *Photogrammetric Eng. Remote Sensing,* 58(10), pp. 1461–1467, 1992.

Lunetta, R.S., R.G. Congalton, L.K. Fenstermaker, J.R. Jensen, K.C. McGwire, and L.R. Tinney. Remote Sensing and Geographic Information System Data Integration: Error Sources and Research Issues. *Photogrammetric Eng. Remote Sensing,* 57(6), pp. 677–687, 1991.

Openshaw, S. Learning to Live with Errors in Spatial Databases, in *Accuracy in Spatial Databases,* M.F. Goodchild and S. Gopal, Eds., Taylor & Francis, London, 1989, pp. 263–276.

SDTS Task Force. The Spatial Data Transfer Standard: Guide for Technical Managers. U.S. Dept. Interior, 1996, <ftp://sdts.er.usgs.gov/pub/sdts/articles/pdf/mgrs.pdf>

Shearer, J.W. Accuracy of Digital Terrain Models, in *Terrain Modelling in Surveying and Civil Engineering*, G. Petrie and T.J.M. Kennie, Eds., Thomas Telford, London, 1990, pp. 315–336.

Smith, T.R., D. Andresen, L. Carver, R. Dolin et al. A Digital Library for Geographically Referenced Materials. *Computer*, 29(7), pp. 54, 1996.

Unwin, D.J. Geographical Information Systems and the Problem of 'Error and Uncertainty.' *Prog. Human Geogr.*, 19(4), pp. 549–558, 1995.

USGS. DEM/SDTS Transfers, in *The SDTS Mapping of DEM Elements*, U.S. Dept. Interior, 1996, <ftp://sdts.er.usgs.gov/pub/sdts/datasets/raster/dem/demmap3.ps>

USGS, National Mapping Program Technical Instructions, Standards for Digital Elevation Models, U.S. Dept. Interior, 1995.

CHAPTER 15

Formulation and Test of a Model of Positional Distortion Fields

C. Funk, K. Curtin, M. Goodchild, D. Montello, and V. Noronha

INTRODUCTION

Hunter and Goodchild (1996) and Kiiveri (1997) have proposed a simple model of positional distortion in geographic data sets to account for the differences between true and measured locations (throughout this chapter *true* should be taken to mean *as determined from a source known to be of higher accuracy*). A point's true location (x,y) is distorted by the addition of a small vector displacement (e_x, e_y) to give the measured location of the point $(x+e_x, y+e_y)$. The displacements vary, but are assumed to be strongly spatially autocorrelated; collectively they form a vector field **e** whose components at any point are (e_x, e_y). In order to avoid folding or ripping, it is necessary that the components of **e** be constrained such that there are no cliffs in either surface, and such that their derivatives with respect to x and y, respectively, are always greater than -1.

In order to detect positional distortion it must be possible to match point locations in the data set in question to points in another data set of higher accuracy (see Church et al., 1998, for an extensive discussion of this issue). This is often difficult, particularly if there is ambiguity in the definition of mapped features, such as often exists in the natural resources area. For example, it would be very difficult to match points between two versions of a soil map, or even to agree that one was of higher accuracy than the other. Higher accuracy is often associated with a larger scale, but if the scale is different between the two data sets there would be no reason to expect pairs of features to match at all. On the other hand, location matching is much more

reasonable in the case of well-defined cultural features, particularly when two sources exist at the same scale for the same area but with independent lineages. We consider such a case in this chapter.

In most industrial countries it is now possible to obtain so-called *street centerline* databases, containing representations of the approximate centerlines of the road and street network. Such databases are of immense value for emergency response, road maintenance, delivery and collection services, routing and navigation, and many other functions. Their importance is such that a significant commercial production and dissemination sector has emerged. In the United States, this industry initially added value to the public-sector TIGER database (produced for the 1980 census through a collaboration between the U.S. Bureau of the Census and the U.S. Geological Survey), and many products still have a TIGER legacy. More recently, however, companies have tended to build from scratch, beginning with aerial photography or vehicle-mounted GPS, in order to obtain higher positional accuracy. At this time, therefore, it is possible to obtain several databases covering a given area of the U.S., with identical features in varying positions. The research described in this chapter is based on an analysis of six such databases for an area of Goleta, California.

There are many practical motivations for wanting to determine the positional distortion between two databases. If one knew **e**, one could in principle correct one database to the other, removing any positional differences. Our research is driven by the problem of interoperation between street centerline databases in *in-*

telligent transportation systems (ITS). Consider the following scenario: a vehicle is being driven through an area, and information is being provided to the driver, including driving directions, from a system in the vehicle that includes a copy of Database A. The system is receiving updates on the state of the road system from a central server, maintained by the Department of Transportation, working off Database B. The position of an accident is broadcast from the server, as a coordinate location. The system in the vehicle attempts to match the location to its own street network, but the match fails because the positional difference between the two databases is sufficient to allocate the accident to the wrong street. Our analyses of the databases in the test area show that differences are frequently this large.

In this chapter we present an analysis of the differences between two of these databases, and test a model of **e**. We show how this model might be used to support error-free interoperation of the two databases using the concept of an *ITS Datum*, a proposed sparse network of control points.

SAMPLING THE ERROR FIELD

Figure 15.1 shows two of the databases for a small part of the study area. The error field was sampled by determining its two components at street intersections, after matching such intersections by the names of the intersecting streets. As in all such databases, it was necessary to deal with differences in the naming and spelling of streets, and with problems caused by multiple intersections between the same pairs of streets. In some areas, particularly the lower left, the magnitudes of the positional differences clearly approach the distances between adjacent streets. The two sets of streets are shown superimposed on an interpolated field of distortion magnitude, or $(e_x^2 + e_y^2)^{1/2}$.

While this method of intersection matching works reasonably well in this case, it clearly will cause problems in rural areas where intersections are few and far between, or in cases where the names of streets are not available. In such cases it may be possible to find match points entirely from geometric information, but this seems inherently more risky and was not pursued in this study.

MODELS OF THE ERROR FIELD

Node matching provides an irregularly-spaced sample of evaluations of **e**. In order to make a complete adjustment of one database to fit the other, we

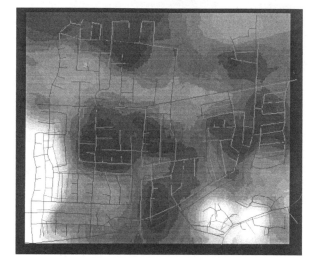

Figure 15.1. Plot of a section of the two databases, superimposed on an interpolated field showing the magnitude of the distortion vector.

need to have exhaustive rather than sample knowledge of **e**. Kiiveri (1997) argues that **e** must be everywhere smooth, and fits simple polynomial and trigonometric functions of the coordinates to both components of **e**. Church et al. (1998) argue that any of a wide range of spatial interpolation methods might also be used; they argue that they are to be preferred over Kiiveri's functions because we have no reason to suppose that **e** is a simple function of the coordinates. In this chapter we focus on the smoothness condition, and argue that it may be unwarranted for two reasons.

Consider first the processes by which such databases are made, and the likely implications of those processes for positional error. Three methods are considered here. First, a street centerline database can be constructed from existing pieces, such as digitized subdivision maps, digitized topographic maps, or existing databases. When such fragments are merged, one can expect differences of registration to show up as failure to match along the fragment edges. But edgematch problems will only appear in areas where streets or other features cross fragment edges; where features are absent such problems may be invisible, or may simply be ignored. In effect, the result is a mosaic of fragments, with distortions that persist within fragments and change sharply at fragment edges.

Second, consider a database constructed from aerial photography. Here the significant sources of positional inaccuracy relate to registration, and to the practice of merging fragments of aerial photographs into a single mosaic. Again, we can anticipate that misregistration errors will persist over areas that depend on the same set of registration points, and change sharply at edge boundaries; and that errors at edge boundaries will be detected only when edges are crossed by features.

Finally, consider a database constructed by driving streets with in-vehicle GPS. Errors in GPS positions are known to be strongly autocorrelated through time and to change sharply when the set of observed satellites changes. While the result will clearly depend on the strategy used to traverse the street network, it seems that this method also is capable of producing sharp discontinuities in the error field. In summary, there are good reasons to believe that the positional error field of a street centerline database will show jumps and discontinuities under each of the standard methods of production.

Consider now the implications of discontinuities in the error field. A discontinuity is defined by the condition that $e(s+\delta s) - e(s)$ tends to some finite vector $\mathbf{a}$ as the magnitude of δs tends to 0; in other words, two points an infinitely small distance apart can have different positional error vectors. It was argued earlier that cliffs could not exist in a distortion field because any such cliff would introduce a sharp break in any linear feature that crossed it. Let $\mathbf{s}$ and $\mathbf{s+ds}$ denote two points on a linear feature a small displacement $\mathbf{ds}$ apart. Then the linear feature will show a sudden horizontal offset or jump with a direction and magnitude defined by $\mathbf{a}$. If this occurred during the production of the database, the producer could adopt one of two strategies: (1) smooth out the jump by editing the feature (Figure 15.2), or (2) revise the production process. But if no feature crossed the cliff, the cliff would go undetected.

The traditional gridiron city of the nineteenth century was highly connected, with interruptions in the network only along rivers, ravines, railroads, or other features too difficult to cross. In the mid-twentieth century, urban planning practice changed to a new model, with high connectivity within subdivisions but low connectivity between them and along major arteries. Access points to subdivisions were limited, either for reasons of security (the gated community being an extreme example), or to limit access to major arteries. The practice is also convenient commercially, if stores can be positioned at or near subdivision access points.

In summary, we propose a model of the distortion field that is very different from that of Kiiveri (1997), or that implied by the interpolated surface of Figure 15.1. In our model, the plane is partitioned into a num-

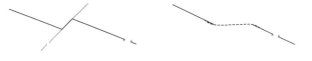

Figure 15.2. Effect of a cliff on a linear feature (left); editing with a smooth line (right).

ber of zones; within each zone, $\mathbf{e}$ is constant. We term this the *piecewise constant* model. The model generalizes readily to the case where $\mathbf{e}$ is a piecewise function of the coordinates, as for example where the distortion within one zone is described by an affine transformation, but we do not consider that generalization here. In the next section we implement the model using a novel clustering technique based on geostatistical principles.

CLUSTERING THE ERROR FIELD

"Regioning" seems to be a natural component of human thought—we speak of this place and that with confidence and regularity. Yet quantitative methods for automatically locating these regions given a set of geographic data are still insufficient to cope elegantly with many real-world problems. Geographic clustering problems are "hard," both mathematically and statistically. Posed as optimization problems, clustering problems are part of the larger class of location-allocation problems whose computational complexity rises exponentially with the number of observations. Posed within a statistical framework, spatial clustering problems immediately inherit the set of difficulties attached to most spatial datasets, chief of which is the lack of independence among nearby observations. Geostatistics is the branch of statistics which has most directly addressed the problem of spatial dependence between observations (e.g., Isaaks and Srivastava, 1989). It thus seems natural that spatial clustering should in some way benefit from the insights of geostatistics. That is the goal we pursue here.

We begin to build the bridge between geostatistics and clustering by making use of a common geostatistical tool, the empirical semivariogram. The semivariogram plots the variance of a pair of spatial observations as a function of their distance apart. The result is a summary statistic that describes the *average* variance between a given point and its neighbors, expressed solely as a function of the distances between that point and its neighbors. Since the variogram is a type of average (defined below), roughly half the data points will have variances above and half below the variogram value. We take advantage of this relationship to define a *ratio of areal dependence* (RAD) statistic, which has an expected value of 0. When calculated for each point and plotted, this statistic is a useful exploratory spatial data analysis technique, helping the analyst to assess the number and locations of clusters, if indeed they exist at all. In addition, we show that it can be used as a type of "variance filter," allowing us to arrive at more meaningful clusters by removing points with high RAD values.

The Angular Semivariogram

The spatial dependence of values may be measured by means of the semivariogram (Cressie, 1993; Deutsch and Journel, 1992; Isaaks and Srivastava, 1989; Matheron, 1963). The semivariogram produces bucket estimates of the variance between pairs of points at varying distances or lags h. The semivariance of the set of points at a given lag, $\gamma(h)$, defined for a stationary random process (Z), is described as:

$$\gamma(h) = \frac{1}{2N(h)} \sum_{(i,j)|h(i,j)=h} \left(Z_i - Z_j \right)^2 \qquad (1)$$

where $\gamma(h)$ is the semivariance for the set of points whose interpoint distances fall within the span of lag h, $(i,j)|h(i,j)=h$ denotes all the pairs of points whose distances place them in lag h, $N(h)$ is the total number of pairs found within this lag, and $(Z_i - Z_j)^2$ is the squared difference in the process between any two points separated by lag h.

The error field $\mathbf{e}$ is a vector field, with two components at any location, whereas the variogram is normally computed for a scalar field Z. In our cluster model, we hypothesize that $\mathbf{e}$ will be constant within each zone. For the purposes of this chapter, we assume that constancy of *direction* within each zone is an adequate test of the model, and ignore the possibility of variation in *magnitude*. But the difference between two angles cannot simply be described by subtracting one angle from the other. We now describe our proposed method for the computation of a semivariogram for angular data, based on principles of circular statistics (see Batschelet, 1981, for an excellent review of circular statistics).

Let $\mathbf{e}_i$ and $\mathbf{e}_j$ denote samples of the error field at two points i and j. We define $\mathbf{r}_{ij}$ as the mean of unit vectors in the directions of the error field at the two points; that is:

$$\mathbf{r}_{ij} = \left\{ \left(e_{xi}/|\mathbf{e}_i| + e_{xj}/|\mathbf{e}_j| \right)/2, \left(e_{yi}/|\mathbf{e}_i| + e_{yj}/|\mathbf{e}_j| \right)/2 \right\} \quad (2)$$

Thus if $\mathbf{e}_i$ and $\mathbf{e}_j$ are the same they will both equal $\mathbf{r}_{ij}$, but $\mathbf{r}_{ij}$ will have no magnitude when the directions are 180 degrees apart.

We define S^2 as one minus the magnitude of $\mathbf{r}$ and substitute it in the computation of the semivariogram, as follows:

$$\gamma(h) = \frac{1}{2N(h)} \sum_{(i,j)|h(i,j)=h} S_{ij}^2 \qquad (3)$$

Applying this equation to the distortion dataset previously discussed yields the empirical angular semivariogram shown in Figure 15.3.

This figure shows how the variation of the distortion angles increases with distance. There is a rather high inherent variability, with a minimum semivariance or nugget of about 0.15. The range is about 800 m; at this distance the semivariance reaches a sill of about 0.35.

The Ratio of Areal Dependence

We began this section by remarking that the semivariogram describes the average variance of observations at different lags. In this section we use this observation to construct a rather simple exploratory data analysis tool called the *ratio of areal dependence*, or RAD statistic. The RAD statistic has an expected value of 0.0. Positive values indicate that the local variance between a point and the set of neighbors found within the range is much higher than the expected variance for a set of points at those distances. Negative values depict the opposite. The RAD statistic is the ratio of the actual local variance at some point (V_i) divided by the expected (or modeled) variance at that point (M_i), or:

$$RAD_i = 1 - \frac{V_i}{M_i} \qquad (4)$$

Note that both V_i and M_i are measured between point i and a set of neighbors within a given range. Let N_r be the set of indices for all neighbors whose distance from the ith point is less than the range, and let n_r be the number of these points. Then the RAD statistic is a measure of actual versus expected local variance, where the actual local variance is defined as:

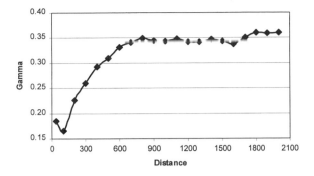

Figure 15.3. Semivariogram of angular distortion values.

Easting

Figure 15.4. A plot of the RAD values associated with angular distortions.

$$V_i = \left(\sum_{j=1}^{N_r} S_{ij}^2 \right) / n_r \qquad (5)$$

and the expected or model variance is:

$$M_i = 2 \left(\sum_{j=1}^{N_r} \gamma\left(d_{ij}\right) \right) / n_r \qquad (6)$$

A plot of the RAD values is shown in Figure 15.4. The filled circles have RAD values of greater than or equal to 0.0, indicating a higher than average spatial contiguity. The points denoted by triangles have relatively low (< 0.0) RAD values, indicating that they are relatively uncorrelated with their neighbors. This plot, and the RAD statistic in general, illustrate a fact sometimes forgotten by practitioners of spatial statistics: the variogram models average variance values, generated from ensembles of points at various lags. Individual points will tend to be more or less correlated with their neighbors than the variogram predicts. Those points which exhibit high spatial autocorrelation will have a *high* RAD, and vice versa. Thus regions with high RAD values designate natural regional patches which share similar values, and regions with high RAD values are dissimilar from their neighbors and are natural "barriers" demarcating one region from another.

Results

The angular variogram provides a statistically valid, somewhat interesting way to measure the spatial dependence of circular data. For our study region, the variogram exhibited a well-defined, highly localized spatial autocovariance structure, with points farther than 800 meters apart showing virtually no correlation. Even nearby intersections exhibited a marked degree of variance, as illustrated by the high nugget of the fitted variogram. Often, nearby streets exhibited dissimilar error vector orientation.

The RAD values plotted above depict the degree of spatial autocorrelation for the angular components of the distortion vectors. Visual inspection of Figure 15.4 indicates how successful any spatial clustering or "patchwise" modeling scheme is likely to be. Some areas, particularly the pockets of homogeneous street intersections in the upper left and lower right of the image, seem like natural "clusters," and are likely to have distortions created by a unique or similar process, which could be modeled with ease. Thus high RAD values suggest areas dominated by sources of "systematic" error. Other regions, such as the pockets of triangles in the top and bottom of the image, show what may be "random" error. Since the purpose of clustering is typically driven by a desire to model or elucidate systematic effects, we explored the effects of screening out the high RAD values in conjunction with an application of Ward's clustering algorithm. The idea is similar to that employed when removing outliers from a dataset, and the hope is that the resulting spatial partition depicts the systematic component of the variation.

As an example, Figure 15.5 shows a clustering of the distortion data. The following six fields were used to cluster the data: the *x* and *y* coordinates, the *x* and *y* components of the error vectors, and the magnitude and angle of the error vectors. Each of these variables was standardized to have a mean of zero and variance of one. The standardized data set was then clustered via Ward's method, a hierarchical clustering technique that seeks to minimize the squared distances (or trace) between the cluster centers and their constituent points (Hartigan, 1975).

This clustering is far from ideal. As a preliminary step in a patch-wise error estimation process, the clusters depicted above would be difficult to interpret and use, since they do not clearly partition the space into mutually exclusive regions. The classes appear to overlap in geographic space, and reappear sporadically throughout the image. There are often situations in which such a pattern is legitimate and even desirable. However, for the problem at hand, we have assumed that errors are, to some degree, spatially determined—nearby points in the same clusters should share the same source of error. This is where the RAD statistic can be of assistance. We suggested earlier that this calculated variable differentiates between points exhibiting "systematic errors" (high RAD values) and "random errors" (low RAD values). Furthermore, in our discussion of the angular variogram, we noticed that the error vectors appeared to exhibit a large nugget effect, from which we inferred that about 42% of the angular variance was not explained by the values' spatial distribution. If these conclusions are valid, then it seems reasonable to exclude points with low RAD values, since presumably these values are dominated by random as opposed to systematic effects. To examine the results of such a clustering we excluded points with RAD statistics of less than zero and reran our clustering on the remaining points. The results appear in Figure 15.6.

This clustering is appealing at an intuitive level. Note that the clusters effectively partition the geographic region, and demarcate well the various patches. Compared to the clustering performed on all the points, the results are much more spatially cohesive, which is not surprising, since we rejected all the data points which differed dramatically from their neighbors. Whether this rejection of some points and retention of others improves our capacity to comprehend and capture the spatial processes under study is difficult to evaluate. Certainly the clusters defined above seem very reasonable. It is difficult to put "space" into clustering algorithms. In general, including the x and y locations, as we have done here, imparts a measure of spatial compactness to the resulting clusters. Their relative importance can be estimated by the sum of their variances compared to the sum of the variances of all the included

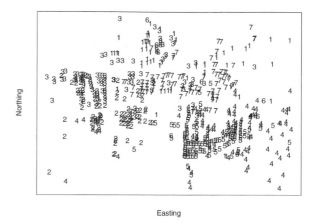

Figure 15.5. Initial clustering based on RAD values.

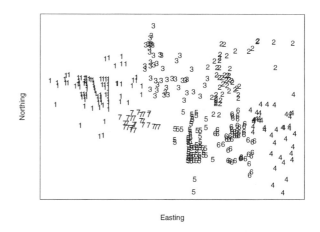

Figure 15.6. Clustering using only high RAD values.

variables. In this example, since the variables were standardized and six variables were used, x and y represent 33% of the total variation. Thus the relative weighting of space in these clusterings is about one-third. By increasing the variance of the x and y variables, we could eventually ensure spatially compact clusters. This compactness comes at a price. As the importance of the x and y variables increases, there will be increased variability within the cluster (in terms of the devaluated variables), as well as the tendency to lose coherent regions which are narrow bands. Clusters will become as round as possible. The RAD filter seems to obviate this undesirable tradeoff. In the clustering above, the clusters seem to be spatially cohesive while still allowing classes to take on complex shapes. One would hope this suggests an increased capability to follow the contour lines of the underlying "truth."

Conclusion

This is clearly a preliminary study. The angular semivariogram can be a useful geostatistical tool for examining the spatial distribution of vector valued data, but the angular component alone is insufficient to describe the errors found in spatial databases. Still, it was our hope that by defining rigorously the way one variable varied spatially over the study site, additional insight and leverage into the clustering problem could be obtained. In this sense, the RAD statistic seems a somewhat promising but still rather ad hoc tool. Simply throwing away outlying points may be a great way to bring out underlying patterns in data, even if they are not really there. Clearly, the RAD statistic must be used with caution as a "variance filter."

We have tried to explore how one might use geostatistics to bring space intelligently into the problem of clustering datapoints into homogeneous units that effectively partition geographic space. The latter two criteria are often at odds. If, however, the spatial data under consideration are in fact the result of a process that really imparts a pattern of spatial correlation, then excluding "variance outliers" may help to reveal the knowledge underneath the noise.

We have argued in this chapter that a piecewise approach to positional distortion may make more sense, in light of what is known about production processes, than an approach that assumes smooth variation. The results appear to bear this out, in a preliminary way, though more testing is clearly needed in a wider range of circumstances.

ACKNOWLEDGMENTS

The National Center for Geographic Information and Analysis is supported by the National Science Foundation. Funding for NCGIA's research on uncertainty in spatial databases comes from the National Imagery and Mapping Agency. Funding for this analysis of street centerline databases comes from Caltrans and Viggen Corporation. The assistance of companies providing data for these analyses is gratefully acknowledged.

REFERENCES

Batschelet, E. *Circular Statistics in Biology*, Academic Press, New York, 1981.

Church, R., K.M. Curtin, P. Fohl, C. Funk, M.F. Goodchild, P. Kyriakidis, and V. Noronha. Positional Distortion in Geographic Data Sets as a Barrier to Interoperation, in *Proceedings, ACSM Annual Convention, Baltimore*, 1998.

Cressie, N.A.C. *Statistics for Spatial Data*, Revised Edition. John Wiley & Sons, New York, 1993.

Deutsch, C.V. and A.G. Journel. *GSLIB: Geostatistical Software Library and User's Guide*, Oxford University Press, New York, 1992.

Hartigan, J.A. *Clustering Algorithms*, John Wiley & Sons, New York, 1975.

Hunter, G.J. and M.F. Goodchild. A New Model for Handling Vector Data Uncertainty in Geographic Information Systems. *J. Urban Reg. Inf. Syst. Assoc.*, 8(1), pp. 51–57, 1996.

Isaaks, E.H. and R.M Srivastava. *Applied Geostatistics*, Oxford University Press, New York, 1989.

Kiiveri, H.T. Assessing, Representing, and Transmitting Positional Uncertainty in Maps. *Int. J. Geogr. Inf. Sci.*, 11(1), pp. 33–52, 1997.

Matheron, G. Principles of Geostatistics. *Econ. Geol.*, 58, pp. 1246–1266, 1963.

CHAPTER 16

Introducing the Concept of Reliability in Spatial Data

M. Azouzi

INTRODUCTION

Uncertainty is a wide concept used very much nowadays to describe information about data. Uncertainty, accuracy, precision, and reliability are some of the words employed indifferently in this context to talk about the general framework of uncertainty. In geodetic networks, only two of these terms are commonly used: accuracy and reliability. They represent a very important set of information about point coordinates resulting from an adjustment. Their definition and mathematical models are different. Accuracy gives information about the closeness of a determined value to the observations used, wile the reliability in information is an indication about the degree of correspondence of the value to what it is supposed to represent in the real world. This conceptual difference of definitions is very important in spatial information, above all because of the diversity of data sets that are stored in a spatial database. In this chapter, an overview of the different types or errors, the concepts of accuracy and reliability, the difference between them and their application to spatial information will be presented.

TYPES OF ERRORS OCCURRING IN GENERAL INFORMATION

There are many types of errors that can occur while acquiring and handling data. They can be classified in two major classes : random errors and gross and systematic errors. In this chapter, both of them will be discussed as well as their influence on the accuracy and the reliability of data.

Random Errors

Many errors occurring during the acquisition of values for an observation are considered to appear randomly, and depend on the normal probability distribution function (p.d.f.), determined by Gauss. These errors, also called *accidental errors,* vary in such a way that the individual errors of an observation can not be precomputed in a significant way. A statistical approach can model them, and an estimation of the closest value to the one wanted can be calculated. The majority of physical observations can have random errors according to the normal p.d.f. Generally, data capture is not always done in a statistical way, and often the observations are assumed to be like random variables so that one can estimate their accuracy. In a spatial database, the cardinal values stored may also have other information, such as attribute accuracy. There are many methods to assess the propagation of such errors, mainly in well-defined mathematical models (Heuvelink, 1993; Azouzi, 1998). However, they can give no indication about the eventual existence of a bias in the original data.

Gross and Systematic Errors

In this category of errors, we find outliers, biases, blunders, and gross errors. Not all of them can be detected by means of numerous observations or measurements. They can be divided into two groups. The first group is the one of gross errors; it contains outliers and blunders. They are characterized by the fact that they do not belong to the data population that is supposed to be representative of certain objects from the real world. These kind of errors do not follow a probability distribution function. They can occur between random errors, for example, when a bad identification of an object is made during the acquisition process. They can also be caused by the presence of wrong information regarding the update of data. The second group is one of biases and systematic errors. They appear while using a model for which one parameter or more is wrong, or an instrument is used with a bad calibration. Such errors can be modeled and corrections be made a posteriori to the observation if some supplementary information can be determined (for example: a missing parameter during the acquisition process).

The presence of gross or systematic errors in a data set can lead to a very bad correspondence between data and reality, or even to no correspondence at all. The results of treatments applied to them (query, spatial analysis, etc.) may have no relation to what is really wanted. In such a context we can talk about reliability of the data. The independent control is the most efficacious way to prevent the presence of such errors in data. The control can be done during the acquisition (*in situ*) or in postprocessing (*a posteriori*).

ACCURACY VS. RELIABILITY IN DATA SETS

Many values can be calculated to represent the uncertainty about information. Among them, accuracy is the most common value and furthermore is often assimilated to the concept of uncertainty. Meanwhile, the term of reliability is often used indifferently to talk about accuracy and precision. A brief presentation of the theoretical fundamentals of both accuracy and reliability follows.

The Accuracy

Accuracy can be defined as the amount of agreement between the resulting value and the accepted reference. It depends directly on the method and the instruments used. Generally there is a probability distribution function that can be applied to the behavior of random errors. From a statistical point of view, the

determination of a random variable x is made by means of a certain number of observations $x_1, x_2, ..., x_n$. Two values are then computed for x:

$$\hat{x} = \frac{1}{n} \sum_{i=1}^{n} x_i \qquad (1)$$

the mean value of x, and

$$s_x = \sqrt{\frac{1}{n-1} \sum_{i=1}^{n} (\hat{x} - x_i)^2} \qquad (2)$$

the standard deviation of x.

The standard deviation represents the accuracy of x; that is, the dispersion of observations around their mean value. The higher n is, the better is the estimation of x. Nevertheless, the mean and the accuracy can be considered meaningful if their estimations are based upon unbiased observations.

The Concept of Reliability

The concept of reliability was developed in the industry to evaluate the failure occurrences in a system, and the degradation properties of its various elements. The theory developed deals with some principles and methods of probability for the determination and analysis of errors that occur in the system and involves failure that makes the system fall in.

The evolution of the industry of electronic components gave the impulse for the development of the concept of reliability some 40 years ago.

The mathematical model of reliability is based on the analysis of the random occurrence of failures in a given system. Let T be the random time to next failure, the reliability function R(t), its cumulative probability function F(t), and the density probability function can be written in the relation below:

$$F(T) = \int_{\infty}^{t} f(t) \cdot dt = \mathrm{Prob}(T \leq t) = 1 - R(t) \qquad (3)$$

In Catuneanu and Miholanche (1989), the reader will find a detailed description of the reliability theory. This theoretical framework can be applied to GIS and databases as softwares, because failures can occur during their use like bugs, for example. But this will not be discussed in this chapter.

In data, the existence of gross and systematic errors, blunders, and biases can be considered equivalent to failure, because their use involves results that do not correspond to what is expected. The existence of large volumes of various types of data, from various sources in a GIS is a sufficient reason to ensure the contents of the database by making provision against gross errors and blunders in the data. If the database already exists, it should be interesting to know and evaluate their frequency within the data.

The use of reliability in geographic information was introduced in the adjustment of geodetic networks. Its main purpose is the detection of gross errors and outliers in measurements, and their influence on the results of adjustment. It can be assimilated to the concept of controllability of information (Dupraz and Stahl, 1995). Figure 16.1 is an illustration of a case where a mean value μ_a is more accurate than the mean μ_c ($\sigma_a < \sigma_c$), but less reliable because of the presence of a systematic error detected by means of a third data set (b).

THE RELIABILITY OF SPATIAL DATA

Reliability of Data

In classical systems, reliability gives an indication about the occurrence of failure in relation to time. In data handling, reliability evaluates the existence of faults; that is, the existence of blunders, outliers, or systematic errors. In such cases, the data cannot be considered representative of reality. All operations based on such data can not give reliable results. The best way to have a reliable set of data is independent control. During the acquisition operations, some of the data should be captured more than once; at least twice to have a double value for the observation; or three times, which is better in case of important difference between the first values. Having a second independent acquisition is necessary, but not for all the data; otherwise, the operation will be too expensive. One way to have independence of control is to carry out primary acquisition, and after that, a second acquisition will be executed by means of other equipment (measurement set, instruments...) or by another observer. Obtaining multiple values for data is not only necessary to get a mean value for which the accuracy is enhanced, it also helps to have a controlled and reliable value.

For example, given a distance **d** that is measured with two different instruments, the first, **d**$_{EDM}$, is made with an accurate electronic distance-meter (EDM), and

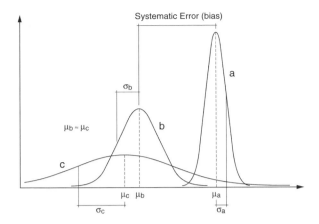

Figure 16.1. Presence of systematic errors detected by control.

the second, **d**$_t$, is made with a tape. We can affirm that **d**$_{EDM}$ is more accurate than **d**$_t$; this is because of the difference of accuracy between the EDM and the tape. However, **d**$_t$ is more reliable than **d**$_{EDM}$ because it is controlled by **d**$_{EDM}$. Effectively, the value of **d**$_{EDM}$ can confirm the one of **d**$_t$, but not the contrary. The smallest digits of **d**$_{EDM}$ do not have their equivalent in **d**$_t$. Example:

$$d_{EDM} = 64.248 \text{ m} \pm 5 \text{ mm}$$
$$d_t = 64.27 \text{ m} \pm 2 \text{ cm} \qquad (4)$$

Taking the wanted accuracy of the distance into account, the distances in meters (64 m), decimeters (2 dm) and centimeters (7 cm) of d_t are confirmed by d_{EDM}, but the millimeter (8 mm) of d_{EDM} is not. This example represents an extreme to illustrate the difference between accuracy and reliability. To do the control with a more accurate instrument is not necessary at all. Independent control is the main requirement for obtaining reliable data.

Reliability in Data and Geodetic Networks

In geodesy and land surveying, the coordinates of the reference points are generally determined by a least-square adjustment method. Currently, for a network, more measurements than needed are made to allow an adjustment. The extra observations allow the enhancement of the accuracy of the results (coordinates of points). Besides, they make possible the assessment of their reliability and the determination of the highest nondetectable errors in the measurements, and their effects upon the coordinates. It would be too much, in this chapter, to develop the mathematical model based on the least squares adjustment for the determination of the reliability parameters. A detailed description is in Pelzer, 1980; Burnand, 1990; and Dupraz and Stahl, 1995. We will only describe some of them, with a brief explanation:

- The first parameter evaluates what is called internal reliability; that is, a percentage representing the degree of certainty of an observation measurement against a gross error. Zero percent is for an observation with no control, and 100% is for an observation with an absolute control; that is, an observation for which a value is already known. Generally, observations should have an internal reliability value within the interval (25%, 60%).
- Another parameter is the ratio between the residual and its standard deviation, which indicates, if above 4, the probable existence of a gross error in the observation. In this case, an estimation of the value of the gross error is calculated.
- Another parameter is the external reliability. It represents, for every result, the effect of the highest nondetectable error occurring in every observation.

All the parameters computed allow the detection of gross errors, blunders, and outliers in the observation, so that the coordinates computed are not only accurate but reliable as far as enough observations are available.

Reliability in Swiss Cadastral Data

For cadastral data in Switzerland, two parameters are required for the coordinates of boundary points : accuracy and reliability. To obtain these parameters, two independent determinations, or one determination and a complementary control (distance or angle) are made for every boundary point. The resulting coordinates are the mean of obtained values. One can say that with the actual accuracy of the different instruments that are in use, there is no need to have more than one determination for one point. This affirmation is correct only for the accuracy of coordinates. Effectively, the actual instruments measure distances and angles with a high accuracy (± 3 mm + 3 ppm for dis-

tances and ± 1 mgon for angles), which permits one to obtain easily a centimetric accuracy for the coordinates. However, an error of identification of the point will always have the same effect, independently of the instrument accuracy; that is, not the right point with the right coordinates. This is an application of the concept of reliability. Thus, all the boundary points stored in a cadastral database are controlled, mainly to have reliable data for them.

DETECTING AND HANDLING GROSS ERRORS

As mentioned above, the independent control of data remains the only way to detect deviations and disturbance in data, and to make the data reliable. For existent data, which data has no control at all, an a posteriori control can be applied. An approach using samples can be used. Controlling all the contents of the database is not realistic at all. Such work will increase heavily the number of tasks, and the cost will be too high with regard to the quality information required. Therefore, executing the control by sampling can give in many situations a good indication about the quality state of the database. The sample should be as representative as possible of contents of the database.

For metric values x, a new determination x_N is made, and compared to the existing one, x_{DB}. Information about the accuracy of the data controlled is needed. Two alternatives are usable for the test: the statistical and the empirical approach.

Statistical Determination

The first approach is to be used if the values are determined statistically. A ratio can be calculated to compare the difference between the values and the standard deviation .

$$y = \frac{\hat{x}_N - x_{DB}}{\sigma_x} \qquad (5)$$

where: $\hat{x}_N$ is the new estimated value of x and σ_x its standard deviation, and

x_{DB} is the value of x existing in the database

A statistical test can be made to determine whether if the value of y can be considered as null, according to a confidence level ($\alpha = 5\%$, 10%, etc.). A confidence interval can be calculated:

- $\left[\hat{x}_N - c_{\alpha/2} \cdot \sigma_x; \hat{x}_N + c_{\alpha/2} \cdot \sigma_x\right]$ if the variance σ_x is known, $c_{\alpha/2}$ is given by the table of the Gauss p.d.f. $[P(X > c_{\alpha/2}) = \alpha/2]$. For example for $\alpha = 5\%$, $c_{0.025} = 1.96$.
- $\left[\hat{x}_N - t_{\alpha/2} \cdot s_{\hat{x}}; \hat{x}_N + t_{\alpha/2} \cdot s_{\hat{x}}\right]$ if only an estimation $s_{\hat{x}}$ of σ_x in known, $t_{1-\alpha/2}$ is given by the table of the Student p.d.f.

If both of the values have their own standard deviation, the χ^2 p.d.f. can be used to compare them and to conclude if they can be considered to be a statistical representation of the same value according to a confidence level α.

This method of comparison is described in many books of statistics in detail. The final result will be the number of values newly determined and considered equal to their equivalent value in the database.

Empirical Approach

The second method used is empirical. The difference between the existing value and the new value is compared to the standard deviation. If it is more that three times the standard deviation, one can conclude that the difference is too large to be accepted. According to the Gaussian distribution function $N(\mu,\sigma)$ for a random variable x, we have the relation:

$$P(-3\sigma_x < x - \mu_x < +3\sigma_x) = 99.73\% \qquad (6)$$

That means that the new value is within the interval $[x_{DB}-3\sigma_x; x_{DB}+3\sigma_x]$ with a probability of 99.73% (Figure 16.2); that is, with a confidence level $\alpha = 0.27\%$. This interval can be used as a reference for the comparison of the values in the database, and those for the

control. The value of $3\sigma_x$ is commonly called the tolerance for x.

Unknown Standard Deviation and Nonmetric Data

If there is no information about the standard deviation of the data in the database and for nonmetric data, an interpretation of the difference and of its possible source must be done. This depends mainly on what data are destined to. This area of research needs more examination for the assessment of the reliability of such a type of data.

Reliability Estimation by Sampling

As described before, a systematic control of all the data stored in a database is a very difficult task. Therefore, control by sampling techniques is essential for existing databases; that is, only some of the data is controlled. The control must be done independently of the first acquisition. It must be executed in other conditions than the first acquisition.

The number of gross and systematic errors can be considered as an indication for the rate of reliability. Let **n** be the total number of entities in the sample, **n₀** the number of entities for which the new value *is not* equal to the existing one, and **n₁** for which the new value *is* equal to the existing one. We have: $n = n_0 + n_1$. We calculate the rates:

$$\hat{p} = \frac{n_0}{n} \quad \text{rate of values accepted;}$$

$$\hat{q} = \frac{n_1}{n} \quad \text{rate of values not accepted;} \qquad (7)$$

The value of $\hat{p}$ is an indication of the degree of reliability of the sample. If the sample is made according to sampling techniques, a confidence interval can be determined for $\hat{p}$ which is :

$$\left[\hat{p} - c_{\alpha/2} \cdot \sqrt{\frac{\hat{q}\hat{p}}{n}}; \hat{p} + c_{\alpha/2} \cdot \sqrt{\frac{\hat{q}\hat{p}}{n}}\right] \qquad (8)$$

This shows that the estimation of the proportion of data that have gross errors in the sample can be generalized to the whole contents on the database if we take a confidence level α. That is, there is approximately a

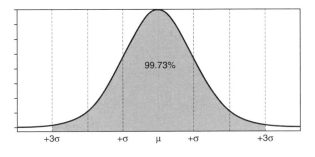

Figure 16.2. Confidence interval for the normal p.d.f ($\alpha = 0.27\%$).

probability of $(1-\alpha)$ that the rate of data not affected by gross errors in the database is in the interval determined above. So we obtain an estimation of the degree of reliability of the database.

CONCLUSION

Two major kinds or errors were examined: random errors and gross and systematic errors. The concept of accuracy and reliability and their application to spatial data were examined as well. It was shown that the random errors influence mainly the accuracy, while the presence of gross and systematic errors in a data set influence the reliability of data. This difference between accuracy and reliability is very important to qualify data. For one value, the repetition on the measurement enhances its accuracy, while the independent control gives an information about its reliability. The control can be done during the acquisition process (in situ) or even a posteriori, if the data are already in a database, by sampling. The presence of gross and systematic errors, blunders, and outliers in a data set can lead to random occurrence of failure. While surveying, digitizing a map, or collecting some information for a database, nobody is protected against a bad identification of a point or an object, nor assured that there will be no confusion between some values.

Independent control remains the main way to assure good reliability for a data set, even if it remains inaccurate.

REFERENCES

Azouzi, M. Error Propagation in Spatial Analysis for the Evaluation of Traffic Noise, in *Data Quality in Geographic Information: From Error to Uncertainty,* R. Jeansoulain and M. Goodchild, Eds., Hermès, Paris, 1998.

Burnand, T. Le Modèle de Fiabilité dans la Mensuration Nationale Suisse, in *Fiabilité dans la mensuration,* Swiss Federal Institute of Technology-Zurich, Zurich, 1990.

Catuneanu, V.M. and A.N. Mihalanche. *Reliability Fundamentals,* Elsevier, New York, 1989.

Dupraz, H. and M. Stahl. *Théorie des Erreurs 3,* Lecture Notes, IGEO/TOPO, Swiss Federal Institute of Technology-Lausanne, Lausanne, 1995.

Heuvelink, G.B.M. *Error Propagation in Quantitative Spatial Modeling,* Thesis, Faculteit Ruimtelijke Wetenschappen Universiteit Utrecht, The Netherlands, 1993.

Pelzer, H. Beurteilung de Genauigkeit und der Zuverlässigkeit Geodätischer Netze, in *Geodätische Netze in Landes- und Ingenieurvermessung,* H. Pelzer, Ed., Konrad Wittwer, Stuttgart, 1980.

Part IV
Representing Spatial Uncertainty

Once spatial uncertainty information has been characterized and described, it remains that it must be communicated to users. A common statistical method of representing and communicating uncertainty in aspatial data is the standard error of the mean for a phenomenon that can be described by measurements made on an interval-ratio measurement scale. However, there is no agreement on how spatial uncertainty information should be represented in order to be communicated reliably. The chapters in this part describe a number of ways to represent spatial uncertainty. The diversity of the methods and the associated target groups is yet another indication of the complexity of the spatial uncertainty domain.

The first chapter, by Fisher, discusses a framework for conceptualizing the uncertainty associated with the assignment of a given object to a category. The second chapter by Bastin et al. extends these ideas by discussing spatial representations that may be effective in communicating spatial uncertainty to nonexpert decision makers. The third chapter, by Vauglin, is concerned with linear rather than areal features, and positional rather than thematic uncertainty; the author employs a geostatistical construct—the variogram—to represent associated spatial uncertainty. Finally, the fourth chapter, by Drummond et al., moves the discussion from the general to the specific by describing a method for communicating uncertainty information to operators of photogrammetric workstations during map preparation, with the goal of extracting better information from aerial photographs.

CHAPTER **17**

Set Theoretic Considerations in the Conceptualization of Uncertainty in Natural Resource Information

P.F. Fisher

INTRODUCTION

This series of conferences highlights a continuing concern with the uncertainty associated with both natural resource information, and its spatial extent and digital representation. The vast majority of papers in the two previous conferences have been concerned with two forms of uncertainty: that associated with measurement error (including interpolation error), and error in the Boolean assignment of an object to a class. These are, however, only two of an expanding number of possible views of uncertainty. In this chapter I do not present any empirical investigation based on use of one of these alternatives. Rather, I attempt to place the two aforementioned methods in the context of the wider set theoretic possibilities, indicating that work in uncertainty in natural resource information actually has a rich variety of opportunities in which to explore uncertainty and error.

MEASUREMENT ERROR

The primary observation in much spatial information is the measurement of a parameter: the color, clay content, pH, species frequency, etc. For any such measurement, there is a probability that they are correctly measured allowing the assignment of a confidence interval; a range of values within which we can be certain (at some probability level) that the real value lies. A special aspect of measurement error is interpolation error, when measurements at a set of locations are used to imply values for the same variable at other measured locations. Again there is a probability of the interpolated value being correct.

Measurement error and its associated statistics are well researched and understood (Taylor, 1982). The properties of interpolation errors, especially in spatial information, are more problematic and are the subject of continuing research (Wood and Fisher, 1993). Most of these methods assume that the point-based measurements are accurate in the first place, however, and do not treat and interpolate measurement errors in that point data.

The most common version of measurement error within GIS is that of the Root Mean Squared Error (RMSE) of a DEM. RMSE is a quantification of the error in the measurement of elevation. It corresponds to the standard deviation, assuming that there are no systematic effects in the error, either bias or trend, and has been used in a variety of studies (Fisher, 1991; Hunter and Goodchild, 1997; Veregin, 1997). Fisher (1998) has also outlined a approach grounded in geostatistical, which accommodates nonstationarity, bias, and trend in the error.

BOOLEAN SETS

Other spatial information is collected in a categorical form; that is, any location belongs to a class of object or theme—a map unit. These classes can be associated with measured attributes, either singly or in a multivariate classification (e.g., climate types). Alternatively, they may be derived from the opinion of an expert (soil and geology). Such measurements and opinions are usually based on visited locations from which measurements may be made or samples examined. In either case, there is commonly a risk that any visited location may be incorrectly assigned to its class, but when a class is assigned to unvisited locations, again interpolation errors invariably ensue.

In a set theoretic context, we are usually saying that any one location may belong to a set of candidates, and any particular location is assigned to one of those candidates and only one. There is therefore a probability that the location may be incorrectly assigned to a class when it is actually some other class. Such class assignment errors have been widely examined in the literature, and are the principal concern of the confusion matrix so beloved in remote sensing (Congalton, 1992). As with the RMSE, however, the confusion matrix implies a number of unwarranted assumptions about the distribution of the errors, especially that they are stationary across any study area. The problems of the confusion matrix are highlighted by attempts to visualize the outcomes (Fisher, 1994a).

In many instances of remote sensing and other areas where data is determined over a grid, then an acceptable argument can be put forward for the phenomenon in question being Boolean in its spatial extent (it can be conceived of as having sharp boundaries). In this instance, the granularity of the cells causes uncertainty, but the cell no longer belongs to any one Boolean set since it is equally true to state that the cell belongs completely to any of the Boolean categories which occur within the grid cell. In this instance, the granularity of the information is the cause of the imprecision in assignment, and the uncertainty may more correctly be described by the recently developed theory of Rough Sets (Pawlak, 1982, 1991). These were specifically developed to address uncertainty caused by granularity in information, and are based around objective methods for defining the upper and lower boundaries (including spatial) of the set. Rough sets are a developing area in information science, and new concepts are being introduced and explored, including appropriate logics. Rough sets may also be appropriate if the objects are not well defined (Boolean). Dubois and Prade (1990) present definitions of Fuzzy Rough sets and Rough Fuzzy sets. The subtle distinctions here may be particularly appropriate for set theoretic problems in, for example, remote sensing information (Fisher, 1997).

FUZZY SETS AND SEMANTIC VAGUENESS

The measures of uncertainty discussed thus far are all based around the concept of the information to be measured or the object to be analyzed being unambiguously defined, in both its attribute domain and its spatial domain. That definition is not necessarily clearcut. There are ample instances, especially among natural resource information where information is profoundly ambiguous. What, for example, is a soil? The actual set of objects which constitute a soil are different in the United States and Britain (Soil Survey Staff, 1975; Avery, 1980), and so the spatial extent of the soil cover must be different. Similarly, in spite of the definition of soils as crisp sets within most modern handbooks of soil classification (Soil Survey Staff, 1975; Avery, 1980), soils are sufficiently variable that positioning the boundary cannot be precise. Several investigations have found that as many boundary patterns can be interpreted in a landscape as there are people surveying the soil (Pomerening and Cline, 1953). Similar problems have been found in land-use mapping from aerial photography (Middelkoop, 1990). Where the classes of objects (sets) to be mapped cannot be conceived of as crisp and well defined, then the appropriate formalization of the sets may be as Fuzzy Sets (Zadeh, 1965), instead of as Boolean sets.

Fuzzy Sets are now well-known in spatial information processing. They define a membership (in the real number range from 0 to 1) which describes the degree to which a phenomenon may match the characteristics of a prototype concept. There are a number of studies which have examined the applicability of the methods to spatial information processing (Altman, 1994; Burrough et al., 1992; Burrough and Frank, 1996), although Robinson and coworkers pioneered the subject (Robinson, 1988, 1990; Robinson and Strahler, 1984; Robinson and Thongs, 1986).

There has been considerable confusion over the distinction between probability and fuzzy memberships (Zadeh, 1980; Fisher, 1994b). The resolution of such problems lies in the conceptualization of the information. If the classes to be examined are clearly defined, then the objects are Boolean sets; if they are not, then they may be fuzzy sets. Reference to the classic Sorites Paradox (the paradox of the heap) from Ancient Greek philosophy can clarify any doubt as to the conceptualization of a fuzzy set in terms of vague concepts, although a fuzzy set is only an approximation of a vague concept (Sainsbury, 1995; Williamson, 1994). There is no conceptual question of a probability being a fuzzy membership. Pragmatically, and because probability and fuzzy membership are both measured on a real number scale from 0 to 1, it may be possible to use the probability yielded by a likelihood classifier as a fuzzy membership, but this does have problems (Bastin, 1997).

HIGHER ORDERS OF UNCERTAINTY

A fuzzy set can be shown to be made up of a large (infinite) number of Boolean sets. More practically, any thresholding of the fuzzy membership function will yield a crisp, Boolean set. Since there is a probability that any location or object is correctly assigned to a Boolean class, when a Boolean class is generated by thresholding a fuzzy set then there is still a probability that locations are correctly assigned. Furthermore, because any particular value of the fuzzy membership itself is a measurement, there is a probability that that membership is correct. In short, associated with any estimate of fuzzy set membership, there is naturally a probability that the estimate is correct. This could be related to the measurement or interpolation error of the spatial parameters used in the membership function (Burrough, 1989). It may also be related to the membership function itself, particularly membership is derived by Similarity rather than Semantic relations, derived by pattern recognition.

In short, Boolean and Fuzzy sets are comparable, and any estimation of set membership attracts a probability of being correct.

Williamson (1994) points out that there are many higher orders of vagueness associated with any real vague concept. Thus if a threshold is applied to a fuzzy membership, there is actually a fuzzy membership function associated with the threshold. The probability of the membership being correct can be viewed as a first empirical estimate of the higher orders of uncertainty is one aspect of this higher level uncertainty. Indeed, the mean, standard deviation, skewness, and kurtosis reported by Hughes et al. (in this volume), are all higher orders of uncertainty associated with the fuzzy membership.

CONCLUSION

In this chapter, the relation of Boolean and Fuzzy sets and probability have, in particular, been examined. Rough sets have been introduced as a formalism which may be well attuned to much grid-cell spatial information, when that information is associated with Boolean classes, although the further Rough Fuzzy sets and Fuzzy Rough sets make the concepts appropriate to poorly defined classes of information. As long as there is doubt over the estimation of any membership values (they may be wrong), then probabilities can be associated with all set memberships, Boolean set, Fuzzy set, Rough set, etc.

In conclusion, I have tried to highlight a number of alternative models for uncertainty which are used in other areas of information science and should all be considered in the analysis of uncertainty in spatial information. Some have been examined extensively in spatial information processing (probability of Boolean set membership and memberships of fuzzy sets), but some confusion may actually have arisen from application of poorly conceptualized set theories. The other set theories outlined here, and still more, are available from the information sciences and should be explored for spatial information processing. The varied models, possibly used in cooperation, may provide a more complete model of uncertainty than any one individually, or in use currently.

Finally, the understanding of uncertainty depends ultimately on one thing only, and that is the conceptualization of the information by the investigator, and its defensibility to a wider group of peers. If a Boolean concept of the information can be defended, then associated probabilities are appropriate, and no recourse to more complex set theories is necessary. More complex conceptualizations of information employing inherent vagueness or granularity may necessitate the use of alternative set theories.

REFERENCES

Altman, D. Fuzzy Set Theoretic Approaches for Handling Imprecision in Spatial Analysis. *Int. J. Geogr. Inf. Syst.,* 8, pp. 271–289, 1994.

Avery, B.W. *Soil Classification for England and Wales (higher categories).* Soil Survey Technical Monograph 14, Harpenden, 1980.

Bastin, L. Comparison of Fuzzy c-Mean Classification, Linear Mixture Modelling and MLC Probabilities as Tools for Unmixing Coarse Pixels. *Int. J. Remote Sensing,* 18, pp. 3629–3648, 1997.

Burrough, P.A. Fuzzy Mathematical Methods for Soil Survey and Land Evaluation. *J. Soil Sci.,* 40, pp. 477–492, 1989.

Burrough, P.A. and A. Frank, Eds. *Spatial Conceptual Models for Geographic Objects with Undetermined Boundaries.* Taylor & Francis, London, 1996.

Burrough, P.A., R.A. MacMillan, and W. van Deursen. Fuzzy Classification Methods for Determining Land Suitability from Soil Profile Observations and Topography. *J. Soil Sci.,* 43, pp. 193–210, 1992.

Congalton, R.G. A Review of Assessing the Accuracy of Classifications of Remotely Sensed Data. *Remote Sensing Environ.,* 37, pp. 35–46, 1992.

Dubois, D. and H. Prade. Rough Fuzzy Sets and Fuzzy Rough Sets. *Int. J. Gen. Syst.,* 17, pp. 190–209, 1990.

Fisher, P.F. First Experiments in Viewshed Uncertainty: The Accuracy of the Viewshed Area. *Photogrammetric Eng. Remote Sensing,* 57, pp. 1321–1327, 1991.

Fisher, P.F. Improving Error Models for Digital Elevation Models, in *Data Quality in Geographic Information: From Error to Uncertainty,* R. Jeansoulin and M.F. Goodchild, Eds., Hermes, Paris, 1998.

Fisher, P.F. Visualization of the Reliability in Classified Remotely Sensed Images. *Photogrammetric Eng. Remote Sensing,* 60, pp. 905–910, 1994a.

Fisher, P.F. Probable and Fuzzy Models of the Viewshed Operation, in *Innovations in GIS 1,* M. Worboys, Ed., Taylor & Francis, London, 1994b, pp 161–175.

Fisher, P.F. The Pixel: A Snare and a Delusion. *Int. J. Remote Sensing,* 18, pp. 679–685, 1997.

Hughes, M., J. Bygrave, L. Bastin, and P.F. Fisher. High Order Uncertainty in Spatial Information: Estimating the Proportion of Cover Types within a Pixel, in *Spatial Accuracy Assessment: Land Information Uncertainty in Natural Resources,* K. Lowell and A. Jaton, Eds., Ann Arbor Press, Chelsea, MI, 1999, pp. 319–323.

Hunter, G.J. and M.F. Goodchild. Modeling the Uncertainty of Slope and Aspect Estimates Derived from Spatial Databases. *Geogr. Anal.,* 29, pp. 35–49, 1997.

Middelkoop, H. Uncertainty in a GIS: A Test for Quantifying Interpretation Output. *ITC Journal,* 1990, pp. 225–233.

Pawlak, Z. Rough Sets. *Int. J. Comp. Inf. Sci.,* 11, pp. 341–356, 1982.

Pawlak, Z. *Rough Sets: Theoretical Aspects of Reasoning about Data.* Kluwer Academic Publishers, Dordrecht, 1991.

Pomerening, J.A. and M.G. Cline. The Accuracy of Soil Maps Prepared by Various Methods that use Aerial Photograph Interpretation. *Photogrammetric Eng.,* 19, pp. 809–817, 1953.

Robinson, V.B. Some Implications of Fuzzy Set Theory Applied to Geographic Databases. *Com., Environ. Urban Syst.,* 12, pp. 89–98, 1988.

Robinson, V.B. Interactive Machine Acquisition of a Fuzzy Spatial Relation. *Com. Geosci.,* 16, pp. 857–872, 1990.

Robinson, V.B. and A.H. Strahler. Issues in Designing Geographic Information Systems Under Conditions of Inexactness, in *Proceedings of the 10th International Symposium on Machine Processing of Remotely Sensed Data,* Purdue University, Lafayette, IN, 1984, pp. 198–204.

Robinson, V.B. and D. Thongs. Fuzzy Set Theory Applied to the Mixed Pixel Problem of Multispectral Landcover Databases, in *Geographic Information Systems in Government,* B. Opitz, Ed., A Deerpak Publishing, Hampton, VA, 1986, pp. 871–885.

Sainsbury, R.M. *Paradoxes,* 2nd ed. University Press, Cambridge, 1995.

Soil Survey Staff. *Soil Taxonomy: A Basic System of Soil Classification for Making and Interpreting Soil Surveys.* USDA Agricultural Handbook 436, Government Printing Office, Washington, DC, 1975.

Taylor, J.R. *An Introduction to Error Analysis.* University Press, Oxford, 1982.

Veregin, H. Effect of Vertical Error in Digital Elevation Models on the Determination of Flow-Path Direction. *Cartography Geogr. Inf. Syst.,* 24, pp. 67–79, 1997.

Williamson, T. *Vagueness.* Routledge, London, 1994.

Wood, J. and P.F. Fisher. Assessing Interpolation Accuracy in Elevation Models. *IEEE Com. Graphics Appl.,* 13(2), pp. 48–56, 1993.

Zadeh, L.A. Fuzzy Sets. *Inf. Control,* 8, pp. 338–353, 1965.

Zadeh, L.A. Fuzzy Sets Versus Probability. *Proc. IEEE.* 68, p. 421, 1980.

CHAPTER 18

Visualization of Fuzzy Spatial Information in Spatial Decision-Making

L. Bastin, J. Wood, and P.F. Fisher

INTRODUCTION

There is considerable potential in the use of fuzzy sets to represent geographical landcover data, and two particular contexts are relevant to this chapter. First, "mixed pixels" in multispectral remotely-sensed imagery are not adequately described by traditional Boolean methods, but can more profitably be represented as having membership in several fuzzy landcover sets. Fuzzy classifications can be used to "unmix" spectral data, and to gain more information on ground cover (Foody, 1996). Secondly, fuzzy sets can be used to represent attribute and positional uncertainty within spatially-referenced datasets. Calculation and representation of uncertainty is not a standard technique in GIS analysis, but is an important consideration in the use of geographic data for practical decision-making (Jensoulin and Goodchild, 1998).

The software toolkit described in this chapter is designed for the following uses: (a) verification of fuzzy classifications against ground data; (b) introduction to the concept of fuzzy classification, using standard visualization methods which can be used to explore the products of a fuzzy classification; and (c) combination and analysis of data layers, which may include uncertainty information, in the process of decision support.

The work described is part of project FLIERS (Fuzzy Land Information from Environmental Remote Sensing). The project aims are (a) to produce and to verify fuzzy classifiers for retrieving land-cover and land-use information from remotely-sensed data, and (b) to visualize the classification results in ways that allow exploration of the data and of the classification process itself. This chapter addresses the latter aim.

Project FLIERS involves large spatial data sets, consisting of multispectral satellite images, multilayered verification data, and fuzzy membership maps produced by thematic classification, as well as supplementary layers representing texture, positional, and attribute uncertainty. The primary value of data visualization with such a large data set is to use the pattern recognition abilities and the expert knowledge of the user to interpret images delivered by computer processing of selected layers. The strength of computer data manipulation lies in pulling out and displaying data which are not immediately obvious in a large and complex data set—for example, animated surfaces representing classification entropy, warped vector meshes representing positional uncertainty, or 3D visualizations of statistical space. The user, with knowledge of the original land area which is being represented, and with prompting from aids such as vector map overlay, can spot spatial patterns or anomalies in these visualizations, and earmark areas for further investigation. The aim of the toolkit is not to automate the interpretation process, but to make a variety of clear and useful visualizations available for informed exploration and analysis.

Many end-users will be familiar with multispectral thematic classification, and with some data manipulation techniques. For these users, the priority is to make the toolkit accessible and flexible, so that they can easily adapt a fuzzy classification and compare it to more familiar classifications.

SOFTWARE LANGUAGE AND INTERFACE

The tools produced are modular, and dynamically linked, so that changes in one part of an analysis sequence show their results in the visualizations. Visualizations are relatively simple and are based on familiar graphs, plots, and maps. The software is written in Java, using the Java Beans standard (Sun, 1997). It uses a graphical programming interface similar to "data flow packages" (Rhyne, 1993) such as AVS; in other words, separate but linkable modules for data import, analysis, display, and manipulation (Figure 18.1). The sequence of actions is constantly visible, parameters can be retrieved or set by clicking on the beans themselves, and any changes are propagated between linked beans. The software components are also serializable, allowing specific analysis "runs" to be saved.

REDUCING DATA DIMENSIONALITY

The fuzzy classifications used in this project produce a separate membership map for each landcover class considered, as compared to the single "likeliest class" map of a traditional thematic classification. This means that there are many more dimensions of data to be handled and explored. A visualization which attempts to display too many data dimensions at once will become incomprehensible to the user (DiBiase et al., 1994). It is therefore a priority to condense the data set into fewer layers for visualization purposes, while retaining useful information.

Single layers of *summary statistics* can be calculated from a combination of data layers. Some of these statistics are well documented.

$$H = -\Sigma p(x) \log_2 p(x) \; / -\log_2(1/n) \qquad (1)$$

Scaled *entropy*, H (Equation 1) describes the conflict between probabilities that an image pixel belongs to n alternative classes (or, in the case of fuzzy sets, between membership values in those classes) (Klir, 1994; Maselli et al., 1994). Fuzzy land cover sets are not mutually exclusive, and so high entropy within any one pixel may actually imply a highly accurate classification of a genuinely mixed area on the ground. A useful index of classification accuracy in this case is *cross-entropy* (H_c)

$$H_c = -\Sigma p(x) \log_2 p'(x) + \Sigma p(x) \log_2 p(x) \qquad (2)$$

(Klir and Folger, 1988; Foody, 1995), where p(x) represents the fuzzy memberships in each land cover class, and p'(x) the ground cover proportions in those classes. This produces a single index of correspondence between the classification and the verification data. In a pixel covering a mixture of land cover types, good correspondence between the membership values produced by a fuzzy classification and the actual pixel fractions covered by different classes will produce a high cross-entropy value, indicating a good classification.

SOFTWARE FUNCTIONS

Verification of Fuzzy Classifications Against Ground Data

The first purpose of the software is to enable evaluation of fuzzy land cover classification techniques, and statistical and visual comparison of their results with field verification data. The visualization toolkit is designed to aid this verification process by calculating validity indices such as cross-entropy. A pixel-by-pixel comparison of predicted and actual membership gives a distributed map of agreement for each land cover class, and can highlight spatially contiguous areas of high or low classification accuracy. The visualization tools also assist the user in picking out spatial patterns in statistical data, using such techniques as scatterplot brushing (Monmonier, 1989), serial animation, and parallel coordinates plots.

Representation of Uncertainty

The second aim is to model positional and attribute uncertainty in an accessible and informative way. Few geographic databases are currently designed to contain any extensive uncertainty information, although various workers have recommended that GIS tackle the issue (e.g., Chrisman, 1991; Hootsmans et al., 1992). In spatially-referenced data sets, a guarantee of positional accuracy, such as an RMSE estimate, is often provided. However, a single statistic such as this can cover considerable variation in accuracy across the image. A preferable approach is a map which gives an estimate of data quality at any selected point or region in the geographic area (van der Wel and Hootsmans, 1993). Distributed maps representing uncertainty can be generated using this toolkit, and easily queried, viewed, and incorporated into the decision process.

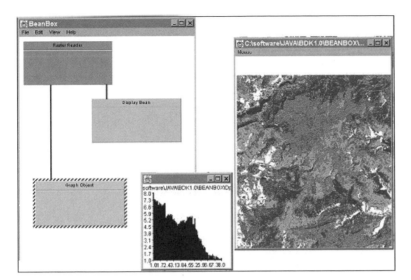

Figure 18.1. A ReadRasterBean reads in a data file and converts it to a multi-layered raster set. The linked Display-Bean and GraphBean produce visual output based on selected layers of the raster.

Standard Tools and Operations for the Visualization Software

Using these tools, single data layers can be viewed as colored or grey-scale flat rasters. Two or three such data layers can be combined by the user and displayed as a single RGB or IHS image. Standard graphs can also be generated to show statistical relationships between different data layers. These methods are all familiar, and are necessary for the visual exploration process. However, there are more novel ways in which to visualize the results of fuzzy classification; for example.

3D Visualization of Data Uncertainty

Raster images can be *draped* over 2.5D surfaces representing real elevation data, or other continuous variables. For example, an image of "most likely landcover" can be draped over a grid showing classification entropy, where vertical value indicates the uncertainty of the hard classification. The numerical data underlying the representation are retrievable by on-screen query of the surface. Such surface representations allow the user to shift viewpoint and to "fly through" the data. However, such representations may also be misleading in that perspective effects distort those areas closest to the viewer, and obscure other features (Wood and Fisher, 1993). The use of contours and of a linked "plan" in this representation could allow a perspective to be maintained on position and context, while exploring the surface at close range (Moore et al., 1997).

Animation

Animation is a useful technique for viewing data representing change through time, or the results of stochastic simulations. Within the context of data exploration and decision support, animation has been used in several interesting ways; to represent data uncertainty in DEM data as a smooth sequence of colored images (Ehlschlaeger et al., 1997); to show uncertainty in soil class membership for fuzzy series analysis (Hootsmans, 1996); and to move through alternative slope stability surfaces, generated by Monte Carlo simulations based on uncertainty "envelopes" (Davis and Keller, 1997). Sets of fuzzy membership maps can also be visualized using random animation based on pixel membership values (Fisher, 1994). Alternative surface realizations are generated and rendered in quick succession, so that some areas of the image are relatively constant in their color, while others can be observed to flicker. When this technique is used to visualize the results of a fuzzy land cover classification, the areas of high entropy (which may be transition zones, highly mixed pixels, or uncertain classifications) draw the user's eye by their apparent motion.

Fuzzy Spectral Signatures

As well as traditional line graphs, scatter plots, and histograms, less conventional graphs can be easily produced. An example is the "fuzzy spectral signature," which represents the fuzzy membership functions underlying the classification process.

Plots of spectral signature (Figure 18.2) are commonly used in thematic classification. Figure 18.3 shows a "fuzzy spectral signature plot." This is an abstraction that shows something of the scatter which is really seen within a spectral cluster representing one land cover class. Like the graphs in Figure 18.2, this is a parallel coordinates plot (Wegman, 1990), where spectral bands are logically ordered by wavelength, and lines interpolated between them, to give a more intuitive feel for the pattern. There is a central reflectance value at which most memberships are clustered, and a spread of values around that center, fading toward the boundaries of the spectral cluster. Sometimes the spread is broad (A), implying that a fuzzy membership function for this spectral band and this class would be broad (i.e., its *dispersion* would be high). Sometimes a much tighter spread is seen (B) implying that in this band, memberships for this class are much more strongly determined by spectral score. A graphical summary of the same data is shown in Figure 18.4. The error bars represent a measure of the dispersion of the fuzzy membership function for this class, in this band, and can be interactively selected and viewed.

To illustrate more clearly what a fuzzy membership function really means, it is useful to take a single spectral band, and to plot a line graph for each land cover class, where, for every reflectance score between 0 and 255, the mean class membership for pixels with that reflectance value is shown (Figure 18.5). The dispersion of the fuzzy membership function has been measured here (A and B) by recording the level at which mean memberships are 50% of the maximum mean membership.

The above example shows the use of a traditional visualization metaphor (the spectral signature plot) to introduce the concepts of a fuzzy classification which is based on clustering in spectral space, and on the assignment of image pixels to classes with non-Boolean boundaries. The "blurred" spectral signatures illustrate the fact that a plotted image pixel can fall anywhere within the boundaries of a class, giving a continuous range of membership "possibilities" (see Figure 18.6).

End Users and Decision Support

The software tools provided should make it possible to carry out a clear, well-documented sequence of procedures from classification to decision, and to manipulate the results in a variety of ways; for ex-

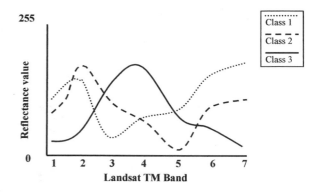

Figure 18.2. A traditional spectral signature plot showing the characteristic reflectances of three landcover classes.

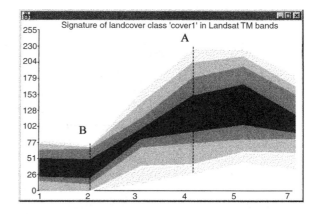

Figure 18.3. A "fuzzy spectral signature" showing the variation that is present within a single land cover class.

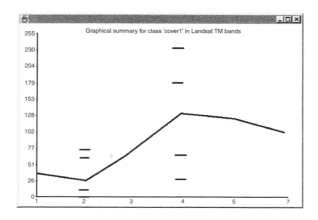

Figure 18.4. A graphical summary of a fuzzy spectral signature showing error bars in two spectral bands selected by the user.

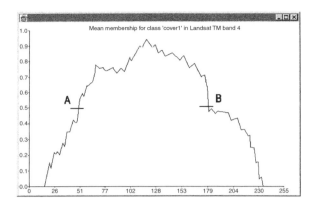

Figure 18.5. A plot of mean membership in a single landcover class, "cover1," against pixel reflectances in a single spectral band (Landsat TM band 4). The spread of values is a rough representation of the fuzzy membership function for that class.

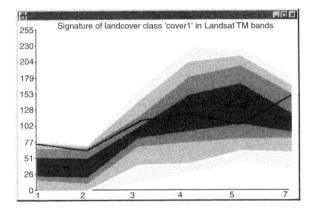

Figure 18.6. Here, the spectral signature of a single pixel, selected by the user, has been plotted over a fuzzy spectral signature, demonstrating the pixel's membership in class "cover1." A multidimensional clustering in feature space is represented simply by a two-dimensional graph, which draws on the traditional metaphor of the spectral signature.

ample, the splitting or collapsing together of qualitative land cover classes, followed by a reclassification, or statistical analyses such as searching for correlations between different data layers.

The analysis tools provided must be flexible enough to allow users to design their own sequences of rules and analyses for specific purposes, while the visualization tools are necessary for viewing and exploring the results of such analyses. To give one example, an ecological worker may be interested in the occurrence of bracken, even at very low densities within a pixel, while having no interest in distinguishing different types of grassland or heath. Using these tools and a modicum of field data, they should be able to specify easily the particular signature and textures which characterize bracken, and to select any pixels which have membership in the class "bracken." These pixels can then be compared to ground verification data, to assess the weighting that should be applied to fuzzy memberships in order to give an estimate of bracken cover. Finally, a variety of visualizations of the area could be produced, to assist in the search for visual patterns or trends over time.

Another user might be interested in the general effect which geocorrection has had on the spatial accuracy of their satellite data. By inputting the RMSE at each of the GCPs used, they should be able to produce a distributed surface showing the estimated spatial error at any point, and from this to predict the qualitative uncertainty introduced by these spatial shifts (Hughes et al., this volume). In this case, 3D representations, with uncertainty or variance as an elevation grid and landcover as a drape, would be useful.

CONCLUSIONS

Thematic classification involves the clustering of data in multidimensional feature space, in order to derive information about a landscape which may itself be familiar. Visual exploration of this multidimensional space can help to explain the process of classification, and to highlight areas for further investigation. Fuzzy classifications in particular multiply the data dimensions which can be explored, increasing the need for standard methods of data combination and summary in order to create clear and comprehensible visualizations. In addition, fuzzy sets and memberships may be unfamiliar to many users of remotely-sensed imagery, but can be more easily introduced and explained by the use of familiar metaphors such as the spectral signature plot. The software described in this chapter is still in development, and recent details can be found at

http://www/geog.le.ac.uk/fliers/vis_software.html

ACKNOWLEDGMENTS

This work is supported by the European Union, DG XII, under the Environment and Climate Programme, Contract No: ENV4-CT96-0305. For more information on the project, and the other European institutions involved, please see *http://www.geog.le.ac.uk/fliers/*

REFERENCES

Chrisman, N.R. The Error Component in Spatial Data, in *Geographical Information Systems: Principles and Applications, Volume 1*, D.J. Maguire, M.F. Goodchild, and D.W. Rhind, Eds., Longman Scientific and Technical, Harlow, United Kingdom, 1991.

Davis, T.J. and C.P. Keller. Modelling and Visualizing Multiple Spatial Uncertainties. *Com. Geosci.,* 23, pp. 397–408, 1997.

DiBiase, D., C. Reeves, A. MacEachren, M. von Wyss, J.B. Krygier, J.L. Sloan, and M.C. Detweiler. Multivariate Display of Geographic Data: Applications in Earth System Science, in *Visualization in Modern Cartography,* A.M. MacEachren and D.R. Fraser Taylor, Eds., Pergamon Press, Great Yarmouth, United Kingdom, 1994.

Ehlschlaeger, C.R., A.M. Shortridge, and M.F. Goodchild. Visualising Spatial Data Uncertainty Using Animation. *Com. Geosci.,* 23, pp. 387–395, 1997.

Fisher, P.F. Visualization of the Reliability in Classified Remotely Sensed Images. *Photogrammetric Eng. Remote Sensing,* 60, pp. 905–910, 1994.

Foody, G.M. Cross-Entropy for the Evaluation of the Accuracy of a Fuzzy Land Cover Classification with Fuzzy Ground Data/ *Photogrammetry and Remote Sensing*, 50, pp. 2–12, 1995.

Foody, G.M. Approaches for the Production and Evaluation of Fuzzy Land Cover Classifications from Remotely-Sensed Data. *Int. J. Remote Sensing,* 17, pp. 1317–1340, 1996.

Goodchild, M.F. and G.J. Hunter. A Simple Positional Accuracy Measure for Linear Features. *Int. J. Geogr. Inf. Sci.,* 11, pp. 299–306, 1997.

Hootsmans, R.M. Fuzzy Sets and Series Analysis for Visual Decision Support in Spatial Data Exploration. Thesis presented to the University of Utrecht, Netherlands, in partial fulfillment of the requirements for the degree of Doctor of Philosophy, 1996.

Hootsmans, R.M., W.M. de Jong, and F.J.M. van der Wel. Knowledge-Supported Generation of Meta-Information on Handling Crisp and Fuzzy Datasets. *Proceedings of the 5th International Symposium on Spatial Data Handling, Charleston,* 2, 1992, pp. 470–479.

Hughes, M., J. Bygrave, L. Bastin, and P. Fisher. High Order Uncertainty in Spatial Information: Estimating the Proportion of Cover Types Within a Pixel, in *Spatial Accuracy Assessment: Land Information Uncertainty in Natural Resources,* K. Lowell and A. Jaton, Eds., Ann Arbor Press, Chelsea, MI, 1999, pp. 319–323.

Jensoulin, R. and M.F. Goodchild, Eds. *Data Quality in Geographic Information: From Error to Uncertainty*, Hermes, Paris, 1998.

Klir, G.J. On the Alleged Superiority of Probabilistic Representation of Uncertainty. *IEEE Trans. Fuzzy Syst.,* 2, pp. 27–31, 1994.

Klir, G.J. and T.A. Folger. *Fuzzy Sets, Uncertainty and Information*, Prentice-Hall International, London, 1988.

Maselli, F., C. Conese, and L. Petkov. Use of Probability Entropy for the Estimation and Graphical Representation of the Accuracy of Maximum Likelihood Classifications. *ISPRS J. Photogrammetry and Remote Sensing,* 49, pp. 13–20, 1994.

Monmonier, M. Geographic Brushing: Enhancing Exploratory Analysis of the Scatterplot Matrix. *Geographic Analysis,* 21, pp. 81–84, 1989.

Moore, K., J. Dykes, and J. Wood. Using Java to Interact with Geo-Referenced VRML within a Virtual Field Course, *http://www.geog.le.ac.uk/mek/usingjava.html*, (accessed 5/2/1998), 1997.

Rhyne, T. Developing an AVS Based Training Program for Environmental Researchers at the U.S. EPA, *Proceedings of the 2nd Annual International AVS User Group Conference, Orlando*, 1993.

Sun. Java Beans Standards, *http://java.sun.com/beans/docs/spec.html*, (downloaded 5/2/1998), 1997.

Van der Wel, F.J.M. and R.M. Hootsmans. Visualisation of Quality Information as an Indispensable Part of Optimal Information Extraction from a GIS, *Proceedings of the 16th International Cartographic Conference, Cologne,* 2, 1993, pp. 881–897.

Wegman, E.J. Hyperdimensional Data Analysis Using Parallel Coordinates. *J. Am. Stat. Assoc.,* 85, pp. 664–675, 1990.

Wood, J. and P.F. Fisher. Assessing Interpolation Accuracy in Elevation Models. *IEEE Com. Graphics Appl.,* 13, pp. 48–56, 1993.

CHAPTER 19

Use of Variograms to Represent Spatial Uncertainty of Geographic Linear Features

F. Vauglin

GEOGRAPHICAL DATA QUALITY

In geographical databases, geographical features are usually divided into two components: geometric data and semantic data. A description of the shape and position of such features is given by points, polylines, and polygons. As digitizing shape and position involves numerous and complex steps, the data quality is never perfect. Some basic principles for assessing geographical data quality are given in this chapter.

Concepts and Definition of Quality

Assessing data quality can be done by comparing the data to a reference data set of higher accuracy. From the geographical data producer's point of view, the quality can be assessed by measuring the discrepancy between what should have been produced and what has actually been produced.

"What has actually been produced" is the data set.

"What should have been produced" is called "*terrain nominal*" in French—"nominal ground." It can be interpreted using a metaphor: if one calls "universe" the region of the world that must be captured, the nominal ground is the universe seen through the filter constituted by the specifications of the database (David and Fasquel, 1997).

Figure 19.1 represents graphically the important steps for assessing data quality. On the left part of Figure 19.1, quality controls are a practical way to actually measure data quality. They involve data sets of higher accuracy that are considered to be very close to the nominal ground, which is an abstract concept. That

way, assessing the data quality is done by measuring the discrepancy between two data sets: the data set itself, and the control data set of higher accuracy, covering the same area and supposed to have been captured with the same specifications. In practice, any data set of higher accuracy may be useful.

Throughout this chapter, two linear features are compared: linear feature A coming from the data set, and its nominal counterpart B coming from a data set of higher accuracy (Vauglin, 1997). Geographical data quality is often subdivided into five, six, or seven components to make its assessment easier and more detailed. These components include positional accuracy, semantic accuracy and completeness, temporal accuracy, logical consistency, and genealogy (Goodchild and Gopal, 1989; CEN, 1996; David and Fasquel, 1997). The focus is put below on positional accuracy of linear features.

Quality is complex information that can be assessed through measuring each of the quality components. The corresponding measuring of components is made using parameters like standard deviation or mean value. In addition to these generic parameters, specific parameters are developed for each of the components of quality.

Positional Accuracy

Positional accuracy is a major component of geographical data quality. Its assessment cannot be done through a global value because positional accuracy must also be divided into absolute positional accuracy and relative positional accuracy (ISO, 1996). For in-

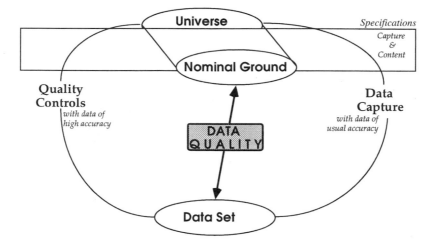

Figure 19.1. Metaphor of "nominal ground" and definition of data quality.

stance, planimetric accuracy and altimetric accuracy should be assessed using the same techniques and parameters, but they should be presented separately for a finer description of positional accuracy. Similar recommendations are made by ISO (International Standardisation Organisation) and other major authors (Goodchild and Gopal, 1989; CEN, 1996; David and Fasquel, 1997).

In the ISO document (ISO, 1996), absolute planimetric accuracy of a geographic feature is assessed by the distance between the feature A and its position in the nominal ground (feature B). For raster data, the distance is computed between pixel centers.

Points

Parameters for points or for raster data that can be computed from the distance in X and Y or Z: mean, variance and standard deviation, matrix of variance-covariance, and interval containing 95% of the measures. This interval is represented by its two boundaries or by a confidence ellipsoid.

Lines and Surfaces

Parameters for polylines or surfaces are based on an epsilon band: an epsilon band is built as the buffer of width epsilon around the feature, and a rate of inclusion is chosen by the user. This rate of inclusion is the percentage of length of the feature that is inside the epsilon band. A function giving the rate of inclusion in function of epsilon is defined in that way.

From the rate of inclusion that has been chosen by the user, the mean value for epsilon is computed as well as the variance and the standard deviation, and the interval containing 95% of variation of epsilon.

Relative accuracy is often assessed by comparing the distance between the location of two features in the data set and the distance between the corresponding two features in the nominal ground. Parameters are then defined in the same manner as for absolute accuracy. Most of the research efforts have been carried out on absolute accuracy of points until now. This chapter gives a focus on relative accuracy of linear features.

REPRESENTATION OF POSITIONAL UNCERTAINTY OF LINEAR FEATURES

The production of geographical data is a long and complicated process that potentially leads to many errors, faults, or other accuracy and shape problems. In this chapter, the positional uncertainty of a line is considered as a statistical field between the line A and its nominal counterpart B. In such statistical context, "uncertainty" will be used instead of "accuracy."

Positional Uncertainty

Abbas (1994) and Goodchild and Hunter (1993) have defined methods to measure positional uncertainty between linear features. Abbas proposes a technique based on the Hausdorff distance. Goodchild's measure is defined on an epsilon buffer zone. These attempts provide complex measures from which statistical information on relative accuracy is hardly deducible.

A New Parameter for Positional Uncertainty: "Geometrical Discrepancy"

Vauglin (1997a) defines another measure, called "Geometrical Discrepancy" that is better suited for assessing a statistical field of positional uncertainty of a line. Like Abbas's technique, this measure derives from Hausdorff distance.

Hausdorff distance d_H between two features A and B is defined as follows:

$$d_{AB} = \sup_{M_a \in A}\left(\inf_{M_b \in B} \left| M_a - M_b \right| \right)$$

$$d_{BA} = \sup_{M_b \in B}\left(\inf_{M_a \in A} \left| M_a - M_b \right| \right) \quad (1)$$

$$d_H(A,B) = \max(d_{AB}, d_{BA})$$

d_{AB} and d_{BA} are also called "Hausdorff distance's components."

In Figure 19.2, both Hausdorff distance's components d_{AB} and d_{BA} are computed and the Hausdorff distance d_H is found to be d_{AB}.

Assessing Hausdorff distance implies computing the expressions

$$\inf_{M_b \in B} \left| M_a - M_b \right|$$

and

$$\inf_{M_a \in A} \left| M_a - M_b \right|$$

for each position on both lines: Hausdorff distance is the maximum of all values of these expressions. Instead of keeping just the very special value (1), "Geometrical Discrepancy" is made of all possible values of d_{AB} and d_{BA}. That measure is represented by the whole set of arrows in Figure 19.2 and gives a very detailed representation of positional uncertainty.

In a more formal way, Geometrical Discrepancy is defined for each point of the line A by its closest Euclidean distance to the line B—and reciprocally. The infinite set of all these closest Euclidean distances can be presented as a function f_A of the curvilinear

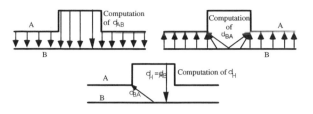

Figure 19.2. Calculation of Hausdorff distance between polylines A and B.

abscissa of A—and reciprocally for function f_B. f_A and f_B are the functional representation for Geometrical Discrepancy.

Use of "Geometrical Discrepancy"

Histograms of functions f_A and f_B can be computed and fitted to classical statistical functions (probability density functions). This is a useful method to model absolute positional uncertainty, leading to mixes of Gaussian and Exponential models. This approach has been developed in Vauglin (1997).

The aim of this chapter is to put a stress on relative positional uncertainty. The use of Geometrical Discrepancy for this focus is presented in the next section.

Relative Positional Uncertainty of Lines

As stated before, positional uncertainty of a line is considered as a statistical field between the line itself and its nominal counterpart. Geometrical Discrepancy is a measure on that field (see Figure 19.2) and will be used to represent the structure of spatial correlation of uncertainty (in other words: the structure of relative uncertainty).

Geostatistics has been developed as a mathematical theory to study fields in geological problems. Its favorite tool about spatial autocorrelation is the variogram. In this chapter, variograms are used to visualize the structure of spatial autocorrelation and to assess parameters representing the relative positional uncertainty of geographical lines.

Definition of Semivariogram

Let Var and E be the variance and the expectation of a random variable. Let h be the lag of the curvilinear abscissa: $h = |x - y|$. The *semivariogram* or *variogram* of γ a regionalized variable Z is defined for $h \geq 0$:

$$\gamma(h) = \frac{1}{2}\text{Var}(Z(x+h) - Z(x)) \qquad (2)$$

If the trend $m(h) = \text{E}(Z(x+h) - Z(x))$ is negligible, one can simplify γ:

$$\gamma(h) = \frac{1}{2}\text{E}\left[(Z(x+h) - Z(x))^2\right] \qquad (3)$$

This definition is taken from Matheron (1978). Isaaks and Srivastava (1989) and Cressie and Hawkins (1980) give further details on variograms. Practical methods to compute variograms are described in Clark (1982); this technique applied to the positional uncertainty field of geographical linear features is detailed in Vauglin (1997).

Practical Use of Variograms

Practical implementation of variograms of Geometrical Discrepancy has been tested on a road network on Amplepuis, France. This data set comes from IGN's cartographic database BDCarto®[1] (scale: about 1:100.000) and the nominal ground is assessed by the IGN's topographic database BDTopo®[1] (scale: about 1:20.000).

Amplepuis is a small city (5000 inhabitants) and the landscape is cut into by small valleys. There are about 120 road sections delimited by crossroads.

Variograms can be described by specific parameters, presented in Figure 19.3. The *sill* is the limit value of γ at higher values of h. The *range* is the value of h at which 95% of the sill is reached. The *nugget effect* is the value of γ when $h=0$. The *slope at the origin* is computed after the nugget effect (Clark, 1982).

Experimental variograms are computed from the theoretical definition 2 or 3 and fitted to existing theoretical variograms from literature (Clark, 1982; Isaaks and Srivastava, 1989). The best fit can be obtained by choosing appropriate parameters for the theoretical variograms. In this study, the validation of the "best fit" has been done visually because there is no definitive method of fitting variograms and of validation of fitting. Some research has been conducted on that theme by Cressie and Hawkins (1980).

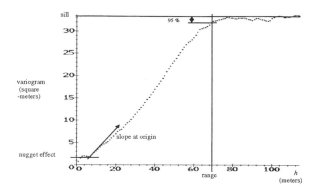

Figure 19.3. A variogram and its parameters.

Theoretical Models of Variograms

Experimental variograms that have been computed on Amplepuis's road network have been found to have a nugget model behavior (Figure 19.4), a spherical model behavior (Figure 19.5), and a power model behavior (Figure 19.6). Some of them also have a periodic model behavior (Figure 19.7). Interpretations of these results are given after the presentation of theoretical models. Many other theoretical models exist in the literature (Clark, 1982; Isaaks and Srivastava, 1989; Cressie and Hawkins, 1980), but these four models have enough variety to describe all experimental results in this study.

Analysis of Experimental Variograms for Positional Uncertainty of Polylines

Besides some exceptions, most experimental variograms are best described by a mix of several theoretical models. A nugget effect could be found for every experimental variogram. The magnitude of this effect is about 5 to 10% in most cases, and sometimes lower than 4%. This effect is intricate to a "starting bump" effect: a bump can be noticed for the lowest values of h (between 0 and 30 meters and mostly about 15 m). These cumulated effects could be a sign of generalization of small geometric details. The width in h is connected to the size of the generalized details and the maximum value of the starting bump is connected to the global amount of details that have been generalized.

The value of the sill can be linked to the parallelism of the lines. A high value for the sill of the variogram means that lines are not parallel at all and a low value means they have a good parallelism. This

---

[1]   Registered Trademark of Institut Géographique National, Paris, France.

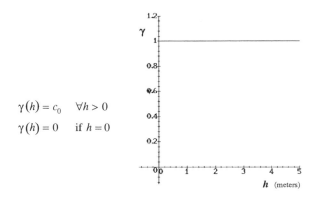

$$\gamma(h) = c_0 \quad \forall h > 0$$
$$\gamma(h) = 0 \quad \text{if } h = 0$$

Figure 19.4. Nugget model.

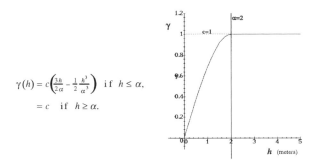

$$\gamma(h) = c\left(\frac{3h}{2\alpha} - \frac{1}{2}\frac{h^3}{\alpha^3}\right) \text{ if } h \leq \alpha,$$
$$= c \quad \text{if } h \geq \alpha.$$

Figure 19.5. Spherical model.

$$\gamma(h) = p.h^{\alpha}, \qquad \forall h \geq 0$$

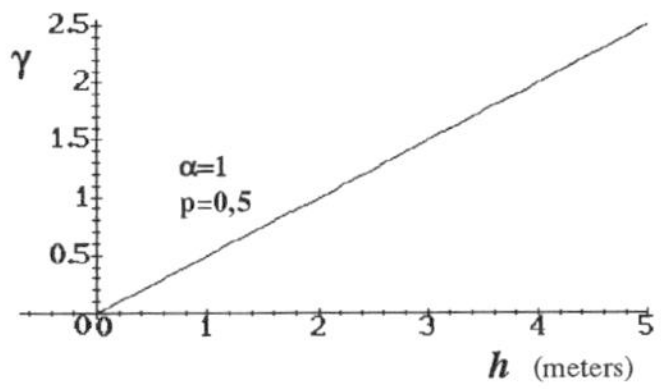

Figure 19.6. Power model.

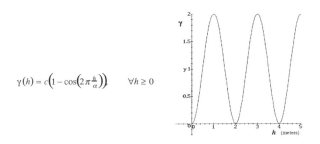

$$\gamma(h) = c\left(1 - \cos\left(2\pi\frac{h}{\alpha}\right)\right) \qquad \forall h \geq 0$$

Figure 19.7. Periodic model.

effect is visible in Figures 19.8 and 19.9. Coordinates of lines in X and Y in Figure 19.8 are meters.

The dotted line of Figure 19.8 is taken from BDCarto®, and the black line is considered as a nominal contour because it is coming from BDTopo® and has a higher accuracy. In the middle of Figure 19.8 is the variogram of Geometrical Discrepancy computed from the BDCarto® line to the BDTopo® line; on the right is the converse (see Figure 19.2).

Similarly to Figure 19.8, the dotted line of Figure 19.9 is taken from BDCarto® and the black line is its BDTopo® counterpart. In the middle of Figure 19.9 is the variogram from the BDCarto® line to the BDTopo® line; on the right is the converse (see Figure 19.2).

Some experimental variograms have been found to have a periodic behavior as in Figure 19.10: this a priori unexpected result can be explained by the particular shape of the line: a variogram of Geometrical Discrepancy is periodic when vertices of the line have a periodical distribution on the curvilinear abscissa or when the lines intersect periodically. See further examples and explanations in Vauglin (1997).

Two other parameters that are presented in Figure 19.4 can also have a physical meaning: the slope at the origin of the variogram has been linked to the regularity of the lines. A formal link can be built between the slope and the fractal dimension of the lines for some types of shapes of geographic lines.

The range of the variograms represents the distance one has to move on a line to keep as much as possible away from autocorrelation problems. More precisely, points of a line that are closer than the range have a high correlation in their positional uncertainty, whereas points that are further than the range cannot have less correlation on this line.

CONCLUSION AND FUTURE RESEARCH

The computation of variograms of positional uncertainty is possible when measured by Geometrical Discrepancy. As shown in this chapter, such experimental variograms compared to theoretical models lead to three main types of variograms: Power model, Spherical model, and Periodic model (plus nugget effect).

Variograms prove to be an efficient tool to represent spatial autocorrelation from which a lot of parameters describing the shape of lines can be assessed. To be more general, parameters of variograms that have been presented here could be used on other data sets, like rivers or land use or houses to test their interpretation given here.

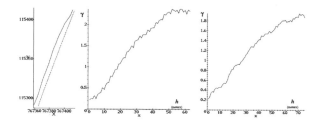

Figure 19.8. A line parallel (dotted) to its nominal counterpart (black) and the corresponding variograms (sill=2 m²).

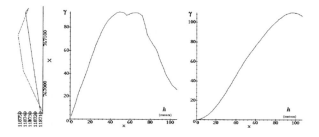

Figure 19.9. A line nonparallel (dotted) to its nominal counterpart (black) and the corresponding variograms (sill or maximum≈100 m²).

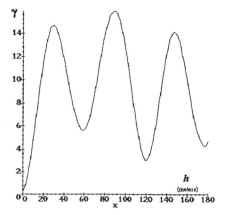

Figure 19.10. Example of a periodic experimental variogram.

The results presented here are a function of the definition of a "road section." "Road sections" are delimited by two crossroads in this chapter, and another definition could lead to different results. This still has to be tested.

The range of a variogram is the most connected parameter to spatial autocorrelation of positional uncertainty. It gives valuable information on positional accuracy autocorrelation within the line. A link still has to be made to have a description of spatial autocorrelation between different lines, and the Geometrical Discrepancy tool may be of some use for that purpose.

Better knowledge of spatial autocorrelation is a step toward heteroscedasticity of positional uncertainty, which implies the definition of regions of homogeneous statistical behavior for positional uncertainty.

REFERENCES

Abbas, I. *Base de Données Vectorielles et Erreur Cartographique: Problèmes Posés par le Contrôle Ponctuel; Une Méthode Alternative Fondée sur la Distance de Hausdorff: le contrôle linéaire,* thesis presented to Paris VII University, Paris, France, in partial fulfillment of the requirements for the degree of Doctor of Philosophy, June 1994.

Comité Européen de Normalisation. TC287. *Geographic Information—Data Description.* Provisory document for public enquiry during 1996–97, 1996.

Clark, I. *Practical Geostatistics,* Applied Science Publishers Ltd., London, 1982.

Cressie, N. and D.M. Hawkins. Robust Estimation of the Variogram, *J. Int. Assoc. Math. Geol.,* 12(2), pp. 115–125, 1980.

David, B. and P. Fasquel. Qualité d'une base de données géographique: concepts et terminologie, *Bulletin d'Information #67 of National Geographic Institute* (IGN), J. Poulain and S. Motet, Eds., IGN-SR Classification Number 970004/R-LIV, Saint-Mandé, France, February 1997.

Goodchild, M.F. and S. Gopal, Eds. *Accuracy of Spatial Databases,* Taylor & Francis, New York, 1989.

Goodchild, M.F. and G.J. Hunter. A Simple Positional Accuracy Measure for Linear Features, Technical Communication, *Int. J. Geogr. Inf. Syst.,* 1993.

Isaaks, E.H. and R.M. Srivastava. *An Introduction to Applied Geostatistics,* Oxford University Press, New York, 1989.

International Organisation for Standardisation (ISO). *Geographic Information—Quality Evaluation Procedures,* Draft, June 1996.

Matheron, G. *Estimer et Choisir,* École des Mines, Paris, France, 1978.

Vauglin, F. Statistical Representation of Relative Positional Uncertainty for Geographical Linear Features, in *Data Quality in Geographic Information—From Error to Uncertainty,* M. Goodchild and R. Jeansoulin, Eds., Hermès, Paris, France, 1997, pp. 87–96.

CHAPTER 20

Assessing and Visualizing Accuracy During 3D Data Capture at Digital Photogrammetric Workstations

J. Drummond, N. Burden, P. Gray, and D. Tait

INTRODUCTION

As we move increasingly toward three-dimensional GIS and incorporate the work already done in 3-D visualization; e.g., Kraak (1994), not only will effective means need to be developed for storing, retrieving, and manipulating these data, but also for gathering them. For several decades the bulk gathering of 3-D terrain data has involved stereophotogrammetry, which has always required considerable skill whether the stereoplotters were optical, mechanical, or analytical. Although the concepts emerged in the 1960s (Hobrough, 1968), it is only within the last four years that digital photogrammetric procedures have begun to dominate in mapping and other geospatial data gathering organizations. Digital photogrammetry is a source of three types of GIS data, namely:

1. Digital elevation models;
2. Digital orthophotos; and
3. Vector data resulting from stereo or mono on-screen orthophoto digitizing.

The three products can be considered part of the same *orthophotomapping production line,* summarized in the next section.

The democratization of photogrammetry may be presented as an outcome of digital photogrammetry (Leberl, 1991). What this means is addressed in this chapter; although the consequences are mainly related to the quality of the end result and the ease of use of the system—which is also addressed. The concluding section of this chapter suggests areas for future effort, with particular reference to data quality.

THE ORTHOPHOTOMAPPING PRODUCTION LINE

Minimum input to the orthophotomapping production line are two files resulting from digitally scanning a stereo-pair of aerial photographs and within the overlap three ground control points whose Eastings (E), Northings (N), and elevation (h) are known.

Although much actual data processing takes place in the photocoordinate system most orthophotomapping tasks require the operator to provide locations in the screen coordinate system. The screen coordinate system's origin is usually the top right corner of the screen; the origin of the photocoordinate system is the fiducial center (subsequently shifted to the principal point of the photograph for computations). The alignment the screen coordinate system's axes is the columns and rows of the screen's pixels, whereas the alignment of the photocoordinate system's axes is defined by the intersection of each photograph's fiducial lines. Transformation is achieved by determining the screen coordinates of the fiducial marks, knowing the fiducial marks' locations in a calibrated aerial camera and the principal point's offset from the fiducial center in the photocoordinate system. Many elementary photogrammetry texts; e.g., Wolf (1994) amply explain the photocoordinate system, principal point, fiducial center (center of collimation) fiducial lines and depict fiducial marks.

The location of each photograph's fiducial center (and hence principal point) is usually achieved by measuring, on screen, the location of the four or eight fiducial marks around the edge of the photo and is (a part of) the process referred to as *interior orientation.*

Other details of the camera have to be known, including focal length and radial lens distortion characteristics. This information will be in a camera calibration report—usually provided by the aerial photography suppliers.

Certain characteristics of the aircraft at the time of photography have to be known, namely:

1. the position of the camera station (Xo,Yo,Zo); and,
2. the orientation (customarily represented by the rotation angles ω,ϕ,κ) of the camera at the time of photography,

with respect to the axes of the ground coordinate system.

Determination of these is referred to as *exterior orientation,* although also referred to as triangulation (or space resection, if one photo at a time is dealt with). Once the orientation elements are determined, new left and right image files can be generated which have had tilt distortion effect removed, leaving only relief distortion effect. Referred to as correction for tilt, rectification, or normalization it relies on photogrammetry's "Collinearity Condition" (well described in Kraus, 1993, pp. 3–16). The resulting digital images are thus "normalized."

A variety of image matching procedures are available, but most are based on correlation, and will result in the most likely pairing of a left-normalized-image pixel with a right-normalized-image pixel. It is assumed the paired pixels will represent the same photographed terrain feature, and the difference in their x-photocoordinates—called x-parallax, is a function of the feature's elevation—exploiting photogrammetry's so-called "Parallax Equations" (see Wolf, 1994). Having determined a feature's elevation, one can return to the "Collinearity Condition" to determine also the features X,Y ground coordinates (or Eastings, Northings).

Image matching is highly automated. Maybe 48 million pixels of the left photograph will be matched to 48 million from the right photograph, in a few minutes. The result is a dense but irregular DTM, which may be resampled to provide a regular DTM, and will be used to correct for relief distortion's effect in the (e.g.) left-normalized-image, giving an image free from both tilt and relief distortion—i.e., the digital orthophoto. A digital orthophoto has exactly the same geometric properties as a topographic map, and can be used for 2-D digitizing, but contains all the additional attribute information of a photo rather than a map. An additional product, the stereomate, can be produced and viewed with the orthophoto in stereo, to support 3-D digitizing.

The orthophoto production line is presented in Figure 20.1.

DEMOCRATIZATION OF PHOTOGRAMMETRY

Before the emergence of digital photogrammetry, measurement was performed directly on glass/film diapositives of the aerial film negative, and correction for tilt distortion was carried out within the viewing system. Computer programs supporting stereoplotting determined what this tilt correction should be, but prior to their existence, removal of tilt distortion was one of the most highly skilled techniques in mapping. Also before the advent of digital photogrammetry the determination of x-parallax (to determine height) was achieved point by point (although this may have appeared continuous) by the operator, and was also highly skilled. Stereoplotter operators were a small group of mapping specialists requiring long and expensive practical (although not theoretical) training. A significant disadvantage was that stereoplotter operators, although skilled at, e.g., plotting contours, rivers, and roads, might not also be able to interpret soil, vegetation, or geomorphology.

Digital photogrammetry, having automated both tilt correction and the determination of x-parallax, presents an environment where the soil, vegetation, or geomorphological expert lacking photogrammetric skills may, nevertheless, extract features of interest from aerial photographs with a high degree of accuracy, while viewing in stereo. This apparently presents an attractive liberation of the general GIS user from the shackles of photogrammetry! Particularly the geomorphologist, for whom digital photogrammetry can provide a very dense representation of the terrain (e.g., a microrelief DTM) to be processed using available terrain-analytic techniques.

But how can the, (e.g.), geomorphologist be assured of the quality of the digital photogrammetric product—which surely must happen if important scientific conclusions or planning decisions are taken based on the DTM? This question will be addressed in the next section, but first, one or two more pitfalls arising from the "democratization of photogrammetry" will be indicated.

Digital photogrammetry is a software-based approach; the photogrammetric past required complex

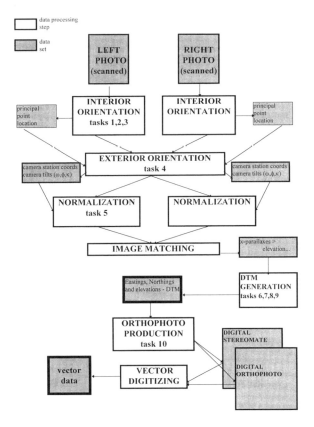

Figure 20.1. Digital orthophoto production line.

and high precision optical and mechanical instruments. Software may be expensive to create but is cheap to reproduce—therefore digital photogrammetric systems are thus available. These systems require scanned aerial photos. Desktop scanners can be bought for a few hundred dollars. For a few dollars, paper print copies of archived diapositives can be bought. Ground control point coordinates can be read off topographic map sheets—also a few dollars each. So with a few skills and a few dollars, photogrammetry is now a 3-dimensional GIS data-gathering tool available to all.

The resolution of low-cost desktop scanners continues to improve and may soon provide pixels as small as the 10 μm recommended by Kolbl et al. (1996). However, such scanners are not stable and introduce geometric distortions which are both rarely and with difficulty modeled (Drummond, 1990). A rather trivial further problem with A4 format desktop scanners is that they may not quite reach and scan the fiducial marks of the standard format 230 mm aerial photograph. Diapositives cost considerably more than paper prints—so paper prints are popular. Like low-cost scanners they,

too, introduce little-understood geometric distortion. Furthermore, some (but by no means all) suppliers of aerial photography are unable or unwilling to provide the calibration reports from which accurate information on camera characteristics can be determined. And, finally, ground control. Using professional instrumentation, easy-to-use GPS procedures can now provide ground control point coordinates having standard deviations of one or two centimeters; the inherent cartographic error of a map is about 0.5 mm at map scale—or 5 m at ground scale off a 1:10,000 map. Thus, low-budget methods can introduce large errors.

QUALITY OF THE DIGITAL PHOTOGRAMMETRIC PRODUCT

Considering Figure 20.1, the data products of concern to the GIS user are the DTM and the Vector Data (shown in heavier boxes). It would be appropriate if any photogrammetric system feeding data into a GIS provided information on the quality of these data. The two systems (Helava-Leica DPW 600 and Virtuozo) most fully investigated by the authors do, indeed, do so.

Quality Reports

Helava-Leica provides a report file of quality statistics (checkpoint residuals and overall RMSE values) using a *checkpoint file* compared to either the DTM File or a Measured Point File (the vector data of selected features; i.e., the checkpoints as they appear in the orthophoto). The checkpoint file represents "the truth." Although truth is never known, a version of it can be generated using surveying methods of a higher accuracy than those being tested—e.g., land survey, or photogrammetry using more precise methods. Parts of an example report are provided in Table 20.1; in this case the "truth" was derived from conventional stereoplotting of the diapositives scanned for the digital procedures. We considered the 20 μm scanning resolution used had degraded the digital photogrammetric procedures compared to traditional steroplotting procedures and we had confidence in the photogrammetric skills of the student performing the experiment. Thus the traditional stereoplotting coordinate determinations were, in this case, considered better than the Digital Photogrammetric ones and could represent the "truth."

Likewise, Virtuozo also provides a Processing Information Report File, showing data quality information. However, these files are very different. The Helava-Leica file is a report of checks on the end prod-

Table 20.1. Part of a Checkpoint File vs. Measured Points.[a]

Point#	X Diff	Y Diff	Z Diff
3	−1.691	0.487	1.223
4	−0.193	−1.059	−2.392
5	−2.016	0.925	−1.640
7	1.110	−0.552	1.523
.			
.			
	rmsX	rmsY	rmsZ
	1.492	1.249	1.905

[a] Wake, 1998.

uct (DTM or orthophoto) and not of any preceding processing steps, whereas the Virtuozo report is of the quality of the each of the processing steps and not the end product.

Quality at Each Processing Step

The three processing steps that determine parameters used in subsequent processing are: interior orientation, exterior orientation, and image matching.

In each system, interior orientation requires the determination of the image coordinates of the fiducials (whose photocoordinates are known from the camera calibration report) to derive the unknown transformation constants to convert all image coordinates to photocoordinates. Based on these constants, the photocoordinates of the fiducials can be redetermined, and the *residuals between the calibrated and redetermined fiducial photocoordinates* considered—thus providing information on the quality of the transformation constants. This is important because if the transformation from image to photocoordinates is poor, any of the subsequent and frequent processing relying on the collinearity condition (which uses photocoordinates) will also be poor. In the Helava-Leica system these residuals are presented as RMSE values in pixels and in the case of Virtuozo, in ground units. In neither case is the operator required to consider these RMSE values or—for example—to repeat the task if they are poor. Interior orientation is highly automated, employing pattern recognition techniques because any standard photogrammetric camera's fiducials' images can be part of an archive. Thus, errors arising at this stage are likely to be blunders, and so extremely serious if not detected and corrected by the operator. If metric cameras have been used, RMSE values greater than 1 pixel at interior orientation are unacceptable; **when this arises the operator should only be able to proceed after (enforced and) careful consideration**. Stated in pixels rather than meters, the magnitude of the problem is clear.

Exterior orientation determines the unknown exterior orientation elements; i.e., camera rotation angles (ω, ϕ, κ) and camera station coordinates (Xo, Yo, Zo) in ground units. Once determined, these too are used in the many digital photogrammetric processing steps which exploit versions of the collinearity condition. Exterior Orientation requires about three or more ground control points whose Eastings, Northings, and elevation values are known. As with interior orientation, once the unknowns have been determined, the ground control point coordinates can be redetermined, and residuals calculated to provide RMSE values for Eastings, Northings, and elevation. One cannot expect the magnitude of these RMSEs to be less than the standard deviation associated with the original coordinates of the ground control points, but it can be suggested that if the RMSE exceeds twice the typical standard deviation of a control point, then consideration should be given to repeating exterior orientation. Possibly a bad control point contributes to a poor result, and so a point with high residuals may need to be removed from the determination of the exterior orientation elements. Both Helava-Leica and Virtuozo present the residuals at each point and the overall RMSE values to the operator, but again there is no **requirement to repeat the step when results are bad—or any indication as to what might be good or bad**.

Many image matching algorithms are available—usually involving correlation. Pixel patches in the left image are matched with pixel patches in the right in order to pair pixels representing the same terrain feature. Paired pixels then have their x-parallax determined—from which can be derived the terrain feature's elevation and then Eastings, Northings, and thus a DTM. A correct image match is important. Image matching is a batch process, initiated after the selection of appropriate parameters; expertise in image matching is required. To provide information on the magnitude of the, e.g., correlation coefficient and thus, perhaps, the success of the match might be considered useful to the operator. This is addressed in the next section.

Visualizing the Quality of the DTM

If one considers the changes in illumination across a pair of photographs arising from moving clouds or nearness to the edge of the photograph, a low correlation coefficient may occur when pixels have nevertheless been correctly matched and vice versa. So the correlation coefficient may not indicate correctness. However, both Helava-Leica and Virtuozo visualize these correlation coefficients. In the case of Helava-Leica, a grid of points on a user-selected spacing can be displayed in the colors red, yellow, green, cyan, blue, magenta, and white to represent increasing correlation coefficients from <0.15 to >0.99. In the case of Virtuozo, the traffic light colors red, amber, and green are used. Furthermore, Helava-Leica will also display any generated contours in the same set of colors—for example, a red contour indicates some points having a correlation coefficient of <0.15 have been used in its interpolation.

As at other steps in the process, the **values of the statistics on the quality of the image matching (and hence of the derived digital elevation model) never prevent the operator from proceeding to the next stage in the orthophoto production line.** Editing tools are available to correct the derived DTM at, e.g., the "red points"; however, their use requires the traditional photogrammetric skill usually referred to "as setting the floating mark on the ground," which is actually a determination of/compensation for x-parallax. This is not a skill generally available outside the photogrammetric community, as its acquisition requires several months of stereophotogrammetric experience.

The derived DTM is used in orthophoto production, and as mentioned elsewhere, Helava-Leica supports procedures for assessing both the DTM and orthophoto using checkpoint files.

USER FRIENDLINESS OF DIGITAL PHOTOGRAMMETRIC SYSTEMS

If digital photogrammetry has democratized photogrammetry, then a geomorphologist with only photointerpretation skills but operating a digital photogrammetric system should be able to identify and measure features of interest, unaided.

At the end of Figure 20.1, production of the digital orthophoto and its stereomate are shown. The orthophoto has tilt and relief distortion removed and has the geometric properties of a map. On-screen digitizing can be performed on the orthophoto to record features of interest, but stereoviewing improves interpretation. The stereomate is the orthophoto with relief distortion reintroduced—permitting stereoviewing. To ensure the stereoview is not lost, a measuring mark on

the stereomate tracks the operator's movement on the orthophoto (possible because the exact amount of relief distortion is known). The operator thus digitizes off the orthophoto while viewing in stereo, without having to compensate for relief distortion. In the Helava-Leica environment, this **terrain tracking, does not require photogrammetric expertise.** The problem is getting through the preceding steps of the production line to ensure DTM, orthophoto, and the gathered data quality meet requirements!

A comparative study carried out from the viewpoint of HCI (human computer interface) looked at the user-friendliness of several digital photogrammetric workstations' production lines (Burden, 1997). The systems investigated were the Helava-Leica DPW 600 and Virtuozo already mentioned, and also the DMS system of R-Wel Inc. In that investigation a method of evaluating digital photogrammetric workstation HCI was developed and tested on the Helava-Leica and DMS systems (both of which were available at Glasgow University), and then the evaluation method tested independently on the company SDS's Virtuozo system. Although the test looked at systems' functionalities, usability was the investigator's main concern—under the headings: Ease of Use; Ease of Learning; and, Support for Intermittent Use. A set of 10 specified tasks, identified approximately (as 1–10) on Figure 20.1, was carried out. During this work image matching was not "a task," as the same image matching algorithm was always used, becoming in effect a default. In reality, the selection of an appropriate image matching algorithm is a **skill required by those who choose to execute photogrammetric tasks at a digital photogrammetric workstation.**

Two operator groups were tested: (1) those familiar with the specific system, and (2) those generally understanding geomatics. Times to complete each task, actual processing mistakes, and correction times were noted. Table 20.2 shows, partially, the result of one session.

Although the thrust of Burden's work was to develop a method for evaluating the user-friendliness of digital photogrammetric workstations, because the evaluation was tried on three systems, some deductions were made. Execution of the tasks was not considered to be easy. In particular, the Helava-Leica system required considerable photogrammetric understanding—although its greater functionality compensated. Neither error messages nor quality statistics were understandable. Burden's evaluators were all exclusively "mapping people," so her conclusion that only

Table 20.2. Example of HCI investigation Observations.[a]

Helava-Leica Task Times: User 1
Total Time: 57.03 minutes

Task	Time (min)	No. of Errors	Error Type	Error Time (sec)
1	3.00	1	m	20
2	5.02	0		
3	14.36	3	c,m,b	5,73,127
4	14.34	2	m,m	30,40
5	2.43	0		
6	7.13	1	m	205
7	0.47	0		
8	2.28	0		
9	3.00	0		
10	2.40	0		

[a] Burden, 1997.
[b] Where error "m" is a simple mistake, e.g., mistyping; "e" the system response is unexpected; "c" the system response is inexplicable; "b" big error, system may need restarting.

people with a photogrammetric background would use such systems and that a knowledge of photogrammetry being required was not detrimental to the systems' use may not stand.

The systems are difficult to use and it has to be assumed that the geomorphologist mentioned at the start of this section can exist. We are also considering error. A system that is difficult to use will not encourage a user to repeat steps when the quality indicators suggest they should—assuming the user understands the quality indicators.

A useful suggestion emerging from Burden's work was that a production flow be mapped before a project begins, so the range of user options can be reduced.

CONCLUSIONS AND RECOMMENDATIONS

That there is no obligation on the operator to consider quality statistics at the various processing stages of digital photogrammetry and that no guidance as to their meaning is provided may be a characteristic of digital photogrammetric systems at this stage in their development. It is hoped that future systems will take into account the requirements of the nonphotogrammetrist and oblige consideration of data quality at relevant stages in the production line. To consider quality only at the end of the production line—as with the

report file approach—is not a procedure used in other industries (e.g., food) where at several stages quality is checked and products may be rejected.

Returning to the fictional geomorphologist, such a person wants to easily measure in stereo; digital photogrammetry allows this. But the geomorphologist also wants reliable measurements. If a user is required to state at the beginning of the project what the maximum acceptable error in measurements taken off an orthophoto can be, then not only can the optimal RMSE be calculated at each of the steps—interior orientation; exterior orientation; image matching/DTM generation—and when these are breached the user told very clearly, but also a running total of error can be kept as production proceeds. Then the user can also be told if the maximum has been reached before getting to the end of the orthophoto production line. Finally, at the end of the orthophoto production line the user can be informed whether acceptable measurements are likely to be made or not.

ACKNOWLEDGMENTS

For their technical contribution to our work we are grateful to the following: Christine Waugh, David Crawford, Neil O'Gorman, Susanne McMahon, and Conor Burns of SDS, Bo'ness, Scotland; the Glasgow University staff members Anne Dunlop (Topographic Science) and Gary Tompsett (Archaeology); and, our students Manolis Papoutsakis, Naief Al Rousan, and Daniel M. Wake.

REFERENCES

Burden, N. *The Evaluation of User Interfaces to Digital Photogrammetric Systems*, Unpublished dissertation for Master of Science degree in Information Technology, University of Glasgow, 1997.

Drummond, J. *Automatic Digitizing*, A report submitted by a Working Group Commission D (Photogrammetry and Cartography), OEEPE Report Nr. 23, Pub IfAG, Frankfurt a.M., Germany, 1990.

Hobrough, G.L. Automation in Photogrammetric Instruments, *Photogrammetric Eng. Remote Sensing*, 31(4), 1968.

Kolbl, O., M. Best, A. Dam, J. Douglass, W. Mayr, R. Philbrik, P. Seitz, and H. Wehrli. Scanning and state-of-the art scanners, Ch. 2 in *Digital Photogrammetry; An Addendum to the Manual of Photogrammetry*, C. Greve, Ed., American Society for Photogrammetry and Remote Sensing, Bethesda, MD, 1996.

Kraak, M.J. Interactive Modelling Environment for Three-Dimensional Maps: Functionality and Interface Issues, in *Visualisation in Modern Catrtography,* D.R.F. Taylor and MacEachern, Eds., Pergamon, Oxford, 1994, pp. 269–286.

Kraus, K. *Photogrammetry, Volume 1, Fundamentals and Standard Processes,* Dummler, Bonn, 1993.

Leberl, F. The Promise of Softcopy Photogrammetry, in *Digital Photogrammetric System,* H. Ebner, D. Fritsch, and C. Heipke, Eds., Wichman, Karlsruhe, 1991, pp. 3–14.

Wake, D.M. *The Quality of Orthophotos Using Helava-Leica DPW,* Unpublished dissertation for Bachelor of Science in Topographic Science, University of Glasgow, 1998.

Wolf, P.R. *Elements of Photogrammetry,* 6th ed., John Wiley & Sons, New York, 1994.

Part V
Spatial Uncertainty Methods

Closely linked to methods of characterizing and representing spatial uncertainty (Chapters 3 and 4, respectively) is a need for appropriate data models. Once one understands how to characterize and represent error, one needs a framework or a structure on which to hang this information. The chapters presented in this Part indicate many possible ways that one can interpret the words "data model." The first two papers in the chapter by Gottsegen et al. and Edwards present conceptual constructs concerning how spatial uncertainty can be handled. The third and fourth papers by De Groeve et al. and Alesheikh et al. present concrete examples of ways that thematic and positional error can be modeled. The fifth paper by Hunter et al. demonstrates the practical implications or error models by presenting the use (rather than the conception) of an error model. Finally, the sixth paper by Duckham does not focus on how particular types of error can be modeled, but rather discusses an error model that has been implemented as computer software.

CHAPTER 21

A Comprehensive Model of Uncertainty in Spatial Data

J. Gottsegen, D. Montello, and M. Goodchild

INTRODUCTION

There is a developing interest in the problem of uncertainty as compared to accuracy or error in spatial data (Unwin, 1995). The notion of uncertainty is broader than error or accuracy and includes these more restrictive concepts. While accuracy is the closeness of measurements or computations to their "true" value or some value agreed to be the "truth" (Unwin, 1995), uncertainty can be considered any aspect of the data that results in less than perfect knowledge about the phenomena being modeled. Thus it is a statement of doubt and distrust in results and is a form of "unknowing" (Thrift, 1985). Researchers of uncertainty in nonspatial data have sometimes used the term "imperfect" data (Motro, 1997).

Uncertainty in data also encompasses data quality described as "fitness for use" by Chrisman (1982). While the terms error and accuracy connote judgments of the appropriateness or usability of data based solely on the comparison of data values to some other set of values, data quality begins to consider the needs of the data user as important in determining the adequacy of data. Issues such as scale, level of aggregation, or classification scheme are critical in a user's assessment of whether a particular set of data is useful for a given task.

This chapter represents an effort to develop a comprehensive view of uncertainty in spatial data. It considers the user a critical component in the definition of uncertainty, and it identifies the processes in the use and creation of data that may contribute to uncertainty. It presents a general model of uncertainty as a framework for this identification and as a possible basis for relating the work being done on the assessment and management of uncertainty in data.

This chapter proceeds by first considering the objectives of defining a conceptual model of uncertainty. It then discusses some of the specific aspects of spatial data development and data use that lead to uncertainty. It then presents the uncertainty model, and identifies where literature in the field fits into the model. It concludes with a consideration of insights gained from the model and the research suggested by the model.

REQUIREMENTS OF A GENERAL MODEL OF UNCERTAINTY

A model of uncertainty should serve several functions. It should expose potential sources of uncertainty and provide ideas for managing or addressing these various sources of uncertainty. This includes identifying sources of uncertainty that cannot be managed. It should also serve as a framework for designing empirical research that can inform efforts resulting from the first function. Finally, it can be used as a benchmark to assess how spatial data researchers are progressing in addressing the topic of uncertainty.

A model of uncertainty needs to address all of the aspects of the use and development of spatial data that constrain a user's knowledge of appropriate uses for the data and the phenomena represented by the data. Such a model must represent the relationship between the data user's knowledge of the world and the data or information maintained by a system (Smets, 1997). It must also integrate the well-accepted definition of data quality as "fitness for use" and the ideas of accuracy and error that have been the subject of considerable research over the past several years. For example, uncertainty in the use of spatial data is partially inversely related to spatial data quality, but it can be considered a distinct concept. That is, high quality spatial data implies a degree of knowledge (i.e., low degree of uncertainty) about the data, the methods used to create it, and its underlying data model or characteristics. The converse is not necessarily true, however. One may have a great deal of metainformation about data that are of low quality simply because they are inappropriate for a certain use. Similarly, accuracy and error have a component that relates directly to uncertainty. The random component of error is, by definition, impossible to know and measure. At best it can be estimated or inferred based on some hypothesized model of its distribution. Bias is deviation from a benchmark that is known to follow a pattern. Hence if error is composed only of bias, there is no uncertainty in terms of the locations (or other measurements) comprising the data.

The model of uncertainty should also describe aspects of a user's query, retrieval, and decision-making process that result in a loss of information about the appropriateness of a given use for particular data. This is the component of uncertainty that is often neglected in the spatial data accuracy literature. However, if one talks about uncertainty in spatial data or the use of spatial data, one must consider the person who may or may not be certain. Another way to view this is that a data user inherently assumes some risk of "incorrect" results from using data. An acceptable level of uncertainty can be considered the amount of this risk that a decision-maker is willing to accept. Information that reduces uncertainty reduces this risk or makes it identifiable (Stinchcombe, 1990). The ability to assess fitness for use implies not only sufficient knowledge about the data, but also a well-defined conception of the use to which the data is applied and the validity of the constructs underlying the application of the data. For example, are the methods for assessing ecosystem health valid? This includes a detailed understanding of how the data that are queried and used represent or correspond to the phenomena that the user is interested in analyzing.

Both of the components described above result from the fact that spatial data and access to them entail formalizations and abstractions. A user must distill her interest in ecosystem health to a set of structured relationships between specific measurements. Additionally, to retrieve data to analyze system health, she must produce a set of discrete valued conditions. Of course, spatial data is also a discrete representation of a model of given phenomena.

The model must also include a consideration of the representation of uncertainty. While considerable research in GIS has concentrated on various techniques of representing error in spatial data, most of it has proceeded based on the assumption that such representation is valuable. It has not developed any consistent guiding principles about how and why such representation would be useful to the user of spatial data. The efficacy of the representation depends on the process by which a data user defines the query or information request, evaluates the appropriateness of the data for a particular use, and uses the data. The interpretation of a representation depends on the user's map schema (MacEachren, 1995). Certain types of representations may facilitate specific decisions to a greater degree, and the user may engage different decision-making methods or heuristics in response to the portrayals of error. In addition, data representations themselves introduce potential uncertainty.

SPECIFIC COMPONENTS OF UNCERTAINTY

Given the comments above, uncertainty in spatial information and its use has several aspects:

1. Uncertainty includes the degree to which the data representation differs from the world or some higher quality representation of it. This may be identifiable or measurable or not. This encompasses the traditional definition of error. The deviation of a representation from reality may be due to (a) measurement error (bias and random), (b) insufficient knowledge to measure a precisely defined concept (e.g., map precisely bounded categories), (c) concepts that cannot be precisely defined, and (d) precision limitations of the measurement or storage device.

2. Uncertainty also relates to the compatibility and consistency between the formally specified need of the user (i.e., data query or information retrieval request) and the data representation. This includes (a) lexical (naming) differences, (b) semantic differences (different associations or meanings for same terms), (c) differences in classification, and

(d) geometric and topological characteristics of the data (e.g., scale, resolution).

3. Uncertainty also depends on the match between the formal specification of a user's query or information request, her subsequent use of the data, and the phenomena of interest. This notion is similar to the idea of construct validity in social sciences (Rosenthal and Rosnow, 1991). That is, do the data and relationships between them specified by the user adequately represent the phenomenon of interest. In some cases where the phenomenon of interest is complex, such as ecosystem modeling, it is extremely difficult to identify and specify the important relationships.

In a sense, uncertainty in data use is the degree to which the formalized structure of data is incompatible with the concepts that a data user is trying to analyze. Together the three aspects listed above influence the applicability of data for a certain use. Other considerations can add to the uncertainty of decisions made with the data, for example, the robustness of the application in which the data is used. However, these do not pertain to uncertainty in spatial data per se.

THE MODEL

The following model describes the different components mentioned above in detail. It identifies several steps in the formalization of user's conceptions of an issue into a query and in the formalization of a conception of a phenomenon into a digital spatial data set. Understanding these steps allows us to see where information is lost or changed in this process. This loss of or change in the information results in the mismatches described above.

The general structure of the model is shown in Figure 21.1. It has two sides, one representing the data user and one representing the data development component and an arrow indicating data representation. The data user and development components begin with a person's conceptualization of the world and proceed to a formal specification that is entered into a computer either as a query or a data set. The database system or query processor matches the two specifications and returns the query results. Each of these components involves a sequence of transformations from less to more formal specifications, and each transformation can entail a potential loss, alteration, or creation

of information. The details of the transformations for the data user and development components are shown in Figures 21.2 and 21.3, respectively. The result of these transformations is that a perfect match of formal specifications may not represent a close match to conceptualizations of a phenomenon.

Data User

The process of choosing which specific set of data is required for a given use involves a continual refinement from a rough identification of a problem or question to a formal specification of the specific data elements of interest. The rough identification of the problem is often simply the recognition that a problem or question is important. The formal specification is often a query in a specific query language.

Step 1

The user of spatial data often begins with a perceived problem or question to be addressed. This may be in the form of a query (e.g., "I wonder where wetlands are in this county?"), or it may be a more complex need, such as assessing the impacts of land-use change on riparian habitats. In either case, the problem or question begins as informal perception of a need.

Step 2

After perceiving a need for spatial data, the user must formulate a conceptualization of the problem in more detail and with greater structure. This formulation begins to specify the parameters of the decision and therefore the necessary aspects of data used for it. For example, in our case of the hypothetical land-use change assessment, the potential data user will start identifying the possible types and magnitudes of change that may be detrimental. She will also identify the spatial characteristics of these changes that are important (e.g., distance from sensitive habitat, pattern of changes, intervening land uses between the habitat and land-use change, etc.).

Step 3

From the formulation of the problem, the data user then identifies the type of data and the aspects of them that she is interested in (e.g., attributes, classification, spatial and attribute resolution, etc.). This eventually results in an informal expression of a data query. That

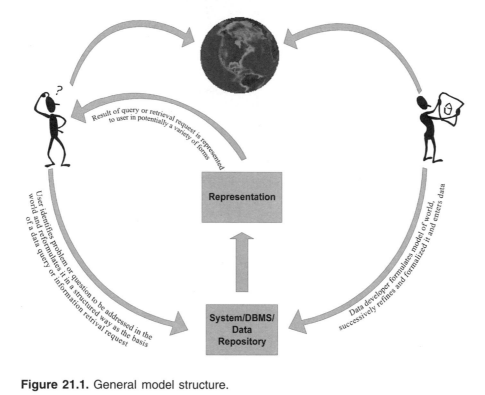

Figure 21.1. General model structure.

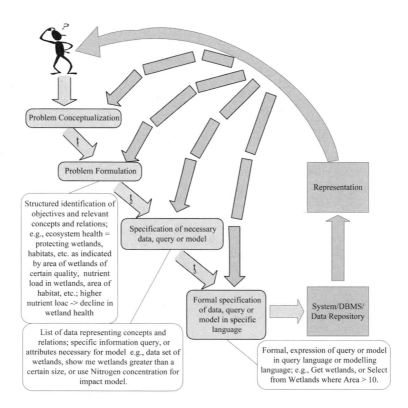

Figure 21.2. Data user component.

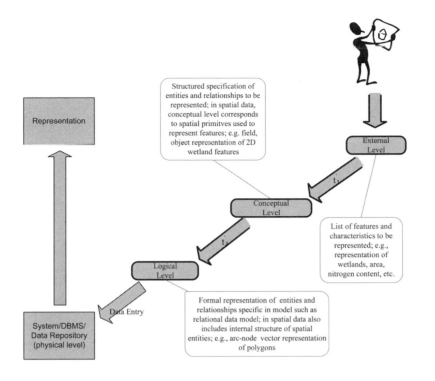

Figure 21.3. Data development component.

is, it is an informal specification of the information that the user requires from the database.

Step 4

The informal specification of the required information leads to a formal query of a database. This can be a query of a data repository where the desired result is a data set, or it can be a query of an individual data set where the desired result is a set of features that match a given condition. Of course, this query is most often expressed in a formal query language.

Data Development

The steps in this component of the model correspond to the data modeling process outlined by many authors (Peuquet, 1984; Ullman, 1988). The process of developing a data set entails defining a data model on which the data set is based. This begins as a fairly informal decision about the features to be represented in the data set and moves to a rigorous, formal specification of the way features and their relationships are represented and stored.

Step 1

The first step in developing a data set is deciding what features and what information about them should be included in the data set. This is often known as the external data model. Thus a data developer may decide that land cover classified in a specific way (e.g., Anderson level I) is important or that streams and stream order numbers are important. In addition, a data developer may also decide that the data set should represent connectivity between stream reaches.

Step 2

The data developer has to make a decision about how the data should be represented in a data set. This is known as the conceptual data model. For example, will land cover be represented as a set of continuous objects or a field of land cover? There are other possibilities, but the field-object differentiation has occupied much of the literature in spatial data modeling (Couclelis, 1992; Goodchild, 1992). We will not dwell on this difference at this point since the reader may peruse one of the papers that treats this difference in detail.

Step 3

The conceptual model is then transformed into a computational or logical model. This specifies the types of data structures that will be used to store the information required. This includes models such as the geo-relational model used in Arc/Info or others. This logical model determines the input of data into a spatial database. In some cases, it also determines the method of measurement or data collection; however, in other cases it may be partly determined by the data collection method used.

Step 4

As mentioned above, the next step is measurement or data input. The notion of accuracy is the comparison of the results of this process to the world. Thus accuracy must incorporate random and systematic sources of differences between the data and the actual characteristics of the features.

Data Representation

The results of a query may be represented to the user in a variety of forms or modes. Clearly, the most common mode is a cartographic representation or textual representation in the form of a table. This representation is then the basis of either decisions or analyses or of refining or altering the initial request or problem formulation. If the latter is the case, then the user proceeds through the steps outlined in the model again to produce a new query.

Thus the representation of data can influence any of the steps in the model. Representations of metadata may cause the data user to change her formulation of the problem to use the data in different ways (or not use the data at all).

LITERATURE

Considerable attention has been directed to estimating, modeling and representing error in spatial data. Veregin (1989) provides a fairly comprehensive, albeit now dated, survey of the literature in this area. This body of research relates to the data component of our model. That is, it concentrates on the relationship between the values stored in the database and characteristics of the world. In some cases (e.g., Chrisman, 1982), these efforts have started by enumerating the possible sources of error in the data development pro-

cess. The representation of error research clearly applies to the representation arrow in our model.

The problem of uncertainty in data has also been addressed by computer science and artificial intelligence researchers (cf., Motro and Smets, 1997). In most cases, these efforts have also concentrated on methods for modeling uncertain data and the mathematical techniques for manipulating the representations. Such techniques include interval mathematics, probability theory, Bayesian statistics, fuzzy set theory, and Dempster-Shafer Theory, among others (Bandemer, 1992). These methods represent one particular aspect of data (e.g., imprecision, vagueness, error) in a database. They also focus on the data side of our model.

Literature on decision making and judgment under uncertainty is prevalent in cognitive psychology. Much of it refers to a classic paper by Tversky and Kahneman that described several biases and heuristics people use when making judgments under conditions of uncertainty (Tversky and Kahneman, 1974). Some research has attempted to identify whether certain conditions determine the use of certain heuristics by people (Payne, Bettman, and Johnson, 1993). This research is critical for the data user component of our model, although its exact position in the model remains to be determined. For example, representing uncertainty in data may motivate the use of heuristics or biases in decision making. However, the substantive decision contains uncertainty introduced from other aspects of the decision environment in addition to uncertainty components deriving from data (Jordan and Miller, 1995).

CONCLUSION AND FUTURE RESEARCH

The model described here helps identify aspects of data uncertainty that have not been included in the consideration of data quality or accuracy. We see that there is a gap in our understanding of the result of representing uncertainty because we do not know how a data user might respond to information about uncertainty. These questions depend on the way a user proceeds through the steps of formalizing her request for data. Thus future research is necessary to investigate the types of information that users employ when formulating the problem and how metainformation such as uncertainty information would affect this process. We are proceeding with such research in this project by interviewing users to ascertain their conceptions of uncertainty in data, what kinds of uncertainty they consider in problem solving, and what kinds of uncertainty would change their approach to problem solving. We will integrate these questions by investigating the changes in their decision making resulting from changes in the representation of uncertainty of a data set.

Certain mathematical methods of modeling and manipulating uncertainty imply specific epistemological conditions. For example, if an interval mathematics approach is used, this implies that users of the data consider value intervals the most appropriate representation of uncertain values. Conversely, fuzzy set theory is more compatible with graded categories in perceptions of phenomena. Categories in cognition have been the subject of research in cognitive science for several decades. A common model of categories is the prototype model which is similar to a graded fuzzy category. Thus there is a question about whether a fuzzy representation of uncertain data is consistent with the cognitive categories of a user. The authors are also completing a paper that compares the representation of fuzzy regions to perceptions of regions.

The model is also useful in directing developments in information retrieval systems such as fuzzy queries or natural language interfaces that attempt to provide a more flexible interaction with a data repository. Similar to the needs for representation, the specific nature of an interface (i.e., what kind of interaction it provides) depends on the types of uncertainty that concern users.

REFERENCES

Bandemer, H. General Introduction, in *Modelling Uncertain Data,* H. Bandemer, Ed., Akademie Verlag, Berlin, 1992.

Chrisman, N.R. A Theory of Cartographic Error and Its Measurement in Digital Data Bases, in *Proceedings of AutoCarto 5, Crystal City, VA,* American Society for Photogrammetry and Remote Sensing, 1982.

Couclelis, H. People Manipulate Objects (but Cultivate Fields): Beyond the Raster-Vector Debate in GIS, in *Theories and Methods of Spatio-Temporal Reasoning in Geographic Space, Lecture Notes in Computer Science 639,* A.U. Frank, I. Campari, and U. Formentini, Eds., Springer-Verlag, Berlin, 1992.

Goodchild, M. Geographical Data Modeling. *Com. Geosci.,* 18, pp. 401–408, 1992.

Jordan, C.F. and C. Miller. Scientific Uncertainty as a Constraint to Environmental Problem Solving: Large-Scale Ecosystems, in *Scientific Uncertainty and Environmental Problem Solving,* J. Lemons, Ed., Blackwell Science, Cambridge, 1995.

MacEachren, A. *How Maps Work: Representation, Visualization, and Design,* Guilford Press, New York, 1995.

Motro, A. Sources of Uncertainty, Imprecision, and Inconsistency in Information Systems, in *Uncertainty Management in Information Systems: From Needs to Solutions,* A. Motro, and P. Smets, Eds., Kluwer Academic Publishers, Boston, 1997.

Motro, A. and P. Smets. *Uncertainty Management in Information Systems: From Needs to Solutions,* Kluwer Academic Publishers, Boston, 1997.

Payne, J.W., J.R. Bettman, and E.J. Johnson. *The Adaptive Decision Maker,* Cambridge University Press, Cambridge, 1993.

Peuquet, D. A Conceptual Framework and Comparison of Spatial Data Models, *Cartographica,* 21, pp. 66–113, 1984.

Rosenthal, R. and R. Rosnow. *Essentials of Behavioral Research: Methods and Data Analysis,* McGraw-Hill, New York, 1991.

Smets, P. Imperfect Information: Imprecision and Uncertainty, in *Uncertainty Management in Information Systems: From Needs to Solutions,* A. Motro and P. Smets, Eds., Kluwer Academic Publishers, Boston, 1997.

Stinchcombe, A.L. *Information and Organizations,* University of California Press, Berkeley, 1990.

Thrift, N. Flies and Germs: A Geography of Knowledge, in *Social Relations and Spatial Structures,* D. Gregory and J. Urry, Eds., Macmillan, London, 1985.

Tversky, A. and D. Kahneman. Judgment under Uncertainty: Heuristics and Biases, *Science,* 185, pp. 1124–1131, 1974.

Ullman, J. *Principles of Database and Knowledge-Base Systems: Volume 1,* Computer Science Press, Rockville, MD, 1988.

Unwin, D.J. Geographical Information Systems and the Problem of 'Error and Uncertainty,' *Prog. Hum. Geogr.,* 19, pp. 549–558, 1995.

Veregin, H. *Accuracy of Spatial Databases: Annotated Bibliography,* Technical Paper 89-9, National Center for Geographic Information and Analysis, University of California, Santa Barbara, CA, 1989.

CHAPTER 22

Toward a Theory of Vector Error Characterization and Propagation

G. Edwards

INTRODUCTION

Goodchild has elaborated a coherent theory for understanding and modeling spatial error within the framework of category coverages in a raster environment (Goodchild et al., 1992). This theory treats spatial data (e.g., a map) as a single sample taken from a population of possible samples (i.e., possible maps) where variations are expressed via the error and spatial autocorrelation structure of the data set. Furthermore, the approach assumes that the dominant factors which affect the production and nature of spatial data are both natural and human-associated processes which operate in the real world. Goodchild and his colleagues argue that this is the only effective means of understanding and propagating spatial autocorrelation, because the latter cannot be derived analytically, directly from a given data set. The approach has been applied successfully to a number of applications, but despite the arguments in favor of this approach, there are a number of problems with it.

First of all, it has been recognized that the spatial autocorrelation structure of a data set may be nonstationary, and hence the spatial properties of a distribution may vary strongly as a function of position. This has been called the problem of "heteroscedicity" (Vauglin, 1997). The second problem is that computing costs of producing stochastic simulations are high, and for large sets of input data, there may be a combinatory explosion problem. The third problem is that the method proposed by Goodchild is not particularly suited to highly generalized data, such as that encountered in a typical vector polygon coverage.

In this chapter I propose a different model for error. I suggest that many spatial data sets are affected more strongly by the perceptual and cognitive processes which pertain to their creation than by the real-world processes they are designed to portray, and that a more appropriate error model would therefore be based on characteristics of these perceptual processes. Most vector or polygonal data sets are produced by photointerpretation, by map generalization procedures, or by a combination of both. For these data, I believe an error model along the lines of what I propose to be more appropriate than the Goodchild error model. The Goodchild model would be more applicable for data sets derived from remotely sensed image products which use little human intervention, or other data sets of like nature, such as some digital elevation models.

In what follows, I shall make a fairly subtle distinction between the concept of error and that of uncertainty. I shall assume that error is the more general concept, but that uncertainty is associated with the error component which arises when cognitive processes interact with data of interest. Hence when discussing error in the context of cognition, I shall use error and uncertainty interchangeably. Although this approach is nonconventional, I believe it should make the chapter more readable than would have been the case for a more rigid distinction between these terms. Furthermore, there does not appear to be widespread consensus yet on the way these terms should be used, and therefore I feel additionally justified in taking this approach.

A THREE-COMPONENT ERROR MODEL

The error model I propose is made up of three semi-independent submodels, related to three different characteristics of the perceptual field and its processing. Submodels include a model for aggregation error, a model for curve or boundary error, and a model for orientation and directional error.

In a purely geometrical environment, these three elements are fully coupled. Hence the boundary of a region is fully determined by the geometry of a region and vice versa, while any directionality associated with either regions or lines is determined by the geometric properties of the latter two features. However, in a perceptual environment, there are strong indications that such is not the case. In the visual field, so-called "surfaces" are not strictly defined by boundaries, nor boundaries by surface regions, and directional fields may be derived and represented independently of boundaries and regions.

Vision appears to operate at multiple scales simultaneously (Bergen, 1991; Cantoni et al., 1997). Hence characteristic and primitive image elements are examined across several scales. Aggregation of image elements into groups appears to be built into low-level vision. Boundaries, however, are not determined at the same time or by similar processes. Boundaries appear to consist of a contour-following action which occurs independently of the multiscale aggregation function (Tversky, 1998). Contour-following may draw on information concerning the aggregates on one side or the other, but the focus of attention for curve-following may shift across scales and may be affected by information already stored in memory. Finally, as well as being multiscale, vision appears to be multidirectional, often expressed through its relation to symmetry principles (Reisfeld et al., 1995). Directionality is a property which is omnipresent in perception, but which has been understudied in geomatics and geographic information science. Within perception, directionality appears to play a role as important as aggregation, but which operates independently of the latter. Curve tracing may use information in the directional field, but will be affected by other criteria, just as for aggregation.

Because these three processes are only partially coupled, their associated errors and uncertainties will be likewise only partially coupled. Take, for example, the problem of boundaries. In several studies which have been reported in the literature (Edwards and Lowell, 1996; Edwards, 1994; Aubert, 1995), my colleagues and I have studied multiple interpretations carried out on the same data. These provide interesting insights into the nature of the interpretation process and how boundary error should be quantified. One of the first results of these studies was to note that the local information provided by the textured regions on either side of a perceived boundary provides only partial information on the uncertainty across the boundary zone (Edwards and Lowell, 1996). In effect, less than half the variability of the latter can be explained by a local texture model. The rest of the uncertainty in the boundary zones appears to be related to the use of contextual information provided by neighboring regions, or by previous information used by the interpreter.

Because the uncertainty of the boundary zones of these regions as determined by photointerpretation cannot be derived directly from the textural characteristics on either side of the boundary, I introduce the term "sloppy boundary" to describe this partial relationship between the two. Hence a "sloppy boundary" is a boundary associated with a region (or pair of regions) under the constraint that the boundary location is only partly determined by the geometry of the region(s). Sloppy boundaries seem to prevail in many forms of spatial data, and are poorly represented by the strict geometric model found in existing GIS software.

Aggregation issues are likewise related to the "sloppiness" of boundary zones. Hence not all interpreters "see" a given boundary zone or label it as important. As a result, polygons interpreted by one interpreter will be distinct, while for another they will be fused. Once again, the difference between these two cases does not arise from geometry, but from perception. Aggregation affects the perception of textures and their corresponding scales, but as I have already indicated, perception of aggregation alone is not enough to determine the presence of a boundary. An aggregation model can be built by using the visual field as a model. Hence the visual field is characterized by a lower limit (albeit variable) on resolution, and larger blocks are built out of smaller ones to form the interpreted visual field (see Figure 22.1). A full model of this aggregation process might consist of the set of all possible incremental chains of partitions formed by merging units from the most primitive scale up to the final interpretation (Figure 22.1).

Although from a given "final" interpretation, many different finer-scale partitions may be consistent with the interpretation, some of these may be favored over

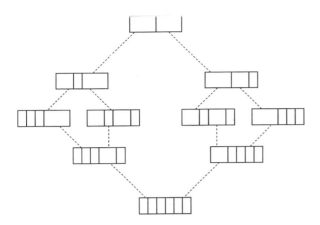

Figure 22.1. Subsample of the aggregation structure of a six-unit, one-dimensional object.

others, based on likely mergings following human perceptual processes. Studying the properties of cognition and visual perception should help identify permissible variations. Such studies might lead to a stochastic simulation of perturbed aggregate partitions, à la Goodchild, which accommodates knowledge of perceiton, and which could then serve as the basis for an error model of aggregation processes.

Hence different nesting structures might correspond to different perceptions of aggregation structure. Large regions which break up quickly into lots of small regions of similar size correspond to regions perceived as uniformly textured. Regions which break up into a smaller regions surrounded by larger regions correspond to zones perceived as characterized by intrusions.

The problem of directionality and orientation has been identified within many fields concerned with spatial data. However, although these play a crucial role in theories of perception, current GIS do not offer much in the way of automated determination of directional fields. In a recent paper (Edwards, 1997), I showed how directional information, when combined with the Voronoï diagram and the Medial Axis Transform, may yield a more powerful shape-modeling capability for polygons than has been achieved by standard (radial) methods. Directions are used as clues on "where to look" by human interpreters. Images contain a great deal of directionality. Polygon coverages contain less directional information, but even they contain enough that the directionality can be used to reason about the map. Map generalization procedures appear to exploit directionality within polygon coverages. For example,

generalizing a base map at 1:25,000 to a tourist map at 1:50,000 appears to preserve the coarse-grained directionality of the map at the expense of other elements (paper in preparation). Furthermore, although humans apparently have the ability to make fine distinctions in directions on a relative or immediate basis, directions stored in memory appear to be characterized by a much coarser graininess (Tversky, 1998). Hence directional uncertainty is related to the way directional information is used, and a perception-based error model for directional uncertainty will need to take such differences into account.

Spatial autocorrelation is expected to play a role in all three error models. However, rather than arising from real-world processes, it arises from the generalization procedures used by interpreters and map makers. This would suggest that standard spatial statistics such as Moran's I and Geary's coefficient along with local indicators of spatial autocorrelation (LISA statistics) which operate along similar lines (Anselin and Getis, 1992; Getis and Ord, 1992), are more appropriate for measurement than point-based sampling strategies such as the semivariogram. Certainly the problem of spatial autocorrelation should be reexamined within the framework of this new theory of error. To the extent that the perceptual and cognitive processes which underly interpretation and generalization can be modeled, it may be possible to decompose spatial autocorrelation into contributions from different cognitive processes and hence improve understanding of its role in spatial data.

ADDITIONAL CONSTRAINTS AND IMPLICATIONS

In the previous section, we outlined the three components of what may be treated as a new error model for polygonal coverages. We now go on to discuss some related issues and to indicate implications of the new approach for spatial data in general.

One of the difficulties encountered in studying different kinds of error and how they relate to each other is that spatial data in one form must be related or "matched" to spatial data from a different coverage, which is not always in the same form. Even matching across data sets with the same form is not trivial. This matching problem is well known in many specialized applications, such as stereo vision, the georeferencing of map and image data, many problems in computer vision, and so forth. Typically, the problem consists of isolating the appropriate data subset to use during a matching procedure. When the match is not exact, identifying the relevant data subset to perform such matching is a very difficult problem. Even when an exact match can be expected, the problem may be related to the identification of a "needle in a haystack," that is, a very large number of tests must be performed before the right match can be found. For inexact matches, there is no simple means of determining the limits of the data subsets to be tested against the target.

In fact, matching is simply a variant on the more general problem of pattern recognition. Pattern recognition is a task at which humans excel, but at which computers still perform poorly, after more than three decades of effort. Because "matching" is such a ubiquitous procedure in the processing of cartographic and image data, and because, aside from a few specialized domains, humans still outperform algorithms, this is another area where human perception affects the ultimate relationships found in spatially referenced data more than the real-world processes these data are designed to model.

Having said this, one shouldn't abandon hope for obtaining better algorithms for matching. Indeed, the point of this article is not to say that because cognition and perception affect data we should give up all hope of quantifying these data and their uncertainty, but rather that to the extent to which we know the effects of cognition and perception on spatial data, we will be able to better quantify their effects.

Some progress on algorithms for matching has been made. For linear and point features, two kinds of matching metrics have been proposed: Hausdorff distances (Vauglin, 1997; Hausdorff, 1919) and Fréchet distance parametrizations (Devogele, 1998; Fréchet, 1906). The Hausdorff distance is a point-based metric which assumes both sets of data are characterized by the same spatial scale, whereas the Fréchet measure allows the determination of a curve-to-curve distance metric which may accommodate scale changes and displacements (a more general metric to that proposed by Edwards in 1996). Hence for nonorthogonal displacements, and for curves which undergo systematic and continuous distortions due to transformations in the data such as generalization procedures, the Fréchet metric seem more appropriate. For different interpretations of a common image, or for discontinuous distortions in spatial data, however, the Hausdorff metrics seem to provide the better measure. For areal features, overlapping area measures seems the most common matching metric (Devogele, 1998), although a procedure for extending such methods to nonoverlapping

regions, based on the use of displacement vectors which must be applied to obtain maximal overlap has also been proposed (Edwards and Rioux, 1996).

A second problem which must be addressed is that of combining the three models into a single error characterization of a data set and propagating this error through several transformations into a final data set. Because these error submodels are partially coupled, care must be taken when combining them. For example, the uncertainty associated with a boundary segment may be partially due to uncertainty in the aggregation level of the polygons on either side of the boundary, partially due to the uncertainty in the directional field in the neighborhood of the boundary, and partially due to previous experience with similar boundaries. Furthermore, the submodels are all characterized by the presence of spatial autocorrelation, but of different types—linear, areal and directional. The presence of three partially correlated types of error may help explain why the spatial autocorrelation of the error structure is so difficult to determine. The spatial autocorrelation structure of any one error component will be affected partially by those of the other two. If the latter two error components are not measured, then it will be difficult to get a handle on the structure of the spatial autocorrelation. Untangling these different effects will likely require a better handle on the perceptual processes involved.

Developing numerical techniques for characterizing and propagating these different error models is also required. It should be possible to adapt Goodchild's simulation approach, consisting of perturbing a given map randomly, as a function of different spatial autocorrelation scenarios, to the three-error submodels discussed above. Hence, for example, boundaries may be treated as sample realizations of a range of boundaries within a given uncertainty range and with a given (linear) spatial autocorrelation structure. Aggregates may likewise be treated as samples of sets of aggregates compatible with a given aggregate, corresponding to the map being considered. Directional vectors may also be modeled by simulating their variability. However, the complexity of these models is likely to be significantly greater than is currently the case.

An alternative consists of the use of analytical propagation methods, such as proposed by Heuvelink et al. (1989). However, these methods fall afoul of the difficulties in propagating spatial autocorrelation. Indeed, the analytical propagation of spatial autocorrelation is not possible without the use of additional information not quantified by the spatial autocorrelation itself. This can be readily seen in the following example. Combining two checkerboard data sets (characterized by negative spatial autocorrelation—see Figure 22.2) will lead to different spatial autocorrelation characteristics in the output data, depending on the relative intensities and offsets of the two data sets. In fact, the spatial variation of the autocorrelation structure is what controls the output, and not the structure itself.

A useful analogy can be made with electromagnetic signal propagation. If the spatial autocorrelation structure of two data sets are coherent (i.e., they are both characterized by the same, single spatial frequency) and phase-matched, the resulting spatial autocorrelation structure will be similar to the input data sets. If their spatial autocorrelation structure is mismatched in phase, however, and the amplitude of the spatial autocorrelation is similar, then we expect the output data set to be characterized by little spatial autocorrelation. This would suggest a possible means of carrying out the propagation of spatial autocorrelation when combining several data sets—i.e., it may be possible to characterize spatial autocorrelation structures as superimposed wave patterns and use wave theory methods to carry out the propagation.

DISCUSSION AND CONCLUSIONS

The broad outlines of a theory of uncertainty based on human perception and cognition of spatially referenced data has been proposed in this chapter. This approach assumes that the dominant source of error in spatial data arises from human interpretation, matching procedures and generalization. Most spatial data in use today may be characterized in this way. A three-component model of uncertainty is proposed, based on three critical aspects of human visual perception—aggregation uncertainty, curve or boundary uncertainty, and directional uncertainty. These models are assumed to be only partially dependent on each other.

Hence perceived boundaries are what has been called *sloppy boundaries* of the regions they supposedly fence in. They are sloppy, because their location is partially dependent on nongeometric considerations. Algorithmic methods for characterizing and representing these different submodels already exist, with the possible exception of directional error, which has been less extensively studied. Hence boundary error has been studied via the analysis of multiple interpretations, which produce mean or median boundaries and boundary distributions around these, while aggrega-

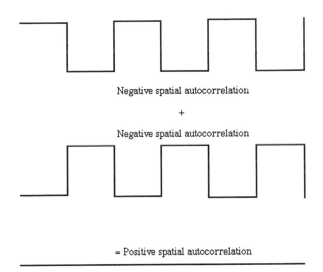

Figure 22.2. Combining two negatively spatially autocorrelated surfaces may yield a positively spatially autocorrelated surface, as a result of particular values of the phase and coherency of the two patterns.

tion uncertainty may be characterized as a subset of all possible aggregation scenarios from primitive visual entities to a given interpreted partition. However, no direct means has been proposed to relate these models to each other, except on the basis of purely geometric techniques.

Might there be more than three error submodels? Of course, this is possible. The human perceptual system is complex, and the characterization of the perceptual process into three partially related systems is a simplification. It might be possible to divide the perceptual field into other relevant features. However, the identification of these three components constitutes a sort of minimalist approach. The inclusion of directionality is already a departure from standard conceptions which circulate in geomatics and geographic information science, which assume that boundary error and aggregation levels are linked phenomena. It is proposed to limit perception-based error models to the three components outlined for the time being, and introduce new components only as they are needed.

The theory as outlined deals with different issues than those generally addressed, based on the assumption that real-world processes are described by spatial data and that the dominant source of error or uncertainty is measurement error. Error models based on the latter approach assume that random perturbations of a given map which are consistent with an assumed spatial autocorrelation structure can be used to characterize the error field. Some maps may indeed be characterized in this way, but the majority are not. A full theory of error therefore needs to combine both approaches, using perturbation-based simulations to study measurement error for process-based data, and perception-based models to study the uncertainties which arise from the transformations data undergoes.

Finally, the difficulties of identifying the spatial autocorrelation structure of data sets are briefly overviewed, along with the problem of (analytically) predicting the outcome of combining such structures when data from different sources are merged. It is argued that information on the spatial variability of spatial autocorrelation is required in order to perform such a task. However, it may be possible to borrow techniques from interferometry in order to carry out such procedures.

In conclusion, although a full theory of error is not laid out here, many of the elements which may characterize such a theory are discussed. The chapter is meant to be thought-provoking more than it is meant to provide a rigorous theory.

ACKNOWLEDGMENTS

This research has been funded by the Canadian Natural Sciences and Engineering Research Council (NSERC) and Québec's Association des industries forestières (AIFQ) through the establishment of an Industrial Chair in Geomatics Applied to Forestry.

REFERENCES

Anselin, L. and A. Getis. Statistical Analysis and Geographic Information Systems, *Ann. Reg. Sci.*, 26, pp. 19–33, 1992.

Aubert, E. *Quantification de l'incertitude spatiale en photo-interprétation forestière à l'aide d'un SIG pour le suivi spatio-temporel des peuplements,* thesis presented to the Université Laval in partial fulfillment of the requirements for the degree of Master of Science, 1995, p. 92.

Bergen, J.R. Theories of Visual Texture Perception, *Spatial Vision*, 10, pp. 114–134, 1991.

Cantoni, V., S. Levialdi, and V. Roberto, Eds. *Artificial Vision: Image Description, Recognition and Communication,* Academic Press, New York, 1997, p. 306.

Devogele, T. *Processus d'intégration et d'appariement de bases de données géographiques: Application à une base de données routières multi-échelles,* thesis presented to the Université de Versailles in partial fulfillment of the requirements for the degree of Doctor of Philosophy, 1998.

Edwards, G. Reasoning about Shape Using the Tangential Axis Tranform (TAT) or the Shape's "Grain," *Proceedings of the AAAI-97 Workshop on Language and Space*, Rhode Island, July 27–28, 1997.

Edwards, G. Characterising and Maintaining Polygons with Fuzzy Boundaries in Geographic Information Systems, *Sixth International Symposium on Spatial Data Handling*, vol. 1, Edinburgh, September, 1994, pp. 223–239.

Edwards, G. Aggregation and Disaggregation of Fuzzy Polygons for Spatial-Temporal Modelling, *Netherlands Geodetic Commission*, 40, pp. 141–154, 1994.

Edwards, G. and K.E. Lowell. Modeling Uncertainty in Photointerpreted Boundaries, *Photogrammetric Eng. Remote Sensing,* 62(4), pp. 337–391, 1996.

Edwards, G. and S. Rioux. A Detailed Assessment of Relative Displacement Error in Cutover Boundaries Derived from Airborne C-Band SAR, *Can. J. Remote Sensing*, 21(2), pp. 185–197, 1995.

Fréchet, M. Sur quelques points du calcul fonctionnel, *Rendiconti del Circolo Mathematico di Palermo*, 22, pp. 1–74, 1906.

Getis, A. and J.K. Ord. The Analysis of Spatial Association by Use of Distance Statistics, *Geogr. Anal.*, 24, pp. 189–206, 1992.

Goodchild, M.F., S. Guoqing, and Y. Shiren. Development and Test of an Error Model for Categorical Data, *Int. J. Geogr. Inf. Syst.*, 6(2), pp. 87–104, 1992.

Harvey, F. and F. Vauglin. Geometric Match Processing: Applying Multiple Tolerances, *Proceedings of the 7th International Symposium on Spatial Data Handling*, Delft, The Netherlands, Volume 1, 1996, pp. 4A13–4A29.

Hausdorff, F. Dimension und ausseres Mass, *Mathematische Annalen*, 79, pp. 157–179, 1919.

Heuvelink, G.M.B., P.A. Burrough, and A. Stein. Propagation of Errors in Spatial Modeling with GIS, *Int. J. Geogr. Inf. Syst.*, 3(4), pp. 303–332, 1989.

Reisfeld, D., H. Wolfson, and Y. Yeshurun. Context Free Attentional Operators: The Generalized Symmetry Transform, *Int. J. Com. Vision*, 14, pp. 119–130, 1995.

Tversky, B. Memory for Pictures, Maps, Environments, and Graphs, Erlbaum, in *Practical Aspects of Memory*, D. Payne and F. Conrad, Eds., Hillsdale, NJ, 1998.

Vauglin, F. *Modèles statistiques des imprécisions géométriques des objets géographiques linéaires,* thesis presented to the Université de Marne-la-Vallée in partial fulfillment of the requirements for the degree of Doctor of Philosophy, 1997.

CHAPTER 23

Super Ground Truth as a Foundation for a Model to Represent and Handle Spatial Uncertainty

T. De Groeve, K. Lowell, and K. Thomson

INTRODUCTION

The key question in forestry management is *where* the forest should be harvested in order to obtain the largest wood volume. To respond to this question, accurate information concerning the forest is needed in order to estimate wood volume with as high a precision as possible. This requires highly accurate map data covering a large area—data that are traditionally acquired by human interpretation of aerial photographs.

The classical method of obtaining such information attempts to identify homogeneous forest areas (forest stands) using subjective interpretation of aerial photographs. In this process, an interpreter of aerial photographs tries to detect the boundaries of the forest stands that are considered homogeneous relative to vegetative parameters such as species, the mean height of the trees, forest density, and the mean age of the forest stand. This information is then compiled to create a thematic forest map which forms the basis for a forest inventory and a number of other forest management activities.

The information offered by this classical interpretation method is, unfortunately, neither accurate nor precise enough to satisfy existing operational forest planning needs. In practice, highly intensive and highly expensive ground-based inventories are necessary in the final phases of harvest planning to obtain sufficiently accurate information for a given area. Relative to the uncertainty inherent in the identification of the forest stands, a large number of the boundaries between forest stands would seem to be highly uncertain (Edwards and Lowell, 1996).

Such spatial uncertainty is basically caused by two factors. First, there is the subjectivity of the interpretation process of the aerial photographs. Second, the boundaries between the forest stands are not discrete changes, but can more accurately be described as transition zones which can be rather large (Lowell, 1995). Correct local information about the accuracy of individual boundaries (and the visual representation of that accuracy) is essential in order to increase the confidence of the user in thematic forest maps, and also to possibly eliminate the costly inventories conducted just before harvesting.

Despite the need for information about spatial uncertainty, the acquisition and the description of such information is a complicated problem. Recently, research has been conducted to develop and evaluate a method for assessing the boundary uncertainty inherent in the subjective interpretation of photographs by overlaying multiple interpretations of the same image (Joy et al., 1994; Aubert et al., 1997). A key element in this multiple interpretation process is a recognition of the same forest stand boundaries on the different maps. If one succeeds in identifying the consistent (over all interpretations) forest boundaries (or stands) on the different forest maps, information about the uncertainty of boundaries between two given forest types can be obtained (Aubert et al., 1997).

Although the method of multiple interpretations for estimating boundary uncertainty offers useful results, some serious drawbacks have to be overcome in order to use it. First, displacement and deformation of the boundaries of a polygon in different interpretations (causing only partial overlap of the polygons) often makes the recognition of the original forest stands difficult (Figure 23.1). Moreover, boundaries can be missing in some interpretations : often a first interpreter will draw a boundary between two forest stands where a second interpreter concludes that there is actually only one forest stand (Aubert et al., 1997). Relative to these spatial problems, the forest type appellation of a given forest stand can be different in each interpretation (especially when values of the classification parameters are close to the class boundaries of the classification system used). Also, scale and thematic category definition can vary between maps. Moreover, if maps of different dates are used (instead of multiple interpretations from the same date), then natural changes or human intervention in the terrain must be taken into account by using growth models or at least a determination of ecologically plausible changes (Joy et al., 1994).

In order to minimize these drawbacks in the use of this method of multiple interpretations, a close examination of spatial uncertainty of forest maps is needed. Aubert et al. (1997) noted that the greatest certainty exists in polygon cores and the least certainty exists at forest stand polygon boundaries. Further characterization of these boundary zones was accomplished for artificial images by Edwards and Lowell (1996), who showed that the width of the uncertainty zone is, among other things, dependent on differences in the visual texture of polygons, and on local information like the length of the boundary, spatial context, etc. Moreover, the distribution of the uncertainty across interpreted boundaries seems to be Gaussian in the uncertainty zone. In spite of these and other related studies, a comprehensive effort designed to examine these methods and other issues of spatial uncertainty on real data has yet to appear.

This chapter attempts to contribute to the understanding of spatial uncertainty in forest maps by attempting to describe the influence of the subjectivity of photo-interpretation, classification errors, and temporal changes of three real data sets of a forest region where most of the real (i.e., on ground) boundaries between forest stands are not sharp lines. In doing this, initially an overlay of multiple interpretations will be examined to characterize the above influences. Subsequently, a Super Ground Truth will be created, which consists of all areas that are interpreted consistently by all interpreters.

DATA AND SITE CHARACTERISTICS

Three forest inventory maps of Montmorency Forest—the research and teaching forest of Laval Univer-

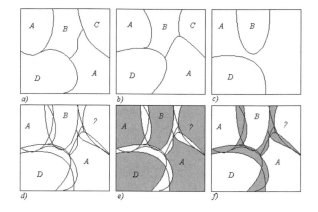

Figure 23.1. In different interpretations of aerial photographs (a, b, and c), boundaries can be displaced or even missing (boundary between B and D in (c)). The greatest certainty exists in polygon cores (d). In (e), the Super Ground Truth is highlighted, while (f) represents boundary uncertainty zones.

sity—of the same scale (1:10,000) were available for this study, covering an area of 6,625 ha. The maps are based on human interpretation of black-and-white aerial photographs obtained for forest inventories conducted in 1973, 1984, and 1992. For each stand polygon, tree species, mean height, mean age, and forest density are identified in a context of three different classification systems. The forest is located 80 km north of Quebec City and is dominated by boreal forest. The predominant species are balsam fir [*Abies balsamea* (L.) Mill.], white spruce [*Picea glauca* (Moench) Voss], and white birch (*Betula papyrifera* Marsh.). The forest has an intensive harvesting regime, and has suffered from a epidemic of the spruce budworm (*Choristoneura fumiferana*). Although most of the forest has been protected, some areas have been affected severely.

The three maps were digitized, and then overlaid—a process that showed a significant geometric transformation among the maps. Using 20 corner points of forest boundary segments of the maps as ground control points, the maps were adjusted to fit the 1992 forest map. This procedure created a maximum error in the center of the map which was estimated (comparing intersections of linear hydrology) to be less than 25 m.

Methodology

A comparison of forest maps from different years is only possible if the associated photographs are interpreted in a comparable fashion over all years. Although all three classification systems (from 1973, 1984, and 1992) offer information on species, height, density, and the age of trees, the definition of the individual classes for each factor is different. In order to make the maps comparable, a common classification system must be created. Density classes were the same in all maps, so the definitions of these could be retained over all years. The definition of the height classes was also identical for all three dates, with lower class boundaries at 22, 17, 12, and 7 meters. However, two supplementary classes for relatively short vegetation (below 7 m) were added after 1973. In the common classification system for height, this class was merged with the lowest class of 1973. Age classes were not directly comparable : boundaries between qualitative classes (young, mature, ripe) in 1973 were later replaced by quantitative classes (10, 30, 50, 70, 90, 120). Because of an inability to reconcile these differences, we discarded age from the analysis. Species composition classes have also been adapted for all classification systems. However, most of the adaptations have split classes, and consequently have caused class boundaries for species to be defined consistently over all years. In our common system to encompass all three dates, we retained nine species classes : three young classes (softwood, hardwood and mixed, without specification of composing species), four mixed classes (mix of fir and birch and other hardwoods in different proportions) and two softwood classes (spruce and fir). Two of these nine classes dominate the forest: balsam fir, and a mix of dominant fir and secondary white birch (which occupy, respectively, 70% and 10% of the total forest area).

Even for factors for which the classification systems are comparable over all years, temporal changes have modified the forest stands on the different maps. In order to compare the same forest stands over time, one must eliminate temporal changes from the analysis. A first step is to exclude areas where changes have been documented. Harvested areas (clear-cuts) and areas affected by the spruce budworm epidemic are excluded from the study territory, which leaves an «undisturbed» territory of 29% of the total area. That is, only 29% of the area had not been subjected to some sort of natural or human disturbance.

After eliminating all such disturbances, growth models need to be applied to the forest stands of this undisturbed territory, in order to age the stands on the 1973 and 1984 forest maps. We considered growth models for species composition and for mean of the height of forest stands. (Models for density changes

are not yet included, because of the high complexity of the evolution of density.) First, considering forest species, we assumed that mature forest stands will not change their composition over a period of 20 years. On the other hand, the species composition of young forest stands does evolve. For example, fast-growing white birch can later be surpassed by (initially) slower growing fir. However, these complicated mechanisms are not estimable from map data only. Consequently, in our model we accepted all transitions from young forest stands to mature forest stands as ecologically plausible. Next, considering mean height of trees in a forest stand, we assumed that trees can only remain the same height or increase in height. Traditional models of height in relation to tree age show a maximum growth of 1 m/year at ages below 20 and an exponentially decreasing growth later on (Boudoux, 1978). Consequently, shorter height classes will change at the fastest rate, while taller classes are not expected to change much. Considering all this, we derived a model which accepts an increase, over a period of 20 years, of two height classes for the shorter heights and only one for the tallest classes. Over a 10-year period, only single class increments are considered plausible for both young and mature stands.

One last step is to study class interpretation uncertainties, which can offer us a "fuzzy classification system" that incorporates these uncertainties. Considering height classes, we accepted an uncertainty of one class, because forest stands having a mean height between two classes might be put in either. However, since trees grow, even when the height of a forest stand was overestimated, it will grow into the taller class. Consequently, a decrement of class is unlikely, and we considered height interpretation uncertainty to be covered by an increment of one class. Next, considering species composition, we compared confusion matrices among the species classes for all possible map pairs. These confusion matrices showed that there was a large amount of confusion among classes within those classified as young. However, there was little confusion of any of the young classes with the mixed and/or softwood classes. The same was also true for the mixed and softwood classes. Species composition uncertainties are probably caused by similar visual textures on the aerial photographs and properties of the classes in these three groups. Consequently, we accepted a fuzzy classification in these groups. In other words, we grouped the nine species composition classes to retain only three classes : young, softwood, and mixed forest stands.

Table 23.1. Percentage of Area Covered by Different Super Ground Truths. (Difference is made between coverage of undisturbed forest and coverage of total map area. SGTs for species, for height and their combination are showed.)

	Undisturbed Area	Total Area
SGT - species	65%	38%
SGT - species, fuzzy	**70%**	19%
SGT - height	34%	23%
SGT - height, grown 1 class	56%	38%
SGT - height, grown 2 classes	**58%**	43%
SGT - species and height, both fuzzy	**41%**	11%

Taking into account these class interpretation uncertainties, the above growth models and the fuzzy classification system, maps from the three years can be compared and analyzed. The physical units of the forest, the forest stands, can be deduced from the overlay: all the consistently interpreted areas on the overlay form the Super Ground Truth. This Super Ground Truth should represent the cores of undisturbed forest stands, or in other words, the area of the map without uncertainty. Inconsistencies between maps (the areas not included in the SGT) are now due to either changes not included in our (simple) growth models, or spatial uncertainty. In this way, the Super Ground Truth offers information about the spatial uncertainty of conventional forest maps.

RESULTS

The application of this method of combining maps resulted in the creation of different Super Ground Truths (SGTs) based on different combinations of parameters considered. Three SGTs were created: a SGT for forest species, a SGT for height, and a SGT for the combination of both. For forest species, the SGT was derived from a "consistency map," showing the number of consistent species interpretations. The SGT covers the area of the consistency map where all three interpretations were consistent for species, i.e., 70% of the undisturbed area (Table 23.1).

For the quantitatively defined height classes, "difference maps" were created for each possible pair of maps, indicating the increment of height class for each polygon. For each difference map, a SGT was created using the growth models mentioned earlier (increment of one class tolerated). Next, the difference maps were overlaid to generate a SGT based on all three maps.

Accepting two classes of growth in 20 years (differences of 0, +1, and +2 tolerated) provides a SGT covering 58% of the undisturbed area (Table 23.1). Note that a lower tolerance for change (maximum difference of one class over the 20-year period) does not affect the results much (56% instead of 58%), meaning that most of the forest stands change only one class over 20 years.

For the overall SGT combining species and height information, the fuzziest classification offered a SGT covering 41% of the undisturbed area (Table 23.1). This is the intersection of the SGT for species using the fuzzy classification and of the SGT for height using a two-class tolerance. This means that 41% of the undisturbed area can be considered as accurate or consistent, considering three interpretations and two classifying parameters (height and species composition).

In addition to these global figures, local behavior of a typical site was studied. Since harvesting of Montmorency forest occurs everywhere over the territory, the undisturbed forest is rather dispersed over the total map area. However, in some regions there are clusters of undisturbed forest stands. One of these clusters is presented in Figure 23.2. On visual inspection, boundary uncertainty zones can be detected as well as inconsistencies caused by classification uncertainty.

The Super Ground Truth covers 41% of the undisturbed forest area, i.e., 22% of the total area in the typical site. Also the SGT is mostly situated in the polygon cores of the 1992 map, as was hypothesized a priori. The same observation can be made for the two other maps.

Finally, we studied some boundary uncertainty zones that have been visually identified. Measuring the widths of these uncertainty bands showed a variation between 5 and 40 meters, with a mean value of

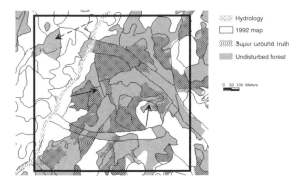

Figure 23.2. A typical site of Montmorency forest. Undisturbed forest is gray shaded, SGT is hatched. Stand boundaries are from the 1992 map. Plain arrows indicate clearly visible boundary uncertainty zones. The broken arrow indicates a classification uncertainty zone.

20. No noticeable differences were apparent between sites at the center of the map and sites in the map boundary area.

DISCUSSION AND CONCLUSION

From the preceding results, we can conclude that highly consistent individual forest stands can be identified using a Super Ground Truth. The application of growth models, fuzzy classification, and the elimination of naturally or artificially changed areas show good results relative to the intelligent comparison of maps of different dates. The area included in the SGT can be accepted as the most accurate representation possible, since all types of uncertainty are accounted for.

The area covered by the Super Ground Truth covers 41% of the undisturbed forest. This figure seems rather low considering the fuzzy classification, the growth models, and the consideration of only undisturbed forest areas. Clearly, the inconsistent area (59%) is not caused by spatial uncertainty alone. Classification uncertainty accounts for a large proportion of this inconsistent area. Consequently, refinement of fuzzy classification methods is necessary to obtain a SGT which covers a larger proportion of the forest. Including other information such as slope could also be necessary: a correlation between uncertain areas and steep topography has been noted, but has not, as yet, been quantified.

A growth model for the evolution of density in the forest stands and a fuzzy classification for density are still required in order to be able to include density information in the construction of the Super Ground Truth. Considering the results with only two variables (species composition and height of stands), we can expect that the SGT will provide an even smaller area considered to be consistent over all interpretations. Nonetheless, though polygon cores will be smaller, the general results will be similar.

Finally, we can conclude that Super Ground Truth offers a way to extract spatial accuracy information using classical forest inventory maps. Moreover, local information on accuracy of map data is easily presented to a user with the Super Ground Truth. As such, the Super Ground Truth can be the basis of a model to represent and manage spatial uncertainty.

ACKNOWLEDGMENTS

The research for this project is conducted within the context of the Industrial Chair in Geomatics applied to Forestry and the Center for Research in Geomatics (CRG) of Laval University. The authors are grateful to the Association of Quebec Forest Industries and the Natural Sciences and Engineering Research Council for financing this work.

REFERENCES

Aubert, E., G. Edwards, and K. Lowell. A Raster Method for Determining Temporal Boundary Change in the Presence of Uncertainty, *Int. J. Geogr. Inf. Syst.,* in revision, 1997.

Boudoux, M. *Empirical Yield Tables for Black Spruce, Balsam Fir, and Jackpine in Quebec.* Government of Quebec, Ministry of Land and Forests, COGEF, Quebec, 1978 (in French).

Edwards, G. and K. Lowell. Modeling Uncertainty in Photointerpreted Boundaries, *Photogrammetric Eng. Remote Sensing,* 62, 4, pp. 337–391, 1996.

Joy, M., B. Klinkenberg, and S. Cumming. Handling Uncertainty in a Spatial Forest Model Integrated with GIS, in *Proceedings of GIS '94,* Vancouver, British Columbia, 1994, pp. 359–365.

Lowell, K. A Fuzzy Surface Cartographic Representation for Forestry Based on Voronoi Diagram Area Stealing, *Can. J. For. Res.,* 24, pp. 1970–1980, 1995.

CHAPTER 24

Rigorous Geospatial Data Uncertainty Models for GISs

A.A. Alesheikh, J.A.R. Blais, M.A. Chapman, and H. Karimi

INTRODUCTION

Geospatial Information Systems (GISs) play an active role in decision-making processes in many disciplines that have planning, research, and/or management using spatial data, at the global, continental, state, or municipal level. More effective use of GIS, however, requires explicit knowledge of the uncertainty inherent in the data, particularly positional data. This becomes more important as the potential impact of decisions based upon GIS increases. The widespread use of GIS as a decision-support tool is therefore dependent on the development of formal modeling of errors in spatial databases.

This chapter focuses on positional errors that refer to the discrepancy between the true location of certain geometric primitives and their measured or expressed locations. If true values are not available, uncertainty is substituted for error. The current uncertainty models of geometric elements are critically evaluated using simulation methods. Particular emphasis is placed on modeling uncertainty of line and polygon objects. A rigorous analytical uncertainty model of a line object is determined by the summation of observational errors and modeling errors affecting points on the line segment. The effects of observational errors on points along a line segment are computed using the covariance law, while the modeling error may be approximated by a Gaussian function. The model has been formulated in such a way that the curve uncertainty model may also be derived. The model has been tested with various data sets to prove its applicability and reliability. The rigorous model has also been used to address the probability of locating points within a polygon.

EVALUATION OF THE CURRENT GEOMETRICAL UNCERTAINTY MODELS

The point uncertainty problem is well understood in disciplines such as surveying, geodesy, and photogrammetry (Mikhail, 1976), and as such it will not be investigated further. Comprehensive knowledge about point uncertainties may be derived from the covariance matrix associated with a point, and presented by an error ellipse in 2-D (Figure 24.1a). A number of models for spatial uncertainty of lines exist. The most commonly used are the epsilon band model (Figure 24.1b) and the error band model (Figure 24.1c). Polygon uncertainty models may be represented by the accuracy associated with the area, the center of gravity of the polygon, or by the combination of the boundary segments uncertainty (Figure 24.1d).

The Epsilon Band Model

The basic concept of the epsilon band model is based upon the principle that a cartographic line is surrounded on each side by an area of constant width, epsilon (ε), similar in appearance to a buffer zone. The concept may be visualized as the effect of a ball with a radius ε rolling along the line as shown in Figure 24.1b. The model was designed to provide users with a measure of the error associated with digitizing cartographic lines. Developed and enhanced by researchers such as Perkal (1966), Blackemore (1984), and Chrisman

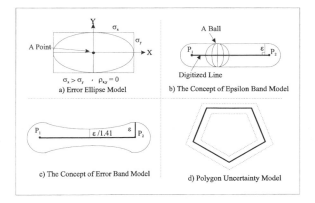

Figure 24.1. Different uncertainty models.

(1989), many other authors have since applied the idea in a variety of ways.

Though many interpretations of the epsilon band exist, they can be categorized in two groups: deterministic and probabilistic. In the deterministic case, the true position of the line is considered to lie somewhere in the buffer zone. Deterministic interpretation of the epsilon band is questionable, because:

- it provides no model of error distribution inside the band, and
- it proposes that the true line lies within the epsilon region.

In the probabilistic interpretation of the epsilon band, the width of the zone is assumed to be a function of different parameters such that their uncertainties accumulate to the final stage. For instance, Alai (1993) assumed scale, digitization, slope, and attributes of the polygons adjacent to the lines are the related variables, while Blackemore (1984) related the band width to the digitizing error, round-off error, and generalization error. However, probabilistic interpretations of the epsilon band are apparently inconsistent with what analytical (Caspary and Scheuring, 1992; Shi, 1994) and simulation (Dutton, 1992) procedures determine.

The Error Band Model

In the error band model, the digitized endpoints of a straight line are drawn from a random sample of possible positions, having a circular normal distribution to form a population of connected line segments. Unlike the epsilon band that assumes constant width

along a line segment, the error band model (Figure 24.1c) suggests a narrower band in the middle of the line segment (Shi, 1994). The error band model is based on the circular error assumption at the endpoints of a line segment. Therefore, the error band model:

- neglects the correlations between endpoints coordinates,
- assumes the magnitudes of uncertainty in perpendicular directions for each endpoint are equal, and
- assumes no modeling error implications.

These restrictions may not be realized when the endpoint positions are determined using only similar methods or instruments, or when the linearity of the model is under question; for instance, when the line represents some natural feature boundary (Chapman et al., 1997). Despite its importance, little work has been done on analyzing the uncertainty arising from the use of inaccurate models within GISs (Karimi and Hwang, 1996).

RIGOROUS POSITIONAL UNCERTAINTY MODEL OF LINE SEGMENTS

Simulation and analytical procedures have been used to model the observational uncertainty of a line segment constructed by connecting its endpoints, and to determine the regions of constant locational probability around the line segment. The effects of modeling error on the intermediate points have been determined empirically, and added to the observational error of the points.

Analysis and Line Uncertainty Simulation

Every measured point is to some degree uncertain, as are the line segments generated by connecting the points. Hence, modeling the uncertainty of a line segment may be determined by properly considering (a) the uncertainty of the coordinates of each point, (b) the correlations among the point coordinates, and (c) the line modeling error.

One way to model the variability of points is to assign an error ellipse to each point. At any given location, the ellipse parameters may be derived from the point covariance matrix—if the point is captured by field surveying methods. Varying amounts of correlation between points occur, depending on the surveying methodology used. For example, points positioned by GPS techniques during a single observational session are influenced by the same atmospheric effects, and satellite geometry. The effect of modeling error to the proposed line uncertainty model will be discussed in a later section.

In the proposed simulation method, each endpoint is drawn from a normal distribution conformed to its coordinates' correlation, and to the assigned correlation with the other point. Figure 24.2 illustrates the realization of this model. Once realized at regular intervals, one can generate dispersion statistics to calculate displacements along the line segment (Alesheikh, 1998). Figure 24.2f illustrates this concept for a particular trial.

Figure 24.2 illustrates selected representations of the many experiments performed. Several different shapes and orientations of error ellipses have been examined, once a correlation factor between two endpoints is assigned. As indicated in Figure 24.2a, once the correlation factor approaches one, the shape of the line variation follows the epsilon band model except at the endpoints. When the correlation factor becomes close to zero (no correlation), the uncertainty of the middle point of the line segment gets smaller (Figure 24.2b). In fact, when the two endpoints are independent, the error band model is approximated, except at the endpoints where the model assumes elliptical shapes instead of circles (Figure 24.2c). If the correlation factor moves toward negative one, the variance of the middle point gets smaller than the error band model (Figure 24.2d and 24.2e).

The immediate conclusions that may be derived from these simulation results are: (1) both the epsilon band and the error band models are special representations of general variations that a line segment may assume, and (2) the line uncertainty indicator should be a function of the correlation among endpoints (Alesheikh and Li, 1996). In the next section, the rigorous line uncertainty model, that respects the correlation among all the coordinates of the endpoints, is analytically derived.

Analytical Procedure of Determining Line Uncertainty Model

A line segment may be defined as a combination of points conforming to a linear function. Hence, one may determine the uncertainty of a line segment as the aggregate of uncertainty of points constructing the line. Point uncertainty is usually represented by error ellipses that are computed from the variance-covariance

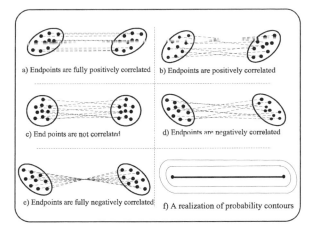

Figure 24.2. Different realizations of line uncertainty model.

matrices of the point coordinates. So the problem of line uncertainty modeling turns out to be the determination of the uncertainty, C_U, of any arbitrary point located along the line segment, based on the statistical information, C_{AB}, of the endpoint coordinates. If U is an arbitrary point located along a line segment AB, its coordinates can be defined as:

$$U = (1-r)A + rB \qquad (1)$$

where

$$r_U = \frac{D_{AU}}{D_{AB}}$$

$$= \frac{\sqrt{(X_U - X_A)^2 + (Y_U - Y_A)^2}}{\sqrt{(X_B - X_A)^2 + (Y_B - Y_A)^2}} \qquad 0 \le r_U \le 1$$

and

$$A = (X_A, Y_A) \text{ and } B = (X_B, Y_B) \qquad (2)$$

The variance-covariance matrix of U may be derived by applying the error propagation law as follows:

$$C_U = J\, C_{AB}\, J^T \qquad (3)$$

where J is Jacobian matrix and C_{AB} is the covariance matrix of the coordinates of endpoints. Once the error matrix is defined, the error ellipse of the arbitrary points may be represented by:

$$\left(\frac{X - X_U}{\sigma_{X_U}}\right)^2$$
$$-2\rho\left(\frac{X - X_U}{\sigma_{X_U}}\right)\left(\frac{Y - Y_U}{\sigma_{Y_U}}\right) + \left(\frac{Y - Y_U}{\sigma_{Y_U}}\right)^2 \qquad (4)$$
$$= (1 - \rho^2)C^2$$

where: σ_{X_U}, and σ_{Y_U} are the diagonal elements of the covariance matrix of U,

$$0 \leq \rho = \frac{\sigma_{x_U y_U}}{\sigma_{X_U}\sigma_{Y_U}} \leq 1$$

is the correlation factor, and
       C is a constant that determines the probability level of error ellipses.

The probability associated with an error ellipse is represented by the volume under the bivariate normal density surface within the region defined by the error ellipse. The probability for various values of C, can be computed from a χ^2 density function with two degrees of freedom.

Once the error ellipses for arbitrary points along the line segment are defined, the region that these ellipses encompass creates the confidence region along the line. It is this region that represents the rigorous uncertainty model of a line segment assuming a perfect known linear model (Figure 24.3).

A very interesting property of this model is that it can easily be extended to determine the uncertainty of a curve, if the curve function, y = f(x), is known (for example, in the case of a road design). The only change that occurs in the method is that the distance from two points on the curve should be computed by the arclength equation:

$$D_{AU} = \int_A^U \sqrt{1 + \left(\frac{dy}{dx}\right)^2}\, dx \qquad (5)$$

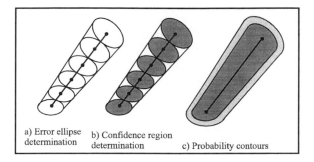

a) Error ellipse determination   b) Confidence region determination   c) Probability contours

Figure 24.3. Line uncertainty model generation.

The Effects of Modeling Errors in the Line Uncertainty Model

Although the effects of modeling errors may be negligible in most man-made structures or cadastral boundaries (Figure 24.4b), they may be of considerable magnitude when the linear function delineates natural feature boundaries (Figure 24.4c).

Modeling error is referred to the deviation between the "true" function representing a boundary and its approximated linear function (Figure 24.4c). Assuming a perfect linear model representing a boundary line causes the uncertainty of the points located along the line to be less than the uncertainty of the endpoints. If the modeling error is under question or the modeling error is of a considerable magnitude, then the proposed line uncertainty model should be investigated further.

To accommodate the modeling error, the uncertainty arises from nonlinearity of the line segment at any point along the line should be added to the observational error of that point. It can be shown that based on the magnitude of the modeling error the shape of the line uncertainty model may well be approximated by a convex shape rather than a parabola (concave) suggested by the error band model. Figure 24.5 demonstrates the differences between line uncertainty models.

Functions representing the modeling error can be determined empirically. As such they may vary from case to case, or from operator to operator. To specify the function for an individual, and for a particular geospatial feature, "true" boundaries can be compared to their linear approximations and the deviations be computed. These deviations present the values of the function at different intervals. Any function that is fitted to the deviations can be used as the representation of the modeling error for the boundaries of that particular object. True boundaries may be determined

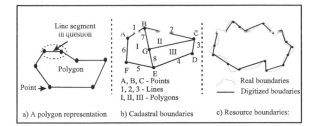

Figure 24.4. The digitized representation of different polygons.

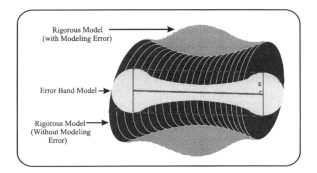

Figure 24.5. Realizations of various line uncertainty models.

using larger scale maps, or other prior knowledge about the function, such as the shapes of the road curves.

Several water bodies have been digitized from two similar maps having different scales. Due the scale reduction, some curvilinear boundaries were represented as straight lines in the small-scale map. The deviations between the curves and the lines were calculated for many line segments. The errors were then averaged and given as the representation of the modeling errors along the line segments. The deviations were approximated by a Gaussian function. The function depicted in Figure 24.5 is the function corresponding to the fitted model which has the form

$$Me \ (d) = \alpha \exp(-\beta d^2) \qquad (6)$$

that is added to the observational uncertainties. Here, d is the separation distance from the center of gravity of the line, α and β are constants that determine the magnitude of modeling error and damping parameter, respectively. For instance, α may be close to zero once the modeling error is negligible; e.g., in the case of cadastral studies, and be higher in the case of digitiz-

ing feature boundaries. Since the modeling in perpendicular direction to the line segment is required, the error is added in that direction to the magnitude of observational error.

POLYGON UNCERTAINTY MODEL

A polygon can be assumed as an areal object defined by its boundary lines. Therefore, the polygon uncertainty is closely related to the previously discussed point and line uncertainties. A comprehensive description of the stochastic variations of a polygon may be derived from the covariance matrix of the vertices of the polygon and the modeling errors of each boundary segment. An obvious indicator of the uncertainty of a polygon object may be defined as the combination of the areas that the boundary line uncertainty models cover over the area of that polygon. In the following, the visualization of the polygon uncertainty will be elaborated using various case studies.

Survey Lot Data Sets

The first case study involves a lot survey consisting of seven points, two of which are fixed. The angular and distance measurements are reported in Mepham and Nickerson (1987). The measurements, together with the approximate coordinates of the points, were used in a least-squares adjustment to determine the estimated coordinates of the points and their covariance matrices. The adjustment results were then imported into the rigorous uncertainty model to illustrate the uncertainty of the line segments and the polygon. Figure 24.6 represents the output of the computations.

Figure 24.6 clearly illustrates that the union of ellipses creates a region similar to the epsilon band model if modeling error is not considered. The reason is attributed to the off-diagonal elements of the covariance matrix that are predominately positive (positive correlation), which supports the results of the simulation method (Figure 24.1b).

Simulated Data Sets

The second case study involves simulation of coordinates of a triangle vertices, together with their covariance matrices. The covariance matrix of the triangle is generated in such a way that the line uncertainty model resembles the error band when the modeling error is not being considered—negative correlation among endpoints are generated. These facts also con-

Figure 24.6. Uncertainty visualization of survey lot data sets (ellipses are not to scale).

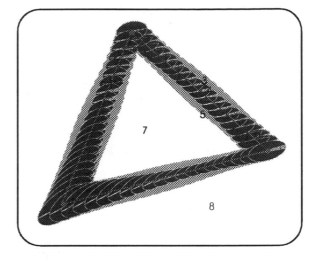

Figure 24.7. Uncertainty realization of the simulated triangle (ellipses as well as the location of test points are not to scale).

firmed the results of the simulation procedures, and consequently the generality of the proposed rigorous uncertainty models. The results of the simulated datasets are shown in Figure 24.7.

Figures 24.6 and 24.7 clearly demonstrate the effects of the correlations between endpoint coordinates on the shape of the line uncertainty model. Correlations were mostly positive in the case of survey lot data sets, making the shape of the line uncertainty to be similar to the epsilon band model, while the endpoints were negatively correlated among the triangle vertices, causing the shape of the line errors to approximate the error band model.

Comparison of Polygon Accuracy Using Different Models

A comparison of the different measures of accuracy of a polygon is given below. The area that is covered by epsilon band, A_{ep}, for a line with length D and width of ε (Figure 24.1b) can easily be determined by:

$$A_{ep} = 2D\varepsilon + \pi\varepsilon^2 \qquad (7)$$

The error band model encapsulates less area, A_{er}, than what the epsilon band model covers (Figure 24.1b). Caspary and Scheuring (1992) showed that this area can be computed by:

$$A_{er} = 2D\varepsilon \int_0^1 \sqrt{1 - 2x - 2x^2} \, dx + \pi\varepsilon^2$$
$$\approx 1.62D\varepsilon + \pi\varepsilon^2 \qquad (8)$$
$$= A_{ep} - 0.38D\varepsilon \approx 0.8A_{ep}$$

The area covered by the proposed rigorous polygon uncertainty model depends not only on the covariance matrix of the endpoints but also on the function representing the modeling error. Table 24.1 shows the difference in the area covered by various line uncertainty models for the two case studies. For the case of the epsilon band and error band models, it is assumed that

$$\varepsilon = \sigma = \{[\sigma^2_{xA} + \sigma^2_{yA} + \sigma^2_{xB} + \sigma^2_{yB}] / 4\}^{1/2} \qquad (9)$$

A single Gaussian function is used to approximate the modeling error for both case studies, and is characterized by: $\alpha = 0.88$ and $\beta = 3$.

THE POINT-IN-POLYGON PROBLEM

The point-in-polygon problem has been addressed by several algorithms, such as the half-line algorithm. However, the quality of the solution to the point-in-polygon problem has yet to be addressed properly. A simple computation of the quality of the point-in-polygon solution could, for example, avoid displaying buoys on dry land or rivers outside their floodplain. Based on the proposed polygon uncertainty model, the solution to the point-in-polygon problem may be more accurate using probability statements.

The procedure used is twofold in nature: (a) determine whether a point is inside, outside, or on the boundary of a polygon, and (b) determine in which uncertainty model the point falls. The shape of the probability contours varies with the uncertainty model applied. Table 24.2 presents the results of a few tests performed on the simulated triangle that is shown in Figure 24.7. Indeed, the probability level of each model may also be changed. Table 24.2 presents the results of the one σ probability level. Table 24.2 clearly shows the different answers to the point-in-polygon problem using various line uncertainty models.

Table 24.1. Comparison Between Areas Covered by Various Line Uncertainty Measures.

	Survey Lot Case (m²)	Triangle Case (m²)
Sum of ε-bands A_{ep}	4.155	44.61
Sum of error bands A_{er}	3.324	35.69
Sum of rigorous bands (without modeling error)	3.983	38.72
Sum of rigorous bands (with modeling error)	—	53.37
Area	599.86	6820.8

CLOSING REMARKS

The role of GIS in decision-based processes will likely improve if uncertainty estimates are provided along with GIS products. The uncertainty estimates may be derived from the information of the uncertainty of their geometric elements. Geometric primitives of a two-dimensional vector GIS are points, lines, and polygons.

Various measures of accuracy for geometrical entities have been assessed. Rigorous models for error measures of lines and polygons that accompany the correlation between endpoints and modeling errors have been derived using simulation and analytical approaches. The results enable a comparison between the frequently used simplified measures and the proposed rigorous models. The approach to extend the model for curvilinear objects is indicated. Conceptual implementation of uncertainty information in GIS databases is recommended. This integration will help the inclusion of uncertainty manipulation for GIS analysis. By applying the uncertainty models, the solution to the problem of point-in-polygon has been augmented using a probability statement.

REFERENCES

Alai, J. *Spatial Uncertainty in a GIS,* thesis presented to the Department of Geomatics Engineering, The University of Calgary, Alberta, Canada, in partial requirements for the degree of Master of Science in Engineering, 1993.

Alesheikh, A. and R. Li. Rigorous Uncertainty Models of Line and Polygon Objects in GIS, *Proceedings of GIS/LIS '96*, Denver, CO, 1996, pp. 906–920.

Table 24.2. Results of Point-in-Polygon Tests, Probability Level is Fixed.[a]

No.	Epsilon Model	Error Model	Rigorous (*)	Rigorous (**)	In	Out	Bound (***)
1	*	*	*	*			*
2	*	*	*	*		*	
3	*			*		*	
4	*		*	*	*		
5				*	*		
6	*	*	*	*		*	
7					*		
8						*	

[a] * without modeling error, ** with modeling error, *** boundary line.

Alesheikh, A. *Modeling and Management of Uncertainty in Geospatial Information Systems,* thesis presented to the Department of Geomatics Engineering, The University of Calgary, Alberta, Canada in partial fulfillment of the Degree of Doctor of Philosophy, (under preparation) 1998.

Blackemore, M. Generalization and Error in Spatial Databases, *Cartographica,* 21(2), pp. 131–139, 1984.

Carver, S. Adding Error Handling Functionality to the GIS Toolkit, *Proceedings of the Second European Conference on GIS (EGIS '91)*, Brussels, Belgium, 1991, pp. 187–196.

Caspary, W. and R. Scheuring. Error-Bands as Measures of Geometrical Accuracy, *Proceedings of the Third European Conference on GIS (EGIS '92)*, Munich, Germany, 1992, pp. 227–233.

Chapman, M.A., A.A. Alesheikh, and H. Karimi. Error Modeling and Management for Data in Geospatial Information Systems, *Proceedings of Coast GIS '97,* August 29–31, Aberdeen, Scotland, 1997.

Chrisman, N.R. Error in Categorical Maps: Testing Versus Simulation, *Proceedings of the Ninth International Symposium on Computer-Assisted Cartography (Auto-Carto 9),* Baltimore, MD, 1989, pp. 521–529.

Dutton, G. Handling Positional Uncertainty in Spatial Databases, *Proceedings of the Fifth International Symposium on Spatial Data Handling,* August 3–7, 1992, Charleston, SC, 1992, pp. 460–469.

Karimi, H. and D. Hwang. Towards Managing Model Uncertainty in GISs: An Algorithm for Uncertainty Analysis of Air Quality Advection Models, *Geomatica,* 50(3), pp. 251–259, 1996.

Mepham, M.P. and B.G. Nickerson. Preanalysis, in *Papers for the CISM Adjustment and Analysis Seminars,* E.J. Krakiwskey, Ed., 1987, pp. 150–181.

Mikhail, E.M. *Observations and Least Squares,* IEP, New York, 1976.

Perkal, J. *On the Length of Empirical Curves*, Discussion Paper Number 10, Michigan Inter-University Community of Mathematical Geography, 1966.

Shi, W. *Modeling Positional and Thematic Uncertainties in Integration of Remote Sensing and GIS,* Ph.D. Thesis, ITC Publication No. 22. ITC, The Netherlands, 1994.

CHAPTER 25

Application of a New Model of Vector Data Uncertainty

G.J. Hunter, J. Qiu, and M.F. Goodchild

INTRODUCTION

In previous papers by Hunter and Goodchild (1996) and Hunter et al. (1996), a model of vector data uncertainty was proposed and its conceptual design and likely manner of implementation were discussed. The model allows for probabilistic distortion of point, line, and polygon features through the creation of independent positional error fields in the x and y directions. These are overlaid with the vector data so as to apply coordinate shifts to all nodes and vertices in the data set to establish new, but equally likely, versions of the original data. By studying the variation in the family of outputs derived from the distorted input data, an assessment may be made of the uncertainty associated with the resultant information product. The model has now been developed and tested, and the purpose of this chapter is to report on its application in practice.

DEVELOPMENT OF THE UNCERTAINTY MODEL

The uncertainty model involves the creation of two independent, normally distributed, random error grids in the x and y directions. These grids are combined to provide the two components of a set of simulated positional error vectors regularly distributed throughout the region of the data set to be perturbed (Figure 25.1). The assumptions made are (a) that the error for each node or vertex has a circular normal distribution, and (b) that its x and y components are independent of each other. The grids are generated with a mean and standard deviation equal to the estimate for positional er-

ror in the data set to be perturbed (a prerequisite for use of the model). These error estimates, for example, might come from the residuals at control points reported during digitizer setup, or from an associated data quality statement.

By overlaying the two grids with the data to be perturbed, x and y positional shifts can be applied to the coordinates of each node and vertex in the data set to create a new, but equally probable, version of it. Thus, the probabilistic coordinates of a point are considered to be (x + error, y + error). With the distorted version of the data, the user then applies the same set of procedures as required previously to create the final product, and by repeating the procedure a number of times the variability residing in the end product may be assessed. Alternatively, several different data sets may be perturbed (each with its own error estimate) before being combined to assess final output uncertainty. While the model does require an a priori error estimate for creation of the two distortion grids, it is the resultant uncertainty arising from the use of perturbed data due to simulation which is under investigation—hence its label as an "uncertainty" model.

As discussed more fully in Hunter and Goodchild (1996), the first step in implementing the model is to determine an appropriate error grid spacing. If it is too large, the nodes and vertices of small features in the source data will receive similar-sized shifts in x and y during perturbation and the process will not be random. Conversely, if the grid is too small then processing time is increased as additional grid points are needlessly processed. Experience to date suggests that an appropriate spacing be selected from one of the following:

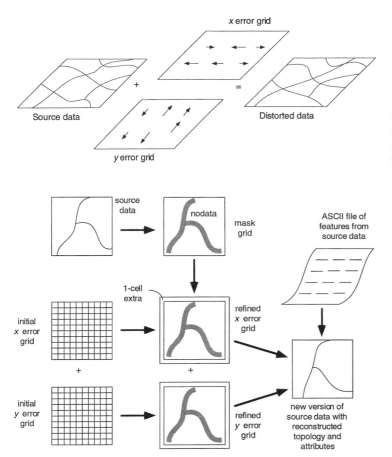

Figure 25.1. The model of vector data uncertainty uses normally distributed, random error grids in the *x* and *y* directions to produce a distorted version of the original data set.

Figure 25.2. Development diagram for the uncertainty model.

- the standard deviation of the horizontal positional error for the source data; or
- a distance equal to an established standard—for example, 0.5 mm at source map scale where the data has been digitized; or
- a threshold value smaller than the user would care to consider, given the nature of the data to be processed.

Using Figure 25.2 as a guide, the second stage is to generate the *x* and *y* error grids. To ensure the grids completely cover the extent of the source data, their dimensions are predetermined by setting a window equal to the data set's dimensions, and the cell size equivalent to the chosen grid spacing. The grids are created automatically with these parameters and populated with randomly placed, normally distributed values having a mean (usually zero) and a standard deviation as previously defined. It should be noted that selecting the standard deviation as the grid spacing has no effect on the random population of the grids. These

two grids are only temporary and will require further refinement before being used to perturb the source data.

To optimize processing time, the number of cells in the error grids needs to be reduced, since unless the data set is extremely dense there will be many unwanted cells processed during the operation. To achieve this the original vector data set is converted to grid format to form a temporary masking grid that only contains "live" cells—that is, those which the source data either lie within or pass through. Polygons are processed as line strings since only their boundaries are perturbed. Cell attributes that are maintained during rasterization are unimportant, given that the grid is only used for masking purposes and all other noncontributing cells are given a null or "nodata" value.

At this point there is a potential problem with using a masking grid that contains 1-cell-wide strings representing line or polygon features (Figure 25.3). As mentioned in Hunter and Goodchild (1996), there is some likelihood (although small) that the magnitude and direction of adjacent *x* and *y* error grid shifts may cause

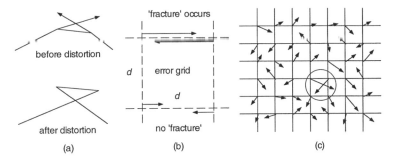

Figure 25.3. In (a), uncontrolled shifts between neighboring error grid points may cause unwanted "fractures" or transposition of features. In (b), "fractures" occur when the difference between neighboring shifts is larger than their separation (*d*). In (c), a "fracture" is circled, requiring filtering on the basis of neighboring shifts.

them to overlap, resulting in possible loss of topological integrity in the source data when applied to it (for example, unwanted loops caused by the transposition of adjacent vertices or nodes). The solution is to filter any "offending" pairs of shifts, which first requires that a spread function be applied to the mask grid for a distance of at least three standard deviations either side of the initial mask cell (up, down, left and right). A new masking grid is created during the process and any previously null cells affected by the operation are automatically returned to active status. The masking grid is then overlaid with the initial error grids to provide reduced versions of the *x* and *y* grids that contain only shift values surrounding features in the source data. Finally, the error grids are expanded by the width of a cell on all sides (with null values) to support the treatment of edge effects during processing.

To test for possible "fractures" between neighboring shifts (Figure 25.3a), a routine was developed to test the difference between consecutive cells (in horizontal or row sequence for the *x* grid, and vertical or column sequence for the *y* grid) to determine whether the absolute value of the difference between them was greater than their separation distance (Figure 25.3b). If so, then a "fracture" has the potential to occur at that location if there are data points nearby and a filter must be applied to average out the shift values on the basis of their neighbors. The procedure is iterative and proceeds until no "fractures" exist in either error grid.

In the final step of the model's development, values in the error grids must be transferred to the data set being perturbed. Naturally, it will be rare for nodes and vertices in the source data to coincide exactly with the error grid points, and a method was required for calculating *x* and *y* shifts based on the neighboring values in the grid. To achieve this, a bilinear interpolation procedure was used in which the *x* and *y* shifts assigned to each point are calculated on the basis of the respective shifts of the four surrounding grid points.

An ASCII feature file containing the identifier and coordinates of each data point was automatically derived, and the four surrounding error shifts were determined for each point then used to interpolate the shift values to be applied. A proximity threshold was also applied to ensure that data points close to a grid point would automatically receive that points' *x* and *y* shifts without computation. The distorted coordinates were then written to an output file and the file topology was rebuilt. Finally, the attributes belonging to each feature in the original data set were rewritten to their parent features in the distorted version of the data set.

The entire process runs as an Arc/Info AML script which calls a random number generator written in C. The AML program prompts the user for the name of the file to be perturbed, its data type (point, line, or polygon), the error grid size, the standard deviation of the horizontal positional error in the features, and the number of perturbations required. The code is freely available at the primary author's website given at the end of this chapter.

APPLICATION OF THE UNCERTAINTY MODEL

Polygon Area Estimation

The first application of the model is a simple one—estimation of the areal uncertainty of a set of polygons. In this case we took a group of six polygons that had been digitized from a source map at a scale of 1:50,000. We estimated that the digitizing was performed with a standard deviation of 25 m which was also the error grid spacing chosen—given that any polygon boundary segment length less than this value would have no significant impact upon subsequent application of the data. The set of polygons was perturbed 20 times and the results of overlaying the 20 realizations can be seen in Figure 25.4. Then, by appending the 20 polygon sets and statistically analyzing the areas for each of the six polygon identifiers, we were able to easily construct a table of mean polygon areas and their standard deviations (see Table 25.1).

Point-in-Polygon Overlay

In the next application we took a set of 30 point features and overlaid them with the set of six polygons used before (see Figure 25.5). The points were deliberately placed near polygon boundaries and junctions. In the first instance we held the polygon boundaries fixed (that is, we assumed they had high positional accuracy), and perturbed the point set 20 times (with an error grid spacing again of 25 m and a standard deviation of 25 m). As each perturbed point set was overlaid with the fixed polygon boundary file, we recorded the identifier of the polygon in which each point was deemed to lie and appended the point identifiers and their associated polygon numbers to an output file. When the 20 overlays were completed, a frequency count was taken and the results were summarized in Table 25.2.

We then perturbed both the points and polygons 20 times each and overlaid them a total of 400 (20 × 20) times—a process that was automated quite simply with a short AML script. While the two data sets once more employed an error grid spacing of 25 m and a standard deviation of 25 m, these parameters are easily varied by a user and need not be the same—which would enable perturbation of different data sets with different errors. The results of the 400 overlays are shown in Table 25.3.

Polygon to Grid Conversion

In the final application we took the same set of six polygons, perturbed them 20 times using the same error grid size and standard deviation as previously, and converted them to grid cells in order to estimate the variation associated with both the allocation of polygons to grid cells and total class areas. After each polygon to grid conversion, the number of cells belonging to the six polygons were counted (since polygon IDs were maintained during conversion), and the mean and standard deviation of the number of cells formed from each parent polygon were recorded. As expected, the mean number of cells remained within one or two of the number recorded when the unperturbed polygon set was converted. However, we believe the standard deviation of the number of cells is a useful statistic that could be put to further use as described later. The results are shown in Table 25.4.

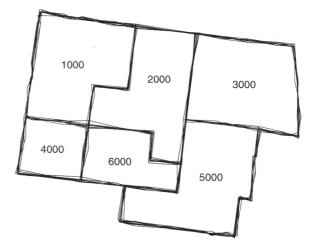

Figure 25.4. Showing the results of 20 perturbations of the polygon data set when overlayed.

Table 25.1. Showing the mean and standard deviation for the area of each polygon after 20 perturbations (using a standard deviation of 25 m for the horizontal positional error of the polygon boundaries).

Polygon ID	Mean Area (sq. m)	Standard Deviation (m)
1000	891858.3	5419.6
2000	890108.5	9920.3
3000	945221.7	3889.6
4000	358774.9	5407.7
5000	980114.9	6748.4
6000	459806.7	7175.6

DISCUSSION OF RESULTS

From these examples, there are several comments that can be made with respect to applying the vector data uncertainty model in practice.

Clearly, it has the potential to help educate users about the meaning of metadata items that are attached to a data set. For example, in conjunction with a statement of the standard deviation of positional error, a diagram such as Figure 25.4 could be included in a data quality report showing how the data may probably vary in position according to the meaning of that error descriptor.

Some useful statistics also arise from the model. For instance, the class membership frequencies shown

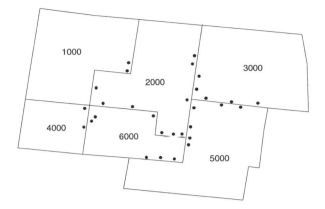

Figure 25.5. Showing the set of 30 points that were overlayed with the set of six polygons.

Table 25.2. Showing the observed frequencies for which each point lies in any of the six polygons after 20 perturbations (for clarity, points lying only in the one polygon are not listed). Asterisks indicate the polygon in which the point was assigned prior to perturbation of the data.

Point ID	Poly 1000	Poly 2000	Poly 3000	Poly 4000	Poly 5000	Poly 6000
2	16*	4	–	–	–	–
9	–	11*	–	–	–	9
13	–	3	17*	–	–	–
17	–	–	7	–	13*	–
18	–	–	12*	–	8	–
19	–	–	6	–	14*	–
23	–	–	–	–	5	15*
27	–	–	–	–	8	12*

Table 25.3. Showing the observed frequencies for which each point lies in any of the six polygons after 20 perturbations each of the point and polygon data sets, then overlayed 400 (20 x 20) times (for clarity, points lying only in the one polygon are not listed). Asterisks indicate the polygon in which the point was assigned prior to perturbation of the data.

Point ID	Poly 1000	Poly 2000	Poly 3000	Poly 4000	Poly 5000	Poly 6000
2	290*	110	–	–	–	–
9	–	260*	–	–	–	140
13	–	27	373*	–	–	–
17	–	–	156	–	244*	–
18	–	–	280*	–	120	–
19	–	–	129	–	271*	–
23	–	–	–	–	108	292*
27	–	–	–	–	140	260*

in Tables 25.2 and 25.3 represent quantities that until now have been quite difficult to define. Certainly, they have been able to be computed in certain cases—for example, when classifying remotely sensed imagery—but there has been no ready solution for vector overlay operations. Furthermore, the standard deviations computed for polygon and cell class areas can serve as input to formal error propagation computations. For example, when calculating population densities we could combine the standard deviation of the area with that of the population count to yield the density variance.

The model has the added ability to indicate portions of a data set that may be highly sensitive to perturbation—thereby warning users that the data set is potentially unsuitable for use in a particular region. For instance, in the polygon-to-grid conversion example above it is suspected that there is little variance in the number of polygons formed after each perturbation due to the fairly regular N-S, E-W boundaries of the polygons. On the other hand, polygons with direction trends at 45° to the cardinal axes might prove highly variable when perturbed and subsequently converted.

Importantly, the model has the capacity (in certain cases) to be able to turn simulated error in position into measurable attribute uncertainty—for example, the transformation of polygon boundary error into polygon area uncertainty. There is also the potential to assess the uncertainty of a final information output after a sequence of spatial operations, in which all data sets have had their positions perturbed to varying degrees according to their individual accuracies. At the same time, we believe that some problems are ill-posed and not well suited to this model—for instance, perturbing closely set contour lines where there is a likelihood of the perturbed contours crossing each other.

CONCLUSION

This chapter has described the development and application of an uncertainty model for vector data which operates by taking an input data set of point, line, or polygon features and then applying simulated positional error shifts in the *x* and *y* directions to calculate new coordinates for each node and vertex. In effect this produces

Table 25.4. Showing the mean and standard deviation of the number of cells formed from each of the six polygons after perturbation 20 times and subsequent conversion to grid format.

Polygon ID	Mean No. of Cells	Standard Dev. (cells)
1000	355.4	2.7
2000	356.1	3.5
3000	376.5	2.3
4000	145.3	2.2
5000	385.6	3.4
6000	182.8	2.6

a distorted, but equally probable, representation of the data set that can be used to create a family of alternative outputs, usually in map form. Assessment of the variation in the outputs can be used to provide an estimate of the uncertainty residing in them, based on the error in the source data and its propagation through the subsequent algorithms and processes employed. The model was tested in several applications, viz: (a) perturbing polygon boundaries to determine a mean and standard deviation for the area of each polygon; (b) perturbing point and polygon data sets prior to point-in-polygon overlay, which yielded class membership frequencies for each point; and (c) perturbing polygon boundaries prior to polygon-to-grid conversion, to generate a standard deviation for the number of cells in each polygon as a result of the conversion algorithm.

FURTHER INFORMATION

For further information, readers are directed to the principal author's home page where a tutorial containing AML and C source code exists to implement the uncertainty model and automatically perturb point, line and polygon files. The URL is:

http://www.geom.unimelb.edu.au/people/gjh.html

ACKNOWLEDGMENTS

The authors acknowledge funding support received under Australian Research Council (ARC) Large Grant No. A49601183—"Modeling Uncertainty in Spatial Databases."

REFERENCES

Hunter, G.J. and M.F. Goodchild. A New Model for Handling Vector Data Uncertainty in GIS, *J. Urban Reg. Inf. Syst. Assoc.,* 8(1), pp. 51–57, 1996.

Hunter, G.J., B. Höck, M. Robey, and M.F. Goodchild. Experimental Development of a Model of Vector Data Uncertainty, *Proceedings of the 2nd International Symposium on Spatial Accuracy Assessment in Natural Resources and Environmental Sciences*, Fort Collins, CO, 1996, pp. 214–224.

CHAPTER 26

Implementing an Object-Oriented Error-Sensitive GIS

M. Duckham

INTRODUCTION

Despite a long history of research into data quality, commercial geographical information systems (GIS) still have little or no data quality management capabilities (Aspinall, 1996; Brunsdon and Openshaw, 1993). Geographical information (GI) science has produced a limited number of implementations which exhibit some of the important capabilities (e.g., Ramlal and Drummond, 1992), but the production of what has been termed an *error sensitive* GIS (Unwin, 1995) is still some way off. The first step in implementing any error sensitive GIS must be to design a model for storing quality statistics and to implement this model in a database. There are a number of barriers to effective design and implementation of such a quality storage model which have hampered the uptake of data quality management within mainstream GIS.

First, there is only limited agreement on the quality statistics which should be stored. Most quality models use at least the five elements of data quality proposed by the United States' National Committee on Digital Cartography and Data Standards (NCDCDS) in 1988: lineage, logical consistency, completeness, positional accuracy, and attribute accuracy (NCDCDS, 1988). Generally, these five quality elements are taken as a basis for any model of data quality. However, subsequent studies have often found that additional quality elements are desirable (e.g., Aalders, 1996; Morrison, 1995), while it is also worth noting that there may be little agreement on the actual definitions of a number of these elements (Drummond, 1996).

Second, it follows that quality statistics carry a data volume overhead. In most models of data quality every feature in the database can be annotated with at least four separate quality statistics (at most, one of positional accuracy or attribute accuracy will apply to a given feature). The elements of data quality can vary widely in size, some studies calling for very large quality statistics. For example, lineage is defined by Clarke and Clark (1995) as "source observations or materials, data acquisition and compilation methods, conversions, transformations, analyses and derivations that the data has been subjected to, and the assumptions and criteria applied at any stage of its life." Consequently, efficiency of storage is crucial to the success of any error sensitive GIS database.

This chapter aims to illustrate how object-oriented (OO) technology can address some of these problems and aid the design of an implementation-independent data quality storage model. Based on such an OO data quality storage model, two implementations are discussed highlighting some of the possible pitfalls of the OO analysis and design method. Finally, an example of how such pitfalls might be addressed using a mathematical formalization of object-oriented systems is explained.

OBJECT-ORIENTED ANALYSIS

The fundamental features of OO are abstraction, encapsulation, and inheritance. Abstraction deals with complexity in the real world from the top down and focuses on the features of an object in the real world which distinguish it from other objects in the real world. Complementary to abstraction, encapsulation approaches complexity from the opposite direction and aims to hide the detailed mechanisms and features of an object in the real world. Finally, inheritance is a way of structuring classes of abstracted, encapsulated objects in a hierarchy which allows blueprint features to be described only once; inheriting objects can then incrementally specialize the blueprint.

Object-orientation has been enormously successful and is increasingly the default choice for a wide range of information systems designs. Object-oriented analysis (OOA) is a method for creating OO representations of real world systems in an implementation-independent way. Object-oriented analysis is important to data quality and GIS because it presents a mechanism for structuring the complexity of the real world in a way which should be supportable in any OO GIS. A modified version of OOA has been demonstrated to significantly improve the process and results of GIS development (Kösters et al., 1997).

To be successful, an OOA approach to data quality storage needs to be able to combat the barriers to quality management (data volume and lack of agreement on quality statistics) presented in the introduction. Furthermore, the provision of an implementation-independent analysis, or at least one where the implementation dependence is clearly outlined, should provide a more secure basis for a general approach to data quality within GIS. A more detailed exposition of OO and OOA is given in the literature (e.g., Booch, 1994; Coad and Yourdon, 1991; Worboys, 1995).

DATA QUALITY STORAGE MODEL ANALYSIS AND DESIGN

An OOA of data quality was undertaken, with a view to creating an implementation-independent data quality storage model. Using the resultant object schema as a basis, two implementations in different OO systems were also completed. An exhaustive exploration of the process and results of the OOA and of the implementation of those results would be too lengthy to present here. Instead, the version of the class diagram shown in Figure 26.1 details enough of the results to illustrate the key successes and difficulties encountered in the production and implementation of the quality storage model. The notation used for this diagram is *not* consistent with either of the two most popular notations, that of Coad and Yourdon (1991) and of Booch (1994), since both of these notations are too rich for the concise version of the results presented here. Consequently, the author's own, simpler notation is used.

A Flexible Data Quality Model

The first steps of any OOA involve looking at the system under analysis in a highly abstracted manner, picking out the most general features and in effect "taking a step back." Starting the analysis with such a wide view of data quality results in a model of data quality storage which places very few restrictions upon the types and structures of quality statistics which can be represented. The model presented here aims to be flexible enough to represent any data quality statistic which has been proposed in the literature or conceivably might be sensible to propose. The analysis took the view that all data quality statistics are essentially collections of conceptually atomic quality attributes. The class *quality element* is proposed as the super-class of all quality statistics. A *quality element* can refer to any number of *quality attributes,* which are classes representing the atomic data values that make up a quality statistic. It is worth noting that while these data values are conceptually atomic, they are not necessarily atomic data types. Abstract data types, such as misclassification matrix, are just as valid *quality attributes* as more conventional data types, such as integer and real.

Additionally, some quality statistics are only meaningful when the data to which they refer is of a particular data type or metric. The quality storage model needs to be able to restrict the scope of a given quality statistic to defined metrics. To represent positional accuracy, for example, a new class *positional accuracy* inheriting from *quality element* would be created. Objects of that class would refer to *x standard deviation* and *y standard deviation* objects, which in turn would inherit from *quality attribute.* The *metric scope* behavior of the *positional accuracy* class would need to be set such that only objects which belonged to a spatial class could refer to *positional accuracy.*

A notable exception to this broad classification of data quality is logical consistency. Logical consistency has been defined as the "fidelity of relationships encoded in the data structure" (NCDCDS, 1988). Therefore, logical consistency is expected to highlight when values fall outside acceptable ranges or when the topology of the data set fails to conform to the topological model being used. However, in the OO world, this type of expected behavior can be more properly encapsulated within the object itself. That the height of a contour must lie somewhere between its topological neighbors can be encoded in the OO representation of a contour. Any attempt to violate this behavior results in the database rejecting the contour, thereby guaran-

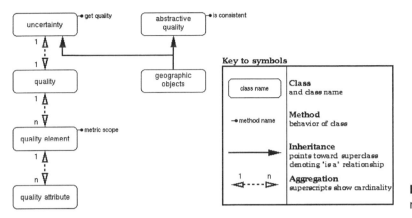

Figure 26.1. Class diagram of OOA results.

teeing the consistency of all objects in the database without the need for a logical consistency statistic. To describe the concept of quality statistics, such as logical consistency, which originate through an incomplete abstraction of the object schema from the real world rather than through some observational error, the term *abstractive quality* will be used here. Abstractive quality generally and logical consistency specifically are represented by the class *abstractive quality*. The *geographic objects* class is intended as the generic superclass of all the geographic objects in the database. All *geographic objects* inherit from *abstractive quality* and so possess the *is consistent* method, the behavior of which would usually be redefined for each subclass and would allow any individual object in the database to report on its own consistency.

Efficient Data Storage

The importance of focusing on discrete objects and the relationships between objects in the OOA process helps maximize the efficiency of data storage. Quality elements can refer to geographical data in a hierarchy of interdependencies at any level. Any information below this level in the hierarchy can automatically infer quality elements. This is achieved using the *uncertainty* class, from which all *geographic objects* inherit. The *uncertainty* class supports a reference to an object in the *quality* class. An object of class *quality* is simply a collection of *quality elements*. The *get quality* method on the *uncertainty* class will return a list of *quality elements* of a specified type by looking at the contents of the *quality* object referenced by that object. Consequently, all objects in the database will be able to store and retrieve references to quality elements defined in the database. If such a reference does not

exist, the *get quality* method is free to infer a *quality* object from its parent object. For example, in an OO database a river system might constitute a large number of individual rivers, which in turn might be made up of a number of line vectors. If it is known that, say, the lineage of the river system is constant across all the information in the database, a single lineage object can represent this situation, since the lineage of river and line objects related to the parent river system object is implied by the hierarchy, shown in Figure 26.2. Similarly, individual rivers or lines can have particular lineage objects inserted at the highest node in the hierarchy that conveys the desired information. The result, in this simple situation at least, is an highly efficient packing of metadata round a hierarchy where the maximum quality information is conveyed for the minimum of stored metadata.

Java and Gothic Implementation

The generic model of data quality in Figure 26.1 was implemented first as a prototype in Java OO programming language and subsequently within Laser Scan Gothic OO GIS. The two OO environments contrast strongly; Java is a general purpose, powerful, fast development language able to provide a working proof of concept very quickly. Gothic is one of the few commercial GIS operating under a fully OO database, so while the implementation took considerably longer to program, it was able to benefit from the range of GIS functionality already available in the Gothic database.

The implementations were able to store successfully a wide range of quality statistics which exist in the literature. An example of how a particular set of quality statistics can be implemented is given in Table 26.1, where the quality elements, quality attributes,

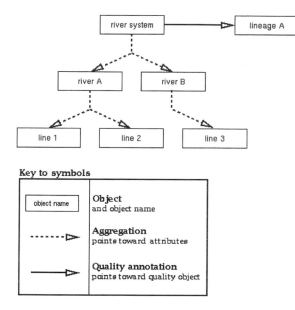

Figure 26.2. Object diagram showing efficient quality storage.

and metric scope used to represent a particular quality statistic are given in addition to the literature source of that statistic. The implementations will support not only those statistics in Table 26.1, but any element of spatial data quality which can be represented by the OOA model; i.e., as long as it can be represented as a named collection of conceptually atomic values. However, the process of implementation of the OOA was not a completely mechanical process. In attempting to produce two separate implementations, a number of problems were encountered during implementation due the difficulty in creating an implementation-independent OOA.

Design and Implementation Problems

Despite general agreement on the principles of OO, outlined in the Object-Oriented Analysis section, there is considerable disparity when examining exactly how and which principles are employed in a programming system. As already mentioned, in principle the results of any OOA are implementation-independent. In reality, no OOA analysis will be truly implementation-independent since different OO system do not all employ the general features of OO which may be used in an OOA. Most authors acknowledge the existence of a "continuum of representation" (Coad and Yourdon, 1991) where the implementation indepen-

dent OOA process blurs with the implementation-dependent design process.

There are numerous examples of OO features which may or may not be supported by a particular implementation. Polymorphism, typing, persistence, and exception handling are all general features of OO (see Booch, 1994) which are not supported by both Java and Gothic. These features will usually be supported differentially dependent on the particular implementation, and any designer wishing to make use of the features in the OOA and OO design process must be aware of this.

Particular problems were encountered here with multiple inheritance, the ability of subclasses to inherit from any number of superclasses. Multiple inheritance is often very useful during OOA, and generally OOA makes does not proscribe its use (Coad and Yourdon, 1991; Booch, 1994). In contrast, most OO programming languages, such as Java, only permit single inheritance where all subclasses must inherit from only one superclass. The results of this OOA made use of multiple inheritance; all *geographic objects* inherit from both *abstractive quality* and *uncertainty*. Consequently, in order to implement the analysis in Java, the class diagram had to be reformulated to remove multiple inheritance. No such difficulties were encountered for the Gothic implementation.

Class-based properties, also termed static properties, are fields and methods which can be defined for a class, meaning that all objects of that class will share identical copies of these properties. Altering a static property means that alteration is seen by every object of that class. Static behavior is implied by the *is consistent* method of the *abstractive quality* class. The test for the consistency of objects in a class will be the same for all those objects. The behavior of the *is consistent* method would therefore best be defined statically. Gothic, however, does not implement static behavior and while Java does, a technical problem in Java meant that static properties are not inherited in a transparent way. Consequently, it was not possible to implement static consistency properties in either system.

These sorts of problems are often encountered during the OOA process, making it harder to design an object schema which could be implemented on any OO GIS. The lack of certainty surrounding a particular object schema and its suitability for implementation in a given OO system *before* programming actually starts, can only serve to destabilize and fragment attempts to produce an error sensitive GIS. Re-

Table 26.1. Example Quality Statistic Implementations.

Quality Element	Quality Attributes	Metric Scope	Reference
Source	Source organization Reason for creation Creation date	Any	CEN/TC287, 1996
Usage	User name Usage type Usage date Usage comments	Any	CEN/TC287, 1997
Lineage	Process name Process description Process date	Any	Aalders, 1996
Positional accuracy	X-Standard deviation Y-Standard deviation	Spatial data only	NCDCDS, 1988
Continuous accuracy	Standard deviation	Aspatial qualitative data only	NCDCDS, 1988
Categorical accuracy	Percentage probability	Aspatial categorical data only	NCDCDS, 1988
Completeness	Percentage present	Any	NCDCDS, 1988
Textual fidelity	Percentage correct spelling	Textual data only	CEN/TC287, 1996
Detail	Mathematical precision	Numerical data only	Goodchild and Proctor, 1997
Currency	Last update Temporal validity	Any	Aalders, 1996

cent developments in computing science have, however, produced a mathematical formalization of object systems, called object calculus, which offers a possible resolution to such difficulties, explored in the following section.

APPLICATION OF OBJECT CALCULUS

The previous section showed that it may not be possible to design an object schema using OOA that can be guaranteed to be implementation-independent. One possible route to providing such a guarantee, or at least detailing exactly what general properties of OO an OO system must support to allow a particular object schema to be implemented, is object calculus. Object calculus was devised by Martín Abadi and Luca Cardelli in 1995 as an object-oriented extension to the established lambda calculus which underpins much of computing science today. It would be impossible to present a significant proportion of the object calculus here, so only enough of the calculus is explained to support an example application. As far as possible the notation used is consistent with the work of Abadi and

Cardelli, to which the reader is referred for more information (Abadi and Cardelli, 1995; 1996).

An OOA schema which uses multiple inheritance cannot be directly translated to an OO system which does not support multiple inheritance, such as Java. However, by representing the object schema using the object calculus it is possible to show under what conditions the translation will preserve the properties of the original, multiply-inheriting schema. Again, such a proof is beyond the scope of this chapter, but a flavor of such a proof can be given by representing the translation of the multiple inheritance in Figure 26.1 to single inheritance.

Multiple Inheritance

Objects within the object calculus are represented as collections of methods l_i each with bodies b_i. The symbol ζ (sigma) is used to bind the postfixed "self" parameter (conventionally s or z) with occurrences of that parameter in the body of the method. Using untyped object calculus a class C can be represented as an object with a *new* method, as in Equation 1. Here M gives the number of methods in that class.

$$C \triangleq \left[\text{new} = \varsigma(z)[l_i = \varsigma(s)z.l_i(s)^{i \in 1...M}], \right.$$
$$\left. l_i = \lambda(s)b_i^{i \in 1...M} \right] \quad (1)$$

The *new* method produces a new object where each of the template methods $l_i = \lambda(s)b_i^{i \in 1...M}$ in the class C are bound to methods in the object being created. The use of λ (lambda) in the template methods is a consequence of the development of object calculus from lambda calculus; however, the λ performs essentially the same binding function as ς. Using this notation, the classes *abstractive quality* and *uncertainty* shown in Figure 26.1 can be represented as follows.

$$abstractive\_quality \triangleq$$
$$[new = \varsigma(z)[is\_consistent =$$
$$\varsigma(s)z.is\_consistent\_template(s)], \quad (2)$$
$$is\_consistent\_template = \lambda(s)b_1]$$

$$uncertainty \triangleq [new = \varsigma(z)[get\_quality =$$
$$\varsigma(s)z.get\_quality\_template(s)], \quad (3)$$
$$get.quality\_template = \lambda(s)b_2]$$

The multiple inheritance of the class *geographic objects* from both *abstractive quality* and *uncertainty* can then be represented by the term in Equation 4.

$$geographic\_objects \triangleq$$
$$[new = \varsigma(z)[is\_consistent =$$
$$\varsigma(s)z.is\_consistent\_template(s),$$
$$get\_quality = \varsigma(s)z.get\_quality\_template(s)],$$
$$is\_consistent\_template =$$
$$abstractive\_quality.is\_consistent\_template, \quad (4)$$
$$get\_quality\_template =$$
$$uncertainty.get\_quality\_template]$$

In order to reformulate this schema into single inheritance it is necessary first to inherit *uncertainty* from *abstractive quality* or vice versa. Neither of these options makes much sense from a semantic point of view, illustrating why multiple inheritance was so useful in the first place. However, the goal is to show how such a change will make no difference to the *working* of an implementation which takes this approach. The object calculus term for, say, an *uncertainty* class which inherits from *abstractive quality* is given in Equation 5.

$$uncertainty \triangleq [new = \varsigma(z)[is\_consistent =$$
$$\varsigma(s)z.is\_consistent\_template(s),$$
$$get\_quality = \varsigma(s)z.get\_quality\_template(s)],$$
$$is\_consistent\_template =$$
$$abstractive\_quality.is\_consistent\_template, \quad (5)$$
$$get\_quality\_template = \lambda(s)b_2]$$

Now the object calculus term for *geographic objects* which singly inherits from *uncertainty* can be written as below.

$$geographic\_objects \triangleq$$
$$[new = \varsigma(z)[is\_consistent =$$
$$\varsigma(s)z.is\_consistent\_template(s),$$
$$get\_quality = \varsigma(s)z.get\_quality\_template(s)],$$
$$is.consistent\_template =$$
$$uncertainty\_quality.is\_consistent\_template, \quad (6)$$
$$get\_quality\_template =$$
$$uncertainty.get\_quality\_template]$$

Under the object calculus, the body of the *is_consistent_template* method in Equation 6 can be substituted for the body of the *is_consistent_template* method in Equation 5. This substitution results in a term which is identical to that in Equation 4. Equality between objects (and so classes) is defined in the object calculus where two objects have exactly the same methods in any order, termed syntactic equivalence. Therefore, the *geographic_objects* class resulting from the single and multiple inheritance from *abstractive_quality* and *uncertainty* are syntactically equivalent. This in turn implies that the reformulation of multiple to single inheritance of the schema in Figure 26.1 required by implementations, such as Java, which don't support multiple inheritance, will not affect the working of the *geographic objects* class.

CONCLUSIONS

The implementation of an OOA of data quality storage in Java and Gothic proved successful in that the storage schema was able to store quality information in the form of a wide variety of quality statistics from the literature. Further, the object schema produced by the OOA is not restricted only to those elements of data quality tested or even currently in the literature; any quality statistic which is of the general form shown in the class diagram can be stored in an implementation of that schema. There is considerable scope for further work testing the implementations and integrating the quality storage model with the other elements of a error sensitive GIS, such as error propagation and visualization.

During the implementation in Java and Gothic, difficulties arose when very few of the general properties of OO were supported equally. This situation is undesirable, since ideally the results of OOA are implementation-independent, but unavoidable, since it is impossible to draw a clear line between analysis and design, between model and implementation. The use of mathematical formalizations in object calculus, however, offers the possibility of encoding object schema in an implementation-free language. The example application of object calculus presented earlier illustrated how changes in the design of the class diagram in Figure 26.1, dictated by the implementation in Java, would have no effect on the working of the resultant implementation. The example use of the object calculus further suggests how the calculus might be used as a general tool within OOA to explore the properties of an object system and inform the process of implementation prior to commencement of any actual programming.

ACKNOWLEDGMENTS

This research is funded as part of a Natural Environmental Research Council CASE studentship, the CASE sponsors being Survey and Development Services (SDS). Supervision for the studentship is from Dr. Jane Drummond and Dr. David Forrest of Glasgow University Geography and Topographic Science Department and from John McCreadie at SDS. Gothic software was kindly supplied under a development license by Laser Scan, UK. Finally, the help, interest, and advice of Dr. Tom Melham of Glasgow University Computing Science Department is gratefully acknowledged.

REFERENCES

Aalders, H. Quality Metrics for GIS, in *Advances in GIS research II; Proceedings of the Seventh International Symposium on Spatial Data Handling*, vol. 1, M.J. Kraak and M. Molenaar, Eds., 1996, pp. 5B1–10.

Abadi, M. and L. Cardelli. An Imperative Object Calculus, *Theory and Practice of Object Systems,* 1(3), pp. 151–166, 1995.

Abadi, M. and L. Cardelli. A Theory of Primitive Objects: Second-Order Systems, *Sci. Comput. Prog.,* 25(2-3), pp. 81–116, 1995.

Abadi, M. and L. Cardelli. *A Theory of Objects,* Springer, New York, 1996.

Abadi, M. and L. Cardelli. A Theory of Primitive Objects: Untyped and First-Order Systems, *Inf. Comput.,* 125(2), pp. 78–102, 1996.

Aspinall. R. Measurement of Area in GIS: A Rapid Method for Assessing the Accuracy of Area Measurement, in *Proceedings of the GIS Research UK 1996 Conference*, 1996, pp. 135–142.

Booch, G. *Object Oriented Analysis and Design with Applications,* 2nd ed., Benjamin-Cummings, CA, 1994.

Brunsdon, C. and S. Openshaw. Simulating the Effects of error in GIS, in *Geographical Information Handling: Research and Applications,* P. Mather, Ed., John Wiley & Sons, Chichester, 1993.

CEN/TC287. Draft European standard; geographic information – quality, Technical Report prEN 287008, European Committee for Standardisation, 1996.

Clarke, D. and D. Clark. Lineage, in *Elements of Spatial Data Quality,* Chapter 2, S. Guptill and J. Morrison, Eds., Elsevier Science, Oxford, 1995, pp. 13–30.

Coad, P. and E. Yourdon. *Object-Oriented Analysis,* Yourdon Press, NJ, 1991.

Drummond, J. GIS: The Quality Factor, *Surveying World,* 4(6), pp. 26–27, 1996.

Goodchild, M. and J. Proctor. Scale in a Digital Geographic World, *Geogr. Environ. Modelling,* 1(1), pp. 5–23, 1997.

Kösters, K., B.-U. Pagel, and H.-W. Six. GIS-Application Development with GEOOOA, *Int. J. Geogr. Inf. Syst.,* 11(4), pp. 307–335, 1997.

Morrison, J. Spatial Data Quality, in *Elements of Spatial Data Quality,* Chapter 1, S. Guptil and J. Morrison, Eds., Elsevier Science, 1995.

National Committee for Digital Cartographic Data Standards. The Proposed Standard for Digital Cartographic Data, *Am. Cartographer,* 15(1), pp. 11–142, 1988.

Ramlal, B. and J. Drummond. A GIS Uncertainty Subsystem, in *Archives ISPRS Congress XVII,* 29, B3, pp. 356–362, 1992.

Unwin, D. Geographical Information Systems and the Problem of Error and Uncertainty, *Progress in Human Geography,* 19(4), pp. 549–558, 1995.

Worboys, M. *GIS: A Computing Perspective,* Taylor and Francis, London, 1995.

Part VI
Generalization and Aggregation

It has long been recognized that aggregation and generalization may lead to certain types of spatial uncertainty. For example, at one level of aggregation—a city block, for example—it may be apparent that human populations tend to cluster. However, at a different aggregation level—states or countries, for example—such effects may not be measurable at all. Hence all spatial analysis may provide an imperfect understanding of how spatial phenomena "truly" behave. Indeed, the "truth" of how something behaves spatially may depend entirely on the level of aggregation and generalization of the spatial data used.

The chapters in this Part all discuss some aspect of the problem(s) associated with spatial generalization and aggregation. The first chapter by Heuvelink provides a conceptual basis for understanding the problem of aggregation by using an applications example. The second and third articles by Dzur, and Scrinzi and Floris describe methods for handling line generalization in different situations. The final three papers deal with some aspect of the effects of generalization and aggregation. Handcock et al. examine the effects on nitrogen-loss estimates, Gesch considers the effects of generalizing a digital elevation model on derived lays (e.g., slope and aspect), and Canters et al. considers aggregation problems relative to satellite imagery.

CHAPTER 27

Aggregation and Error Propagation in GIS

G.B.M. Heuvelink

INTRODUCTION: DAILY LEAD INGESTION IN THE GEUL VALLEY

To introduce the problem addressed in this chapter, consider the following example: The floodplain of the Geul River valley, located in the south of The Netherlands, is strongly polluted by heavy metals deposited with the stream sediments. Historic metal mining has caused the widespread dispersal of lead, zinc, and cadmium in the alluvial soil. Figure 27.1 shows a map of the study area. The pollutants may constrain the land use in these areas, so detailed maps are required that delineate zones with high concentrations. Leenaers (1991) used point kriging on measurements of soil samples of 100 grams at 101 sites in the area to derive maps of the predicted lead concentration in the topsoil (0–10 cm depth) and of the standard deviations of the prediction errors. The maps are given in Figure 27.2.

For the general assessment of health risks to children playing in the Geul valley, it could be sensible to make maps of potential daily lead ingestion. This can be done if it is known how much soil a child is likely to ingest per day. Van Wijnen et al. (1990) provide experimental data on the daily ingestion of soil by young children playing on a camping ground. After making correction for "background" ingestion, these data fit a lognormal distribution with mean 0.120, median 0.052, and standard deviation of 0.250 g/day. Maps of the potential daily ingestion of lead (I) can now be obtained by multiplying the lead concentration of the soil (PB) by the amount of soil consumed (S):

$$I = PB \cdot S \qquad (1)$$

However, uncertainty in both the lead concentration of the soil and the daily soil ingestion will propagate to the daily lead ingestion. This implies that even when a site is on average safe, there can be incidences in which the children's health is at risk.

Because in this case the output is a simple multiplication of two inputs whose errors are uncorrelated, the propagation of errors can be analytically computed, yielding the following expressions for the mean and variance of the daily lead ingestion (Heuvelink, 1998):

$$mean(I) = mean(PB) \cdot mean(S) \qquad (2)$$

$$var(I) = var(PB) \cdot (mean(S))^2 \\ + var(S) \cdot (mean(PB))^2 \\ + var(PB) \cdot var(S) \qquad (3)$$

The maps of the mean and standard deviation of the daily lead ingestion are given in Figure 27.3.

To identify areas where there is a substantial risk that critical values of lead ingestion are exceeded, the lead ingestion should be compared with the Acceptable Daily Intake (ADI), which for lead is given by 50 µg/day (Leenaers et al., 1991). Judging from Figure 27.3a, the ADI would only be exceeded in a relatively small area in the south of the study area. However, when the uncertainty in the predictions is taken into account, then a much larger area that is po-

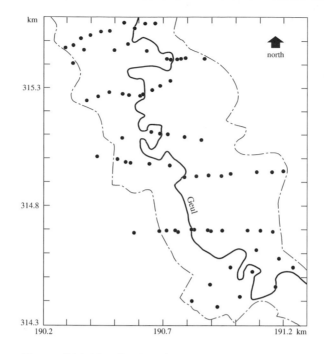

Figure 27.1. The Geul study area showing sampling points.

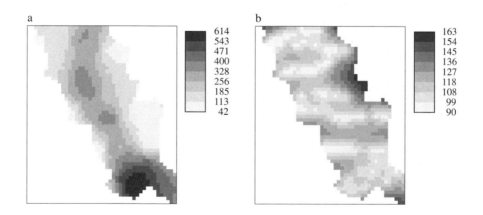

Figure 27.2. Lead concentration of topsoil (mg/kg) for the Geul study area: (a) point kriging mean, (b) point kriging standard deviation.

tentially at risk may result. For instance, if a location is only judged safe when one is at least 95% confident that the ADI is not exceeded, then a Monte Carlo uncertainty analysis shows that almost the entire study area is unsafe (Heuvelink, 1998).

Change of Support

In the analysis above we have overlooked a crucial factor: the *support* (i.e., aggregation level) of the

measurements. We used point kriging to predict the lead concentration of the soil at unsampled locations. This means that the mapped predictions and prediction error standard deviations refer to the lead concentration of soil samples of 100 grams. But the average daily soil ingestion is only 0.120 grams. Can we reasonably assume that the uncertainty about the lead concentration in a 100 g sample is the same as in a 0.120 g sample? The answer is no. It is likely that the variability in the lead concentration of 0.120 g

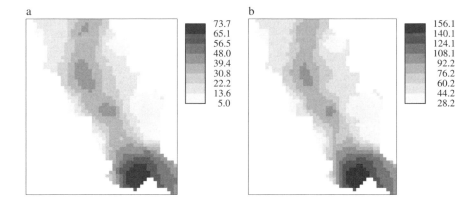

Figure 27.3. Error propagation results for potential daily lead ingestion (μg/day) in the Geul study area: (a) mean, (b) standard deviation.

samples is larger than that of 100 g samples, simply because lead will not be homogeneously distributed in a 100 g sample. However, it is hard to calculate how much larger the uncertainty about the predicted lead concentration in a 0.120 g sample should be. This requires that we know how the lead is spatially distributed within a 100 g sample, but this information obviously cannot be inferred from measurements with a 100 g support.

Even if we would be able to determine the uncertainty about the lead concentration of a 0.120 g sample, then this still would not solve all our problems. Clearly, the soil ingested during the day is a composite of many even smaller samples, taken at various locations on the camping ground. This causes spatial variability to average out, leaving less uncertainty about the lead concentration of the ingested soil. Again it will be very difficult to quantify the reduction of uncertainty.

Yet another problem concerns the temporal support used in the analysis. The uncertainty in the daily soil ingestion is based on measurements with a temporal support of one day. Temporal variation will average out for longer time periods, leaving less uncertainty about the soil ingestion. Again it will be very difficult to predict how much variability averages out over the longer time period, because the variability in children's behavior remains.

From these considerations we cannot but conclude that the computed uncertainty as presented in Figure 27.3b is *only* valid for the support that was used for collecting the data. These are a spatial support of 100 grams and a temporal support of one day. Particularly the spatial support has little practical relevance. Consequently, the intrinsic value of the computed uncer-

tainty should be questioned. A change of support is required to obtain more meaningful results. However, in this particular case it turns out to be extremely difficult to carry out the change of support and to determine how this would affect the uncertainty about the daily lead ingestion.

Implications for Uncertainty Analysis in Environmental Modeling

The crucial role of the support is not a unique property of the Geul example. The support will play an important role in many more applications of uncertainty analysis in environmental modeling. Are the research and user communities sufficiently aware of this fact? This is an important question, because recently, with the advent of theory and algorithms for error propagation and spatial stochastic simulation, we see an increase in spatial error propagation applications (e.g., Loague et al., 1989; Kros et al., 1992; Gotway, 1994; Lafrance and Banton, 1995; Leenhardt, 1995; Finke et al., 1996; Rogowski, 1996; Burrough and McDonnell, 1998; Heuvelink, 1998; Kros et al., 1999). It is important that we realize that the results presented in these studies refer to the chosen support only and that the results might well have turned out quite differently if a different support had been used.

Since the results of an uncertainty analysis are support-dependent, a change of support will be required when the obtained output support differs from the desired output support. The next sections discuss how a change of support may be carried out for situations less complicated than the Geul example.

THEORY OF CHANGE OF SUPPORT

A change of support may be directed upward (aggregation) or downward (disaggregation). Here we restrict ourselves to aggregation, which is usually easier and in addition more frequently needed in practice (Heuvelink and Pebesma, 1999). We will only consider *spatial* aggregation, but note that *temporal* aggregation can be dealt with in a similar manner. In this section we also restrict ourselves to *linear* aggregation; i.e., our objective is to compute the arithmetical mean of all "point" values (the values at a small support) that are contained in a "block" (the larger support).

In mathematical terms, the problem is as follows. Given a spatial attribute $Z(\cdot)=\{Z(x)|x \in D\}$ defined on a spatial domain D. We consider attributes that are uncertain to some degree and so $Z(\cdot)$ is represented as a random field (Heuvelink, 1998), with mean $\mu(\cdot)$ and variance $\sigma^2(\cdot)$. $Z(x)$ refers to the point value of $Z(\cdot)$ at location x, but rather than in point values we are interested in the average of $Z(\cdot)$ for some spatial block B: $Z(B)=\int_B Z(x)dx/|B|$, where $|B|$ is the area of B. Although use of the word "block" may suggest that B is rectangular, its shape may well be irregular. Also, use of the word "point" suggests that the $Z(x)$ has zero support but in fact what we assume is that the support of the $Z(x)$ is negligible compared to that of $Z(B)$.

A straightforward result now is that the mean and variance of $Z(B)$ are given as:

$$\mu(B) = \frac{1}{|B|} \int_B \mu(x)dx \qquad (4)$$

$$\sigma^2(B) = \frac{1}{|B|^2} \int_B \int_B C(x,y)dxdy \qquad (5)$$

where $C(x,y)$ is the covariance of $Z(x)$ and $Z(y)$, $C(x,y)=E[(Z(x)-\mu(x))(Z(y)-\mu(y))]$. Thus the uncertainty of the block average can be easily computed provided the uncertainty at point support and the associated within-block spatial correlation are known.

Although $\mu(B)$ will not systematically differ from the $\mu(x)$, $\sigma^2(B)$ typically is much smaller than the $\sigma^2(x)$. This is because within-block uncertainty is averaged out, leaving less uncertainty about the block-average. In fact this is a well-known result from geostatistics; i.e., recall that the block kriging variance is usually much smaller than the point kriging variance (Cressie, 1991). The variance reduction will be substantial when the within block-variability is large.

When multiple attributes are considered, a change of support may also affect the correlations between the attributes. This point was recently addressed by Cressie (1998), and it is observed in standard textbooks as well (Burt and Barber, 1996, section 12.3). A change of support thus not only affects the variance but it may affect other statistical parameters as well.

Although the theory of linear aggregation is relatively straightforward, in the general nonlinear situation there remain many fundamental problems not yet completely resolved (Cressie, 1991; Cressie, 1998). The problem is that Equations 4 and 5 cannot easily be generalized to the case of nonlinear aggregation. One way around the problem is through numerical simulation (i.e., Monte Carlo simulation). The next section gives an example.

MODELING SOIL ACIDIFICATION AT THE EUROPEAN SCALE

The soil acidification model SMART2 (Kros et al., 1999) is a fairly simple, strictly vertical, one-layer dynamic model that predicts for natural and semi-natural areas the long-term response of aluminium (and nitrate) concentration below the root zone to changes in atmospheric deposition. The model was developed to analyze how atmospheric deposition causes soil acidification and how this in turn affects the groundwater quality. The model was deliberately kept simple to be able to be used on a regional (and European) scale (De Vries et al., 1998).

At the European scale, the inputs to SMART2 have to be derived from general-purpose soil maps and land-use maps that are available at the European level. These maps are not very accurate and in addition the procedure involves the use of transfer functions which further deteriorates the quality of the model input. In order to judge whether application of SMART2 at the European scale yields results that are sufficiently accurate, a study was carried out to analyze how the uncertainty in the source maps and the transfer functions propagate to the model output. This was done using a Monte Carlo approach. The exact procedure is described in Kros et al. (1999). Here we focus on the change of support methodology.

Change of Support with SMART2

Although SMART2 is intended to be used at the regional scale (De Vries et al., 1998), it still is a model operating at the point support. In order to assess

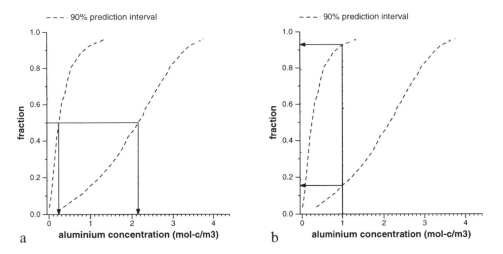

Figure 27.4. (a) the 90% prediction interval for median concentration value in the block is [0.23, 2.15], (b) the 90% prediction interval for the areal fraction where point concentration values do not exceed 1 mol-c/m³ is [0.155, 0.930].

groundwater quality at the European scale (5 km × 5 km and larger blocks), the model is therefore applied to many point locations within each block. After the model is run at these point locations the model outputs are aggregated to yield a single block value. Block aggregation is typically nonlinear, because from legal considerations the interest is not so much in the block mean aluminium concentration but rather in the block median concentration or the block areal fraction where the aluminium concentration exceeds a critical value.

For a single block, a single Monte Carlo run gives the output at all points in the block. These can be represented in a single graph by a cumulative distribution function. However, due to input uncertainty there will be a population of graphs, since each Monte Carlo run produces a different graph. In Figure 27.4, the population of graphs is represented by the sides of a 90% prediction interval. From it, we can construct 90% prediction intervals for quantiles, such as the median (Figure 27.4a) or for the block areal fraction exceeding a fixed concentration (Figure 27.4b). Because the procedure yields as an intermediate result the point support output at all point locations in the block, nonlinear aggregation is as easily coped with as linear aggregation.

An interesting property of the procedure presented here is that it nicely separates spatial variability from uncertainty. Typically, increasing the block size moves one from a situation with a small within-block spatial variation (steep distribution curves) and large uncertainties (wide prediction intervals) in Figure 27.5a to a situation with a large within-block spatial variation (flatter distribution curves) and small uncertainties (narrower prediction intervals) in Figure 27.5b. This effect that had been anticipated on theoretical grounds was confirmed by the case study (Kros et al., 1999). Thus aggregation to larger blocks reduces uncertainty at the expense of a loss of resolution.

CONCLUSIONS

The results of a spatial uncertainty analysis strongly depend on the support of the data used in the analysis. Therefore, the support of the output of an analysis should always be stated. If it is not, then the reported output accuracies effectively become meaningless. In fact, meta-information about the accuracy of data stored in spatial databases should always include the support, regardless of whether the data is the output of a model, a map derived from direct observations, or the observations themselves. Therefore, in our appeal to map makers that they should routinely convey the accuracy of the maps they produce (Heuvelink, 1998), we should extend this call by requiring that the associated support is reported as well.

In environmental modeling practice, it may frequently occur that the support of the obtained output does not match the desired output support. This calls for a change of support, which is more often in the form of an aggregation than a disaggregation. Existing techniques can be used to perform the aggregation, although nonlinear aggregation will usually

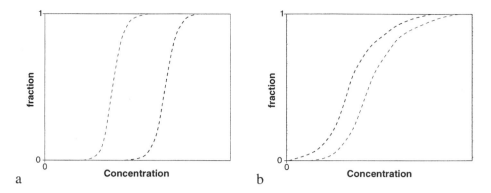

a   b

Figure 27.5. Prediction intervals for cumulative distribution of point concentration values within (a) a small block (small within-block spatial variability and large uncertainty), (b) a large block (large within-block spatial variability and small uncertainty).

prohibit the use of an analytical approach so that one has to rely on numerical methods (such as Monte Carlo simulation). Use of aggregation techniques does presuppose that the within-block spatial variability of the attribute is known. A complete characterization of the accuracy of spatial data should therefore also include the spatial correlation of (the uncertainty of) the spatial attributes stored in the spatial database.

In the early days of spatial accuracy assessment we might have been very happy with a summary of the accuracy of a map by means of a single Root Mean Squared Error. We now know that this is not enough but that we need much more information to completely characterize the accuracy of spatial attributes. Specification of the accuracy of a spatial attribute must include the support of the attribute as well as a complete definition of the stochastic error model associated with the uncertain spatial attribute.

ACKNOWLEDGMENT

I thank Dr. H. Leenaers for permission to use the Geul data set.

REFERENCES

Burrough, P.A. and R.A. McDonnell. *Principles of Geographical Information Systems,* Oxford University Press, Oxford, 1998.

Burt, J.E. and G.M. Barber. *Elementary Statistics for Geographers*, 2nd ed., Guilford Press, New York, 1996.

Cressie, N. *Statistics for Spatial Data,* John Wiley & Sons, New York, 1991.

Cressie, N. Aggregation and Interaction Issues in Statistical Modeling of Spatiotemporal Processes, *Geoderma,* 85, pp. 133–140, 1998.

De Vries, W., J. Kros, J.E. Groenenberg, G.J. Reinds, and C. Van Der Salm. The Use of Upscaling Procedures in the Application of Soil Acidification Models at Different Spatial Scales, *Nutrient Cycling in Agroecosystems,* 50, pp. 225–238, 1998.

Finke, P.A., J.H.M. Wösten, and M.J.W. Jansen. Effects of Uncertainty in Major Input Variables on Simulated Functional Soil Behaviour, *Hydrological Processes,* 10, pp. 661–669, 1996.

Gotway, C.A. The Use of Conditional Simulation in Nuclear-Waste-Site Performance Assessment, *Technometrics* 36, pp. 129–161, 1994 (with discussion).

Heuvelink, G.B.M. *Error Propagation in Environmental Modelling with GIS*, Taylor & Francis, London, 1998.

Heuvelink, G.B.M. and E.J. Pebesma. Spatial Aggregation and Soil Process Modelling, *Geoderma,* 89, pp. 47–65, 1999.

Kros, J., W. De Vries, P.H.M. Janssen, and C.I. Bak. The Uncertainty in Forecasting Trends of Forest Soil Acidification, *Water, Air Soil Pollut.,* 66, pp. 29–58, 1992.

Kros, J., E.J. Pebesma, G.J. Reinds, and P.F. Finke. Uncertainty in Modelling Soil Acidification at a European Scale: A Case Study, *J. Environ. Qual.,* 28, pp. 366–377, 1999.

Lafrance, P. and O. Banton. Implication of Spatial Variability of Organic Carbon on Predicting Pesticide Mobility in Soil, *Geoderma,* 65, pp. 331–338, 1995.

Leenaers, H. Deposition and Storage of Solid-Bound Heavy Metals in the Floodplains of the River Geul (The Netherlands), *Environ. Monitoring Assessment,* 18, pp. 79–103, 1991.

Leenaers, H., M.C. Rang, and D.M.C. Rang. Coping with Uncertainty in the Assessment of Health Risks, in *Heavy Metals in the Environment,* J.G. Farmer, Ed., Page Bros, Norwich, 1991, pp. 286–289.

Leenhardt, D. Errors in the Estimation of Soil Water Properties and their Propagation through a Hydrological Model, *Soil Use Manage.,* 11, pp. 15–21, 1995.

Loague, K., R.S. Yost, R.E. Green, and T.C. Liang. Uncertainty in Pesticide Leaching Assessment in Hawaii, *J. Contam. Hydrol.,* 4, pp. 139–161, 1989.

Rogowski, A.S. GIS Modeling of Recharge on a Watershed, *J. Environ. Qual.,* 25, pp. 463–474, 1996.

Van Wijnen, J.H., P. Clausing, and B. Brunekreef. Estimated Soil Ingestion by Children, *Environ. Res.,* 51, pp. 147–162, 1990.

CHAPTER 28

Evaluation of a Procedure for Line Generalization of a Statewide Land Cover Map

R.S. Dzur

INTRODUCTION

Line generalization procedures are as numerous as the motivations for generalizing. Considerable effort has been devoted to developing both a procedural and a theoretical framework for map generalization (McMaster and Shea, 1992). While the generalization process has enjoyed extensive theoretical and analytical exploration, guidelines for assessing the results are sparse (Richardson, 1994). Assessing the results of the generalizing procedures may be difficult. In the operational (project oriented) setting of many geographic information systems (GIS) organizations, general issues of accuracy and generalization may not be as well understood as proximity or overlay analysis (Muller, 1989).

For the purpose of this chapter, the term map generalization refers to an overall process of altering mapped information across scale dimension. Line generalization refers to the actual task of altering individual line features that make up a map. Whether map generalization focuses on text, line, or area elements, most researchers agree that generalization of map data by employing differing methodologies is necessary because the objectives are different in each case.

Rule-based and phenomenon-based methodologies often are categories which define approaches to line generalization (Mark, 1989; Muller, 1989). Those categories have been discussed in terms of statistical and graphical generalization (Mark, 1989). They have also been labeled as geometric generalization and conceptual generalization (Muller, 1989). The former emphasizes generalization strictly through the statistical or geometric definition of cartographic features, while the latter concentrates on the semantics of cartographic features and as a result may apply a certain generalization approach to railroad features and another to forest-cover polygon features. These are important concepts for the future of generalization in operational GIS environments, especially as image (raster)-based information increasingly comes into usage in dominantly vector-based GIS systems.

Naturally, generalization implies process. Geospatial data undergo (Li and Su, 1995) transformations in scale that simplify their representation for effective communication of the reality those mapped features depict. Shea and McMaster (1989) emphasized process in cartographic generalization by organizing their theoretical context for generalization around questions of why, when, and how. Line smoothing, in their vocabulary (how), is a spatial operator altering the arrangement of verticies to effect a more graphically appealing line. It has been suggested that algorithm selection may be constrained by data type and/or scale (Shea and McMaster, 1989). This is certainly the case for enhancing the characteristic appearance of raster data in a vector form.

Automated generalization raises map accuracy concerns in the realm of digital geospatial databases. There appears to be little empirical work describing those concerns and the consequences of such generalization. Congalton (1997) recently presented a procedure for testing the accuracy of vector-to-raster-to-vector transformations on simulated geometric surfaces. Error matrices described changes in area due to raster-to-vector conversion. Outlining the process of raster-to-vector conversion, Congalton (1997) identified three data processing steps: preprocessing, enhancement, and postprocessing. Those steps are designed to carry remotely sensed information through the generalization process toward an efficient GIS information product. Bury (1989) described a methodology for raster-to-vector conversion of a 21-class land cover map. Included were preprocessing steps to remove small pixel groups below a specified minimum mapping unit (MMU) threshold and postprocessing to smooth raster stair-stepping to produce a more traditional cartographic vector product. Accuracy in that study was measured according to distance from original vector. Results were considered acceptable when the generalized line was within one pixel's distance of the original line.

Objective

This chapter explores changes in area between a source raster-based vector product and a final smoothed (postprocessed) vector product. The objective of this investigation is to identify what effects the procedure had on informational accuracy of a statewide land cover map generated as part of the Arkansas Gap Analysis Project (AR-GAP). Other issues, including classification, positional, and aggregation accuracy will not be addressed in this study.

METHODS

Study Area and Data

AR-GAP implemented line smoothing on an aggregated (preprocessed) statewide land cover database with a 100 ha MMU with allowable 40 ha inclusions for special features (e.g., lakes, palustrine land cover) originally classified from 30-meter Landsat TM imagery (Dzur et al., 1995). The 100 ha level of detail was primarily an application-based standard that was expected to satisfy general applications related to the broad ecoregional terrestrial vertebrate modeling goals that corresponded to gap analysis' principal objective (Scott et al., 1993). When converted to vector form, resultant polygons were defined by the regular (30 m) grid of the input raster data that defined the raster array. A more generalized depiction of the data was desired to ensure scale/map utility consistencies at the required 100 ha MMU level in a vector format (Muller, 1989). Resolution of 30 m created the appearance of a very detailed and precise boundary between land cover classes. At that scale (100 ha MMU), boundaries are

not necessarily as detailed as the raster might indicate. AR-GAP considered the representation of those boundaries inconsistent with the scale at which the data had been aggregated.

Line Smoothing Approach

One of the problems with implementing line generalization is a lack of experience. Iterative testing is often required to identify appropriate parameters that meet an application's specific generalization requirements (Richardson, 1994). Benton County, Arkansas was selected for testing a variety of line smoothing algorithms both in ARC/INFO and Intergraph MGE Map Finisher. Because of the need to process the entire state of Arkansas, algorithms were initially tested for computational efficiency. Results indicated that Map Finisher moving-averages method had superior processing speed. More importantly, the algorithm also appeared to have greater capacity for maintaining the general shape of input polygons.

Next, an extensive testing schedule was established to determine threshold values that generated the most aesthetically pleasing output from the moving-averages line smoothing filter. To smooth features, the algorithm moves or reduces the number of vertices in the linework utilizing a weighted average of five points: a center point, the first two neighbor points, and the second two neighbor points (Intergraph, 1995). Based on the evaluation exercise, equal weights of 20 were the parameters selected for the center, first neighbor, and second neighbor vertices. Vectors output from that method achieved a generalized line shape while maintaining most of the original character of the polygon. Visual inspection of the results led to the assumption that the new line traced an average route between two endpoints forming the corners of pixel-based polygon boundaries (Figure 28.1).

To overcome file size and processor limitations, AR-GAP land cover map was tiled into 35 30 x 60 minute U.S. Geological Survey (USGS) 1:100,000 scale quadrangles (Figure 28.2) in the raster (GRASS) domain. Those raster tiles were vectorized (r.poly)[1] and transferred to Intergraph MGE for line smoothing. Topology was maintained through the transfer. Unfortunately, maintaining topology through generalization was not a primary concern of the moving

---

[1] GRASS commands are noted in lower case; ARC/INFO commands are noted in uppercase.

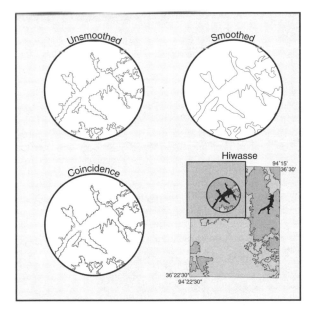

Figure 28.1. Detail of generalization process for Loch Lomond reservoir, Benton County, Arkansas.

Figure 28.2. 30 x 60 Minute quadrangles with randomly selected 7.5 minute quadrangle name below the 30 x 60 minute name.

averages method. Whereas few topological errors were generated by the processing, their existence added extra complexity to the process.

Line smoothing introduced two types of topological problems: duplicate centroids and false intersections. Centroids near polygon boundaries were occasionally located in adjacent polygons after smoothing. To retain proper polygon attribution, those centroids needed to be relocated to their appropriate polygons. Lines were first smoothed, then a text file

containing centroid locations was created. That file allowed for review of centroid placement. Duplicate centroids were readjusted upon visual inspection when they were found to jeopardize polygon attribution. A second topological error existed in areas characterized by narrow polygon features with few vertices defining their narrow peninsular shape. In these situations, line smoothing would collapse a thin feature on itself, subsequently creating a second polygon without a centroid. Those topological errors were corrected by applying line cleaning (intersection processor) tools to flag and fix false intersections where line work crossed itself without the presence of a node at the intersection point. After centroid relocation and intersection processing, topology was again restored for each quadrangle.

Smoothed tiles were edge-matched across their 1:100,000 scale quadrangle boundaries to smoothed linework in neighboring quadrangles. Edge-matched tiles were exported to ARC/INFO where all 35 quadrangles were joined (MAPJOIN) and common boundaries dissolved (DISSOLVE) to form a single topologically-intact map product.

Analysis

The following procedures were set by the large quantity of data to be analyzed. To analyze the relationship between smoothed (generalized) and unsmoothed (source) AR-GAP 100 ha landcover data, a sample 7.5 minute quadrangle was randomly (r.random) selected from each of the 35 30 x 60 minute quadrangles (Figure 28.3). The set of randomly located quads represented nearly 4% of the total number of Arkansas quadrangles. For each compared data set, data were "cookie cut" by a 7.5 minute quadrangle boundary. Unsmoothed data were then unioned (UNION) with smoothed data creating an output data set containing the coincidence between the smoothed and unsmoothed data sets (Congalton, 1997). Small union thresholds were employed to capture the smallest polygons that were either omitted or committed to their respective polygons during overlay. Error matrices were used to quantify performance of the moving-averages line smoothing algorithm for each randomly 7.5 minute selected quadrangle.

RESULTS AND DISCUSSION

Compiled statewide results showed a decrease of 35% in number of arcs comprising the two vector data sets (Table 28.1). A corresponding reduction in file

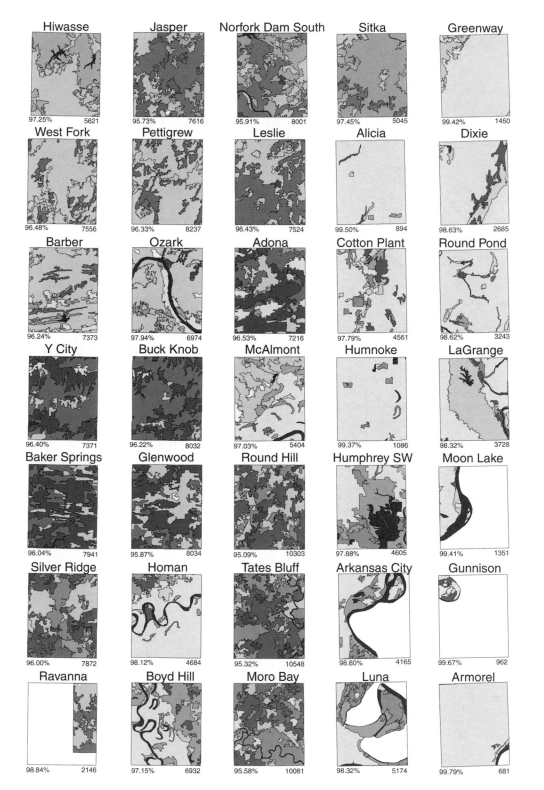

Figure 28.3. Randomly selected 7.5 minute quadrangles with percent area unchanged noted on lower left and number of polygons noted on lower right of each quadrangle. Except for Armorel (lower right), all quadrangles are organized by relative geographic location in Arkansas.

size was also achieved. Total number of polygons increased by 10 polygons in the smoothed land cover map, an increase of 0.05%.

Output of the union between those two data sets provided a very simple and descriptive illustration of error distribution among the two data sets. All quadrangles exhibited little change in area from the line smoothing process. No sampled quadrangle displayed less that 95% overall agreement. Table 28.2 presents the error matrix for the Hiwasse quadrangle: the median value for overall percent area unchanged (97.25%). Distribution of omission and commission errors was remarkably even. Notably, all categories bore less than 1.5% difference in omission and commission rates.

Goals related to computational efficiency were met. One concern with this implementation was that it was not completely respecting of topological relationships. While original intersections remained unchanged through the smoothing process, some false intersections were created. Topology is an absolute necessity when the objective is to maintain the generalized data set in a GIS environment for analysis or modeling. If the objectives are solely related to cartographic output, this may not have particularly strong bearing on algorithm selection.

The error matrices revealed that the smoothed representation of these land cover data did not substantially alter the quality of information. The algorithm did an adequate job of maintaining true polygon area during the transformation from unsmoothed to smoothed land cover vector database.

While parameter selection was based more on the subjective measure of appearance and maintenance of land cover polygon form. No less than 95% coincidence among source and smoothed data in all sampled areas is strong validation of algorithm performance forecast and assumed by visual analysis. A rise in the number of polygons is concerning and deserves further study. Somewhat surprising was that less than 1% discrepancy in total number of polygons would emanate after processing over 19,000 spatially complex polygons.

CONCLUSIONS

The present study focused on a single aspect of map accuracy related to map generalization. Line smoothing was successfully implemented on a statewide land cover map of Arkansas, and results were evaluated using overlay analysis. The study achieved objectives of

Table 28.1. Statewide Line Smoothing Results.

Data	Total Number of Arcs	Total Number of Polygons	File Size (mega-bytes)
Unsmooth	13,757,232	19,738	111
Smooth	8,865,827	19,748	75

data reduction, scale consistency in feature boundary delineation, and graphical appeal.

In the vector domain, raster artifacts in polygon boundaries often unnecessarily increase detail in already intricate spatial data. As a result, unprocessed vector data sets from remotely sensed sources require more disc space and are more difficult to manipulate and display. Intergraph's Map Finisher moving-averages algorithm reduced the risk of materially degrading accuracy of map classes in a GIS database derived from remotely sensed data. The described line smoothing reduced redundant data while retaining a high level of information content. Error matrices proved useful for evaluating the performance of line smoothing and provided a means for assessing distribution of errors ranging across individual classes. The technique was not difficult to apply; however, it was a pioneering effort to smooth such a large and complex map. The task was time-consuming and challenging.

These results could be expanded to look at the influence of polygon shape and possibly to develop strategies for generalizing land cover according to more erudite (phenomenon based) definitions. This sort of investigation might be able to ascertain whether distinct patterns of accuracy might correspond to distinct spatial configurations of land cover on the landscape. It is unknown whether these results exhibit some correlation with physiographic province or are simply a function of geometry and the number of polygon edges. Answering those concerns might be irrelevant in the present application of line smoothing, considering that the objectives were met. Moreover, if map (database) users are willing to tolerate the accuracy of the algorithm, it may not be worth pursuing a less geometrically based approach to line smoothing. With such demonstrably small changes in area, the current sample provided information confirming the assumption drawn upon visual appearance of resulting generalized data. The high level of agreement between the two data sets suggested that there was little change in accuracy due to the application of postprocessing routines.

Table 28.2. Error Matrix for the Hiwasse Quadrangle.

Hiwasse Quadrangle Area (hectares) Smooth	Unsmooth						
	0[a]	8	11	33	36	Total	Commission
0	84.29	0.00	3.63	0.00	0.00	87.92	4.13%
8	0.00	1095.75	37.55	0.26	60.37	1193.94	8.22%
11	2.67	37.73	6364.80	17.75	89.99	6512.95	2.27%
33	0.00	0.18	23.51	286.37	0.61	310.67	7.82%
36	0.00	56.53	96.18	0.62	7280.25	7433.58	2.06%
Total	86.97	1190.20	6525.67	305.01	7431.22	15539.06	Overall
Omission	3.08%	7.94%	2.47%	6.11%	2.03%		97.25%

[a] 0 = Background.
8 = T.1.b.3.a.II; Quercus alba—mixed hardwoods (white oak—mixed hardwoods).
11 = T.1.B.3.a.V; Quercus stellata (post oak).
33 = Water.
36 = Agriculture (pasture).

ACKNOWLEDGMENTS

The described research was supported by the U.S. Geological Survey Biological Resources Division as part of the National Gap Analysis Program: Award #14-16-000901567 RWO #13. Many thanks to P. Teague, J. Wilson, Y. Yuan, and A. Bayard for their patience and hard work. AR-GAP coprincipal investigators were Dr. Kim Smith and Dr. W.F. Limp.

REFERENCES

Bury, A.S. Raster to Vector Conversion: A Methodology, in *GIS/LIS '89 Proceedings, Orlando, Florida,* American Society for Photogrammetry and Remote Sensing and American Congress on Surveying and Mapping, Bethesda, MD, 1, 1989, pp. 9–11.

Congalton, R.G. Exploring and Evaluating the Consequences of Vector-to-Raster and Raster-to-Vector Conversion, *Photogrammetric Eng. Remote Sensing,* 63, pp. 425–434, 1997.

Dzur, R.S., M.E. Garner, K.G. Smith, W.F. Limp, X. Li, and W. Song. Arkansas Gap Analysis: State-Wide Biodiversity Mapping Research, in *1995 URISA Proceedings, San Antonio, Texas,* M.J. Salling, Ed., Urban and Regional Systems Association, Washington, DC, 1995, pp. 74–83.

Intergraph. *MGE Map Finisher (MGFN): User's Guide for the Windows NT Operating System,* Huntsville, AL, 1995.

Li, Z. and B. Su. From Phenomena to Essence: Envisioning the Nature of Digital Map Generalization, *The Cartographic Journal,* 32, pp. 45–47, 1995.

Mark, D.M. Conceptual Basis for Geographic Line Generalization, in *Auto-Carto 9 Proceedings: Ninth International Symposium on Computer-Assisted Cartography, Baltimore, Maryland,* 1989, American Society for Photogrammetry and Remote Sensing and American Congress on Surveying and Mapping, Falls Church, VA, 1989, pp. 68–77.

McMaster, R.B. and K.S. Shea. *Generalization in Digital Cartography,* Association of American Geographers, Washington, DC, 1992, p. 134.

Muller, J.C. Theoretical Considerations for Automated Map Generalization, *ITC Journal,* 3/4, pp. 200–204, 1989.

Richardson, D.E. Generalization of Spatial and Thematic Data Using Inheritance and Classification and Aggregation Hierarchies, in *Advances in GIS Research, Volume 2: Proceedings of the Sixth International Symposium on Spatial Data Handling,* T.C. Waugh and R.C. Healey, Eds., Taylor & Francis, Ltd., London, 1994, pp. 957–972.

Scott, J.M., F. Davis, B. Csuti, R. Noss, B. Butterfield, C. Groves, H. Anderson, S. Caicco, F. D'Erchia, T.C. Edwards, Jr., J. Ulliman, and R.G. Wright. Gap Analysis: A Geographic Approach to Protection of Biological Diversity, *Wildlife Monographs,* 123, pp. 1–41, 1993.

Shea K.S. and R.B. McMaster. Cartographic Generalization in a Digital Environment: When and How to Generalize, in *Auto-Carto 9 Proceedings: Ninth International Symposium on Computer-Assisted Cartography, Baltimore, Maryland,* 1989, American Society for Photogrammetry and Remote Sensing and American Congress on Surveying and Mapping, Falls Church, VA, 1989, pp. 56–67.

CHAPTER 29

ARIANNA: An Experimental Software for Regularization of Lines Surveyed by Differential GPS

G. Scrinzi and A. Floris

INTRODUCTION

Global Positioning System—GPS—(Wells, 1987; Leick, 1990; Hurn, 1989; Hurn, 1993)[1] is used to plot, detect, and map any natural or artificial "object" on land which could be of interest in forestry or agriculture and, more in general, in the thematic mapping of natural environments (projects, inventories, assessment, and land management). The GPS techniques for radiolocalization, in fact, can reliably support and, in certain cases, quite entirely replace other methods, improving precision and efficiency of surveys with a remarkable reduction in costs (Liu and Brantigan, 1995; Marchetti et al., 1995; Floris and Scrinzi, 1995a,b).

As known, two main "families" of procedures allow one to calculate the satellite-receiver distance, which is the basis of the positioning.

1. identification of the time delay in the transmission of a binary code on one of the two carrier frequencies, which in most receivers is the *C/A code* on the L1 frequency (*pseudorange measurement*).
2. computation of the carrier frequency phase cycles (either L1 or L2, or both) and multiplication of these latter by their relevant wavelength. This is the so-called *measurement on the carrier frequency phase* which,

although requiring greater care in its realization, allows one to achieve high-precision results like those required in topographic and geodetic survey.

For the moment, the requirements typical of forest surveys (easiness, low-cost equipment, nonoptimal environmental conditions, tolerability vs. low precision levels, etc.) make us turn quite exclusively to the first modality, *pseudorange measurement* (Floris and Scrinzi, 1995c).

Although in some instances the surveys can still take advantage of the use of stand-alone GPS (e.g., localization of faunal ranges), the use of Differential GPS (DGPS)—a fixed reference receiver placed on a point with known coordinates operating simultaneously with one or more receiver units on field—is necessary to achieve fairly good levels of accuracy and precision. The use of differential correction, in fact, allows one to neutralize the influence of Selective Availability (SA) and, in any case, it further reduces the positioning error affecting nondifferential surveys even in absence of SA. Anyway, the total absence of SA is an exception in the current system.

GPS technology underwent in the last years a considerable development in the field of pseudorange measurements thanks above all to the introduction of the procedure called "carrier phase smoothing" (Van Dierendonck, 1995) and to its use in dynamic surveys.

This procedure improves the "single fix" precision level (CEP) from a few meters to submetric values, frequently with an error of approximately 10–20 cm only. Such performance levels, however, can be ob-

---

[1] We have chosen only few titles within the wide literature concerning all theoretical and technological aspects of Global Positioning System.

tained only on particularly featured environments (with a low-grade orography, and in the absence of natural or artificial shields) as those typical, for example, of rural contexts; these conditions, however, are uncommon in mountain areas or wherever orography is particularly remarkable, as well as in well-developed and thick forests (Scrinzi, 1992).

In these cases orography and tree coverage induce a drop in performance due, on one side, to a non-favorable satellite geometry (summarized in the PDOP index whose integer value is inversely proportional to the uncertainty in positioning[2]) and, on the other side, to the *multipath phenomenon*, caused by the signal reflection on natural or artificial obstacles (rocks, trees, buildings, and so on). These problems (both poor PDOP and multipath) cannot be either eliminated or reduced by differential correction (Hurn, 1993; Gilbert, 1995).

SOME EXPERIENCES OF GPS USE IN FOREST

Limitations of GPS Use in Rough Orography Environments

In 1994, a study to analyze the hypothetical reduction of GPS operative time and precision when the receiver works in rough orography forest scenerios was carried out (Scrinzi and Floris, 1994).

In 52 sites randomly chosen in Trentino (northeastern Italy) the elevation of the horizon along the eight main cardinal directions (N, NE, E ...NW) were measured on 1:25,000 maps (Figures 29.1 and 29.2).

$$E = \operatorname{arctg} \delta h / \delta d$$

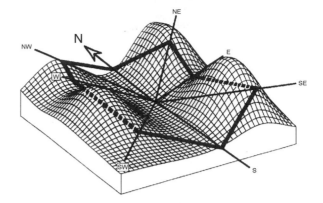

Figure 29.1. Example of horizon elevation along the eight main cardinal directions.

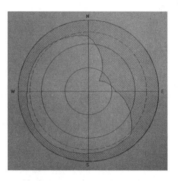

Figure 29.2. Profile of the "true" horizon of a sample point (the broken line represents the elevation mask at 15°).

where δh and δd are, respectively, the altitude difference and the distance between the site and the highest natural obstacle.

Data were processed using the satellite constellations' availability forecast utility supplied from GPSurvey® software. For each site, time reduction (with respect to a working period 9:00 A.M. to 5:00 P.M.) was calculated under three different conditions:

(a) minimum constellation for a 3D positioning (four satellites, low precision survey)
(b) maximum pdop 8 (fair precision survey)
(c) maximum pdop 4 (high precision survey)

Data were compared with those obtained if working, in the same sites, with completely free horizons (elevation mask 15°).

Results, as shown in Table 29.1 and in Figure 29.3, don't point out important limitations to GPS use in

---

[2] It must be remembered here that PDOP is an index of the plotting accuracy. It depends on the geometrical arrangement of the satellites used for the survey in the moment in which the survey occurs (Scrinzi, 1992). The optimal geometry would consist of four satellites placed on the vertices of an imaginary pyramid whose sides are the straight lines between each satellite and the receiver. PDOP value is inversely proportional to the volume of such a pyramid. The greater this volume, the better is the satellite geometry. PDOP results from two other parameters; i.e., HDOP (horizontal) and VDOP (vertical), giving the horizontal and vertical range of the positioning, respectively. The horizontal component, in particular, is of basic importance in planimetric tracings.

Table 29.1. Mean Operativity of GPS (percentage of time during which the positioning is possible) with Free Horizon (el. mask 15°) and with True Horizon (52 sample point).

	4 Satellites Low Precision	PDOP < 8 Fair Precision	PDOP < 4 High Precision
Free horizon	100.0	100.0	95.3
True horizon	96.0	90.5	77.7

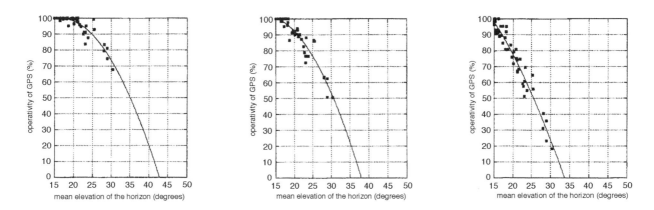

Figure 29.3. Trends in operativity of GPS (percentage of time during which the positioning is possible) as related to mean elevation of the horizon, at three different conditions: four satellites (left), PDOP < (middle), PDOP < 4 (right).

environments with rough orography, if used at low and medium precision.

Nevertheless, the time during which GPS is operational at high precision (pdop < 4) decreases almost 20% on average, in certain cases (mean horizon elevation of about 30°, 10% of our sites) dropping at 20–40% of the total working period.

Tree Coverage Influence on Accuracy and Efficiency of GPS Surveys

Another factor affecting the GPS positioning in natural environments is the tree coverage. In a forest survey, this effect determines an increase of difficulties and errors when operating inside forest stands, particularly in high density conditions (Gilbert, 1994; Lachapelle and Henriksen, 1995; D'Eon, 1995; Floris et al., 1996).

To evaluate the tree coverage influence, 28 forest stands in Trentino were selected in summer 1995, each one neighboring with a "coverage-free" site placed in the same orography conditions (Table 29.2). A series

of 10 survey sessions were executed in each pair of subplots (covered/uncovered), varying the acquisition parameters (Table 29.3).

An 8-channel receiver Pathfinder Pro-XL® was used in the field; all the data thus collected have been differentially corrected in postprocessing, using as reference station a 12-channel Community Base Station®[3] located at 10–80 km far from the surveyed plots.

Each survey session had a three-minute duration, acquiring one position per second (maximum 180 positions). Horizontal coordinates, DOP values, and three acquisition parameters (PDOP mask, SNR mask, and antenna height) were recorded for each position. If no positions were acquired during the first three minutes, we waited a maximum of ten minutes to try to obtain at least one position.

After transferring them in ASCII format, data concerning each session were processed to calculate:

---

[3] Registered trademark of Trimble Navigation Ltd., Sunnyvale, California.

Table 29.2. DGPS Survey under Tree Canopy. Main Site and Forest Features (Characteristics) of the Stands Sampled in Trentino.

Plot No.	Altitude (m m.s.l.)	Mean Slope (°)	Mean Elevation of Horizon (°)	Management System	Composition	Mean Tree Height (m)	Tree Crown Coverage (%)	No. of Trees per Hectare	Mean dbh Diameter (cm)	Basal Area (m²/ha)
1	1045	30	14	high forest	mixed	15	60	849	19.5	32.6
2	1270	5	7	high forest	mixed	20	90	1957	16.1	48.8
3	1300	13	5	high forest	mixed	15	80	1578	17.2	49.7
4	1300	13	6	high forest	mixed	15	70	2246	13.5	45.4
5	1240	40	17	high forest	conifer	20	40	380	27.7	32.6
6	1430	22	12	high forest	conifer	15	70	1481	21.2	57.3
7	1430	14	8	high forest	conifer	10	80	2198	17.1	54.4
8	1430	23	8	high forest	conifer	5	70	1390	13.6	21.6
9	1470	18	4	high forest	conifer	5	50	2217	7.7	13.1
10	1180	20	11	high forest	mixed	15	70	2127	9.8	37.7
11	1280	35	15	high forest	mixed	10	80	3795	9.6	42.4
12	1270	10	8	high forest	conifer	15	80	4376	12.9	30.9
13	1225	17	12	high forest	mixed	15	80	1949	14.8	68.4
14	1170	12	8	high forest	mixed	15	80	832	17.8	26.8
15	1230	18	11	high forest	broadleaved	15	80	2006	19.6	76.8
16	1120	14	11	high forest	mixed	15	80	1082	16.3	41.5
17	1200	18	14	coppice	broadleaved	5	70	8727	4.4	19.6
18	690	7	14	high forest	broadleaved	15	70	582	24.3	28.2
19	470	0	11	coppice	broadleaved	5	100	7894	5.8	25.3
20	350	27	18	coppice	broadleaved	7	100	4851	7.2	28.9
21	456	0	10	coppice	broadleaved	10	90	2674	9.2	26.0
22	466	0	11	coppice	broadleaved	10	90	3947	7.1	21.5
23	1430	28	18	high forest	broadleaved	15	70	29	20.8	57.0
24	1440	32	15	high forest	broadleaved	10	80	881	22.4	36.4
25	1480	24	21	high forest	conifer	40	80	312	54.0	71.8
26	1680	22	13	high forest	conifer	40	70	480	46.7	85.9
27	1300	6	10	high forest	broadleaved	25	60	177	45.7	30.1
28	1310	0	0	high forest	broadleaved	20	50	45	35.1	39.0

Table 29.3. DGPS Survey under Tree Coverage. Different Sets of Acquisition Parameters.

Subplot	PDOP Mask	SNR Mask	Antenna Height (m)
Uncovered	4	6	2
Covered	4	6	2
Uncovered	8	4	2
Covered	8	4	2
Covered	4	4	2
Covered	8	6	2
Covered	4	6	4
Covered	8	4	4
Covered	4	4	4
Covered	8	6	4

- RIS (reception of the signal) at least one position within ten minutes: Y/N;
- FEA (frequence of acquisition): number of positions/180;
- mean PDOP (arithmetic mean of pdop values for each position);
- mean HDOP (arithmetic mean of hdop values for each position);
- SCO (horizontal distance between the mean position of each session and the mean position resulting from the average of all positions collected for each plot).

In Table 29.4 results of this work are presented. They show the very good operativity of GPS in uncovered sites (FEA = 99%, SCO = 40 cm). In covered sites, stems number and size (high-forest stands) show more influence than the crown density (coppices).

The average errors of positioning, however, are included within a good range of values: 1.7 meters in coppice forest and 3.3 meters in high forest, in which we have also found more reception difficulties (RIS = 75.5%).

More elevated errors could occur; fortunately, just in a few cases, because of poor DOP values and multipath phenomena.

Using specific values of acquisition parameters (PDOP mask = 8, SNR mask = 4), raising the receiver antenna up to 4 meters, and averaging a sufficient number of positions for the same point, it is possible to conjugate quickness and accuracy of survey in a balanced manner, as shown also in other studies (D'Eon, 1996).

Dynamic Survey of Forest Trails

The above-described strategies can be easily applied in "static" survey of points. In dynamic survey,[4] on the contrary, the ever-changing operative conditions make the control on the acquisition parameters very difficult (Trimble Navigation, 1992).

In particular, in some cases these parameters may temporarily fall off (the constellation of reference for the receiver gets darkened under the four satellites required for the radiotriangulation, or the PDOP and SNR thresholds have been exceeded), and transient blackouts in survey may occur.

Provided that well-planned survey sessions supported by a software for the forecasting of satellite ephemeris may help to keep these inconveniences under control, the problem can be quite entirely solved anyway by repeatedly carrying out many dynamic surveys on the same line: on forest routes, for example, this can be easily done, as the "return" course almost always coincides with the "outgoing." The several position files obtained in this way are then grouped by software.

This procedure has been followed in a set of surveys carried out on four distinct routes characterized by severe orography and thick forest coverage. Each route was surveyed 30 times using different acquisition parameters (elevation mask, PDOP mask) and recording, on a planning software, the number of available satellites. Some statistics from these surveys are described in Tables 29.5 and 29.6.

In this way, the lines obtained after differential correction are often gap-free even in areas with severe orography; nevertheless, lines are not represented as a more or less regular sequence of points, but rather as a belt of positions surrounding the actual position of the track; such a belt is usually 1.5 to 2.0 times wider than the mean error of the single positioning (in our surveys, the belt width ranges 6–7 m to 10–12 m).

In view of the track mapping, therefore, the problem consists in objectively defining one regular and univocal sequence of points—which could be eventually linked to each other by a continuous line—representing the "most probable mean track" (provided that the "actual" position of the track is unknown).

---

[4] In this context, we call "dynamic" the survey of routes, perimeters, etc., where the current position of the receiver antenna is recorded at short intervals of time during the track, running on foot or by a vehicle.

Table 29.4. Operativity and Accuracy of DGPS under Tree Coverage. Statistics about Surveys Carried Out in 28 Sample Points in Trentino (North-eastern Italy).

Variable	Site Category	Descriptive Statistics	PDOP Mask 4	PDOP Mask 8	SNR Mask 6	SNR Mask 4	Antenna Height (m) 2	Antenna Height (m) 4	Total
						Reception Parameters			
RIS	HIGH FOREST	mean	60.9	90.2	66.3	84.8	77.2	73.9	75.5
		n. obs.	—	—	—	—	—	—	184.00
		max	—	—	—	—	—	—	—
		min	—	—	—	—	—	—	—
		st. dev	—	—	—	—	—	—	—
	COPPICE	mean	75.0	95.0	80.0	90.0	75.0	95.0	85.0
		n. obs.	—	—	—	—	—	—	40.00
		max	—	—	—	—	—	—	—
		min	—	—	—	—	—	—	—
		st. dev	—	—	—	—	—	—	—
	UNCOVERED	mean	100.0	100.0	100.0	100.0	100.0	—	100.0
		n. obs.	—	—	—	—	—	—	48.00
		max	—	—	—	—	—	—	—
		min	—	—	—	—	—	—	—
		st. dev	—	—	—	—	—	—	—
PDOP	HIGH FOREST	mean	3.66	5.30	5.00	4.38	4.78	4.52	4.65
		n. obs.	56.00	83.00	61.00	78.00	71.00	68.00	139.00
		max	4.00	8.00	8.00	8.00	8.00	8.00	7.46
		min	2.50	2.50	2.50	2.50	2.80	2.50	2.55
		st. dev	0.39	1.45	1.60	1.17	1.49	1.29	1.26
	COPPICE	mean	3.51	4.22	4.03	3.79	4.02	3.82	3.91
		n. obs.	15.00	19.00	16.00	18.00	15.00	19.00	34.00
		max	4.00	7.80	7.70	7.80	7.70	7.80	7.21
		min	2.70	2.60	3.10	2.60	2.60	2.70	2.71
		st. dev	0.44	1.51	1.22	1.22	1.18	1.25	1.16
	UNCOVERED	mean	2.70	2.80	2.70	2.80	2.75	—	2.75
		n. obs.	24.00	24.00	24.00	24.00	48.00	—	48.00
		max	3.80	3.80	3.80	3.80	3.80	—	3.80
		min	1.90	2.10	1.90	2.10	1.90	—	1.97
		st. dev	0.53	0.48	0.53	0.48	0.51	—	0.51
HDOP	HIGH FOREST	mean	2.02	2.70	2.58	2.31	2.48	2.36	2.43
		n. obs.	56.00	83.00	61.00	78.00	71.00	68.00	139.00
		max	2.90	4.40	4.40	4.40	4.40	4.40	4.20
		min	1.30	1.40	1.40	1.30	1.60	1.30	1.39
		st. dev	0.29	0.90	0.91	0.67	0.85	0.73	0.74
	COPPICE	mean	1.79	2.07	2.11	1.81	1.93	1.97	1.95
		n. obs.	15.00	19.00	16.00	18.00	15.00	19.00	34.00
		max	2.20	3.60	3.60	3.00	3.60	3.20	3.21
		min	1.00	1.40	1.60	1.00	1.40	1.00	1.23
		st. dev	0.32	0.59	0.54	0.44	0.52	0.50	0.49
	UNCOVERED	mean	1.35	1.45	1.35	1.45	1.40	—	1.40
		n. obs.	24.00	24.00	24.00	24.00	48.00	—	48.00
		max	2.00	2.10	2.00	2.10	2.10	—	2.07
		min	1.00	1.00	1.00	1.00	1.00	—	1.00
		st. dev	0.22	0.29	0.22	0.29	0.26	—	0.25
		mean	3.75	3.02	3.30	3.32	3.62	2.99	3.31
		n. obs.	56.00	83.00	61.00	78.00	71.00	68.00	139.00

Table 29.4. Continued.

Variable	Site Category	Descriptive Statistics	Reception Parameters						
			PDOP Mask		SNR Mask		Antenna Height (m)		Total
			4	8	6	4	2	4	
SCO	HIGH FOREST	max	19.23	20.84	20.84	19.23	11.28	20.84	18.70
		min	0.17	0.17	0.17	0.34	0.38	0.17	0.23
		st. dev	3.37	3.21	3.23	3.35	2.90	3.63	3.28
		mean	**1.78**	**1.60**	**1.62**	**1.73**	**1.84**	**1.55**	**1.68**
		n. obs.	15.00	19.00	16.00	18.00	15.00	19.00	34.00
	COPPICE	max	5.33	4.66	4.66	5.33	5.33	2.93	4.65
		min	0.43	0.39	0.41	0.39	0.41	0.39	0.40
		st. dev	1.39	1.11	1.06	1.38	1.65	0.78	1.21
		mean	**0.43**	**0.36**	**0.43**	**0.36**	**0.40**	**—**	**0.40**
		n. obs.	24.00	24.00	24.00	24.00	48.00	—	48.00
	UNCOVERED	max	1.54	0.87	1.54	0.87	1.54	—	1.32
		min	0.08	0.07	0.08	0.07	0.07	—	0.08
		st. dev	0.32	0.23	0.32	0.23	0.27	—	0.27
		mean	0.33	0.55	0.35	0.55	0.42	0.51	0.46
		n. obs.	56.00	83.00	61.00	78.00	71.00	68.00	139.00
FEA	HIGH FOREST	max	1.00	1.00	1.00	1.00	1.00	1.00	1.00
		min	0.01	0.01	0.01	0.02	0.01	0.01	0.01
		st. dev	0.32	0.33	0.32	0.34	0.33	0.35	0.33
		mean	**0.49**	**0.72**	**0.43**	**0.78**	**0.60**	**0.63**	**0.62**
		n. obs.	15.00	19.00	16.00	18.00	15.00	19.00	34.00
	COPPICE	max	1.00	1.00	1.00	1.00	1.00	1.00	1.00
		min	0.04	0.02	0.02	0.08	0.08	0.02	0.05
		st. dev	0.38	0.32	0.34	0.30	0.34	0.38	0.34
		mean	**0.98**	**1.00**	**0.98**	**1.00**	**0.99**	**—**	**0.99**
		n. obs.	24.00	24.00	24.00	24.00	48.00	—	48.00
	UNCOVERED	max	1.00	1.00	1.00	1.00	1.00	—	1.00
		min	0.51	1.00	0.51	1.00	0.51	—	0.67
		st. dev	0.10	0.00	0.10	0.00	0.07	—	0.06

Table 29.5. Statistics Concerning DGPS Survey of Four Forest Routes Carried Out by Car. (Percentage of route surveyed in one session, with different acquisition parameters.)

PDOP Mask	Mean	Std. Dev.
<= 6	70.6	24.1
<= 9	81.6	21.9
> 9	85.9	14.0
Number of satellites available		
< 7	76.9	22.7
>= 7	89.8	9.3
Elevation mask		
12°	81.4	19.8
15°	79.8	21.5
TOTAL	80.5	20.7

Table 29.6. Statistics Concerning DGPS Survey of Four Forest Routes Carried Out by Car. (Width of the belt covered by 10 repetitions of the same route [meters].)

	Mean	Std. Dev.
Route 1	6.85	2.82
Route 2	7.40	1.90
Route 3	7.15	3.03
Route 4	7.62	3.70
TOTAL	7.25	2.90

Availability of a univocal sequence of points is a prerequisite also for exporting the tracks into GIS environment, a procedure which is ever more used, in particular since most GPS receivers allow one to link information on the nature of the surveyed objects to their geographic location, in this way exploiting the improved opportunities the GIS software offers for an integrated processing of geographic data and the elements they are linked to (features and their relevant attributes).

In view of this rectification, therefore, a specific software application has been arranged and developed to objectively delineate an optimal and univocal track, the distance of which does not exceed a prudentially estimated value of 1–2 m from the actual one (unknown).

ARIANNA

A Software Overview

The software has been developed on the basis of prototypes in which all the required steps of analysis, design, coding, and testing have been iteratively carried out.

The so-obtained version, called ARIANNA (Applied Rectification and Interpolation Algorithm for Navstar Navigation Analysis[5]), consists of different modules or subroutines (batch and on-line) characterized by a high grade of interactivity with the user as well as by functions relatively independent from each other.

First of all, it must be remarked that ARIANNA does not work through regression models on the DGPS position "belt," because this would require the line to be fractionated in each time to subjectively fixed segments which, consequently, would have to be equalized through mathematical models. Moreover, segments would have to be mutually continuous in the link-points, and this approach would result in being too complex for the aim of the research, as well as difficult to implement.

ARIANNA Operative Steps

Figure 29.4 shows the flowchart of the application, with its logic sequence of operations which can be summarized in four main steps.

A preliminary step consists of loading the coordinate file (in ASCII format) according to a structure which could be processed by ARIANNA. Each DGPS point is stored in a different record and each coordinate covers a field the length of which varies according to the chosen measure unit and the required digits.

In a preliminary version, the input module was adjusted to the format in which GPS files are exported from the PFINDER®[6] software; the current version, on the contrary, derives the information by means of data entry form (field length, auxiliary strings, and so on), in this way adapting various file formats to its own structure.

The current version of ARIANNA works on the planimetric coordinates pairs (latitude and longitude). In fact, in our opinion, the point position on plane is the basic factor of mapping since the point altitude can subsequently be deduced from the map (or from DTM, in the case of GIS).

Observing the results of a DGPS survey on a line, one frequently sees a main belt along with some occasional outliers, i.e., positions which are more or less remarkably far from the belt (Figure 29.5).

The first step in data processing consists in identifying and deleting such positions from the DGPS track, so that they do not affect the next steps of the elaboration. Around each DGPS point, the software identifies a "proximity circle" centered on the point and with required radius, in this way detecting how many and what points lie within the circle (Figure 29.6). Due to their isolation, in fact, outliers are characterized by a low incidence of other points in their surroundings.

This procedure therefore allows one to identify both single outliers (points whose the proximity circle does not include other points) and clusters of outliers (points in the proximity circle of which few other points lie, their number being fixed by the user). Software implements this function in both automatic and supervised mode (respectively, deleting all points which have been found to have in their surroundings less than n coordinate pairs within a radius of length y and identifying all the points suspected to be outliers, giving to the user the possibility of choosing whether to delete them or not); this second mode may prove useful when the survey parameters (i.e., a very high pdop and/or a strong incidence of multipath) make it hard to distinguish between "good positions" and outliers,

---

[5] The proper denomination of the GPS system is NAVSTAR GPS, since the satellites of this system belong to the class NAVSTAR (NAVigation Satellite for Time and Ranging).

[6] Registered trademark of Trimble Navigation Ltd., Sunnyvale, California.

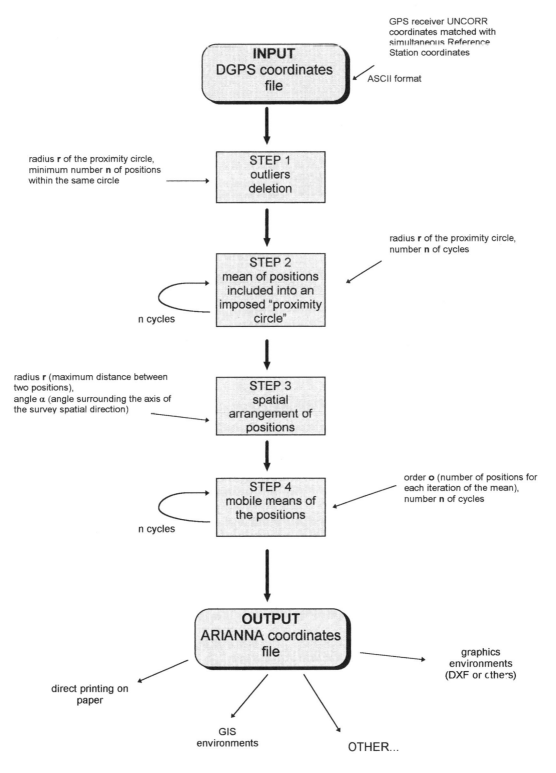

Figure 29.4. Flowchart of ARIANNA.

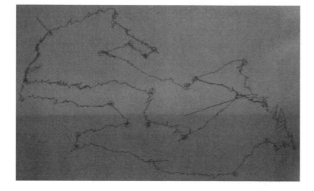

Figure 29.5. DGPS survey of three forest compartments (total extent of about 10 hectares) carried out in three consecutive sessions: positions have been united with lines to put in evidence the outliers.

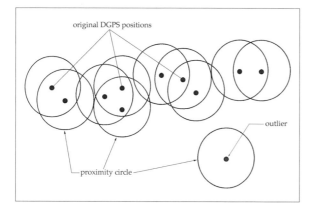

Figure 29.6. ARIANNA's first step. Identification and deletion of outliers.

and whenever an isolated position represents the only information obtained within a gap due to a non-surveyed line.

In the second phase, ARIANNA works on a preliminary rectification of the line through an iteration of the mean for each position left in the file after the outliers have been deleted (Figure 29.7). In other words, the mean is calculated on all the points that are included in a proximity circle of radius fixed by user, the center of which coincides with each position still being present in the file.

Since each position may represent in turn either the center or a peripheral point of a given proximity circle, it enters many times the calculation of the mean. The process output is a new file with the same number of coordinates as before, each one representing the mean

of all the positions detected within the proximity circle. It must be noted that the size of this circle may be different from the size of the circle used in the outlier deletion step.

Some inconveniences affecting this procedure of rectification may be avoided by carefully selecting the proximity radius and comparing the output file with the original one. For example, a too-large radius could cause a translation of positions belonging to strongly bending lines (e.g., winding forest routes) toward the bending center. For this reason, in our elaborations the radius never exceeds the value of 12 m.

In the third step, indispensable for following rectifications, the coordinates must be arranged in a new file in "spatial" order so that their sequence in the file can match the real sequence on the ground (Figure 29.8).

Positions in GPS are in fact recorded in chronological order which corresponds to the actual spatial arrangement only when the line has been surveyed all at once in one session.

When DGPS survey series are not univocal (i.e., when lines are tracked several times in order to fill the reception gaps)—and this, far from being an exception, is a common procedure in forest surveys—this phase is particularly complex. It in fact may be considered the most critical, but at the same time, the most interesting step of the whole process.

First, the operator must fix the sorting starting point and a second point, lying within the proximity circle of the first point, toward which sorting must proceed. In other words, the azimuth between two positions p1 and p2 identifies the initial direction of the spatial sorting.

Now, the software "scans" the angular surrounding of the p1–p2 axis for an angle α and a radius r, both fixed by the operator according to the requirements of survey, finding in this way the closest point to p2 (p3). Then it repeats the procedure on the p2–p3 axis.

During this procedure, the detected positions are recorded in spatial order into a new file, and at the same time they are deleted from the file under processing to prevent them from being reincluded in the sorting: this circumstance could occur, for example, in case of crossing over between tracks.

This phase of the program is carried out with a higher level of interactivity than others, because the sorting process must stop whenever one of the conditions (angle α and radius r) is not fulfilled. If the spatial distance between two positions immediately consecutive in time is greater than r, the signal recep-

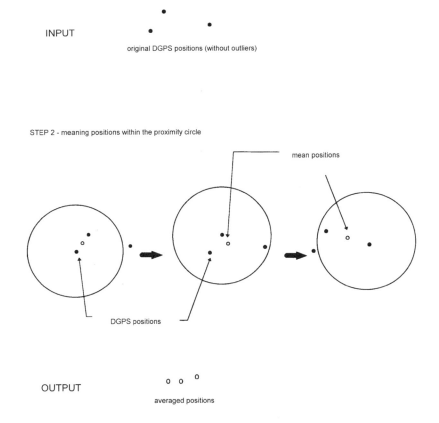

Figure 29.7. ARIANNA's second step. Regularization of DGPS positions by means of the "proximity circle" method.

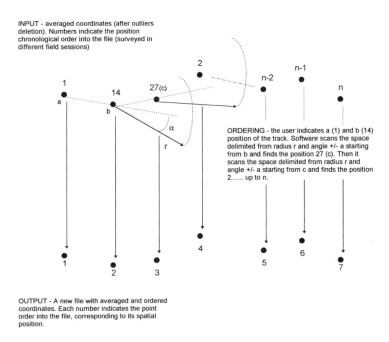

Figure 29.8. ARIANNA's third step. DGPS positions are arranged in a new file in spatial order to match their real sequence on the ground.

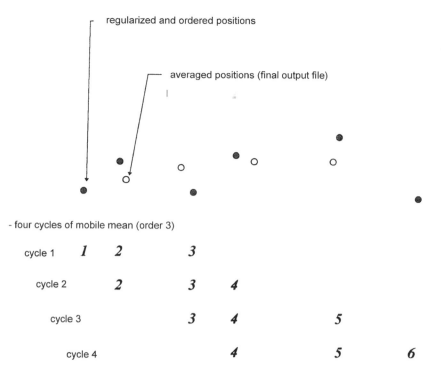

Figure 29.9. ARIANNA's fourth step. Regularized and ordered positions are interpolated by means of mobile mean cycles.

tion can be reasonably thought to have been broken off for a long time. In this case, the track gaps cannot be filled but introduce strongly subjective elements fixed by the operator.

In case the software does not detect any position within the angle α, a sudden change in direction between p_n and $(p_n -1)$ could have been occurred. For this reason, the values of α and r must be accurately selected, taking into account the mean density of the surveyed points. Moreover, during on-field surveys it is of great importance to reduce speed (either on foot or on vehicles) in the bending lines of the track and in particular on windings, in order to make the local points density increasing, with the side effect of minimizing the risk of break-offs in the sorting process.

Shorter radius (10–20 m) is the choice when surveys are carried out on foot, with a high mean density of positions (one position every 1–2 meters), and whenever the nature of the track (e.g., trails) is such that sudden changes in direction can be reasonably expected to occur. Larger radius (50–100 m), on the contrary, is allowed for faster surveys (i.e., on vehicle), with a lesser position density (one position every 10–20 meters), and whenever the nature of the track (i.e.,

roads) excludes the occurrence of sudden changes in direction. After each sorting process break-off, a new sequence automatically starts for sorting the next track.

The fourth and last ARIANNA processing step works on an interpolation with mobile means of n order (Figure 29.9) on the sequences detected in the previous phase of sorting; each sorted sequence is here individually taken to avoid the automatic arbitrary introduction of middle points which don't necessarily comply with the actual track.

This procedure is aimed, on one side, at an additional regularization of the track and, on the other side, at redistributing the coordinates so that they can cover all the gaps which presumably can be reconstructed because of their smallness.

Such processes can be iterated in order to improve rectification. On the grounds of our experience, we suggest n values = 3 or 5, for 2 or 3 mobile mean cycles. With higher values and/or more cycles the risk is that the line track results are excessively simplified.

In Figures 29.10 and 29.11 some ARIANNA typical video outputs compare the original files (DGPS coordinates) with those finally emended by the program (software).

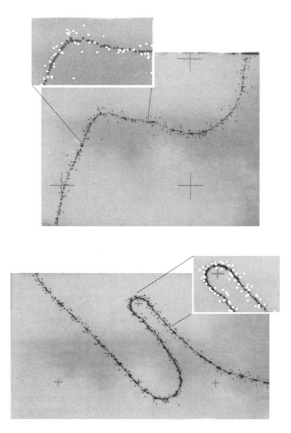

Figure 29.10. Two examples of video output of forest trails processed by ARIANNA, with the original DGPS positions (dark points) and the final track (red points in the main picture, white points in detail).

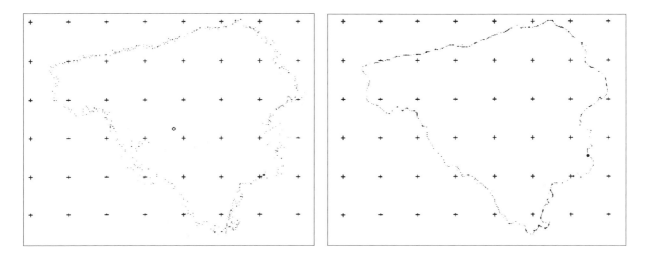

Figure 29.11. An example of video output of a forest compartment processed by ARIANNA, with the original DGPS positions (left) and the final track (right). The distance among marks is 100 meters.

CONCLUSIONS

The first experimental version of ARIANNA was written in Qbasic, a programming language easily accessible to not particularly skilled users; such a version, however, was affected by a number of limitations; in particular, it cannot be "compiled" and always requires source files, apart from being unsuitable to handle files exceeding a given size. For these reasons, a Windows-compatible version in C++ language has been subsequently developed.

This new release, besides the possibility of implementing all the typical Windows functions as dialog-boxes, tool-bars, and so on, allows a substantial bettering of performance, the handling of vectors, arrays, and variables being here improved; moreover, this release has been oriented since its design to the integrated use of widespread GIS packages.

At present, a commercial version of ARIANNA is available on market as an additional module to ArcView[7], under the name ELIT-GPS[8].

REFERENCES

D'Eon, S.P. Accuracy and Signal Reception of a Hand-Held GPS Receiver, *The Forestry Chronicle,* 71(2), 1995.

D'Eon, S.P. Forest Canopy Interference with GPS Signals at Two Antenna Heights, *West. J. Appl. For.,* 13(2), 89–91, 1996.

Floris, A. and G. Scrinzi. GPS in foresta: un aiuto prezioso, *Sherwood, foreste e alberi oggi,* anno 1, 4, 1995a.

Floris, A. and G. Scrinzi. GPS in foresta: esperienze d'uso, *Sherwood, foreste e alberi oggi,* anno 1, 5, 1995b.

Floris, A. and G. Scrinzi. Glossario di terminolgia GPS, Sherwood, *Foreste e Alberi Oggi,* anno I, 6, pp. 46–47, 1995c.

Floris, A., A. Cutrone, and G. Scrinzi. Influenza della copertura arborea su precisione ed efficienza dei rilievi GPS, *Monti e Boschi,* XLVII, 5, pp. 5–14, 1996.

Gilbert, C. Using GPS in the Shade, *Earth Observation Magazine,* May 1994.

Gilbert, C. Sources of GPS Error: What Can Be Fixed and What Cannot, *Earth Observation Magazine,* May 1995.

Hurn, J. *GPS: A Guide to the Next Utility.* Trimble Navigation, Sunnyvale, CA, 1989.

Hurn, J. *Differential GPS Explained.* Trimble Navigation, Sunnyvale, CA, 1993.

Lachapelle, G. and J. Henriksen. GPS Under Cover: The Effect of Foliage on Vehicular Navigation, *GPS World,* March 1995.

Leick, A. *GPS Satellite Surveying.* Wiley Interscience, New York, 1990.

Liu, C.J. and R. Brantigan. Using Differential GPS for Forest Traverse Surveys, *Can. J. For.,* 25, pp. 1795–1805, 1995.

Marchetti, M., F. Campaiola, and G. Lozupone. Esperienze di applicazione del metodo GPS differenziale nel monitoraggio e nella gestione delle risorse forestali, *Monti e Boschi* XLVI, 5, 1995.

Scrinzi, G. GPS: una nuova tecnologia per il rilevamento del territorio, *Monti e Boschi* XLIII, 4, pp. 20–26, 1992.

Scrinzi, G. and A. Floris. Limitazioni operative di GPS in ambienti ad accentuata orografia, *Monti e Boschi,* XLV, 3, 1994.

Trimble Navigation. *GPS Surveyor's Field Guide,* Sunnyvale, CA, 1992.

Van Dierendonck, A.J. Understanding GPS Receiver Terminology: A Tutorial, *GPS World,* pp. 34–44, January 1995.

Wells, D. *Guide to GPS Positioning.* Canadian GPS Associates, Fredericton, N.B., Canada, 1987.

---

[7] Registered trademark of ESRI Inc.
[8] Registered Trademark of Lumisoft Inc., Trento, Italy.

CHAPTER 30

Monte Carlo Sensitivity Analysis of Spatial Partitioning Schemes: Regional Predictions of Nitrogen Loss

R.N. Handcock, S. Mitchell, and F. Csillag

INTRODUCTION

A wide variety of hydroecological models exists to predict phenomena such as plant growth, and the cycling of water, nutrients, and carbon. These models perform well at particular sites, especially with homogeneous land cover, and known land-use history. However, there is a definite need for regional and continental scale modeling, to tackle questions such as the effects of global warming or policy alternatives to combat acid rain (e.g., EPA, 1995). This modeling must either scale up site models, a process which has several inherent problems, or spatially partition the region into smaller and more manageable units (Rastetter et al., 1992).

To perform this scale of prediction, it is normally unreasonable to simply overlay a very large, fine mesh over the region, in effect running a large number of site models and aggregating. This will not work either in terms of practicality (computing power, data volumes), or due to the associated uncertainty; it is unlikely that the model parameters are known for all the locations in the grid, so some method of estimation must be used. However, the uncertainty in this estimation can be considerable and nonintuitive (Lammers, 1998).

The other extreme would be to use spatially large, highly aggregated regions. This leads to questions about whether or not it is reasonable to assume that variability of the model output will stay within reasonable bounds in response to potentially greater variability in the input data. In addition, the inability to measure many ecosystem parameters at scales that correspond to these spatial units (e.g., one can't measure with any certainty the total nitrogen runoff of the Adirondack region) contributes to prediction uncertainty (Aber et al., 1993).

If a fine mesh is not feasible, and we can't make measurements at the highest level of aggregation, it would help to be able to evaluate suitable compromises. There are many possible choices within these limits, including the size, shape, and relative arrangement of the partitioning units, and what, if anything, these have to do with the underlying terrain data. Since the degree of uncertainty of our data will vary from one location to another (Csillag, 1991), we suspect that the most appropriate partitioning methods will pay attention to how this uncertainty varies across the landscape. The "goodness" of any partitioning scheme will depend on a combination of the nature of the data, the problem being investigated, and statistical considerations.

RATIONALE

Our test area, the Adirondack Mountains in northern New York, receives elevated inputs of precipitation and acidic deposition; in combination with the edaphic/geologic characteristics, this places terrestrial and aquatic resources at risk. Comprehensive lake surveys and long-term monitoring have revealed that a significant proportion of lakes are impacted. In spite of decreasing atmospheric inputs, the acid neutralizing capacity (ANC) of most lakes has not increased. Decrease in ANC appears to be due to increases in NO_3 concentrations, which appear to be due to decreases in the retention of N by older stands of forest vegetation (Driscoll and van Dreason, 1993). Thus, modeling the nitrogen dynamics of the region's forests is a priority in understanding and managing the effects of acidic precipitation. There are intensive efforts to develop suitable models of these dynamics (Aber et al., 1991, 1993; Lam et al., 1992), but they all share the challenges of spatial partitioning and uncertainty.

Part of our objective is to meet the needs of environmental managers for such a modeling system in a way that can evaluate partitioning schemes. For example, a New York State resource manager who needs to provide input to the legislation of some level of nitrate loading reductions may not be as interested in the variability of the input data as in the resulting output range. They would probably choose individual regions (partition units) familiar to them, such as counties or watersheds, to determine whether or not the range of predicted output met their tolerances. If the modeling support environment is interactive and designed to facilitate the analysis of uncertainty, the partitioning methods can be tailored to the model and application requirements.

We recognize that this approach may still be problematic, due to attitudes or common practices regarding uncertainty in decision making. For example, existing studies of acid deposition (Dillon et al., 1987; Driscoll and van Dreason, 1993; EPA, 1995; Pardo and Driscoll, 1996) have many tables or maps of suggested loading or predicted nitrogen fluxes for the areas of North America affected by acidic deposition, but have little or no indication of uncertainty or prediction error estimates. The policy maker, if asked, may request zero error tolerance, which is not feasible. Analysis tools such as those outlined in this chapter may help bring about a more informed and confident management of the uncertainty of ecosystem models in regional applications.

TECHNICAL TOOLS

The huge volumes of data both used and produced while running an environmental model are of increasingly fine resolutions and often of increasing spatial and temporal extents. The combination of GIS tools and environmental models is increasingly common (Fedra, 1993; Goodchild, 1993), both for the generation of input data and the analysis of output. Effective visualization and exploratory data analysis (EDA) techniques such as the animation and querying of spatio-temporal data are necessary to extract meaningful information from this data. As this complex visualization of spatio-temporal data is not a GIS specific task,

the GIS's graphics may be supplemented by a data visualization package such as IBM's Data Explorer. Some types of queries of a spatio-temporal series include a one-dimensional probe through time to create a time series graph of attributes, a two-dimensional probe through time to create a time series graph of attribute summary statistics, as well as animation of the spatial distribution of attributes over time.

Deciding on an appropriate partitioning scheme is an interactive process that involves assessing how the model behaves using different schemes. The actual process of running the model is a task that is often nonautomated, undocumented, and difficult to change. An overall framework is needed which will automate and standardize the modeling process, making it repeatable according to principles of the scientific method. Other requirements include the transparent coupling of tools (modeling, visualization, EDA) needed to create model input and analyze output. These requirements are met by STAMP (the **S**patial **T**empor**a**l **M**odeling **P**rogram), which uses the concept of a Model Object to encapsulate modeling data and processes in the one unit.

STAMP—A DYNAMIC MODELING ENVIRONMENT

STAMP is a loosely coupled system incorporating the ESRI's ARC/INFO GIS, the PnET environmental model, and the Data Explorer visualization package, for the purpose of viewing spatio-temporal data and managing environmental models. A graphical user interface (GUI) is used to control the spatial analysis tasks and parameterization needed to generate model input, and control the execution of the model. The spatio-temporal model output is read back into the GIS for spatial analysis. This output is archived along with metadata that describes the model input. Graphing of the time-series output from the model may be handled from within the GIS, or exported to a spreadsheet package for further analysis, while animation and complex visualization is available from the data visualization package. This is all incorporated within a modular (object-oriented) and extendable system framework run from within the GIS. STAMP also handles basic viewing and query of spatial data themes, but discussion of this is beyond the scope of this chapter.

The advantage of STAMP is that the path from the selection of model inputs to the analysis of model outputs can be traced, allowing repeatability and verification of model results. Model input, output, and

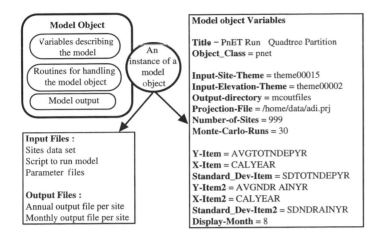

Figure 30.1. The Model Object, and an example of an instance of it.

runtime parameters are encapsulated within a *model-object* (Handcock, 1997) which describes it completely (Figure 30.1). A particular instance of a model object will store both the input data and parameters needed to create a particular run of the model, as well as the output resulting from the model run.

When an environmental model such as PnET is run from STAMP, a new PnET model object is created. The GUI (Figure 30.2) leads the user through the required steps and data selections needed to create the model object. Input parameters and partitioning schemes can be selected from existing STAMP spatial data themes or created "on the fly," and model parameters can be updated from a scrolling list of variables. This menu structure restricts the data that the user can input to the model, and ensures consistency and repeatability between model runs.

STAMP is a multiuser system, so that a number of users can share the modeling environment in a transparent manner. Only one copy of very large data sets is necessary since each user of the system can create individual data themes based on a single unchanged data layer such as a digital elevation model (DEM). In addition, STAMP is easily modified to run different models because the code that handles model objects is itself written in a object-oriented manner. This allows code for general model creation to be reused, while only model input-output routines need to be updated to run a different environmental model (e.g., RHESSys, CENTURY).

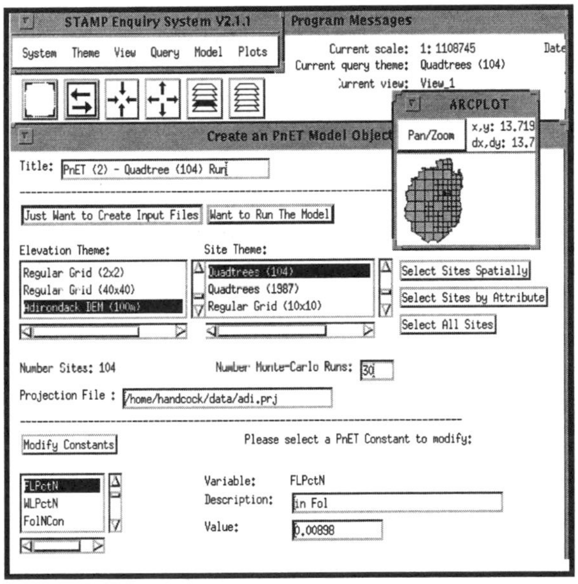

OPERATIONAL DETAILS

Using the STAMP environment and the PnET (Aber and Federer, 1992) model, we investigated the predic-

Figure 30.2. Creating a PnET model object with the STAMP environment.

tion problems of acid deposition and nitrogen saturation in the Adirondack National Park, New York State. PnET is a site-level model that can be run over a spatial extent by assuming that the mean value of input variables across a partition unit represents the site average. To examine how the partitioning methods interact with this assumption to create aggregation effects, we tested a variety of partitioning schemes (Figure 30.3). Some were completely arbitrary, such as a regular grids, which were created by dividing the study area into a square grid of various resolutions. Other partitions were controlled by either landscape, or statistical constraints. Watershed partitions were constructed using the recursive drainage area threshold algorithms of Band (1986), with two threshold levels. The "Lakes/Voronoi Polygon" partitions are Voronoi polygons constructed around lake centroids from the Adirondack Lake Survey. Quadtrees were formed with a variance-based partitioning system (Csillag and Kabos, 1996) which minimizes within-unit, and maximizes between-unit variation, for given numbers of units, or a given residual variance. We used a 100 m DEM from the U.S. Geological Survey as the driving variable for this partitioning algorithm and the watershed delineation.

A version of PnET-CN (Aber et al., 1997) was modified to allow repeated runs with changing inputs, over multiple sites. This model allows for recorded climate data to be used when known, otherwise site characteristics are used to estimate maximum and minimum temperatures, precipitation, and nitrogen deposition. The site characteristics can be long-term averages of climate in that area, or modeled climate. For the Northeastern United States, the model user can choose to use regression coefficients reported by Ollinger et al. (1993), in order to predict the climate, solar radiation, and nitrogen deposition based on latitude, longitude, and elevation. The temporal change in nitrogen deposition since industrialization is modeled based on empirical observations.

Each of the nine partitioning schemes and a deciduous land-cover parameterization was used to develop input data for PnET, as well as some runs using a coniferous land cover (Table 30.1). The source of uncertainty used for Monte Carlo analysis was the variability of elevation within the partition; for each partition unit (to be considered a site by PnET), STAMP uses ARC/INFO routines to calculate the mean and standard deviation of elevation, as well as the latitude and longitude of the partition unit's centroid. The variability of elevation within a partition unit will cause uncertainty in the estimates of temperature and precipitation, which will consequently produce uncertainty in any predictions of quantities which depend on climate. Wet deposition of nitrate is affected by the latitude and longitude of the partition unit's centroid, and the total flux of nitrogen input to a site is affected by elevation variability through its effect on precipitation. Thirty simulations per site were executed, drawing elevation values randomly from the normal distribution defined by the mean and standard deviation of the partition unit. For statistical reasons, 100 Monte-Carlo runs would have been preferred, but

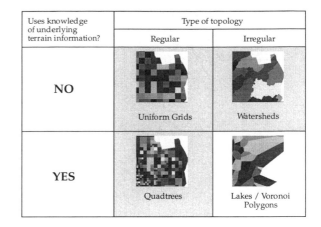

Uses knowledge of underlying terrain information?	Type of topology	
	Regular	Irregular
NO	Uniform Grids	Watersheds
YES	Quadtrees	Lakes / Voronoi Polygons

Figure 30.3. Types of partitioning schemes. Gray shading indicates schemes that are scalable. (Erratum: Watershed-based partitioning uses terrain information, Voronoi polygons do not.)

Table 30.1. Partitions and Land-Cover Parameterizations.

Land-Cover Parameterizations	
Coniferous	Deciduous
Partitions	
Quadtree (104 units)	Regular Grid (2x2) (4 units)
Quadtree (1001 units)	Regular Grid (10x10) (100 units)
Quadtree (1987 units)	Regular Grid (40x40) (1600 units)
Watersheds (6 units)	Voronoi Polygons (1514 units)
Watersheds (120 units)	

there were technical constraints on performing large numbers of runs with large numbers of sites. These model runs will help identify our confidence in the predictions, from the user perspective (N dynamics), and from the modeling perspective (sensitivity).

OBSERVATIONS AND DISCUSSION

Uncertainty/Sensitivity

For the purposes of examining the effects of partitioning schemes in this study, the effects of variability in elevation and nitrogen deposition on the annual losses of nitrogen in drainage water were studied. Figure 30.4 illustrates the general trend of model sensitivity with respect to elevation uncertainty. There is no consistent trend of deposition with respect to elevation, since the effect is indirect, through precipitation (which is also affected by latitude and longitude). However, variability of predicted nitrogen runoff rises with that of elevation, and the highest sensitivity of this prediction occurs when there is high uncertainty in both elevation and nitrogen deposition.

Intersite Variability

The sensitivity to elevation means that with all other factors held equal, the climatic and deposition regime differences caused by changes in location causes differing trends of nitrogen runoff across sites. Figure 30.5 displays a sample of this variability.

Partition Size/Number of Units

The size of partition units within any one scheme will affect prediction uncertainty because as the area of a partitioning unit decreases (i.e., finer units), there is less within-unit variability. Figure 30.6 illustrates the effects of this on predictions, with narrower standard deviation bands around the means of nitrogen fluxes as the number of units increases.

Arbitrary versus Terrain-Based Partitioning

Since the quadtree partitioning was based on the variability of the DEM, we expected model output uncertainty to be generally lower for sites based on quadtrees, than for sites developed from an arbitrary partitioning scheme. Indeed, examination of output across all sites for regular grid and quadtree themes with similar numbers of units showed that this was true. Direct comparison of sites across different partitioning schemes is not possible, since the locations do not correspond well; however, the random selection of sites in Figure 30.7 demonstrates the trend.

Vegetation Parameterization

Elevation is not the only variable being aggregated by the partitioning methods. Nitrogen dynamics are particularly sensitive to interspecific differences in nitrogen use efficiency. Coniferous and deciduous vegetation have very different nitrogen contents and efficiencies, which through feedback mechanisms can cause very different ecosystem nitrogen dynamics

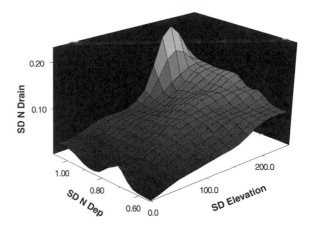

Figure 30.4. Standard deviation of elevation, nitrogen deposition, and nitrogen in drainage. Nitrogen fluxes are annual totals, in g N m^{-2} yr^{-1}. This particular surface was created using half of the 40 x 40 cell regular grid partition and a deciduous vegetation parameterization.

(Vitousek and Howarth, 1991). Work has begun on incorporating this into STAMP's interface to PnET, but a full coniferous parameterization has not been completed at this time. Initial attempts have confirmed the model authors' warning that the model is quite sensitive to these parameters, and more care needs to be taken before we have confidence in STAMP's treatment of a coniferous site. Although the results were not satisfactory enough to include here, the sensitivity at least highlights the potential importance of aggregation effects with respect to land-use data such as vegetation type.

Technical Issues

The process of performing large numbers of PnET runs identified some technical constraints. Model input was restricted to less than a thousand sites due to limitations in the internal GIS routines that calculate grid statistics. Even then, the routines that managed the PnET runs also ran into memory limitations, resulting in the reduction to 30 Monte Carlo runs. Improving the computing efficiency of this experimental code is an obvious target. There will always be the trade-off between memory requirements and the speed requirements of an interactive system; one solution would be to run PnET in parallel processing or load distribution environment, to spread out the processing of the independent PnET sites and runs. However, we believe that in further work we should decrease the

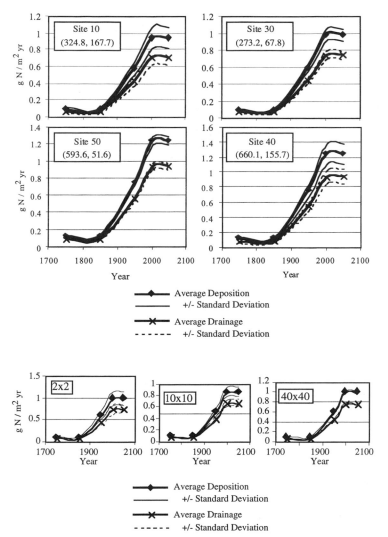

Figure 30.5. Nitrogen fluxes in a random selection of sites in the 120 watersheds partition scheme. Note scales on vertical axes differ. Site 50 has considerable separation between nitrogen input and output fluxes, while this gap is much closer, given the intersecting bands of uncertainty, at Site 10. Numbers in parentheses refer to average, std. dev. of elevation.

Figure 30.6. Nitrogen flux under three levels of aggregation; these samples were taken from the 2 x 2, 10 x 10, and 40 x 40 cell regular grid partitions, and were qualitatively judged to be representative sites of general trends.

number of sites processed at any one time, and increase the number of Monte Carlo simulations. It is difficult to analyze output from the overwhelming numbers of sites which were processed, and we envision a user being mostly interested in certain familiar sites anyway. When more advanced visualization tools are completed, analysis can be done at a landscape level instead of individual sites, but this will require much longer computing times.

CONCLUSIONS

The partitioning schemes and scenarios used in this project to drive the model are a small subset of the huge numbers of possible combinations. To choose the "best" partitioning scheme would involve creating a large number of runs, and narrowing down the choice based on the modeling application. This chapter shows that which partitioning method is chosen will affect the uncertainty of the model output, whether in terms of partition unit size (large, small), or arrangement (regular, arbitrary). These effects are often ignored, or only taken into account once. A partition scheme may be chosen for an initial problem on a new data set and used for all subsequent analyses; however, it may be more suitable to reevaluate partitioning for each problem.

With regard to the particular PnET scenario we have examined, our results show that the model is sensitive to elevation as a source of uncertainty. The other obvious source of relevant variability in this problem is vegetation type. Further work will use dominant veg-

Regular Grid:

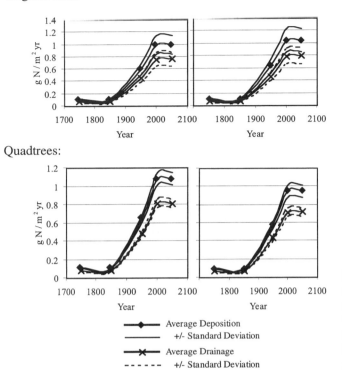

Quadtrees:

Figure 30.7. Nitrogen flux under arbitrary and natural partition schemes: representative sample sites from each of the 10 x 10 regular grid, and the 104 leaf quadtree partitioning. Qualitative analysis revealed that quadtree-derived sites had less variability in nitrogen predictions than their regular grid counterparts.

etation groups as another driving variable to the partitioning, such that within-unit variation of vegetation type is minimized in each PnET site. STAMP could be modified to choose default model parameters based on the vegetation in a partition unit. Another future direction that seems suitable for our hypothetical user would be to add the capability to select different emissions scenarios, and evaluate the effects of different nitrogen loading on the ecosystem dynamics.

We have shown that a system such as STAMP is successful in allowing different model runs and scenarios to be tested in a interactive manner that is both controlled and repeatable. Although it is possible within STAMP to create simple graphs of the uncertainty of the output for each site, it is clear that more complex EDA tools are necessary to examine the spatial and temporal distribution of this uncertainly. Animation was used to observe the model input data, but further development is necessary to make these tools suitable for analyzing spatio-temporal series. The temporal series of model output, with its associated uncertainty, will be mapped back onto the spatial domain.

REFERENCES

Aber, J.D., J.M. Melillo, K.J. Nadelhoffer, J. Pastor, and R.D. Boone. Factors Controlling Nitrogen Cycling and Nitrogen Saturation in Northern Temperate Forest Ecosystems, *Ecol. Appl.,* 1(3), pp. 303–315, 1991.

Aber, J.D. and C.A. Federer. A Generalized, Lumped-Parameter Model of Photosynthesis, Evapotranspiration and Net Primary Production in Temperate and Boreal Forest Ecosystems, *Oecologia,* 92, pp. 463–474, 1992.

Aber, J.D., C. Driscoll, C.A. Federer, R. Lathrop, G. Lovett, J.M. Melillo, P. Steudler, and J. Vogelmann. A Strategy for the Regional Analysis of the Effects of Physical and Chemical Climate Change on Biogeochemical Cycles in Northeastern (U.S.) Forests, *Ecol. Modelling,* 67, pp. 37–47, 1993.

Aber, J.D., W. Currie, C. Driscoll, E. Farrell, C. Federer, C. Goodale, M. Goulden, J. Jenkins, D. Kicklighter, R. Lathrop, G. Lovett, M. Martin, S. McNulty, J. Melillo, S. Ollinger, K. Postek, P. Reich, and R. Santore. PnET: A Simple, Lumped Parameter Model of Forest Biogeochemistry for Regional Applications, [Online] Available URL: http://pyramid.sr.unh.edu/csrc/aber/pnetweb/main/more.html, 1997.

Band, L.E. Topographic Partition of Watersheds with Digital Elevation Models, *Water Resour. Res.,* 22(1), pp. 15–24, 1986.

Csillag, F. Resolution Revisited, in *Auto Carto 10, (ASPRS-ACSM), Baltimore,* 1991. pp. 15–29.

Csillag, F. On Uncertainty in Geographic Databases Driving Regional Biogeochemical Models, in *Auto Carto 12, February 27–March 2, Charlotte, North Carolina,* 1995, pp. 264–270.

Csillag, F. and S. Kabos. Hierarchical Decomposition of Variance with Applications in Environmental Mapping Based on Satellite Images, *Math. Geol.,* 28, pp. 385–405, 1996.

Dillon, P.J., R.A. Reid, and E. de Grosbois. The Rate of Acidification of Aquatic Ecosystems in Ontario, Canada, *Nature,* 329(6134), pp. 45–48, 1987.

Driscoll, C.T. and R. van Dreason. Seasonal and Long-Term Temporal Patterns in the Chemistry of Adirondack Lakes, *Water, Air Soil Pollut.,* 67, pp. 319–344, 1993.

Englund, E.J. Spatial Simulation: Environmental Applications, in *Environmental Modeling with GIS*, Chapter 20, M.F. Goodchild, B.O. Parks, and L.T. Steyaert, Eds., Oxford University Press, Oxford, 1993, pp. 432–437.

Environmental Protection Agency (EPA), Acid Deposition Standard Feasibility Study Report to Congress—Draft for Public Comment, EPA 430-R-95-001, February 1995.

Fedra, K. GIS and Environmental Modeling, in *Environmental Modeling with GIS*, M.F. Goodchild, B.O. Parks, and L.T. Steyaert, Eds., Oxford University Press, Oxford, 1993.

Goodchild, M.F., The State of GIS for Environmental Problem-Solving, in *Environmental Modeling with GIS,* M.F. Goodchild, B.O. Parks, and L.T. Steyaert, Eds., Oxford University Press, Oxford, 1993.

Handcock, R.N. Model Objects—Managing an Environmental Model from within the GIS, *Electronic Proceedings of Environmental Systems Research Institute User Conference, July 8–11 1997, San Diego, CA,* 1997.

Lam, D.C.L., I. Wong, D.A. Swayne, and J. Storey. A Knowledge-Based Approach to Regional Acidification Modelling, *Environ. Monitoring Assess.,* 23, pp. 83–97, 1992.

Lammers, R.B. Hydro-Ecological Simulation Models—From Local to Regional Scales, thesis presented to the Department of Geography, University of Toronto in partial fulfillment of the requirements for the degree of Doctor of Philosophy, January 1998.

Ollinger, S.V., J.D. Aber, G.M. Lovett, S.E. Millham, R.G. Lathrop, and J.M. Ellis. A Spatial Model of Atmospheric Deposition for the Northeastern U.S., *Ecol. Appl.,* 3(3), pp. 459–472, 1993.

Pardo, L.H. and C.T. Driscoll. Critical Loads for Nitrogen Deposition: Case-Studies at 2 Northern Hardwood Forests, *Water, Air Soil Pollut.,* 89(1–2), pp. 105–128, 1996.

Rastetter, E.B., A.W. King, B.J. Cosby, G.M. Horrnberger, R.V. O'Neill, and J.E. Hobbie. Aggregating Fine-Scale Ecological Knowledge to Model Coarser-Scale Attributes of Ecosystems, *Ecol. Appl.,* 2(1), pp. 55–70, 1992.

Vitousek, P.M. and R.W. Howarth. Nitrogen Limitation on Land and in the Sea: How Can It Occur*?, Biogeochemistry,* 13, pp. 87–115, 1991.

CHAPTER 31

The Effects of DEM Generalization Methods on Derived Hydrologic Features

D.B. Gesch

INTRODUCTION

Digital elevation model (DEM) data are being used routinely as a source for deriving hydrologic features such as drainage basins and stream networks. Many of the popular geographic information systems (GIS) and image processing software packages include functions for deriving hydrologic data directly from DEMs. The scientific literature includes many examples of both the background of the methods (Mark, 1984; Band, 1986; Jenson and Domingue, 1988; Fairfield and Leymarie, 1991) and application of the techniques (Band, 1989; Jenson, 1991; Moore et al., 1991; Maidment, 1993).

The increasing availability of global elevation data, such as the GTOPO30 global DEM produced by the U.S. Geological Survey (USGS) (Gesch and Larson, 1996), has allowed usage of the techniques for deriving hydrologic features from DEMs to be extended to continental and global scales (Verdin and Jenson, 1996). Development of continental and global scale DEMs usually involves generalization of higher resolution source DEMs. While the effects of terrain data resolution and DEM aggregation on hydrology and geomorphology applications have been well documented (Chang and Tsai, 1991; Wolock and Price, 1994; Vieux, 1995), there is little documentation on the effects of the method of generalization on derivative hydrologic products. Fisher (1996) has described the effects of generalization methods on another DEM derivative, the viewshed.

DEM GENERALIZATION

Generalization of DEM data may be done for several reasons, including "scale" reduction and data volume reduction. Scale reduction may be done strictly for cartographic display purposes, or it may also be performed to adjust the grid spacing of the terrain data to a resolution appropriate for the features being analyzed. The latter is the case when DEMs are used to derive features for use in large area (continental and global) land surface studies. Higher resolution DEMs are generalized to a level that reduces the data volume to be processed but still portrays topographic features at a level of detail suitable for the applications in which they are to be used. A case in point is the USGS GTOPO30 global DEM which has elevations regularly spaced every 30-arc seconds of latitude and longitude (approximately 1 kilometer). A large portion of GTOPO30 was produced by generalizing higher resolution source DEMs having a grid spacing of 3-arc seconds (approximately 100 meters).

Cartographic generalization is an area in which significant research and development have been conducted, although most studies have concentrated on generalization of linear and polygonal features in vector cartographic data. Generalization of raster terrain data, while similar in concept to generalization of other forms of spatial data, poses unique challenges, due in part to the manner in which the information content is represented in gridded data. In vector cartographic data, the features of interest have already been delin-

255

eated during the compilation process. In gridded terrain elevation data the features of interest; for instance, ridges and stream lines, are included with other features that may not be of interest; for example, broad flat areas. The challenge, then, in generalizing DEMs is to retain the maximum amount of information content (by keeping a proper representation of critical features), but this is made more difficult by not knowing beforehand all the various applications in which the generalized data may eventually be used.

There are several categories of generalization methods for DEM data: statistical, resampling, and morphology-based. Statistical methods use as the aggregated elevation value a representative statistic, such as minimum, maximum, mean, median, mode, or weighted average, calculated from a window of full resolution elevation cells. Resampling methods use a geometric transformation with interpolation to accomplish a change in the grid spacing. Nearest neighbor and bilinear resampling algorithms are often used for raster elevation data. When the generalized cell size is an integer multiple of the full resolution cell size (and there is no change in projection or orientation between the original and generalized grids), nearest neighbor resampling is equivalent to systematic subsampling in which every nth row and column are selected. Statistical and resampling approaches to generalization have the advantage of being widely available in many GIS and image processing software packages. They work well in low relief areas, but smoothing of features can occur in high relief areas and prominent topographic features may not be retained, especially if the degree of generalization is significant.

Morphology-based generalization methods take into account the topographic features present in the DEM. The goal is to retain as much of the significant topographic information (represented by the critical terrain features of ridges, valleys, peaks, and pits) as possible in the generalized DEM. Weibel (1992) presents a very thorough discussion on the types of terrain generalization methods. Morphology-based approaches are similar in concept to the structure line model of generalization described by Weibel. The "breakline emphasis" method of DEM generalization, an implementation of a morphology-based approach, is the subject of the work reported here. The method is named for its emphasis on retaining the topographic breaklines (ridge lines and stream channels) as depicted in the full resolution elevation data.

Breakline Emphasis Generalization Method

The breakline emphasis method of DEM generalization involves two steps: extraction of topographic breaklines from the full resolution DEM, and selection of generalized cell values based on the absence or presence of breaklines. Topographic breaklines are extracted directly from the full resolution data with the widely used methods described by Jenson and Domingue (1988). Figure 31.1 outlines the breakline extraction procedure. Flow direction processing determines the direction of flow from each cell to its steepest downslope neighbor (1 of 8 possible directions). Flow accumulation finds the total number of upslope cells that flow into each cell. Ridges are delineated simply by extracting those cells which have a flow accumulation of zero (there is no flow from any of the neighboring cells). Streams are found by inverting the DEM and then performing the same flow direction and accumulation processing used for ridges. The streams extracted in such a manner may not coincide exactly with the actual stream channels, especially in low relief areas where factors other than just topographic position influence the location of the channel, but use of the inverted DEM allows for complete automation of the procedure. An alternative method of delineating the streams would be to threshold the flow accumulation calculated from the original DEM (where higher accumulation values represent the stream lines where flow accumulates), but such an approach would require intervention by an analyst to interactively select a threshold value.

Figure 31.2 shows how the breaklines are used in selection of a value for each cell in the generalized DEM. If full resolution ridge and/or stream cells are present in the area covered by the generalized cell, a simple majority criteria is applied to classify the generalized cell as either a ridge cell or a stream cell. Ridge cells are assigned the maximum value of the corresponding full resolution elevation values, while stream cells receive the minimum elevation from the area of corresponding full resolution cells. In this manner, the significant topographic structure lines are preserved in the generalized DEM. Generalized cells that do not contain breaklines are simply assigned a representative value from a nearest neighbor resampling of the full resolution DEM.

Breakline emphasis processing was applied to a portion of the 3-arc second source data that were used for the GTOPO30 global DEM. In this case, the generalized 30-arc second data were created for each grid

Ridge processing: Stream processing:

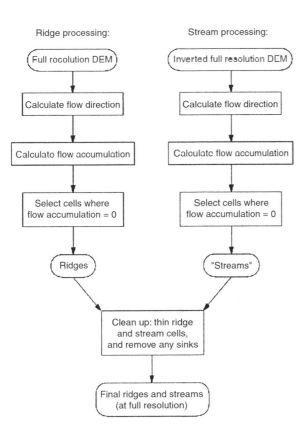

Figure 31.1. Breakline extraction procedure.

cell by selecting one elevation value to represent the area covered by 100 full resolution grid cells (a 10-by-10 matrix).

EFFECTS OF GENERALIZATION ON HYDROLOGIC DERIVATIVES

To investigate the effects of DEM generalization methods, tests were conducted for two drainage basins in the United States. DEM data with a 100-meter grid spacing were generalized to 500-meter, 1-kilometer, and 2-kilometer grid spacings using resampling (nearest neighbor), morphology-based (breakline emphasis), and statistical (mean and median) methods. The James River basin in North and South Dakota and the Allegheny River basin in western Pennsylvania were used for the tests. The basins are part of a national data set of hydrologic unit codes (HUCs) (Seaber and Kapinos, 1987), delineated from 1:250,000-scale topographic maps, that are commonly used for watershed level applications. The HUCs serve as the reference against which DEM-derived basins are compared.

Drainage Basins

Drainage basins were derived from the full resolution and generalized DEMs using the hydrologic analysis tools available in the ARC/INFO[1] GIS software package. Figure 31.3 shows graphically the DEM-derived Allegheny River basins (shaded) compared with the reference HUC (outlined) at the various grid spacings. The percentages refer to the percent of the HUC covered by the derived basin. Table 31.1 lists more comparison statistics for two of the generalization methods, nearest neighbor resampling and breakline emphasis. Note that even in the full resolution DEM a significant portion of the HUC is not covered by the DEM-derived basin. Further investigation has indicated that by slightly modifying just a few elevation values along the sub-basin drainage divide, the derived basin would include the previously missing area. The percent difference in total area of the derived basin against the reference HUC can sometimes be misleading. The percent difference in total area should ideally be zero, but even at zero difference the spatial correspondence of the derived and reference basins could be poor if the "undershoot" area (error of omission) is equal to the "overshoot" area (error of commission). Therefore, when comparing the results, it is useful to examine the percentages of undershoot and overshoot (which should ideally both be zero), as well as a graphic portrayal of the basin shapes.

Figure 31.3 and Table 31.1 indicate that for the Allegheny River basin, a basin with high relief and well defined drainage, the breakline emphasis generalization approach performs better than nearest neighbor resampling, in terms of derived basins matching the reference basin. At the 1-kilometer grid spacing the statistical methods perform equally well, but as the generalized grid spacing increases to 2 kilometers, only the basin derived from the breakline emphasis generalized DEM maintains a match of the reference basin consistent with that derived from the full resolution DEM. At 500 meters all the generalization methods perform about the same, presumably because the level of generalization is less coarse, allowing enough of the significant topographic features to be retained in each case.

---

[1] Any use of trade, product, or firm names is for descriptive purposes only and does not imply endorsement by the U.S. Government.

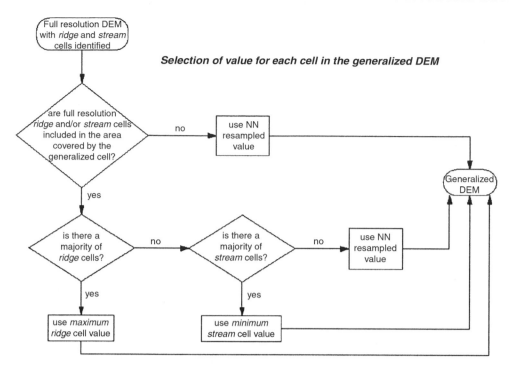

Figure 31.2. Use of breaklines to guide selection of generalized DEM values.

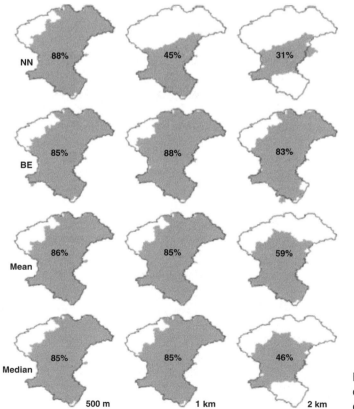

Figure 31.3. DEM-derived Allegheny River drainage basins (shaded) compared with reference basin outline.

Table 31.1. Comparison of Drainage Basins Derived from DEMs Generalized with Breakline Emphasis (BE) and Nearest Neighbor Resampling (NN).

Generalization Method	Grid Spacing	Percent Difference in Total Area: Derived Basin vs. HUC	Percent of HUC Covered by Derived Basin	Percent of HUC Not Covered by Derived Basin (Undershoot)	Percent of HUC in "Extra" Cells (Overshoot)
None	100 m	−13.95%	85.81%	14.82%	0.87%
NN	500 m	−13.54%	85.57%	14.43%	0.89%
BE		−13.41%	85.40%	14.60%	1.19%
NN	1 km	−53.93%	45.40%	54.60%	0.67%
BE		−10.49%	88.36%	11.64%	1.15%
NN	2 km	−67.47%	30.57%	69.43%	1.96%
BE		−15.07%	83.12%	16.88%	1.81%

Stream Networks

Stream lines (as defined by cells of concentrated flow accumulation) also were derived from the full resolution and generalized DEMs using the ARC/INFO hydrologic analysis tools. The reference data set for the James River basin was the U.S. Environmental Protection Agency Reach File (Dewald et al., 1996), a national hydrology data set derived from 1:500,000-scale map sources. Figure 31.4 shows the comparison of stream lines derived from DEMs generalized with nearest neighbor resampling and breakline emphasis approaches. The derived stream lines (shown as shaded cells) are overlaid on the reference stream lines for a portion of the drainage network of the James River basin. A low relief area such as the James River basin is the type of area where the effectiveness of automatic stream extraction procedures is controlled by the quality and resolution of the DEM. The gross mismatches between the derived and reference streams at the 500-meter grid spacing likely occur in nearly flat areas where the resolution of the original source DEM was too coarse to capture the subtle topographic gradients. Overall, the streams derived from DEMs generalized with breakline emphasis correspond much better spatially to the mapped hydrography than do those derived from nearest neighbor resampled DEM's. The drainage pattern is also much more consistent across the three grid spacings of the breakline emphasis DEMs than it is for the nearest neighbor DEMs.

Surface Curvature

Surface curvature (second derivative of the surface) is a topographic parameter that can be used in model-ing erosion, deposition, and sediment transport (Zeverbergen and Thorne, 1987). Curvature was calculated for the Allegheny River basin from the full resolution DEM and from 1-kilometer generalized DEMs. Table 31.2 shows a comparison of the areas classified as flat, concave, and convex in each DEM. Because of the significant difference in cell sizes it is difficult to compare quantitatively the spatial correspondence of the curvature classes from the generalized DEMs with those from the full resolution DEM. This is reflected in the much larger percentage of area classified as flat in the full resolution DEM. The 3 by 3 cell window used to calculate curvature covers a much smaller ground area in the full resolution DEM, thus increasing the likelihood of calculations based on nearly identical elevations. However, the relative proportion of concave and convex areas is a convenient way to compare the performance of the generalization methods. As seen in Table 31.2, breakline emphasis was the only generalization method of the four tested which resulted in a DEM that maintained the correct ratio of concave to convex area (as found in the full resolution DEM).

CONCLUSIONS

Limited tests indicate that breakline emphasis, an implementation of a morphology-based DEM generalization approach, is effective in preserving topographic structure in generalized DEMs, resulting in better derived hydrologic features. Drainage basins and stream networks derived from DEMs generalized by breakline emphasis correspond better to reference data sets than do those derived from DEMs generalized with

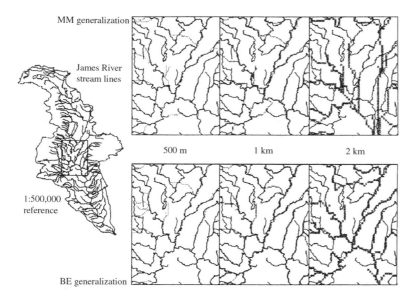

Figure 31.4. Comparison of DEM-derived stream lines.

Table 31.2. Comparison of Land Surface Curvature Classes Derived from Generalized DEMs.

Generalization Method	Grid Spacing	Percent of Basin Classified as:			Ratio of Concave to Convex Areas
		Flat	Concave	Convex	
None	100 m	14.7%	43.6%	41.7%	1.046
NN	1 km	0.7%	49.0%	50.3%	0.975
BE	1 km	1.0%	50.7%	48.3%	1.048
Mean	1 km	0.7 %	48.9%	50.4%	0.971
Median	1 km	0.9%	48.7%	50.4%	0.966

resampling and statistical methods, especially at coarser grid spacings. Further tests resulting in quantitative comparisons of derived features are required on basins in a wide range of terrain relief conditions. The epsilon band concept for assessing the accuracy of area measurements as applied to drainage basins (McAlister et al., 1996) would be useful for quantitative comparison of derived basins. Suitability of the breakline emphasis method for generalizing very high resolution elevation data also should be investigated. Comparisons should also be done against terrain data generalization schemes that use fractal interpolation (Bindlish and Barros, 1996).

The breakline emphasis method offers the advantage that it can be completely automated using the hydrologic analysis tools and raster data processing routines available in many popular GIS and image processing packages. The approach is also flexible, in terms of the criteria that can be used for selection of a generalized grid cell value. As currently implemented, the criteria applied for selection of either the maximum or minimum full resolution elevation value is whether there is a majority of ridge or stream cells (Figure 31.2), but this is just one option. Other options would include taking into account the locations of the breaklines within the generalized cell, or perhaps using a threshold at which a certain percentage of a generalized cell must be classified as ridge or stream before it is handled as such. Selection of the generalized cell value could also be done differently than simply using the maximum or minimum value. For example, a weighted average of the values of the full resolution breakline cells could be used instead of just the maximum or minimum. Another modification to the procedure would be to use an alternative method of identifying the significant topographic features to be emphasized. A more sophisticated landform description that depicts concave, convex, and horizontal and sloping flat areas in addition to ridge and stream lines, such as that described by Blaszczynski (1997), could be used to guide selection of generalized values.

The need for effective DEM generalization techniques will remain strong as better quality and higher resolution data become more available, and as applications of DEM data become more widespread. Newer gridding algorithms, such as that developed by Hutchinson (1989), produce source DEMs that are already optimized for hydrologic applications, so use of morphology-based generalization schemes appropriately retain the increased topographic information inherent to the original DEM. As coverage of higher resolution elevation data increases, DEMs at coarser grid spacings will be routinely produced through generalization. One such example is the data to be generated from the Shuttle Radar Topography Mission (SRTM), scheduled for launch in September 1999. The near global 100-meter terrain data to be produced from SRTM data will be an excellent source for producing generalized DEM's for continental scale hydrologic studies.

ACKNOWLEDGMENT

This work was performed under U.S. Geological Survey contract no. 1434-CR-97-CN-40274.

REFERENCES

Band, L.E. Topographic Partition of Watersheds with Digital Elevation Models, *Water Resour. Res.,* 22, pp. 15–24, 1986.

Band, L.E. A Terrain-Based Watershed Information System, *Hydrol. Proc.,* 3, pp. 151–162, 1989.

Bindlish, R. and A.P. Barros. Aggregation of Digital Terrain Data Using a Modified Fractal Interpolation Scheme, *Comput. Geosci.,* 22, pp. 907–917, 1996.

Blaszczynski, J.S. Landform Characterization with Geographic Information Systems, *Photogrammetric Eng. Remote Sensing,* 63, pp. 183–191, 1997.

Chang, K. and B. Tsai. The Effect of DEM Resolution on Slope and Aspect Mapping, *Cartogr. Geogr. Inf. Syst.,* 18, pp. 69–77, 1991.

Dewald, T.G., S.A. Hanson, L.D. McKay, and W.D. Wheaton. Managing Watershed Data with the USEPA Reach File, in *Proceedings, Watershed '96 Technical Conference and Exposition, Baltimore, Maryland, June 8–12, 1996, Water Environment Federation,* Alexandria, VA, 1996. http://www.epa.gov/OWOW/watershed/Proceed/

Fairfield, J. and P. Leymarie. Drainage Networks from Grid Digital Elevation Models, *Water Resour. Res.,* 27, pp. 709–717, 1991.

Fisher, P. The Effect of Database Generalization on the Accuracy of the Viewshed, in *Spatial Accuracy Assessment in Natural Resources and Environmental Sciences: Second International Symposium,* May 21–23, 1996, Fort Collins, CO, H.T. Mowrer, R.L. Czaplewski, and R.H. Hamre, Eds., General Technical Report RM-GTR-277. U.S. Department of Agriculture, Forest Service, Rocky Mountain Forest and Range Experiment Station, Fort Collins, CO, 1996, p. 272–280.

Gesch, D.B. and K.S. Larson. Techniques for Development of Global 1-km Digital Elevation Models, in *Pecora Thirteen, Human Interactions with the Environment: Perspectives from Space, Sioux Falls South Dakota, August 20–22, 1996, Proceedings,* American Society for Photogrammetry and Remote Sensing, Bethesda, MD (in press). http://edcwww.cr.usgs.gov/pecora/contents.html

Hutchinson, M.F. A New Procedure for Gridding Elevation Data and Stream Line Data with Automatic Removal of Spurious Pits, *J. Hydrol.,* 106, pp. 211–232, 1989.

Jenson, S.K. and J.O. Domingue. Extracting Topographic Structure from Digital Elevation Data for Geographic Information System Analysis, *Photogrammetric Eng. Remote Sensing,* 54, pp. 1593–1600, 1988.

Jenson, S.K. Application of Hydrologic Information Automatically Extracted from Digital Elevation Models, *Hydrol. Proc.,* 5, pp. 31–44, 1991.

Maidment, D.R. GIS and Hydrologic Modeling, in *Environmental Modeling with GIS,* M.F. Goodchild, B.O. Parks, and L.T. Steyaert, Eds., Oxford University Press, New York, 1993, p. 147–167.

Mark, D. Automated Detection of Drainage Networks from Digital Elevation Models, *Cartographica,* 21, pp. 168–178, 1984.

McAlister, E., N. Domburg, T. Edwards, and R. Ferrier. Hydrological Modelling of the River Ythan Using ARC/INFO GRID, in *Proceedings, Sixteenth Annual ESRI User Conference,* May 20–24, 1996. Environmental Systems Research Institute, Inc., Redlands, California, 1996. http://www.esri.com/base/common/userconf/proc96/WELCOME.HTM

Moore, I.D., R.B. Grayson, and A.R. Ladson. Digital Terrain Modeling: A Review of Hydrological, Geomorphological, and Biological Applications, *Hydrol. Proc.,* 5, pp. 3–30, 1991.

Seaber, P.R. and F.P. Kapinos. Hydrologic Unit Maps, *U.S. Geological Survey Water-Supply Paper* 2294, 1987.

Verdin, K.L. and S.K. Jenson. Development of Continental Scale Digital Elevation Models and Extraction of Hydrographic Features, in *Proceedings, Third International Conference/Workshop on Integrating GIS and Environmental Modeling,* Santa Fe, NM, January 21–26, 1996. National Center for Geographic Information and Analysis, Santa Barbara, CA, 1996. http:/

/www.ncgia.ucsb.edu/conf/SANTA_FE_CD_ROM/main.html

Vieux, B.E. DEM Aggregation and Smoothing Effects on Surface Runoff Modeling, in *Wetland and Environmental Applications of GIS*, J.G. Lyon and J. McCarthy, Eds., Lewis Publishers, Boca Raton, FL, 1995, pp. 205–229.

Weibel, R. Models and Experiments for Adaptive Computer-Assisted Terrain Generalization, *Cartogr. Geogr. Inf. Syst.,* 19, pp. 133–153, 1992.

Wolock, D.M. and C.V. Price. Effects of Digital Elevation Model Map Scale and Data Resolution on a Topography-Based Watershed Model, *Water Resour. Res.,* 30, pp. 3041–3052,1994.

Zeverbergen, L.W. and C.R. Thorne. Quantitative Analysis of Land Surface Topography, *Earth Surface Processes and Landforms,* 12, pp. 47–56, 1987.

CHAPTER 32

Estimation of Land-Cover Proportions from Aggregated Medium-Resolution Satellite Data

F. Canters, H. Eerens, and F. Veroustraete

INTRODUCTION

A good knowledge of the spatial distribution of major land-cover types on a global scale is of critical importance for global environmental research. To accurately monitor changes in land cover, sensors on satellite platforms may be used that are able to deliver data at regular points in time for the whole of the globe. The present study is part of a research project aimed at developing a strategy for biome classification at a global scale using data to be delivered by the VEGETATION instrument. The strategy to be developed is to make optimal use of the features offered by VEGETATION and should guarantee that reliable estimates of land-cover proportions can be derived at resolutions that are suitable for global change studies. To accomplish this, attention is focused on the optimization of phenology-based classification techniques, as well as on the development of methods for the removal of areal bias present in coarse-scale estimates of land-cover proportions. The latter issue is the subject of this chapter.

In the last two decades many attempts have been made to characterize land cover over large areas using NOAA-AVHRR data. Almost all strategies that have been proposed are based on the characterization of variations in the Normalized Difference Vegetation Index (NDVI) temporal profile for different types of vegetation (Tucker et al., 1985; Townshend et al., 1987). In most studies, however, land-cover classification is restricted to the continental scale. Also the IGBP-DIS 1 km global land-cover product is created on a continent by continent basis. To obtain a maxi-

mum discrimination between different cover types at the full 1 km resolution, extensive use is made of detailed ancillary data, including elevation, climate, and ecological data (Loveland et al., 1991; Brown et al., 1993). As a result, the classification procedure is time-consuming and highly dependent on the expertise of well-trained image classification personnel.

Not all users of global land-cover data, however, require a strong differentiation in cover types at the full 1 km resolution. A detailed survey, set up by the partners of this project in the spring of 1997, indicated that land-cover data requirements may vary considerably, in terms of spatial resolution, as well as in terms of land-cover classes that are needed or preferred (Canters and Veroustraete, 1997). A considerable part of the respondents proved to be satisfied with up-to-date information about the distribution of a relatively small number of broadly-defined cover types at resolutions equal or comparable to the grid-cell size of their models, on the condition that reliable estimates of subcell class proportions are made available. This justifies further investigation of the potential of global classification strategies for land-cover characterization. While global classification methods may not be able to provide the same level of detail as continent-based approaches, neither spatially, nor thematically, they may offer a valuable and cheap alternative for applications requiring up-to-date information about major biome distributions at a global scale.

One of the issues one is confronted with when developing a global strategy for biome classification is the choice of an appropriate resolution. Spatial degradation of 1 km sensor data prior to classification is an interesting option, since it reduces the amount of data to be classified to more manageable proportions. As has been pointed out, many users of global land-cover data do not require information at the 1 km resolution. On the other hand, one should keep in mind that although global-scale models may not be sensitive to the locational accuracy of land-cover types at the sub-grid scale, they will be sensitive to the proportions of different land-cover types within the cell (Moody and Woodcock, 1994). This raises fundamental questions about the proportional accuracy one may expect to obtain if data are classified at lower resolutions, compared to the results obtained at the full resolution (1 km). The choice of an appropriate method for image degradation is an important issue in this respect. Also, methods to improve the estimation of class proportions at lower resolutions should be examined, since the use of such methods may increase the level of proportional accuracy that can be attained for a given resolution. The same questions can be raised about the proportional accuracy of data sets produced by aggregating higher-resolution land-cover data.

To reduce the loss of proportional accuracy due to spatial aggregation of image data two different approaches might be considered: (a) estimating the proportion of component land-cover classes within each pixel in the coarse resolution image, using spectral mixture models (Pech et al., 1986; Cross et al., 1991; Zhu and Evans, 1992; DeFries and Townshend, 1994); (b) correcting coarse-scale area estimates in a post-classification mode, using calibration models that define the relationship between class proportions at the fine and at the coarse resolution. Mayaux and Lambin (1995) and Moody (1996) showed that calibration models relating fine-scale to coarse-scale proportions can be significantly improved by including the spatial pattern of land cover as an independent variable in the model. In this study a similar approach has been followed to determine to what extent calibration models that make use of information about the spatial structure of the scene can reduce the proportional bias that is present in coarse-scale area estimates derived from spatially-degraded versions of 1 km land-cover data, which have been produced through majority-based resampling.

Results described in this study represent only a first step toward the final goal of our work, which is to find out if structure-based calibration can be used to derive reliable coarse-scale area estimates from land-cover maps, obtained through classification of spatially degraded versions of 1 km resolution sensor data. The idea is that if one would succeed in defining a resampling technique which is able to extract the spectral signature of the most prominent class within the coarse-scale pixel, then classification of the degraded image should produce a result which is very similar to the one obtained by majority-based resampling of the classified image at the original resolution. If a proper resampling technique can be defined, and if structure-based calibration proves to works well in the present design of the experiment, then one may expect that it can be used as well for the correction of area estimates derived from coarse-scale classifications. The problem of defining a proper method of resampling is currently investigated.

While it is our intention to develop and evaluate strategies for area estimation that can be applied to data coming from VEGETATION, our present work is based on NOAA-AVHRR data. Scale-dependent changes in class proportions have been modeled for a multiple-resolution land-cover data set (1 km, 8 km, 32 km) for the entire African continent (16 classes), produced by majority-based resampling of a multi-temporal classification of AVHRR maximum 10-day composites of NDVI values for 1992. The objective of the case study was to use aggregated land-cover data at 8 km and 32 km resolution to produce estimates of land-cover class proportions at grid sizes that are eight times larger (64 km and 256 km, respectively), and to correct these estimates in order to bring them closer to the "true" class proportions that are obtained by counting within each grid cell the number of pixels attributed to each category in the original classification (1 km) (Figure 32.1).

BACKGROUND

When land-cover data are aggregated to lower resolutions through a majority-based resampling procedure, or obtained directly from classification of coarse-resolution sensor data, the proportional accuracy of different land-cover types will drop as the resolution is decreased. In general, classes that dominate the scene will grow larger, while small classes will diminish in size. This is clearly seen from the proportion transition curves depicting "true" (1 km) class proportions against proportions estimated from aggregated data (Figure 32.2). Previous studies have shown, however, that changes in cover-type proportions are not related to the scale of aggregation and the original proportion of each class only. Also, the spatial structure of the landscape seems to play a major role in the way class

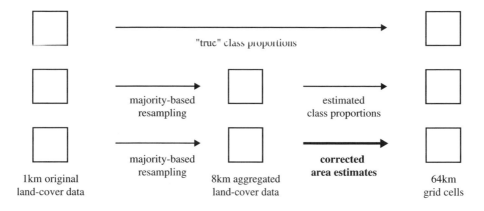

Figure 32.1. Estimation of class proportions for grid cells with a size of 64 km.

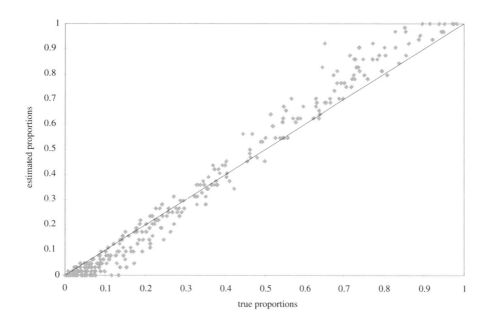

Figure 32.2. "True" class proportions, derived from the original land-cover data set (1 km), against class proportions estimated from 32 km aggregated land-cover data, for grid cells with a size of 256 km.

proportions change due to spatial aggregation (Turner et al., 1989; Moody and Woodcock, 1994).

In a study on the estimation of tropical forest cover from AVHRR 1 km data, Mayaux and Lambin (1995) proposed a two-step calibration model to correct for proportional bias that includes a measure of the spatial fragmentation of the landscape as an independent variable. First of all, they classified Landsat TM and AVHRR data for a nonrandom sample of locations. Both the Landsat TM and AVHRR classifications were coregistered and overlaid with a grid of contiguous

13×13 AVHRR pixel blocks. For each block the proportion of pixels classified as evergreen forest, as well as a number of indices characterizing the spatial structure of the landscape, were calculated at the two resolutions. Then the blocks with the most similar values for a particular landscape index were clustered in equally-sized subsets. For each subset a least-squares regression between the coarse and fine resolution estimates of forest proportion was computed. Once the regression parameters for each subset had been derived, the relationship between the average value of the landscape index for each subset, and the intercept (a) and slope (b) of the regressions between the coarse and fine resolution estimates, calculated in the first step of the procedure, were computed. This produces two equations, one for the intercept and one for the slope of the proportion transition curve, which both include a spatial index as an independent variable.

Mayaux and Lambin experimented with various calibration models, using different spatial measures to control the parameters of the proportion transition curve. They also compared the results obtained with spatial indices calculated at the fine and at the coarse resolution. Calculating spatial indices at the fine resolution yielded a stronger relationship between spatial structure and parameters of the transition curve, since changes in forest cover proportion due to spatial aggregation are, of course, determined by the structure of the detailed scene, not by the spatial characteristics of the aggregated image data. Yet also the calibration models based on coarse-scale spatial measures succeeded in producing much better estimates of fine scale proportions compared to a simple linear correction model that relates proportions at coarse and fine scales without taking account of the spatial organization of the landscape. This proves that structure-dependent backward scaling is feasible; i.e., that fine-scale proportions can be retrieved from coarse resolution data, using coarse-scale spatial measures as a substitute for spatial measures calculated at a fine resolution. Of course, this process of backward scaling requires that the spatial measures which are used in the calibration are relatively scale-invariant. This should be verified before the model is constructed.

Moody (1996) generalized the structure-based calibration approach to situations with more than two land-cover categories. He also included four spatial measures in the calibration model instead of one, each of them measuring a different characteristic of the landscape. Landsat TM data for two forested areas (a calibration and a validation site) were classified into five land-cover categories, and aggregated to 1020 m resolution by majority-based resampling. Changes in land-cover proportions were modeled for 50 randomly chosen sample units consisting of 7×7 1020 m pixels. For each unit the intercept and slope of the proportion transition curve were computed using a linear least-squares fit between the 30 m and 1020 m proportions for all the classes occurring within the unit. Then the relationships between the slope and the intercept of the proportion transition curve and the four spatial measures calculated for each unit were mod-

eled using multiple linear regression. Results of the modeling, using coarse resolution data only, showed an improvement of the total absolute error within each sampling unit for 80% of the validation sample.

Although the results of the previous studies are very promising, the question remains if structure-based calibration models can be used in areas where the number of land-cover classes and the spatial structure of the scene is very different from one location to another, as is the case with global land-cover data covering a region of continental size or the entire world. It is also not clear if the approach is extensible to other scale ranges; more in particular, if it can be used for the correction of very coarse-scale area estimates; i.e., for spatial units with a size comparable to the grid-cell size of global-scale models. The aim of this study is to investigate if structure-based calibration models can be generalized to such situations.

METHOD

To be able to demonstrate the feasibility of developing one general calibration model for the correction of coarse-scale class proportions in areas with strong differences in landscape characteristics, it was decided to work out a case study for the entire African continent. AVHRR maximum NDVI values for four 10-day periods in 1992, derived from Eidenshink's 1 km global data set, were classified into 16 categories with different NDVI-profiles using the ISODATA method. The classification shows very strong local differences in terms of spatial organization, including relatively homogeneous areas where only a small number of classes occur (e.g., deserts), as well as highly fragmented areas where most of the classes are encountered. Two aggregated versions of the classification were produced, one at 8 km and one at 32 km resolution, using a simple majority-based resampling procedure.

The 8 km and 32 km data set were used to derive and correct area estimates for grid cells that are eight times larger in both the x and y direction; i.e., grid cells with an area of 64×64 km for the 8 km data set and with an area of 256×256 km for the 32 km data set. For both resolutions a grid of contiguous 8×8 coarse-scale pixel blocks was overlaid on the aggregated version as well as on the original 1 km version of the classification. Modeling of changes in class proportions was based on a large random sample of cells selected from this grid and covering the entire African continent (687 cases for the 8 km scenario, 456 cases for the 32 km scenario). About 75% of the cells were used for model calibration; the remaining cells were used for validation. For each cell in the sample fine-scale and coarse-scale class proportions were determined, as well as several landscape measures, each characterizing a particular aspect of the structural organization of the cell. Each measure was derived both at the fine scale (1 km) and at the coarse scale (8 km or 32 km).

To decide which characteristics of the cells are most appropriate to be included in the calibration modeling, correlations between any two structural measures as well as cross-scale correlations for each measure were evaluated. Correlation proved to be very high between measures referring to the relative proportion of different classes, as well as between measures describing different aspects of spatial complexity. It was therefore decided to use only two structural measures in the modeling, one describing the distribution of class proportions, and one describing the complexity of the scene. From each group of measures the one with the highest cross-scale correlation (a prerequisite for scale-invariance) was selected. The two measures are: the reciprocal of Simpson's index (Baker and Cai, 1992)

$$SIMP^{-1} = 1 / \sum_{i=1}^{k} P_i^2 \qquad (1)$$

which combines class richness and evenness (k is the number of classes, P_i is the proportion of class i in the cell), and a simple measure of complexity ($COUNT$), which is obtained by dividing the total length of patch boundaries in the cell by the total length of pixel edges, not including the edges which are on the boundary of the cell. The first measure takes a value between 1 (one class, maximum dominance) and k (k classes, all have equal proportions); the second measure takes a value between 0 (one class) and 1 (all adjacent pixels belong to different classes).

Calibration models were developed by applying the two-step procedure proposed by Mayaux and Lambin (1995) (see above). Apart from using a simple linear model to characterize the shape of the proportion transition curve, tests have also been performed with a non-linear model (S-shaped) (Figure 32.3). Inspection of the proportion transition curves for individual cells indicated a maximum deviation from the no-error line for intermediate classes, suggesting that the curve might be better approximated by an S-shaped model than by a straight line. Use was made of the left half of

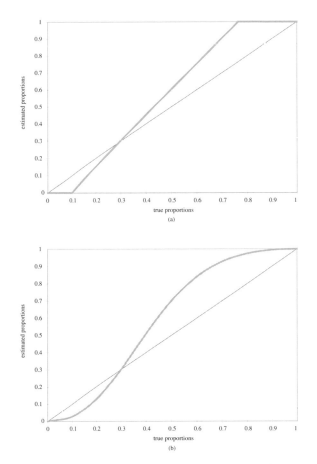

Figure 32.3. Linear (a) and nonlinear modeling (b) of the proportion transition curve.

Dombi's bell-shaped function, known from fuzzy set theory (Jiang and Kainz, 1997). This function has the required shape and boundary conditions and has the advantage that its two most important characteristics, i.e., its sharpness and its inflection point, can be modified by two parameters. When defined on the interval [0,1] the function is written as

$$f(x) = \frac{(1-\upsilon)^{\lambda-1} x^{\lambda}}{(1-\upsilon)^{\lambda-1} x^{\lambda} + \upsilon^{\lambda-1}(1-x)^{\lambda}} \qquad (2)$$

where υ indicates the position of the inflection point along the horizontal axis, and λ determines the sharpness of the curve.

Calculating the two parameters of the proportion transition curve for each cell in the calibration set and using these as the dependent variables in a multiple regression analysis relating each parameter to the structural characteristics of the cell, did not produce regression models with highly significant coefficients. To obtain stronger relationships, the population of cells was partitioned into subsets of cells with a similar structure, as suggested in the study by Mayaux and Lambin (1995). Since we wanted to develop a model based on two structural measures, cells had to be grouped according to the value of both. This was done by a k-means clustering (50 clusters), using the standardized value of each measure as a variable for the partitioning. Parameters of the proportion transition curve were derived for each cluster by determining a least-squares fit between fine- and coarse-scale class proportions for all cells found within a specified distance of the cluster center. Then again multiple regression was applied, using the mean value of the two structural measures for each cluster as the independent variables. Clusters for which the least-squares fit did not produce a significant regression model were excluded from the multiple regression analysis.

Calibration models were developed for both aggregation levels (8 km, 32 km), using structural measures calculated at the fine scale (1 km) as the independent variables in the multiple regression. Applying the models directly allows for the testing of the validity of the modeling approach; in other words, it allows us to find out if the models, as defined, can successfully explain observed changes in class proportions due to spatial aggregation. To be able to apply the models for the correction of coarse-scale area estimates they must be reformulated so that use is made of coarse resolution data only. This can be accomplished in two ways. Either one can repeat the two-stage modeling strategy described above, using structural measures calculated at the coarse scale instead of their fine-scale equivalent, and derive a new calibration model, based on coarse-resolution data only (Mayaux and Lambin, 1995), or one can estimate structural measures at the fine scale from the coarse-scale measures using simple linear regression and then apply the original calibration model. Both strategies have been tested and did not produce a significantly different output. The results listed below are based on the second approach.

RESULTS

To evaluate the performance of the modeling two types of error were calculated for each cell: the total absolute error for all classes (*TOTABS*), and the maximum absolute error (*MAXERR*):

$$TOTABS = \sum_{i=1}^{k} |P_{Ci} - P_{Fi}| \qquad (3)$$

$$MAXERR = \max_{i=1,...,k} |P_{Ci} - P_{Fi}| \qquad (4)$$

where k is the number of classes, P_{Ci} is the proportion of class i at the coarse resolution and P_{Fi} is the proportion of class i at the fine resolution.

Both measures were averaged for all cells in the validation set. Results are listed in Tables 32.1 and 32.2. Use of the calibration models with structural measures calculated at the fine scale (1 km), indicate an improvement of the average total error by 25% for the 8 km estimation and by 27% for the 32 km estimation. The use of a simple linear correction model, not taking account of the structure of the scene improves the average total error only by 15% (8 km) or 8% (32 km). The average maximum error within a grid cell is improved by 35% for the 8 km scenario and by 44% for the 32 km scenario, against 22% (8 km) and 24% (32 km) for the simple correction model. These results prove that the use of structure-based calibration models will, at least theoretically, lead to much better area estimates for grid-cell sizes used in global scale modeling than the use of simple correction models relating directly proportions at coarse and fine scales.

Applying the calibration using only coarse-resolution data reduces the performance of the models, especially for the linear approach. Although the average maximum error is about the same for both models, the average total error is much higher for the linear model and hardly differs from the result obtained with a simple correction model that takes no account of spatial structure. With the nonlinear model much better results are obtained. The average total error is improved by 23% for the 8 km scenario and by 22% for the 32 km scenario, compared to an improvement of only 15% (8 km) and 9% (32 km) for the linear approach. The average maximum error is improved by 30% (8 km) and 35% (32 km).

DISCUSSION AND CONCLUSIONS

The results of the case study demonstrate that nonlinear, structure-based calibration models, as proposed in this chapter, can substantially improve coarse-scale area estimates derived from aggregated 1 km global land-cover data, even for large areas that show drastic differences in scene characteristics. Calibration mod-

Table 32.1. Results for Different Calibration Models (Total Absolute Error).

	TOTABS (%)		Improvement (%)	
	8 km	32 km	8 km	32 km
Original error	15.5	20.1	—	—
Simple correction model	13.2	18.4	14.5	8.4
Linear calibration (fine scale)	12.4	17.4	20.3	13.1
Nonlinear calibration (fine scale)	11.6	14.6	24.9	27.4
Linear calibration (coarse scale)	13.2	18.2	14.6	9.1
Nonlinear calibration (coarse scale)	12.0	15.6	22.6	22.1

Table 32.2. Results for Different Calibration Models (Maximum Absolute Error).

	MAXERR (%)		Improvement (%)	
	8 km	32 km	8 km	32 km
Original error	6.6	8.3	—	—
Simple correction model	5.2	6.3	21.7	23.7
Linear calibration (fine scale)	4.3	4.9	35.4	41.1
Nonlinear calibration (fine scale)	4.4	4.6	34.4	44.2
Linear calibration (coarse scale)	4.6	5.4	30.2	34.0
Nonlinear calibration (coarse scale)	4.6	5.3	30.3	35.2

eling of this kind gives us an idea of the relative loss of proportional accuracy one may expect to encounter when using aggregated versions of 1 km land-cover data sets instead of the original data for the estimation of class proportions at the grid-cell size of global scale models, and may help to evaluate the potential and the limitations of using aggregated land-cover data sets in global change studies.

The calibration models that have been developed in this study may be further improved in different ways. Although they are based on a large number of samples, covering a broad range of scene characteristics, evaluation of both the 8 km and the 32 km model for individual cells showed that the performance of the model rapidly decreases once the characteristics of a cell are outside the range of the set used for model calibration. In other words, the relationship between the structural measures and the parameters of the transition curve cannot be extrapolated outside the range on which the model has been defined. Simulation of artificial land-cover scenes with known structural characteristics might be an interesting alternative to generalize the model to more extreme situations, although one should be very cautious not to produce scenes that differ from "real" landscape conditions in some aspect not captured by the structural measures used in the modeling,

but influencing observed changes in class proportions in a systematic way.

Better results might be obtained also by including other aspects of spatial structure in the modeling. Experimentation with simulated land-cover scenes in the initial stages of model development indicated that the degree of spatial homogeneity observed in the distribution of different classes within the cell has a substantial impact on the shape of the proportion transition curve. The more the classes are homogeneously spread over the area of the cell, the larger the changes in class proportions due to spatial aggregation will be. So far, this element has not been included in the modeling due to the difficulty of describing this particular aspect of spatial structure by a single measure.

As has already been explained in the introduction, one of the objectives of the present project is to find out if an approach similar to the one described in this chapter could be used to derive reliable coarse-scale area estimates from land-cover maps obtained through classification of spatially degraded versions of 1 km resolution sensor data. A crucial point in developing such a strategy is the definition of an appropriate method for spatial image degradation. This is the subject of ongoing research. Results of the study should give us indications on the possible use and limitations of coarse-scale data sets, especially in connection with the development of global classification strategies, and may also help in deciding if it would be appropriate to produce spatially degraded versions of 1 km VEGETATION data, once these data become available.

ACKNOWLEDGMENT

This research is part of the Belgian Scientific Research Programme on Remote Sensing by Satellite— phase four (Office for Scientific, Technical and Cultural Affairs), under contracts T4/DD/002 and T4/DD/003. The scientific responsibility is assumed by its authors.

REFERENCES

Baker, W.L. and Y. Cai. The Role Programs for Multiscale Analysis of Landscape Structure Using the GRASS Geographical Information System, *Landscape Ecol.,* 7, pp. 291–302, 1992.

Brown, J.F., T.R. Loveland, J.W. Merchant, B.C. Reed, and D.O. Ohlen. Using Multisource Data in Global Land-Cover Characterization: Concepts, Requirements, and Methods, *Photogrammetric Eng. Remote Sensing,* 59, pp. 977–987, 1993.

Canters, F. and F. Veroustraete. *Global Mapping of Land Cover Using Medium Resolution Satellite Data,* unpublished survey report, Belgian Office for Scientific, Technical and Cultural Affairs, Contract T4/DD/002, T4/DD/003, 1997, p. 15.

Cross, A.M., J. Settle, N.A. Drake, and R.T.M. Paivinen. Sub-Pixel Measurement of Tropical Forest Cover Using AVHRR Data, *Int. J. Remote Sensing,* 12, pp. 1119–1129, 1991.

Defries, R.S. and J.R.G. Townshend. NDVI-Derived Land Cover Classifications at a Global Scale, *Int. J. Remote Sensing,* 15, pp. 3567–3586, 1994.

Jiang, B. and W. Kainz. Fuzzy Overlay Analysis with Linguistic Degree Terms, in *Advances in GIS Research II, Proceedings of the Seventh International Symposium on Spatial Data Handling,* M.J. Kraak, and M. Molenaar, Eds., Taylor & Francis, London, 1997, pp. 301–318.

Loveland, T.R., J.W. Merchant, D.O. Ohlen, and J.F. Brown. Development of a Land-Cover Characteristics Database for the Conterminous U.S., *Photogrammetric Eng. Remote Sensing,* 57, pp. 1453–1463, 1991.

Mayaux, P. and E.F. Lambin. Estimation of Tropical Forest Area from Coarse Spatial Resolution Data: A Two-Step Correction Function for Proportional Errors Due to Spatial Aggregation, *Remote Sensing Environ.,* 53, pp. 1–15, 1995.

Moody, A. A Calibration-Based Model for Correcting Area Estimates from Coarse Resolution Land Cover Data, in *Spatial Accuracy Assessment in Natural Resources and Environmental Sciences: Second International Symposium, Fort Collins, Colorado, 1996,* General Technical Report RM-GTR-277, H.T. Mowrer, R.L. Czaplewski, and R.H. Hamre, Eds., U.S. Department of Agriculture, Forest Service, Rocky Mountain Forest and Range Experiment Station, Fort Collins, CO, 1996, pp. 83–90.

Moody, A. and C.E. Woodcock. Scale-Dependent Errors in the Estimation of Land-Cover Proportions: Implications for Global Land-Cover Datasets, *Photogrammetric Eng. Remote Sensing,* 60, pp. 585–594, 1994.

Pech, R.P., A.W. Davis, R.R. Lamacraft, and R.D. Graetz. Calibration of LANDSAT Data for Sparsely Vegetated Semi-Arid Rangelands, *Int. J. Remote Sensing,* 7, pp. 1729–1750, 1986.

Townshend, J.R.G., C.O. Justice, and V. Kalb. Characterization and Classification of South American Land Cover Types Using Satellite Data, *Int. J. Remote Sensing,* 8, pp. 1189–1207, 1987.

Tucker, C.J., J.R.G. Townshend, and T.E. Goff. African Land-Cover Classification Using Satellite Data, *Science,* 227, pp. 369–375, 1985.

Turner, M., R. O'Neill, R. Gardner, and B. Milne. Effects of Changing Spatial Scale on the Analysis of Landscape Pattern, *Landscape Ecol.,* 3, pp. 153–162, 1989.

Zhu, Z. and D.L. Evans. Mapping Midsouth Forest Distribution, *J. For.,* 90, pp. 27–30, 1992.

Part VII
Decreasing Spatial Uncertainty

"Everyone complains about spatial uncertainty but no one ever does anything about it."

While most of the papers and presentations at the conference dealt with characterizing, describing, representing, and modeling uncertainty, there are, in fact some practitioners who are examining techniques for decreasing or minimizing spatial uncertainty. The first chapter is the only one to concern itself with categorical data; Thierry and Lowell describe and evaluate a method of aerial photointerpretation that does not require closed polygons, but instead requires that a photointerpreter only identify features of "100% certainty." The second chapter, by Duh and Brown, describes a procedure for passing a spatial filter over a digital elevation model to detect and smooth anomalous areas. The final two chapters are concerned with the improvement of estimates for environmental variables. Viau and Huang examine the linkage of satellite data with ground and point-based weather station information to improve estimates of air temperature. Goovaerts presents a simulation method for studying soil contaminated by cadmium in which the constraining of the simulation to reproduce the histogram and variogram of point data causes local error to be minimized.

CHAPTER 33

A New Method of Photointerpretation to Increase the Overall Reliability of Forest Maps

B. Thierry and K.E. Lowell

INTRODUCTION

Background and Context

In commercial forest management, industry relies heavily on thematic maps to manage their resources. These maps are constructed from aerial photographs which have been processed subjectively through human photointerpretation. However, due to the subjectivity of the photointerpretation process, the resulting map may have high level of associated uncertainty—a problem made even more acute in the case of forest mapping because of the gradual transitions that may actually exist between different forest types on the ground (rather than the abrupt transitions mapped). In fact, the uncertainty in question results primarily from the subjective nature of the interpretation method, which in turn depends mainly on the judgment, the experience, the reasoning, etc. of the human operator—factors that can hardly be verified and/or controlled. For example, if we give the same aerial photograph to two different photointerpreters, they will produce two different maps—both of which can be considered to be completely acceptable. The same has also been found to be true when the same interpreter completes two interpretations of the same photograph at different times (Edwards and Lowell, 1996). These problems of map uncertainty have serious repercussions for forestry since, in managing forested areas, one utilizes maps that have been produced by this subjective photointerpretation process.

Consequently, in order to increase the reliability of the forest mapping process, we need to improve the photointerpretation process by removing—or at least

by significantly decreasing—the uncertainty induced by the human operator. In a previous study, Aubert et al. (1994; see also Aubert, 1995) developed a method based on "fuzzy polygons" (i.e., polygons with buffered boundaries) to address this type of uncertainty. The major drawback of the method is the need for a minimum of three different interpretations in order to estimate the width of the fuzzy boundaries. What if comparable results can be obtained with just a single photointerpretation?

When one examines the photointerpretation process more closely, one sees that it can be broken down into two stages (Lowell, personal communication). First, the interpreter identifies areas for which there is no doubt (i.e., "100% certainty areas"), and then, second, refines this initial rough photointerpretation by dealing with the zones left blank—i.e., those whose classification seems more doubtful. It is reasonable to believe that most of the uncertainty which taints the photointerpretation process comes from this latter stage. Indeed because all photointerpreters are forced to categorize the entire surface of the photograph (rather than identifying only "100% certainty areas"), they tend to extrapolate to the entire image those characteristics (e.g., homogenous textures, or distinct boundaries between two different textures) that are identified unhesitatingly. This further implies uneven levels of reliability across the map produced, whereas the end-user assumes a constant error throughout (Lowell, 1995), thus preventing a fully efficient utilization of the map.

In order to address the problem of subjectivity, the solution we propose is an alternative photointerpretation which consists of two phases. First, the photo-interpreter is asked to classify, on an image, only that which can be identified "without error." This is equivalent to stopping the interpretation at the end of the first stage and provides for a partial, but reliable interpretation. Second, using conventional geomatic software, the rest of the map can be interpolated from the already-classified zones and an uncertainty estimate can be calculated for each point on the completed map. These uncertainty values can then be used to produce a final thematic map in much the same way that a satellite image is classified. It should be noted that the resultant map that the end-user handles looks like a map produced by a more classical photointerpretation—i.e., it is composed of a series of closed polygons. However, an additional companion map is also produced that indicates the reliability of its features.

The purpose of this chapter is to present and evaluate a new method of photointerpretation designed to address the problem of uncertainty in forest mapping. We will first present and explain the methodology, and then will present the results we have achieved thus far.

METHODOLOGY

As mentioned above, our new method of photointerpretation is based upon the assumption that the photointerpretation process can be broken down into two stages. First, the photointerpreter identifies the "100% certainty" areas. Second, he extends the interpretation to the rest of the image which has been left blank. It is this latter stage that is primarily responsible for the uncertainty associated with the final map. We propose to stop the photointerpretation process just before the second stage and, with the aid of a spatial interpolation algorithm, to treat the partial interpretation (i.e., the "100% certainty" areas) in order to achieve a completed map along with an indication of its reliability over the entire map surface.

The Data

In this study, instead of using actual aerial photographs, we utilize a synthetic image (see Figure 33.1) to be photointerpreted since this allows us to know the "ground-truth" corresponding to the image. The synthetic image was constructed from a combination of a polygon mask and a set of textures. [See Edwards and Lowell (1996) for details concerning the construction of this image.] The resultant black-and-white image

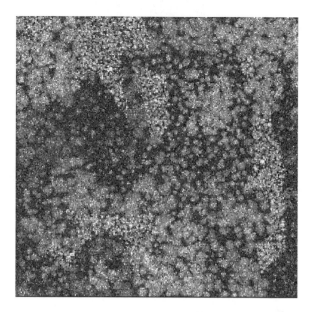

Figure 33.1. The synthetic image used in this study.

resembles a forest enough that the photointerpretation process on real aerial photographs should not be significantly different from that examined in this study.

The polygon mask (i.e., the ground-truth) used to create the synthetic image is also available, as well as the nine classical interpretations (i.e., the entire image is interpreted) produced as part of Aubert's original study.

Image Interpretation

In the present study, the synthetic image was photointerpreted by five different people. These were all students, four of whom had no prior photointerpretation experience. In accordance with the alternative interpretation principles presented in the previous sections, the photointerpreters were asked to categorize only the areas which they were able to identify unhesitatingly and to leave the rest of the image surface blank. They were also allowed to draw portions of lines where they were able to discriminate "without doubt" boundaries between two textures. The resultant interpretations (surfaces and lines of 100% certainty) were then harmonized with respect to each other in order to ensure the same map category for a given "100% certain" area over all interpretations. This harmonization was consistent with the foregoing work of Aubert (1995) and permitted comparisons between alternative and classical photointerpretations.

The final result of each interpretation is a rasterized image, with 399 rows and 399 columns on which only areas and lines that are "100% certain" are present (Figure 33.2).

Certainty Interpolation

The algorithm used to interpolate the partial interpretations made by the five interpreters to "fill in the gaps" between "100% certain" areas is based on the *Area Stealing Algorithm* (Gold, 1992). Briefly, it requires two sets of points: the first is the 100% certainty points from the new method of interpretation, and the second is a regular grid of sample points over the entire image. The algorithm estimates the probability that a sample point belongs to one particular class. To estimate that probability, the program builds a Voronoi diagram around the first set of points, and places in this diagram one of the sample points from the second set and reproduces the Voronoi diagram. The system keeps track of how much area of the sample point (Thiessen) polygon was stolen from polygons of each class from the first Voronoi diagram, and uses these values as the probability that the sample point belongs to a given class (see Figure 33.3 for an illustration of the method). This process continues until all the points in the second set have been treated. If the sample points in the second set are, in fact, each pixel in the final image, the result is a surface showing the probability of an area belonging to any given class

From the resulting surface, we produced thematic maps using three different levels of uncertainty. The first one, the "100% certainty map," is composed solely of the 100% certainty features (areas and lines), i.e., the rough photointerpretation before interpolation. The second map, the "75% certainty map," is produced using only those points having a probability for any class greater than or equal to 75%. The third one, the "50% certainty map," is made of points having a probability for any class greater than or equal to 50%. It should be noted that this last map covers almost—but not all—the surface; on average, only about 5% of the surface is left blank.

RESULTS

We conducted three types of analysis on the three certainty maps produced. First, we studied the consistency of interpretation among the maps; i.e., whether or not two (or more) interpreters identified the same class at the same place. Secondly, we studied the correctness of the interpretations; i.e., whether or not an interpreted class corresponded to the same ground-truth

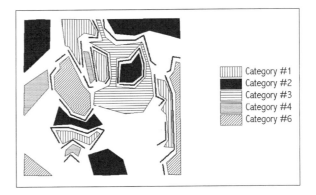
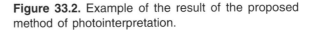

Figure 33.2. Example of the result of the proposed method of photointerpretation.

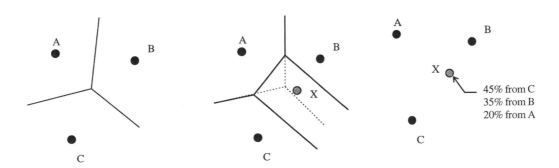

Figure 33.3. Illustration of the Area Stealing Algorithm. Left, the Voronoi diagram around the "100% certain" points; center, the new Voronoi diagram induced by one sample point added; right, the resulting probability of belonging to a given class for the sample point.

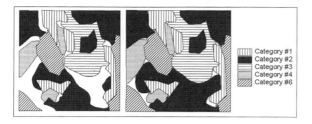

		Category #1
		Category #2
		Category #3
		Category #4
		Category #6

Figure 33.4. Example of 75% certainty (left) and 50% certainty (right) maps produced from the interpretation shown in Figure 33.2.

Table 33.1. Consistency of 100%, 75%, and 50% Certainty Maps. (Values are the mean of all possible combinations.)

	100% Certainty Maps		75% Certainty Maps		50% Certainty Maps	
	Consistent	Inconsistent	Consistent	Inconsistent	Consistent	Inconsistent
2-Map combination	24.3%	2.3%	45.4%	14.3%	56.5%	33.3%
3-Map combination	18.6%	27.0%	36.3%	46.7%	45.1%	52.9%
4-Map combination	15.6%	43.0%	30.7%	59.7%	37.8%	60.5%
5-Map combination	13.8%	53.2%	26.5%	69.0%	32.0%	66.5%

class at the same place. Thirdly, we analyzed the maps resulting from both the alternative and the classical photointerpretation in order to evaluate the maps resulting from the alternative one compared to the classical one.

Consistency of the Maps

We compared each possible combination of 2, 3, 4, and 5 maps and recorded the number of consistent and inconsistent pixels found throughout each combination. A pixel was said to be consistent for a given combination when all maps of that combination give the same class for that pixel. Since blank pixels are not considered in this evaluation, a particular case of consistency arises when all classes but one are blank. Table 33.1 shows the percentage of consistent and inconsistent pixels in the interpolated maps.

One should note that the sum of consistent and inconsistent pixels is not 100% because of the zones which remained blank in the maps over all interpretations; these blank pixels have not been included in the analysis.

Although the very low percentage of inconsistent pixels resulting from the 2-map combination of 100% certainty maps seems to indicate that the interpreters classified almost all of the areas interpreted in the same way, the marked increase in inconsistent pixels in the next three combinations—i.e., 3-, 4-, and 5-map—refutes that conclusion: the interpreters do not produce a highly consistent classification of the image. Instead,

one can explain this low percentage of inconsistent pixels (and the subsequent increases) by the fact that the interpreters did not generally classify the same zones in the image. That is, the classified zones in one map correspond, at least partially, to blank ones on another map. Of course, when the number of maps included in a combination increases, the probability that a classified zone remains unclassified in all the other maps decreases greatly. This explanation is also valid for the increasing inconsistency when one considers the three maps resulting from the selected certainty levels: the surface which remains blank in a 50% certainty map is much lower than that left blank in a 100% certainty map.

Correctness of the Maps

The actual correctness of the interpretations can be determined since we utilized synthetic images generated from a known ground-truth. Accordingly, the maps constructed from the new method of image interpretation were compared with this polygon mask that was used to create that image. We derived from that comparison the confusion matrixes shown in Tables 33.2, 33.3, and 33.4.

In these tables, the term "Null" indicates those pixels left blank in the map produced after the interpolation. In "% Correct," statistics are calculated only for the classified areas and blank pixels are not taken into account. Conversely "% Strictly Correct" accounts for

Table 33.2. Correctness of the 100% Certainty Maps. (Values in each cell are the mean number of pixels over all five interpreters.)

Category	Truth (Mask)						
	1	2	3	4	5	6	Total
Null	8963	17409	10548	10832	24957	7050	79759
1	14941	49	167	253	117	242	15770
2	1280	15706	308	341	5191	220	23046
3	253	1166	13584	31	1864	1	16898
4	72	11	3	3612	684	312	4694
5	7	479	20	18	512	0	1036
6	126	246	385	611	194	14753	16315
Other	541	94	44	981	23	0	1683
Totals	26183	35160	25058	16680	33542	22578	159201
% Correct	86.8%	88.5%	93.6%	61.8%	6.0%	95.0%	69.8%
% Strictly correct	57.1%	44.7%	54.2%	21.7%	1.5%	65.3%	39.6%

Table 33.3. Correctness of the 75% Certainty Maps. (Values in each cell are the mean number of pixels over all five interpreters.)

Category	Truth (Mask)						
	1	2	3	4	5	6	Total
Null	3463	8963	3143	5189	14298	1724	36780
1	18781	675	1009	956	1266	577	23263
2	2027	21733	457	884	7432	516	33048
3	448	1640	17917	48	4003	1	24058
4	354	536	634	7247	2226	1252	12249
5	29	982	94	448	1224	0	2777
6	557	479	1449	1844	2908	18435	25671
Other	523	153	356	64	185	73	1355
Totals	26183	35160	25058	16680	33542	22578	159201
% Correct	82.7%	83.0%	81.8%	63.1%	6.4%	88.4%	65.3%
% Strictly correct	71.7%	61.8%	71.5%	43.4%	3.6%	81.6%	53.6%

Table 33.4. Correctness of the 50% Certainty Maps. (Values in each cell are the mean number of pixels over all five interpreters.)

Category	Truth (Mask)						
	1	2	3	4	5	6	Total
Null	776	1737	337	1525	3889	370	8635
1	20337	2639	1193	1706	2928	752	29556
2	2374	24058	544	1638	10164	582	39359
3	722	2213	19519	98	6388	1	28943
4	576	1333	682	8242	3110	1588	15531
5	58	1768	207	545	2224	0	4803
6	812	1228	1972	2758	4376	19078	30224
Other	527	185	603	167	462	207	2150
Totals	26183	35160	25058	16680	33542	22578	159201
% Correct	80.0%	72.0%	79.0%	54.4%	7.5%	85.9%	60.9%
% Strictly correct	77.7%	68.4%	77.9%	49.4%	6.6%	84.5%	58.7%

both the consistent and blank pixels. In fact, this latter amounts to counting blank pixels as ones being incorrectly classified.

One can see from the above tables that the new method of interpretation and the associated certainty interpolation does not decrease excessively the percentage of pixels correctly classified although the interpolation affects on average about one half of the map surfaces: the "% Correct" percentage falls only from 69.8% for the "100% certainty" map to 60.9% for the "50% certainty" map. One can also see the relatively low percentage of "% Correct" for Category 5; this is clearly the category that was classified with the most difficulty. Indeed, only one interpreter identified it at all, whereas it is one of the most widespread categories on the ground-truth map (with about 20% of the total surface).

It should also be noted that one of the alternative interpretations tended to be more like a classical one. Indeed, only 10% of the entire surface was left blank (compared to 50% to 75% for the remaining interpretations). This single interpretation tended to cause a decrease in the "% Correct" percentage in Tables 33.2, 33.3, and 33.4.

In general, the quality of the certainty maps resulting from the certainty interpolation seems to be tied to the quality of the alternative interpretation: the more the 100% certainty areas are scattered evenly across the image, the better the resulting map. In the same way, large areas of "100% certainty" do not necessarily give better interpolated maps. One should also note that the interpolation may fail to produce pertinent results when it encounters some special cases. For example, a certainty area with a "U" shape and no other 100% certainty feature "within" the "U" shape. Conversely, interpretations comprised principally of compactly shaped areas seem to provide better results.

Comparison of the Alternative and Classical Interpretations

The complete results for the nine classical interpretations—i.e., the full confusion matrixes—can be found in Edwards and Lowell (1996). In this section, we will present only the statistics required to compare the results of the alternative and classical interpretations. Recall that the classical interpretations were produced by nine different interpreters (not the same ones as those who made the alternative interpretations), and, as the classical interpretation method requires, the whole surface of the image was classified.

Table 33.5. Consistency of Maps Produced by the Classical Interpretation Method. (Values are the mean of all possible combinations.)

	Consistent	Inconsistent
2-Map combination	60.6%	39.4%
3-Map combination	45.9%	54.1%
4-Map combination	37.7%	62.3%
5-Map combination	32.3%	67.7%

In order to determine the consistency of the classical interpretations, we compared all possible combinations of two to five maps of the nine produced by the classical interpretation method. Table 33.5 presents the results obtained for these combinations. A pixel is said to be "consistent" when its category is the same for all the maps of the combination; a single pixel of one map belonging to a different category renders that pixel "inconsistent." Since the entire surface of the map is classified, all the pixels belong either to the Consistent or Inconsistent category.

When one compares the results in Table 33.1 with those in Table 33.5, the results attained with the combinations of 50% certainty maps and those of classical ones show very little difference: except for the 2-map combination, the differences between both results are within 1% to 2%.

Relative to ground-truth for the classical interpretation method, Table 33.6 sums up the confusion matrix of the nine classical maps (the table shows the mean values over the nine interpretations). The "% Correct" term that appears in Table 33.6 represents the mean percentage of pixels correct (relative to the ground-truth) for each class over all nine interpretations. Since each classical map is entirely classified, the "% Strictly correct" term found in Tables 33.2, 33.3, and 33.4 is not applicable to Table 33.6.

Once again, the results obtained, respectively, by the 50% certainty maps and the classical ones are very similar. Indeed, the percentage of correct pixels (the "% Correct" term in Table 33.4) reaches an amount of 60.9% for the 50% certainty maps, while the classical maps sustain an average of 60.6% of correctly classified pixels. The most noticeable exception to the similarity between the 50% certainty maps and the classical ones comes from the results for Category 5, which was found to be present by eight classical interpreters, whereas only one of the alternative ones found it.

The main remark that arises when one draws a comparison between the alternative and classical interpre-

Table 33.6. Correctness of Classical Maps. (Values in each cell are the mean of the correctly classified pixels for each of the nine interpreters.)

Category	Truth (Mask)						
	1	**2**	**3**	**4**	**5**	**6**	**Total**
% Correct	72.9%	56.8%	73.2%	62.0%	39.9%	68.1%	60.6%

tations is the similarity of maps judged to be comparable (i.e., the 50% certainty maps and the classical ones) in terms of statistical consistency and correctness. Operationally, if the results attained with a classical interpretation are slightly better than those achieved with the alternative one, it is the latter that is preferable due to a decreased time for interpretation. Consequently, in regard to quality attained relative to effort required, the alternative interpretation appears to be more efficient.

CONCLUSION

The proposed alternative method of photointerpretation described relies on the following hypothesis: the photointerpretation process is a two-stage process that can be broken into, first, a partial but reliable interpretation of the image and, second, an implicit extrapolation by the human operator of the known features to the rest of the image. This latter stage is the one that is primarily responsible for the uncertainty inherent in the map production process. To avoid this problem, we developed an alternative photointerpretation process which leaves the first stage unchanged (the human operator still interprets the areas he identifies unhesitatingly) but entrusts an algorithm with the second stage (the interpolation stage). This way of conducting photointerpretation allows us to evaluate the level of certainty of the produced maps.

In the present study we carried out two types of experiments, one to determine the consistency among the alternative interpretations, and the other to check their correctness against ground-truth. We also compared the results attained from the alternative interpretations with those obtained from the classical interpretations produced in a previous study. The results revealed two main conclusions. First, the alternative interpretations are not as consistent as was hoped, i.e., the areas categorized by different interpreters did not always belong to the same class. Sec-

ond, alternative and classical interpretations produced very similar maps in terms of consistency and correctness, although the classical maps gave slightly better results. Nevertheless, when we compare the time spent to produce a certainty map with the time required to produce a classical one, the operational advantage of the alternative interpretation becomes apparent.

ACKNOWLEDGMENTS

The research for this project was conducted within the context of the Chaire industrielle en géomatique appliquée à la foresterie and the Centre de Recherche en Géomatique (CRG) of Laval University. The authors are grateful to the Association of Quebec Forest Industries and the Natural Sciences and Engineering Research Council for financing this work.

REFERENCES

Aubert, E. Quantifying the Spatial Uncertainty in Forest Photo-Interpretation Using a GIS for the Spatial-Temporal Monitoring of Forest Stands (in French), thesis presented to the Université Laval, Québec City, Canada in partial fulfillment of the requirements for a Master's degree, 1995.

Aubert, E., G. Edwards, and K.E. Lowell. Quantifying Boundary Errors in Forest Photo-Interpretation for the Spatial-Temporal Tracking of Forest Stands (in French), *Proceedings of the Canadian Conference on GIS*, Ottawa, Canada, 1994, pp. 195–205.

Edwards, G. and K.E. Lowell. Modeling Uncertainty in Photo-Interpreted Boundaries, *Photogrammetric Eng. Remote Sensing*, 62(4), pp. 337–391, 1996.

Gold, C.M. Surface Interpolation as a Voronoi Spatial Adjacency Problem, *Proceedings of the Canadian Conference on GIS*, Ottawa, Canada, 1992, pp. 419–431.

Lowell, K.E. A Fuzzy Surface Cartographic Representation for Forestry Based on Voronoi Diagram Area Stealing, *Can. J. For. Res.*, 24, pp. 1970–1980, 1995.

CHAPTER 34

Local Reduction of Systematic Error in 7-1/2 Minute DEMs by Detecting Anisotropy in Derivative Surfaces

J.-D. Duh and D.G. Brown

INTRODUCTION

Using semivariograms and fractal dimensions to examine the presence of anisotropy, systematic errors can be detected in digital elevation data obtained from the U.S. Geological Survey (USGS) as 7-1/2-minute quadrangle maps (Polidori et al., 1991; Brown, 1994). Anisotropy, caused by systematic errors of DEMs, is more noticeable in the curvature surfaces than in the elevation and slope surfaces. Anisotropy is almost insignificant in elevation surfaces, but as systematic errors can be increased through the calculation of derivative surfaces so too does the observed anisotropy increase (Brown, 1994). Brown also pointed out that the use of a 3 by 5 low-pass filter will reduce the systematic errors and eliminate anisotropy effectively in 7-1/2 DEMs.

Other researchers have asserted that smoothing results in a decrease in the spatial variability of elevation (Vieux, 1993), i.e., information loss. Smoothing may also alter elevation values that were actually correct, and thus propagate error when these surfaces are applied to hydrological modeling (Vieux, 1993).

In this study, we tried to minimize the alteration of elevation values in DEMs and reduce the systematic error to an acceptable level by using an anisotropy-detection filter which identifies the anisotropic areas on slope and curvature surfaces. Then, the 3 by 5 LPF is applied only to identified areas in the DEMs. In a second pass, a 3 by 3 LPF smoothes the boundaries of areas which were smoothed by a 3 by 5 LPF to prevent the discontinuity that results from altering the elevation values.

After describing the procedure in more detail, we evaluate the nature of the trade-off between decreasing anisotropy and increased information loss that occur with an increasing proportion of the DEM being filtered. We provide some general guidelines for the use of the procedure.

STUDY AREA AND DATA

The study area is the same area as that used in Brown and Bara's study (1994). It is a portion of Glacier National Park (GNP) in northwestern Montana. The elevations of this rugged landscape range from 1,120 to 3,000 meters. The study area encompasses parts of four 7-1/2-minute quadrangles, including Logan Pass, Many Glacier, Lake Sherburne, and Rising Sun. Each of the DEMs was produced through the TRASTER (manual profiling) procedure at the Rocky Mountain regional office of the USGS (U.S.G.S, 1987) (Brown, 1994). The derivative slope angle and slope curvature surfaces of the unfiltered elevation surface show prominent banding errors (Figure 34.1).

METHODS

In this study, we used PCI, ARC/INFO, and a FORTRAN program written by Brown to independently calculate semivariance as a function of lag distance in the two cardinal directions of all the surfaces. Semivariance, $\gamma(h)$, in each direction is calculated by

$$\gamma(h) = \frac{1}{2(N-h)} \sum_{i=1}^{N-h} (z(i) - z(i+h))^2 \qquad (1)$$

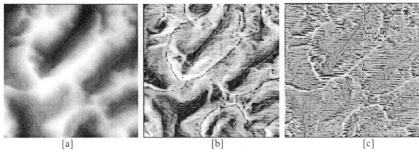

[a]   [b]   [c]

Figure 34.1. The unfiltered elevation and its derivative surfaces of study area: [a] elevation, [b] slope angle, [c] slope curvature.

where N is the number of points on the surface, $z(i)$ is the value of the surface at any point i, and $z(i+h)$ is the value of surface at a distance h units from i.

The following procedures had been established and programmed in an AML script (Appendix) to reduce the systematic errors in 7-1/2-minute DEMs:

1. *Calculating the derivative surfaces of elevation.*

   Slope angles and slope curvatures were calculated using the CURVATURE function in ARC/INFO GRID module. The function calculates the profile curvature, plan curvature, tangential curvature, slope angle, and slope aspect at each cell center from input elevation data. Tangential curvatures are the slope curvatures used for further analysis. Curvature units are 1/100 Z unit. The values of slope angle are in units of degrees and range from 0 to 90.

2. *Detecting the anisotropy in slope and curvature surfaces.*

   Considering that anisotropy is present when the general pattern of variation in one direction (e.g., north to south) is different from the pattern of variation in another direction (e.g., east to west) (Brown, 1994), the anisotropy-detection filter used the following 5 by 5 convolution mask which calculated the difference between the patterns in two cardinal directions:

   Anisotropy - det ection filter =

0	0	0.125	0	0
0	0	0.125	0	0
−0.125	−0.125	0	−0.125	−0.125
0	0	0.125	0	0
0	0	0.125	0	0

The filter was applied to both slope angle and slope curvature surfaces and detects anisotropy only when the pattern of error is parallel to one of the cardinal axes.

3. *Using predefined thresholds to identify anisotropic areas.*

   Thresholds were used to define the magnitude of anisotropy, defined by the above filter, to be treated as systematic error. In this study, thresholds were expressed in terms of the number of standard deviations from the mean of the anisotropic filter surfaces of both slope angle and curvature. Areas with anisotropy values outside the bounds of thresholds were identified as anisotropic areas. In the AML script, the threshold values were defaulted to 1 standard deviation for both anisotropic surfaces derived from slope angle and curvature.

4. *Sieving out the one-pixel areas and the vertical lines with one-pixel width.*

   With the understanding that the systematic errors were horizontal stripes resulting from the manual profiling process, we sieved out the one-pixel anisotropy areas and the vertical lines (i.e., N-S) with one-pixel width from the anisotropic areas identified both from slope and curvature surfaces. Those sieved areas might be the peaks, sinks, or ridges on terrain. Ridges or valleys that were parallel to the horizontal direction were detected as stripes as well, yet they are not common in the DEMs used.

5. *Creating anisotropic areas for the reduction of anisotropy.*

   The anisotropic areas identified from both slope and curvature surfaces were then merged to identify the areas that contained systematic errors in the DEMs (Figure 34.2).

6. *Applying a 3 by 5 low-pass filter to elevation values within anisotropic areas.*

   The elevation values within the anisotropic areas that were defined from previous steps were replaced with the values averaged by a 3 by 5 LPF while other elevation values were unaltered.

7. *Creating edge zones for each anisotropic area.*

   To prevent discontinuity resulting from altering only some of the elevation values, the edges of those altered areas, i.e., anisotropic areas, needed smoothing. We identified the edge cells by merging the ends of line segments in each anisotropy area which were detected vertically and horizontally (Figure 34.3).

8. *Applying a 3 by 3 low-pass filter to elevation values on the edge zones.*

   As a second pass, a 3 by 3 LPF smoothed elevation values along the edge zones detected from the previous step.

We used semivariograms and central tendency measures of the difference surfaces between filtered and unfiltered elevations to evaluate the performance (i.e., the amount of systematic errors reduced and the potential amount of information loss) of the local smoothing algorithm proposed in this study. Those measures were calculated from a series of derivative surfaces which were derived from elevations that were smoothed with different anisotropic thresholds. Because cell values greater than the thresholds were identified as anisotropic areas, the larger the thresholds were, the fewer cells were identified as anisotropic (Table 34.1, Figure 34.4).

FINDINGS AND RESULTS

By applying the anisotropic filter to the slope angle and curvature surfaces, we found that systematic errors in DEMs became significant at high elevations and steep slope areas (Figure 34.5), which supports the notion that the errors were induced by the profiling process over hills. By visual judgment, the derived slope and curvature surfaces from both globally and locally smoothed elevation surfaces had no prominent stripes (Figures 34.6 and 34.7). The locally smoothed surfaces still had some stripes in very low contrast areas if examined carefully, but they retained more subtle features in terrain than the derivative sur-

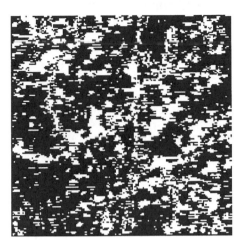

Figure 34.2. The anisotropy mask (white) for areas to be filtered by a 3 X 5 LPF.

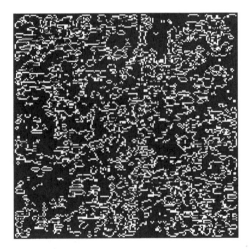

Figure 34.3. The edges (white and gray) of anisotropy areas to be filtered by a 3 X 3 LPF.

faces derived from the globally smoothed elevations. In general, the locally smoothed images give an impression of having higher resolution than the globally smoothed images.

The local smoothing method reduced the anisotropy in the elevation and derivative surfaces to almost the same degree as the global smoothing method. The difference in the semivariograms that were calculated in two cardinal directions (i.e., N/S and E/W) from a surface was used as an indicator of anisotropy. We found that anisotropy can be detected more obviously from the semivariogram of derivative surfaces (i.e., slope angle and slope curvature) than elevation (Fig-

Table 34.1. Percentage of Cells in Which Elevation Values were Altered by Threshold.[a]

Threshold	Altered by 3x5	Altered by 3x3	Total Altered %
>>2.0	0.0	0.0	0.0
2.0	7.7	9.1	16.8
1.8	10.0	11.3	21.3
1.6	13.0	13.7	26.7
1.4	7.2	16.2	33.5
1.2	22.8	19.2	42.0
1.0	29.8	22.4	52.2
0.8	39.7	24.8	64.6
0.6	52.5	25.3	77.8
0.4	68.2	21.8	90.0
0.2	86.2	11.2	97.4
0.0	96.8	1.6	98.4

[a] There are some lakes in the study area that have no variations in elevation.

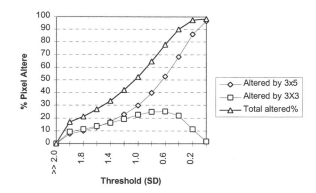

Figure 34.4. Relationship between percentage of pixels altered and the value of threshold.

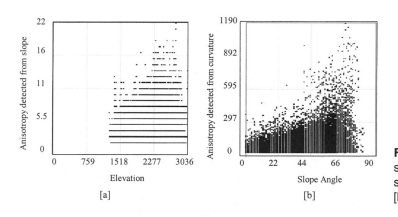

Figure 34.5. Scatter plots of anisotropic surfaces against elevation and slope angle surfaces: [a] slope anisotropy—elevation, [b] curvature anisotropy—slope.

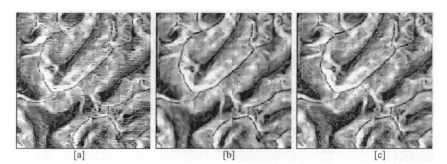

Figure 34.6. Slope angle surfaces derived from elevation surfaces that are: [a] unfiltered, [b] globally smoothed by 3 by 5 LPF, and [c] locally smoothed by 3 by 5 LPF with threshold 1.0.

ures 34.8a, 34.9a, and 34.10a). The global smoothing method reduced the anisotropy in slope angle and slope curvature (Figures 34.9b and 34.10b). The local smoothing method also reduces anisotropy, but not as completely as the global smoothing method (Figures 34.9c and 34.10c).

To observe the potential amount of information loss between the global and local smoothing methods, statistical summaries of the difference and absolute difference between filtered and unfiltered elevation surfaces were calculated. These showed that both methods produced minor effects. The mean differences were

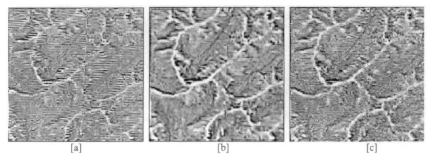

Figure 34.7. Slope curvature surfaces derived from elevation surfaces that are: [a] unfiltered, [b] globally smoothed by 3 by 5 LPF, and [c] locally smoothed by 3 by 5 LPF with threshold 1.0.

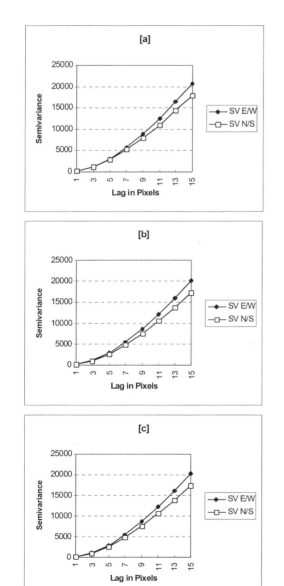

Figure 34.8. Semivariograms of elevation surfaces that are [a] unfiltered, [b] globally smoothed by 3 X 5 LPF, [c] locally smoothed by 3 X 5 LPF with threshold 1.0.

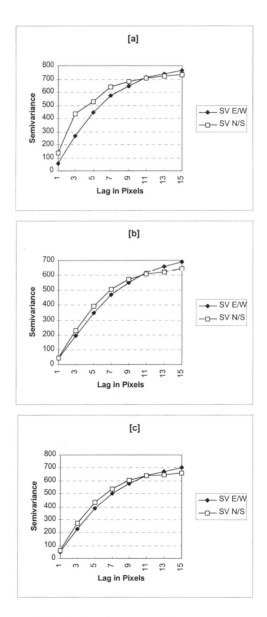

Figure 34.9. Semivariograms of slope angle surfaces that are derived from elevation surfaces that are [a] unfiltered, [b] globally smoothed by 3 X 5 LPF, [c] locally smoothed by 3 X 5 LPF with threshold 1.0.

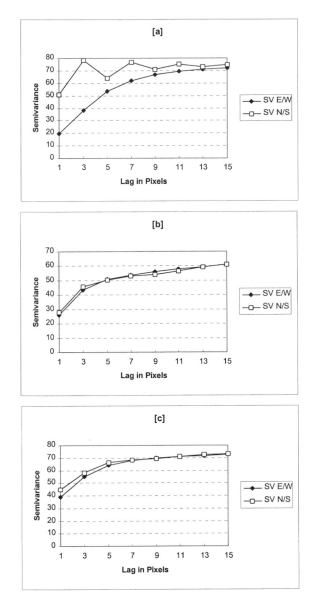

Figure 34.10. Semivariograms of curvature surfaces that are derived from elevation surfaces that are [a] unfiltered, [b] globally smoothed by 3 X 5 LPF, [c] locally smoothed by 3 X 5 LPF with threshold 1.0.

Table 34.2. Mean Values of the Difference and Absolute Difference Surfaces Between Unfiltered and Filtered Elevation Surfaces.

% Pixels Altered	Mean of Diff. Surface	Mean of Abs. Diff. Surface
0	0.00	0.00
17	0.27	1.56
21	0.29	1.81
27	0.30	2.08
33	0.31	2.38
42	0.29	2.73
52	0.27	3.08
65	0.22	3.46
78	0.15	3.83
90	0.09	4.13
97	0.04	4.29
98	0.04	4.45

anisotropy reduced from elevation surfaces and found that the more cells altered by the 3 X 5 LPF, the more anisotropy will be reduced (Figures 34.11 and 34.12) and the more the information loss will be (Figure 34.13). Bias in the elevation values, measured by the mean of the difference surface, resulting from smoothing was always less than one-half the vertical resolution of the DEM (i.e., < 0.5 m). We noticed a significant amount of reduction of anisotropy with a small number of cells altered; i.e., the decrease of anisotropy is exponentially related to the number of altered cells (Figure 34.14), while the potential amount of information loss increased linearly with the number of altered cells measured by the mean of the absolute difference surface (Figure 34.13). This implies that a higher standard deviation threshold for anisotropy (i.e., fewer cells) in the local smoothing method is favorable to achieve an effective reduction of anisotropy and cause less information loss.

CONCLUSIONS

In this study, we found that by using local smoothing, we could reduce the systematic error caused by the manual profiling process in the 7-1/2-minute DEMs as effectively as we could with global smoothing, as proposed in a previous study (Brown and Bara, 1994). At the same time, local smoothing results in less information loss from the DEMs. By detecting local anisotropy in the derivative surfaces of DEMs, the spatial distribution of the systematic errors and its correlation with elevation and slope angle could be dem-

4.45 and 3.08 (with threshold 1.0) meters for the global and local method, respectively (Table 34.2). This implies that the local smoothing method resulted in less information loss than the global method. This should be expected because fewer cells in the DEMs had been smoothed by local method. We calculated the sum of differences of semivariances of the N/S and E/W directions as an indicator for the amount of

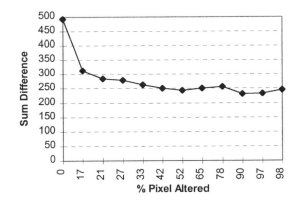

Figure 34.11. Relationship between the amount of anisotropy reduced (i.e., sum of absolute semivariance differences) and % of pixels altered for slope angle surface.

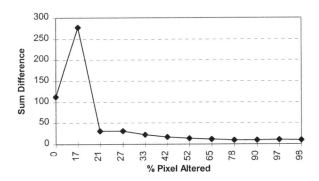

Figure 34.12. Relationship between the amount of anisotropy reduced (i.e., sum of absolute semivariance differences) and % of pixels altered for curvature surface.

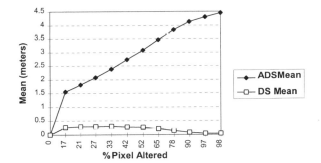

Figure 34.13. Relationship between the potential amount of information loss [mean of difference (DS) and absolute difference (ADS) surfaces] and % of pixels altered for elevation.

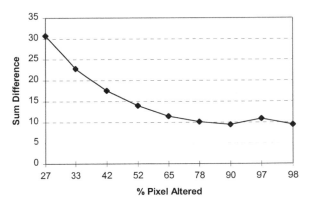

Figure 34.14. Relationship between the amount of anisotropy reduced (i.e., sum of absolute semivariance differences) and % of pixels altered for curvature surface (closeup view).

onstrated. We also suggest that thresholds used to identify anisotropic areas could be set to high values, i.e., only a small amount of cell values will be altered. This will produce an effective reduction of systematic error and less information loss.

REFERENCES

Brown, D.G. and T.J. Bara. Recognition and Reduction of Systematic Error in Elevation and Derivative Surfaces from 7-1/2 Minute DEMs, *Photogrammetric Eng. Remote Sensing,* 60(2), pp. 189–194, 1994.

Brown, D.G. Anisotropy in Derivative Surfaces as an Indication of Systematic Errors in DEMs, *Proceedings: International Symposium on Spatial Accuracy of Natural Resources Data Bases,* Williamsburg, 1994, pp. 98–107.

Polidori, L., J. Chorowicz, and R. Guillande. Description of Terrain as a Fractal Surface, and Application to Digital Elevation Quality Assessment, *Photogrammetric Eng. Remote Sensing,* 57(10), pp. 1329–1332, 1991.

U.S.G.S. *Digital Elevation Models, Users Guide 5,* U.S. Geological Survey, Reston, VA, 1987.

Vieux, B.E. DEM Aggregation and Smoothing Effects on Surface Runoff Modeling, *J. Comput. Civil Eng.,* 7(3), pp. 310–338, 1993.

APPENDIX

File 1: aniso.knl (kernel definition of anisotropy detection filter used by deband.aml)

```
5 5
0  0  0.125  0  0
0  0  0.125  0  0
-0.125  -0.125  0  -0.125  -0.125
0  0  0.125  0  0
0  0  0.125  0  0
```

File 2: deband.aml (ARC/INFO AML script for reducing systematic errors in 7-1/2 minute DEMs)

```
/***********************************************************************
/*   deband.aml - basic implementation of the Local anisotropy detection
/*                    filter
/*
/*   Geoffrey Duh - Department of Geography, Michigan State University
/*   01.30.1998
/*
/*   stdv: threshold that will be used in identifying anisotropy from
/*         slope and curvature surfaces. Enter the number in terms of
/*         standard deviation, i.e., 1 means one stdv.
/***********************************************************************

&args ingrid outgrid stdv_r

&if [show program] NE GRID &then
&do
 &type
 &type Please run DEBAND from GRID
 &type
 &stop
&end

&if [null %ingrid%] &then
&do
 &type
 &type Usage: DEBAND <in_grid> <out_grid> {number of stdv}
 &type
 &stop
&end

&if [null %outgrid%] &then
 &sv outgrid %ingrid%fix

disp 9999 2
mape %ingrid%
maplimits 0 7.5 2.5 10
gridpaint %ingrid% # linear wrap gray

/****************************************************/
/* Calculating the derivative surfaces of elevlation */
/****************************************************/

&sv dcurva = %ingrid%c
&sv dslope = %ingrid%s

&type
&type Calculating the derivative surfaces of input grid...
```

```
%dcurva% = curvature ( %ingrid%, #, #, %dslope%, # )

maplimits 2.5 7.5 5 10
gridpaint %dslope% # linear wrap gray
maplimits 5 7.5 7.5 10
gridpaint %dcurva% # linear wrap gray

/***************************************************/
/* Apply global 5 X 3 smoothing to DEM            */
/***************************************************/

&sv gsmth %ingrid%g
&sv gsmths %gsmth%s
&sv gsmthc %gsmth%c

%gsmth% = focalmean( %ingrid%, RECTANGLE, 3, 5 )
%gsmthc% = curvature ( %gsmth%, #, #, %gsmths%, # )

maplimits 0 5 2.5 7.5
gridpaint %gsmth% # linear wrap gray

maplimits 2.5 5 5 7.5
gridpaint %gsmths% # linear wrap gray

maplimits 5 5 7.5 7.5
gridpaint %gsmthc% # linear wrap gray

/***************************************************/
/* Detecting anisotropy from slope and curvature  */
/***************************************************/

&sv ans = %ingrid%ans
&sv anc = %ingrid%anc

&type
&type Detecting anisotropy for slope and curvature...

%ans% = focalmean( %dslope%, WEIGHT, "aniso.knl", NODATA )
%anc% = focalmean( %dcurva%, WEIGHT, "aniso.knl", NODATA )

/* maplimits 0 2.5 2.5 5
/* gridpaint %ans% # linear wrap gray

/* maplimits 2.5 2.5 5 5
/* gridpaint %anc% # linear wrap gray

/*****************************************************/
/* Get STDV values from anisotropy coverages and use */
/* as thresholds if thresholds didn't set by user    */
/*****************************************************/

&if [null %stdv_r%] &then
&do
 &sv stdv_r 1
&end

&describe %ans%
&sv sthresh %GRD$STDV% * %stdv_r%
&type
&type Using slope threshold of %sthresh%
```

```
&describe %anc%
&sv cthresh %GRD$STDV% * %stdv_r%
&type Using curvature threshold %cthresh%

/*******************************************************/
/* Using thresholds to create anisotropic masks      */
/*******************************************************/

&sv masks %ingrid%ks
&sv maskc %ingrid%kc

&type
&type Using thresholds to create anisotropic masks...

%masks% = con( (%ans% > %sthresh%), 1, (%ans% < -%sthresh%), 1, 0 )
%maskc% = con( (%anc% > %cthresh%), 1, (%anc% < -%cthresh%), 1, 0 )

kill %ans% all
kill %anc% all

/*******************************************************/
/* Sieving out the 1-pixel areas and the vertical    */
/* lines which are 1-pixel width                     */
/*******************************************************/

&sv masks1 %ingrid%ks1
&sv maskc1 %ingrid%kc1

%masks1% = pick( %masks%, focalsum( %masks%, RECTANGLE, 3, 1 ) )
kill %masks% all
%masks% = con( isnull( %masks1% ), 0, %masks1% )

%maskc1% = pick( %maskc%, focalsum( %maskc%, RECTANGLE, 3, 1 ) )
kill %maskc% all
%maskc% = con( isnull( %maskc1% ), 0, %maskc1% )

kill %masks1% all
kill %maskc1% all

/*******************************************************/
/* Creating final mask                               */
/*******************************************************/

%masks1% = con( (%masks% > 1), 1, 0 )
%maskc1% = con( (%maskc% > 1), 1, 0 )

&sv mask %ingrid%mask
%mask% = max( %masks1%, %maskc1% )

kill %masks% all
kill %maskc% all
kill %masks1% all
kill %maskc1% all

&type ***********************************************************
&type *    Mask coverage report: values equal to 1 are masks *
&type ***********************************************************
list %mask%.vat

maplimits 0 0 2.5 2.5
gridpaint %mask%
```

```
/******************************************************/
/* Applying a 3 by 5 LPF to elevation under mask   */
/******************************************************/

&sv patch %ingrid%f1

&type
&type Debanding with a 3 by 5 LPF under mask...

%patch% = con( (%mask% == 1), focalmean( %ingrid%, RECTANGLE, 3, 5 ), %ingrid% )

/******************************************************/
/* Creating edge zones for each anisotropic area    */
/******************************************************/

&sv vedge %ingrid%ve
&sv hedge %ingrid%he
&sv edge %ingrid%edge

%vedge% = con( (%mask% == 0), focalsum( %mask%, RECTANGLE, 1, 3 ), 0 )
%hedge% = con( (%mask% == 0), focalsum( %mask%, RECTANGLE, 3, 1 ), 0 )
%edge% = max( %vedge%, %hedge% )

maplimits 2.5 0 5 2.5
gridpaint %edge%

kill %vedge% all
kill %hedge% all
kill %mask% all

/******************************************************/
/* Applying a 3 by 3 LPF to elevation under edges */
/******************************************************/

&type
&type Creating output grid...

%outgrid% = con( (%edge% <> 0), focalmean( %patch%, RECTANGLE, 3, 3 ), %patch% )
&type *************************************************************
&type *    Edge coverage report: values greater than 0 are edges *
&type *************************************************************
list %edge%.vat
maplimits 0 2.5 2.5 5
gridpaint %outgrid% # linear wrap gray

&sv fixs %outgrid%s
&sv fixc %outgrid%c

&type
&type Creating derivative surfaces for dedanded elevation...
%fixc% = curvature( %outgrid%, #, #, %fixs%, # )

maplimits 2.5 2.5 5 5
gridpaint %fixs% # linear wrap gray
maplimits 5 2.5 7.5 5
gridpaint %fixc% # linear wrap gray

kill %edge% all
kill %patch% all
```

```
&type
&type Done!
&if [query 'Do you want to delete all the derivative surfaces? (y/n) ' .TRUE. ] &then
  &do
    kill %dslope% all
    kill %dcurva% all
    kill %gsmths% all
    kill %gsmthc% all
    kill %fixs% all
    kill %fixc% all
  &end
&type

&type
&if [query 'Do you want to measure the absolute difference surfaces? (y/n) ' .TRUE. ]
&then
  &do
    &sv patchg global
    &sv patch local
    %patchg% = abs( %ingrid% - %gsmth% )
    %patch% = abs( %ingrid% - %outgrid% )
    clear

    &type **********************************************
    &type * Mean ADS report for globally smoothed DEM *
    &type **********************************************
    describe %patchg%
    maplimits 0 5 5 10
    histogram %patchg%

    &type **********************************************
    &type * Mean ADS report for locally smoothed DEM *
    &type **********************************************
    describe %patch%
    maplimits 0 0 5 5
    histogram %patch%
    &type
    kill %patch% all
    kill %patchg% all
  &end
kill %gsmth% all
&return
```

CHAPTER 35

Improving Air Temperature Interpolation Using Satellite Data

A.A. Viau and Y. Huang

INTRODUCTION

Air temperature near the surface of the earth is the most important climatological variable and plays a major role in various scientific disciplines. It is of great interest because most terrestrial life occurs within this zone and air temperature regulates many land surface processes, such as photosynthesis, respiration, and evaporation. Detail measurements of spatial-temporal variations in air temperature across the earth's land areas are critical in the cultivation of many economically important crops (e.g., Tabony, 1985), or more generally in agrometeorology (Söderström and Magnusson, 1995).

The current meteorological stations are generally located more for convenience than for representative sampling. In some areas, such as mountains and remote locations, the density of station, which is already low, is decreasing. There are also areas where data are either unavailable, delayed, or prohibitively expensive (Jones, 1995). Generally, the distribution of climatological stations is only suitable for a rough representation of the spatial variation of this parameter (McClatchey, 1992). Insufficient and spatially changing station density, the nonhomogeneity in the time of observation, the limited intrinsic precision of the measurement, as well as the complexity of the terrain, may result in unprecise information on the spatial variability of air temperature (Karl et al., 1986). This problem becomes more acute when dealing with instantaneous measurements, as compared to daily, decadal, or monthly means (Vogt et al., 1997).

Because the representation of the weather patterns is important to interpolation or regionalization of weather variables, it follows that a definition for the spatial accuracy of the networks is needed. Previous studies have shown that this accuracy depends in large part upon the station density. The purpose of this chapter is to develop a method to quantify the spatial representativity area for near-surface temperature using AVHRR image on board NOAA satellite or meteorological modeling results.

DATA AND METHOD

The Study Area

A 3° latitude by 5° longitude area in southern Quebec, centered on 46.2°N and –72.8°W, was chosen as the study site. The topography ranges from –35 m below sea surface level to 1742 m above sea level. The main land cover and landscape features of the area. The vegetation cover is formed primarily of deciduous, mixed forest (mixture of coniferous and deciduous trees), and closed coniferous forest.

Meteorological Data

In the area over southern Quebec, two meteorological station networks are in operation. One is the manual station network of Ministere de l'Environment et de la Faune du Québec (MEFQ) with daily measurements including minimum and maximum temperatures. The other is the automatic

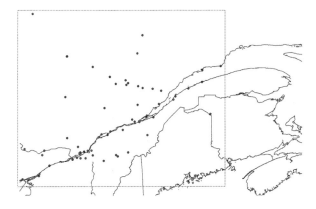

Figure 35.1. CMQ meteorological station network.

station network managed by Centre de Météorologique du Québec (CMQ) with hourly measurements. Sixty-two CMQ stations, as depicted in Figure 35.1, were selected for this study. Hourly measures of air temperature from these stations were obtained. The meteorological data from the station network have undergone a quality control to check for spatial and temporal consistency of the measurements. These temperatures were then used to compare with those derived from nine selected cloud-free satellite images from the overpasses of NOAA-9, 11, and 12.

Satellite Data

Contrary to meteorological station networks which are generally with low density and uncertain representativity, remotely sensed observations can provide better spatial coverage of surface conditions because they are spatially contiguous and available on a regular basis (Seguin, 1991). This provides the potential of using satellite data to improve our knowledge of the spatial patterns of air temperature. Recent studies have been seen exploring such potentials. The variability in land surface response imposes a major forcing to the atmosphere, thus the changes in land surface temperature have a considerable influence on air temperature variability. Based on this hypothesis, Vogt et al. (1997) studied the possibility of using remotely sensed surface skin temperature to map the spatial pattern of air temperature field with a high accuracy. Our study looks at the area of southern Quebec which extends from southwest of Montreal to northeast of Quebec city. The complex landscapes of this area range from agricultural fields, water bodies and urban covers to mountainous areas. Therefore the study is based

on a similar hypothesis as that of the study of Vogt et al. (1997) which establishes the correlation between satellite-derived surface skin temperature and station-measured air temperature. However, in the case of southern Quebec of summer 1994 (May to September), only a limited number (9 out of 130) cloud-free NOAA AVHRR images are available, with acquired times ranging from early morning to early afternoon. Radiometric calibration and atmospheric correction were performed on the images using the AVHRR image processing program developed in CRG (Centre de Recherche en Géomatique, Université Laval). The images were georeference in Lambert Conformal Conic projection. Band 4 (10.334 to 11.252 μm) and 5 (11.395 to 12.342 μm) were used to calculate surface temperature with the use of split window technique outlined by Deschamps and Phulpin (1980) (Table 35.1).

Spatial Representativity of the Meteorological Stations for Near-Surface Temperature

Statistical measures have been seen in literature used to quantify the spatial variabilities of meteorological variables. Spatial variability was expressed by standard error of estimate (Morin et al., 1979); by plotting correlation coefficient against separation distance (Hopkins, 1979); by comparing coefficient of variability with separation distance (Hay and Suckling, 1979); or by using correlation coefficient as a function of separation distance (Harcum and Loftis, 1987).

One of the latest works is by Vogt et al. (1997) for the Mediterranean region of Andalusia in southern Spain. The work involved 31 meteorological stations and 148 AVHRR images from the year 1992. For each of the stations, regression analysis is performed between the daily maximum air temperature and the mean surface skin temperature retrieved for 11 km=AVHRR image windows centered over the station, and strong correlation is found (with mean $R=0.823$). The spatial representativity of each of the meteorological stations is evaluated by a cross-validation regression model.

In this study, since only nine relatively cloud-free images are available for comparison purposes, a different method of estimating the spatial representativity of station network is advocated. This method estimates the spatial representativity of a station in terms of the homogeneity of AVHRR-derived surface skin temperature or MC2-derived screen air temperature. The homogeneity is expressed as the standard deviation

Table 35.1. NOAA-AVHRR Data Used for the Study.

NOAA	noaa-11	Noaa-11	noaa-09	noaa-12	noaa-09	noaa-11	noaa-12	noaa-12	noaa-12
Date (in 1994)	May 23	June 17	Sep. 4	Aug. 6	Aug. 7	Aug. 7	Aug. 7	Aug. 8	Aug. 10
Time (GMT)	20:33	20:28	20:30	12:14	16:30	11:30	11:52	13:11	12:27
Source	SEA[a]	SEA	UQAC[b]	SEA	UQAC	UQAC	SEA	SEA	SEA

SEA: Service de l'Environment Atmosphérique, Ministére de l'Environment du Canada.
UQAC: l'Université du Québec à Chicoutimi.

(SD) of temperature within the area around a station. The higher an allowed SD, the larger the spatial representativity area of a station, shows a procedure of calculating standard deviation in ERDAS IMAGINE software. The input raster is a temperature map. The value of each pixel of the output raster is the SD of the central pixel of the user-defined window. The SD value at the station point will then be extracted using ARC/INFO software.

The discussions on station representation that have been seen so far refer to the representativity of a station as an area centered over the station. If this common method is applied in this study, difficulty appears in measuring the spatial representativity area of a station when the station is located close to delineation of land-cover types.

RESULTS AND DISCUSSIONS

The Difference and Correlation Between Satellite and Station Measurements

Regression analysis is used to evaluate these two measurements. Correlations between AVHRR-derived surface temperature and station-measured air temperature were calculated for all the stations on each of the dates with image overpass, based on the mean surface temperature extracted for different windows centered over the station locations. As can be seen in Table 35.2, the correlation found between AVHRR and station measurements is generally low. About one-third of the stations have residual value higher than ±2.0°C.

Due to the limited number of images used in this study, the spatial representation of individual stations cannot be estimated using such regression analysis. However, the regression calculation provides an opportunity to evaluate the general performance of the station network as a whole. Table 35.3 summarizes the station residuals at three levels for each date. For the nine images under evaluation, an average of 43.6% of the stations have residuals lower than 1°C. Other research shows that the measurement bias of meteorological stations is within the range of 1 K (Vogt et al., 1997). Therefore, this indicates that compared with satellite measurements, less than half of the network has measurement errors within the range of instrumental bias. Reported error of surface temperature retrieved from AVHRR image using split-window algorithm ranges within 2 K (Becker and Li, 1990; Sobrino et al., 1994).

As can be seen in Table 35.4, for the image of May 23, when all the stations are used in regression calculation, a high R square value (0.70) is already found. This value increased to 0.87 after the 11 stations with high residuals are eliminated from the analysis. Except for the dates of August 6 and August 10, R square values increase to a much higher level. It seems that for these two images, there is not much accordance between satellite-observed surface temperature and station-measured air temperature. It is not clear at this time whether the inconsistency is due to an actual difference between air temperature and surface temperature on those particular two dates, or an artifact of the measurements themselves.

Note:

A - All stations are included in the regression calculations.

B - Stations with residuals higher than 2.0 are eliminated from regression calculations.

Station Representativity Area

Standard deviations (SD) of surface temperature around each station in the four directions with different window sizes are calculated for each of the images. The window size starts from 3×3 to 60×60 with five pixels for each step. The spatial resolution of AVHRR image indicates that one pixel represents a square of 1.1×1.1 km. In the calculation of standard deviation of temperature of a station, the processed window is centered around the station. Since SD is measured in four individual directions, the measuring distance ranges from 2 to 33 km away from the station.

Table 35.2. Correlation Coefficient of Station-Observed Air Temperature vs. AVHRR-Derived Surface Temperature.

| Auto. Station | | | NOAA | | | | | |
| No. of Observations | Time of Observation (GMT) | NOAA Image | r with Different Pixel Size | | | | | |
			1x1	3x3	5x5	7x7	9x9	11x11
31	21:00	1	0.83	0.90	0.64	0.67	0.69	0.69
31	20:00	2	0.39	0.26	0.32	0.35	0.25	0.25
40	12:00	3	0.24	0.46	0.49	0.50	0.49	0.48
33	17:00	4	0.76	0.78	0.78	0.79	0.79	0.79
30	11:00	5	0.53	0.43	0.40	0.40	0.39	0.39
40	12:00	6	0.63	0.65	0.66	0.66	0.66	0.66
41	14:00	7	0.63	0.65	0.65	0.66	0.67	0.67
40	13:00	8	0.41	0.43	0.47	0.49	0.51	0.53
41	21:00	9	0.49	0.51	0.40	0.45	0.47	0.50

NOAA image: 1. 23-May-94 noaa-11 20:33
2. 17-Jun-94 noaa-11 20:28
3. 6-Aug-94 noaa-12 12:14
4. 7-Aug-94 noaa-09 16:30
5. 7-Aug-94 noaa-11 11:30
6. 7-Aug-94 noaa-12 11:52
7. 8-Aug-94 noaa-12 13:11
8. 10-Aug-94 noaa-12 12:17
9. 4-Sep-94 noaa-09 20:30

Table 35.3. Summary of Stations with Different Residual Levels.

| | No. of Observations | Station with Residual <1°C | | Station with 1≤ Residual <2°C | | Station with Residual >2°C | |
		No.	%	No.	%	No.	%
23-May-94 noaa-11	31	9	29.0	11	35.5	11	35.5
17-Jun-94 noaa-11	31	7	22.6	10	32.3	14	45.2
6-Aug-94 noaa-12	40	16	40.0	12	30.0	12	30.0
7-Aug-94 noaa-09	33	13	39.4	14	42.4	6	18.2
7-Aug-94 noaa-11	30	17	56.7	5	16.7	8	26.7
7-Aug-94 noaa-12	40	21	52.5	11	27.5	8	20.0
8-Aug-94 noaa-12	41	24	58.5	8	19.5	9	22.0
10-Aug-94 noaa-12	40	19	47.5	10	25.0	11	27.5
4-Sep-94 noaa-09	41	19	46.3	15	36.6	7	17.1
Average			**43.6**		**29.5**		**26.9**

It was discussed before that a variance of 2°C is a criterion we used to evaluate the network representation. The results of SD calculation for all the images shows that within the window size of 60 × 60 pixel, the SD in four directions for most of the stations are within the range of 2.0. This is in accordance with the results of other research which found that 60 km seems to be the maximum distance between stations accept-able to represent the spatial variability of mean temperature (Hubbard, 1994).

SUMMARY AND CONCLUDING REMARKS

This chapter used NOAA AVHRR image as the reference to monitor the spatial representativity of the automatic meteorological station network of southern

Quebec area. Split-window algorithm is used to extract surface temperature from AVHRR image. This temperature is first compared with the air temperature measurement from meteorological stations, based on which the overall representation of the network is assessed. We found that 70.1% of the network can produce measurements within an error range of 2°C.

The satellite measurement is further used to analyze the spatial variability of surface temperature in terms of standard deviation centered over the station locations. Interesting features are found that in different directions of a meteorological station, the variability of surface temperature shows different characteristics. An abrupt change in standard deviation with distance indicates the spatial limit of station representativity. We suggest that different criterion (i.e., threshold value of standard deviation) should be used to measure the spatial representation for different directions of a station.

REFERENCES

Becker, F. and Z.L. Li. Towards a Local Split Window Method Over Land Surface, *Int. J. Remote Sensing,* 11, pp. 369–393, 1990.

Deschamps, P.Y. and T. Phulpin. Atmospheric Correction of Infrared Measurements of Sea Surface Temperature Using Channels at 3.7, 11 and 12 μm, *Boundary-Layer Meteorol.,* 18, pp. 131–143, 1980.

Harcum, J.B. and J.C. Loftis. Spatial Interpolation of Penman Evapotranspiration, *Trans. Am. Soc Agric. Eng.,* 30(1), pp. 129–136, 1987.

Hay, J. and P.W. Suckling. An Assessment of the Networks for Measuring and Modelling Solar Radiation in British Columbia and Adjacent Areas of Western Canada, *Can. Geogr.,* 13(3), pp. 222–238, 1979.

Hopkins, J.S. The Spatial Variability of Daily Temperature and Sunshine Over Uniform Terrain, *Meteorol. Mag.,* 106, pp. 278–292, 1979.

Hubbard, K.G. Spatial Variability of Daily Weather Variables in the High Plains of the USA, *Agric. For. Meteorol.,* 68, pp. 29–41, 1994.

Jones, P.D. Land Surface Temperatures: Is the Network Good Enough?, *Clim. Change,* 31, pp. 545–558, 1995.

Karl, T.R., C.N., Willians Jr., P.J. Young, and W.M. Wendland. A Model to Estimate the Time of Observation Bias Associated with Monthly Mean Maximum, Minimum, and Mean Temperatures for United States Locations, *J. Climatol. Appl. Meteorol.,* 25, pp. 145–160, 1986.

McClatchey, J. The Use of Climatological Observations as Ground Truth for Distributions of Minimum Temperature Derived from AVHRR Data, *Int. J. Remote Sensing,* 13, pp. 153–163, 1992.

Morin, G., J. Fortin, W. Sochanska, and J. Lardeau. Use of Principal Component Analysis to Identify Homogeneous Precipitation Stations for Optimal Interpolation, *Water Resour. Res.,* 18(4), pp. 1269–1277, 1979.

Robeson, S.M. Resampling of Network-Induced Variability in Estimates of Terrestrial Air Temperature Change, *Clim. Change,* 29, pp. 213–229, 1995.

Seguin, B. Use of Surface Temperature in Agrometeorology, in *Applications of Remote Sensing to Agrometeorology,* F. Toselli, Ed., Kluwer Academic Press, Boston, pp. 1991, 221–240.

Sobrino, J.A., Z.L. Li, M.Ph. Stoll, and F. Becker. Improvements in the Split-Window Technique for Land Surface Temperature Determination, *IEEE Trans. Geosci. Remote Sensing,* 32(2), pp. 243–253, 1994.

Söderström, M. and B. Magnusson. Assessment of Local Agroclimatological Conditions—A Methodology, *Agric. For. Meteorol.,* 72, pp. 243–260, 1995.

Tabony, R.C. Relations between Minimum Temperature and Topography in Great Britain, *J. Climatol.,* 5, pp. 503–520, 1985.

Viau A., A. Royer, and C. Ansseau. Integration of Climatological and Remote Sensing Data in a Geographic Information System: Biome-Tel Project, *Proceedings of the Workshop on Camadian Climate System Data,* Québec, May 16–18, 1994, pp. 13–20.

Viau, A.A., J. Vogt, and F. Paquet. Regionalisation and Mapping of Air Temperature Fields Using NOAA-AVHRR Imagery, *Actes du 9e Congrés de l'Association québécoise de télédétection (AQT): La télédétection au sien de la géomatique,* 30 avril–3 mai, Québec, Canada, 1996.

Vogt, J.V., A.A. Viau, and F. Paquet. Mapping Regional Air Temperature Fields Using Satellite Derived Surface Skin Temperatures, *Int. J. Climatol.,* 17, pp. 1559–1579, 1997.

Zemel, Z. and J. Lomas. An Objective Method for Assessing Representativeness of a Station Network Measuring Minimum Temperature Near the Ground, *Boundary-Layer Meteorol.,* 1, pp. 3–14, 1976.

CHAPTER 36

Combining Minimum Error Variance and Spatial Variability in the Mapping of Environmental Variables

P. Goovaerts

INTRODUCTION

Most environmental applications, such as the delineation of contaminated areas, require a prior mapping of the target attribute, say a soil pollutant concentration, over the study area. A common approach consists of estimating the pollutant concentration at unsampled grid nodes using minimum error variance (kriging) interpolation algorithms. The map of such local estimates typically smooths out local details of the spatial variation of the attribute, with small values being overestimated while large values are underestimated. This type of selective bias is called conditional bias (Journel and Huijbregts, 1978), and is a serious shortcoming when one aims at detecting large pollutant concentrations.

Stochastic simulation allows the generation of maps of pollutant concentrations that reproduce the spatial variability of the data without smoothing effect, and it is thus increasingly preferred to estimation (Srivastava, 1996; Goovaerts, 1997a; Kyriakidis, 1997). In so doing, the "minimum error variance" property of kriging is lost. Several studies (Olea and Pawlowsky, 1996; Goovaerts, 1997a, 1998) have shown that the mean prediction error is larger for simulated values than for kriging estimates.

Goovaerts (1998) presented estimation and simulation as two optimization problems that differ in their optimization criteria, minimization of a local expected loss for estimation, and reproduction of global statistics (semivariogram, histogram) for simulation. Maps with intermediate properties in terms of mean square error and reproduction of histogram and semivariogram

were generated by modifying gradually an initial random image using simulated annealing and an objective function that incorporates both local and global constraints. The algorithm was applied to the mapping of permeability values, and flow simulation results showed that accounting for local constraints in stochastic simulation yields, on average, smaller errors in production forecast than a smooth estimated map or a simulated map that reproduces only the histogram and semivariogram.

This chapter presents an application of this optimization algorithm to the mapping of topsoil zinc concentration in a 14.5 km^2 region of the Swiss Jura. The prediction performances of kriging, simulation, and the proposed mixed approach are investigated using a validation set.

THEORY

Consider the problem of determining the value of a continuous attribute z at N grid nodes $\mathbf{u}_j$ discretizing the study area A. At each unsampled grid node $\mathbf{u}_j$, the uncertainty about the unknown z-value is modeled by the conditional cumulative distribution function (ccdf) of the random variable $Z(\mathbf{u}_j)$:

$$F\left(\mathbf{u}_j; z|(n)\right) = \operatorname{Prob}\left\{Z\left(\mathbf{u}_j\right) \leq z|(n)\right\} \qquad (1)$$

where the notation "|(n)" expresses conditioning to the local information, say, n neighboring data $z(\mathbf{u}'_\alpha)$. The function (1) gives the probability that the unknown is no greater than any given threshold z, and it can be established using a variety of algorithms that are classified as parametric and nonparametric (Deutsch and Journel, 1998; Goovaerts, 1997b).

Estimation or simulation can be viewed as the selection, within the range of possible z-values at $\mathbf{u}_j$, of a single value that is "optimal" for some criterion. Three constraints of increasing complexity will be considered:

1. minimization of the local error variance
2. minimization of the local error variance, and reproduction of the target histogram
3. minimization of the local error variance, and reproduction of the target histogram and semivariogram model

Depending on the constraint retained, the number and characteristics of the map(s) of optimal values will greatly differ.

Map of E-Type Estimates

Imposing the first constraint amounts to selecting the value $z^*(\mathbf{u}_j)$ that minimizes the quadratic function of the estimation error $[e(\mathbf{u}_j)]^2=[z(\mathbf{u}_j)-z^*(\mathbf{u}_j)]^2$, which is referred to as a loss (Journel, 1989, pp. 27–28; Christakos, 1992). Because the actual value $z(\mathbf{u}_j)$ is unknown, only the expected loss can be computed:

$$\varphi\left(z^*\left(\mathbf{u}_j\right)|(n)\right) = E\left\{\left[Z\left(\mathbf{u}_j\right)-z^*\left(\mathbf{u}_j\right)\right]^2|(n)\right\}$$
$$= \int_{-\infty}^{+\infty}\left[z-z^*\left(\mathbf{u}_j\right)\right]^2 dF\left(\mathbf{u}_j;z|(n)\right) \quad (2)$$

This expected loss appears as a function $\varphi(.)$ of the estimated value $z^*(\mathbf{u}_j)$. The optimal estimate is shown to be the expected value of the ccdf at location $\mathbf{u}_j$, also called E-type estimate:

$$z_E^*\left(\mathbf{u}_j\right) = \int_{-\infty}^{+\infty} z\, dF\left(\mathbf{u}_j;z|(n)\right) \quad (3)$$

The corresponding expected loss is but the variance of the conditional cdf:

$$\sigma^2\left(\mathbf{u}_j\right) = \int_{-\infty}^{+\infty}\left[z-z_E^*\left(\mathbf{u}_j\right)\right]^2 dF\left(\mathbf{u}_j;z|(n)\right) \quad (4)$$

The unique solution to this optimization problem is the set of N "locally optimal" estimates $\{z_E^*(\mathbf{u}_j), j=1,...,N\}$. The global expected loss associated with the optimal estimation grid is the sum of local expected losses, that is the sum of the variances of the N ccdfs, and is minimal.

Map of Rescaled E-Type Estimates

The minimization of the mean square estimation error yields a smooth map of optimal estimates which does not reproduce the sample histogram. Because of this smoothing effect, the second constraint of histogram reproduction requires the selection of another set of values which is not any more optimal for the single first criterion. Intuitively, one would like to transform (rescale) the set of E-type estimates of type (3) so as to reproduce the target histogram while keeping the global expected loss as small as possible. Goovaerts (1998) showed that there is a unique optimum for the joint constraints of histogram reproduction and minimization of global expected loss. The set of optimal values $\{z_c^*(\mathbf{u}_j), j=1,...,N\}$ is obtained by applying the following rank-preserving transform to the E-type estimates:

$$z_c^*\left(\mathbf{u}_j\right) = F^{-1}\left[F_E\left(z_E^*\left(\mathbf{u}_j\right)\right)\right] \quad j=1,...,N \quad (5)$$

where $F_E(.)$ is the cumulative distribution function (cdf) of the N estimates, and $F(.)$ is the target cdf. Note that to ensure the honoring of data values, correction (5) is not applied at the sampled grid nodes, hence the target histogram is only approximately reproduced, in particular when the proportion of sampled grid nodes is large (Journel and Xu, 1994). The price to pay for the better reproduction of the histogram is an increase in the global expected loss:

$$\sum_{j=1}^{N}\varphi\left(z_c^*\left(\mathbf{u}_j\right)|(n)\right) = \sum_{j=1}^{N}\varphi\left(z_E^*\left(\mathbf{u}_j\right)|(n)\right) + \sum_{j-1}^{N}\left[d\left(\mathbf{u}_j\right)\right]^2 \quad (6)$$

where $d(\mathbf{u}_j)=z_c^*(\mathbf{u}_j) - z_E^*(\mathbf{u}_j)$ is the difference between corrected and initial values at $\mathbf{u}_j$.

Maps of Simulated Values

The last constraint includes the additional requirement that the set of values reproduces the semivariogram model $\gamma(\mathbf{h})$ inferred from the z-data. Reproduction of the semivariogram model is generally limited to a specified number S of the first lags, and the lack of reproduction can be measured as the sum of squares of differences between the model and experimental semivariogram values over the S lags:

$$O_1 = \sum_{s=1}^{S}\left[\gamma(\mathbf{h}_s)-\hat{\gamma}(\mathbf{h}_s)\right]^2 \tag{7}$$

Intuitively, one would like to modify the map of rescaled E-type estimates, which already matches the first two constraints, so as to lower the quantity (7) to a value close to zero.

Since the set of rescaled estimates $\{z_c^*(\mathbf{u}_j), j=1,...,N\}$ already reproduces the sample variance, the sill of the corresponding semivariogram is generally close to the sill of the target model. The problem usually lies in the underestimation (smoothing) of the short-range variability (e.g., nugget effect) by the set of rescaled estimates. Thus, there is no need to change the set of estimated values, but rather their locations on the grid should be modified so as to improve the reproduction of the semivariogram. Of course, one wants to keep the desirable property of "minimum error variance," that is, keep the following quantity as small as possible:

$$O_2 = \sum_{j=1}^{N}\left|\varphi\left(z_E^*(\mathbf{u}_j)|(n)\right)-\varphi\left(z^{(l)}(\mathbf{u}_j)|(n)\right)\right| \tag{8}$$

where $\varphi(z_E^*(\mathbf{u}_j)|(n))=\sigma^2(\mathbf{u}_j)$ is the variance of the ccdf, and $\varphi(z^{(l)}(\mathbf{u}_j)|(n))$ is the expected loss associated with the new value $z^{(l)}(\mathbf{u}_j)$ at location $\mathbf{u}_j$.

Unlike the two previous constraints, there are usually many solutions (i.e., sets of values $\{z^{(l)}(\mathbf{u}_j), j=1,...,N\}$) to this optimization problem since a given semivariogram can be obtained from different arrangements of the same set of values. The set of solutions can be explored using simulated annealing which is a generic name for a family of optimization algorithms based on the principle of stochastic relaxation (Farmer, 1988; Srivastava, 1996). The optimization process amounts to systematically modifying an initial image, say the map of rescaled E-type estimates, so as to get this image acceptably close to the target statistics; that is, decrease the value of the following two-components objective function:

$$O(i) = \frac{\lambda}{O_1(0)}O_1(i)+\frac{1-\lambda}{O_2(0)}O_2(i) \tag{9}$$

where $O_1(i)$ and $O_2(i)$ are the values of the functions (7) and (8) after the i-th perturbation of the image. The image is perturbed by swapping z-values at any two unsampled locations $\mathbf{u}_j$ and $\mathbf{u}_k$ chosen at random: $z_{(i)}^{(l)}(\mathbf{u}_j)$ becomes $z_{(i+1)}^{(l)}(\mathbf{u}_k)$ and vice versa. In this way, the histogram of the initial image remains unchanged, hence there is no need to include an additional component in the objective function to control the reproduction of the histogram. To prevent the component with the largest unit from dominating the objective function, each component O_c is standardized by its initial value $O_c(0)$. The relative importance of each component is controlled by the weight λ, which allows the user to strike a balance between a local criterion (minimization of a local expected loss) and a global criterion (reproduction of a semivariogram model).

Different types of decision rules and convergence criterion can be adopted for the iterative algorithm (Deutsch and Cockerham, 1994). In this chapter, all perturbations that diminish the objective function were accepted, while unfavorable perturbations were accepted according to a negative exponential probability distribution. A fast annealing schedule was used; that is, the initial temperature was set to 1 and lowered by a factor 20 (reduction factor=0.05) whenever enough perturbations ($5 \times N$) have been accepted or too many ($50 \times N$) have been tried.

CASE STUDY

Consider the problem of mapping the topsoil Zn concentration in a 14.5 km^2 region of the Swiss Jura. The information available consists of 259 Zn measurements depicted in Figure 36.1 (top left). A validation set of 100 test locations is available to investigate the prediction performances of the different techniques. The bottom graphs of Figure 36.1 show the sample histogram and omnidirectional semivariogram with the model fitted.

Conditional distributions (ccdf) were modeled using ordinary indicator kriging (Goovaerts, 1997b, p. 294) with nine threshold values corresponding to the deciles of the sample distribution of Figure 36.1. The resolution of the discrete ccdfs was increased by performing a linear interpolation between tabulated bounds provided by the sample cdf (Deutsch and Journel, 1998, p. 136). Figure 36.2 (top graphs) shows the maps of E-type

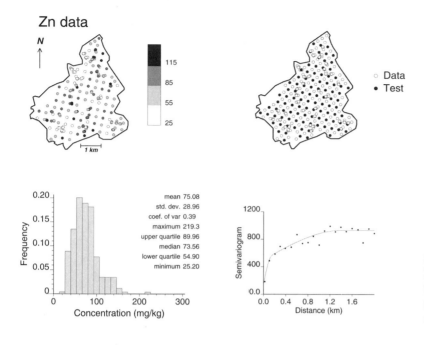

Figure 36.1. Location maps of 259 Zn data and 100 test locations. Bottom graphs show the sample histogram and semivariogram with the model fitted.

estimates before and after rescaling using the transform (5). The rescaling causes a 34% increase of the global expected loss, but the sample histogram is now reproduced. Yet, the experimental semivariogram on Figure 36.3 (right top graph) indicates that the short-range variability of Zn concentrations is still underestimated by the map of rescaled E-type estimates.

To improve the reproduction of the semivariogram, the map of rescaled E-type estimates was postprocessed using simulated annealing. Results were, however, unsatisfactory because of the salt-and-pepper effect displayed by the postprocessed images. Better results were obtained when starting with an initial random image generated by assigning to each unsampled grid node a z-value drawn at random from the target cdf $F(z)$. The random image was postprocessed using simulated annealing and the two-components objective function (9). To investigate the relative influence of both components on the final realization, five different sets of weights were considered (see Figure 36.2). The corresponding semivariograms with the target model are displayed in Figure 36.3.

When all the weight is given to the minimization of local expected loss ($\lambda = 0$), the realization (Figure 36.2, second row) is very close to the smooth map of rescaled E-type estimates, which is the optimum for the joint constraints of histogram reproduction and minimization of local expected loss (Figure 36.2, right top graph). As the weight given to the first component increases, the realization becomes less smooth while deviation from the semivariogram model decreases (see Figure 36.3).

Validation Set

The same approach was used to generate 100 realizations of the spatial distribution of Zn values for the different weighting schemes. Table 36.1 gives a few statistics computed at the 100 test locations where the actual Zn concentration is known. As expected, the global loss decreases as the weight λ given to the first component (7) decreases. When all the weight is given to the minimization of local expected loss ($\lambda = 0$), the global loss is 62% of the loss obtained when the local constraint is ignored ($\lambda = 1$). This score is slightly larger than the score of the rescaled E-type estimates (58%). Consequently, if the objective is to minimize the local expected loss while reproducing a target histogram, better results are obtained by rescaling E-type estimates using a transform of type (5). This unique optimum cannot be reached by simulated annealing: the realizations appear to be trapped in suboptimal situations. In the absence of constraint of histogram reproduction, the optimum is the map of E-type estimates which yields a global loss of 46%.

The mean square error of prediction (MSE) was computed as the arithmetic average of square differences between actual concentrations and values gen-

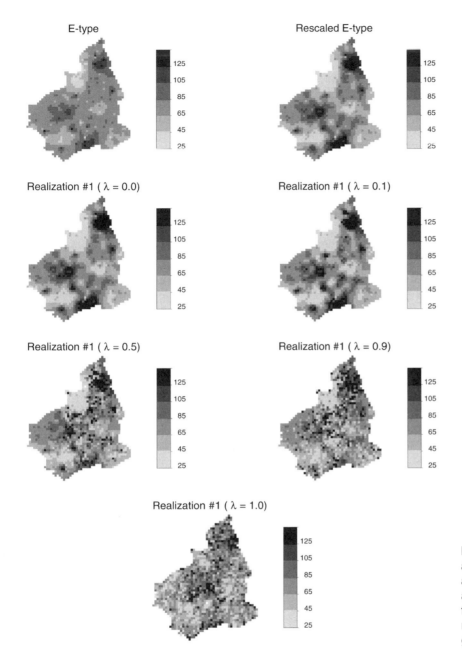

Figure 36.2. Maps of original and rescaled E-type estimates, and the first realization generated using increasing weight λ for the component that controls reproduction of the semivariogram.

erated at the 100 test locations using the different approaches. The smallest errors are obtained using an estimation approach (E-type) that seeks only the minimization of local expected loss. When the constraint of histogram reproduction is included, the prediction error increases. As for the global expected loss, better results are obtained by rescaling E-type estimates (MSE=74%) instead of processing an initial random image using simulated annealing and a weight λ = 0 (MSE=77%). When a third constraint of semi-

variogram reproduction is included, the prediction error increases even more. In other words, reproduction of spatial variability as modeled by semivariogram is achieved at the expense of larger errors of prediction.

Each test location was classified as contaminated if the simulated value or E-type estimate exceeds a threshold taken as 100 mg/kg. Table 36.1 (fourth column) shows that the classification based on smooth E-type estimates yields the largest proportion of misclassified locations. Better results are obtained for simulated

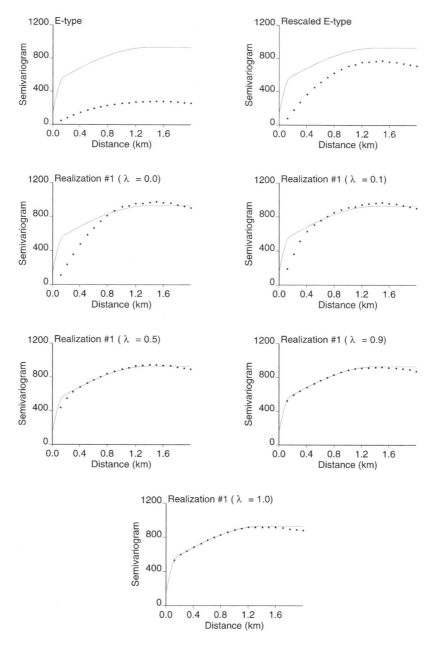

Figure 36.3. Experimental semivariograms computed from the maps of Figure 36.2 with the target model (solid line).

values; in particular, when the weight of the local constraint increases. Unlike the two previous criteria (global loss and MSE), simulated annealing yields better results than a rescaling of E-type estimates.

The last column of Table 36.1 gives, for each weighting scheme, the average variance of the distributions of simulated values, which is a measure of local uncertainty. Giving more weight to the minimi-

zation of local expected loss reduces differences between realizations which become more similar to the unique optimum; that is, the map of rescaled E-type estimates. For E-type estimates, the uncertainty measure is the average variance of ccdfs which is larger than the average variance of the distribution of simulated values.

Table 36.1. Statistics Measuring the Prediction Performances Obtained on Average over 100 Realizations Generated Using Decreasing Weight λ for the Constraint of Semivariogram Reproduction, and for the Rescaled and Original E-Type Estimates. (For the global loss and mean square error (MSE), results are expressed as percentage of the simulation score when local constraints are ignored (1 − λ = 0): best is smallest.)

Algorithm	Global Loss (%)	MSE (%)	Misclassification (%)	Uncertainty ($[mg/kg]^2$)
Simulation				
$\lambda = 1.0$	100	100	66	524
$\lambda = 0.9$	99	100	65	545
$\lambda = 0.5$	91	99	61	451
$\lambda = 0.1$	73	93	54	55
$\lambda = 0.0$	62	77	39	4
Rescaled E-type	58	74	42	0
E-type	46	63	74	579

CONCLUSIONS

Both estimation and simulation approaches can be formulated as the selection of a set of attribute values that are optimal for specific criteria which are typically conflicting. Simulated annealing with a two-components objective function allows one to find a balance between local and global constraints, and to generate maps that minimize the local error variance under the constraints of histogram and semivariogram reproduction.

The case study shows that as the weight of the local constraint increases, the realizations become smoother and more similar to each other, thereby reducing the extent of the space of uncertainty. Also, accounting for local constraints in stochastic simulation reduces the average prediction error and the risk of classifying wrongly contaminated locations as safe. Smallest prediction errors are obtained for the map of E-type estimates, yet the smoothing effect lessens the ability to detect large pollutant concentrations.

In this case study, the reproduction of the pattern of spatial variability (semivariogram) does not appear to bring a clear advantage over the straightforward rescaling of E-type estimates. The reason is that all performance criteria (MSE and misclassification score) are local in that they involve each grid node separately. Decision-making involving jointly many locations should, however, benefit from the constraint of semivariogram reproduction, as demonstrated for flow properties in Goovaerts (1998).

REFERENCES

Christakos, G. *Random Field Models in Earth Sciences,* Academic Press, New York, 1992, pp. 341–343.

Deutsch, C.V. and P. Cockerham. Practical Considerations in the Application of Simulated Annealing to Stochastic Simulation. *Math. Geol.,* 26, pp. 67–82, 1994.

Deutsch, C.V. and A.G. Journel. *GSLIB: Geostatistical Software Library and User's Guide, Second Edition,* Oxford University Press, New York, 1998, p 76.

Farmer, C. The Generation of Stochastic Fields of Reservoir Parameters with Specified Geostatistical Distributions, in *Mathematics in Oil Production,* S. Edwards, and P. King, Eds., Clarendon Press, Oxford, 1988, pp. 235–252.

Goovaerts, P. Kriging vs. Stochastic Simulation for Risk Analysis in Soil Contamination, in *geoENV I—Geostatistics for Environmental Applications,* A. Soares, J. Gómez-Hernández, and R. Froidevaux, Eds., Kluwer Academic Publishers, Dordrecht, 1997a, pp. 247–258.

Goovaerts, P. *Geostatistics for Natural Resources Evaluation,* Oxford University Press, New York, 1997b, pp. 259–367.

Goovaerts, P. Accounting for Estimation Optimality Criteria in Simulated Annealing, *Math. Geol.,* 30, in press, 1998.

Journel, A.G. *Fundamentals of Geostatistics in Five Lessons,* Volume 8 Short Course in Geology. American Geophysical Union, Washington, DC, 1989, p. 40.

Journel, A.G. and C.J. Huijbregts. *Mining Geostatistics,* Academic Press, New York, 1978, p. 458.

Journel, A.G. and W. Xu. Posterior Identification of Histograms Conditional to Local Data, *Math. Geol.,* 26, pp. 323–359, 1994.

Kyriakidis, P.C. Selecting Panels for Remediation in Contaminated Soils via Stochastic Imaging, in *Geostatistics Wollongong '96,* E.Y. Baafi and N.A. Schofield, Eds., Kluwer Academic Publishers, Dordrecht, 1997, pp. 973–983.

Olea, R.A. and V. Pawlowsky. Compensating for Estimation Smoothing in Kriging, *Math. Geol.,* 28, pp. 407–417, 1996.

Srivastava, M.R. An Overview of Stochastic Spatial Simulation, in *Spatial Accuracy Assessment in Natural Resources and Environmental Sciences: Second International Symposium,* H.T. Mowrer, R.L. Czaplewski, and R.H. Hamre, Eds., General Technical Report RM-GTR-277, U.S. Department of Agriculture, Forest Service, Fort Collins, CO, 1996, pp. 13–22.

Part VIII

Decomposing Digital Images to Improve Classification and Describe Uncertainty

Among the first digital spatial data available were remotely sensed imagery. Since its introduction in the early 1970s, such imagery has been widely used and studied. Much of this effort has involved attempting to improve image classifications, and to be able to extract more information from this imagery than was initially possible. The chapters in this Part all describe some aspect of increasing the information content of images, or documenting where there are weaknesses in the information content.

The first chapter discusses new techniques for image classification: Griffith and Fellows describe a supervised classification procedure based on principal components analysis (PCA) but linked to techniques for describing spatial autocorrelation. The second chapter by Hughes et al. describes a fuzzy set-based method for obtaining subpixel information about land-cover types. The third chapter by Myers describes a method of condensing the information in a multi-band image with little or no loss of information content in the image. Finally, the fourth chapter by Gabrosek et al. touches on the future by addressing issues of information extraction from data that will be collected by the Earth Observing System (EOS) satellite to be launched in 1998.

CHAPTER 37

Pixels and Eigenvectors: Classification of LANDSAT TM Imagery Using Spectral and Locational Information

D.A. Griffith and P.L. Fellows

INTRODUCTION

On the early morning of July 15, 1995, the Adirondack Park of New York State was struck with a fierce combination of wind, rain, thunder, and lightning. Hundreds of thousands of acres of trees were uprooted, snapped off, or defoliated. The problem that this research examines may be stated as follows: can an accurate and precise classification of the severity of the blowdown in the Adirondack Park be generated? Specifically, this research explores a new approach to image classification using principal components analysis with spectral and locational information derived from the images.

The research summarized in this chapter utilizes two recently developed approaches in satellite image analysis. First, the utility of a spatial statistical extension is explored. It provides additional data that may enhance a classification. This extension attempts to quantify significant spatial trends in satellite data by identifying significant map patterns using eigenvectors and spatial lag variables. These significant map patterns provide additional locational information for each pixel and improve the classification of the spectral data. Second, multivariate image analysis (MIA), a modified supervised classification methodology based on principal components analysis (PCA) and corresponding components scatterplots, is employed to generate the classification. Spectral data and locational data are used in MIA. These two methods are expected to provide a straightforward, accurate approach to image classification.

BACKGROUND

Multivariate Image Analysis

Esbensen and Geladi (1989) and their associates at the Research Group for Chemometrics, Umea University, Umea, Sweden and the Norwegian Computing Center, Oslo, Norway, have published extensively on the use of multivariate image analysis in a variety of applications (see Geladi et al., 1989). ERDAS®[1], Inc. has produced a software module for Multivariate Image Analysis (MIA) following the strategies developed by these researchers. The methods implemented in this chapter parallel the plan outlined in the manual that accompanies the ERDAS® MIA module 19 for Revision 7.4. (ERDAS®, Inc., 1990).

Esbensen and Geladi's (1989) seminal paper, entitled "Strategy of Multivariate Image Analysis," describes their methodology applied to LANDSAT thematic mapper (TM) data. They demonstrate how principal components analysis (PCA) and the related use of principal components (PC) scatterplots are used to classify satellite imagery. Geladi et al.'s earlier paper (1989) introduces the concept of PC scatterplots and their utility in the classification of satellite imagery. These two papers appeared in the chemometric literature where image analysis is common in spectroscopy applications. Several researchers in chemometrics have extended Esbensen and Geladi's work. Wienke et al. (1995) explore the identification of plas-

---

[1] Registered trademark of ERDAS®, Inc., Atlanta, Georgia.

tics in mixed household and industrial waste through multivariate image rank analysis. These researchers used images generated from the component scores instead of component scatterplots themselves to uncover hidden relationships in the data. The team then classified the PC images.

Esbensen and Geladi's (1989) utilization of a LANDSAT TM dataset is very fortuitous for the remote sensing community. However, the remote sensing literature offers few examples of MIA applications. Wang (1993) used PCA to transform a LANDSAT multispectral scanner (MSS) dataset to four components and then, through an iterative process, clustered the component values into general brightness classes using a minimum distance to means algorithm. These classes were used later in the analysis as a change detection mask. The literature does discuss related uses of PCA as a preprocessing tool. Crosta and McMoore (1989) outline a methodology called Feature Oriented Principal Components Selection (FPCS). These researchers utilized eigenvector loadings to identify which PC images concentrate certain information about a target.

Spatial Autocorrelation and Map Patterns

Much of the literature relating to spatial statistics and remote sensing is concerned with the problem of highly autocorrelated (correlation within one variable or band) data comprising training samples used in classification algorithms. Due to the high positive spatial autocorrelation latent in individual cover types in satellite imagery, many training areas underestimate the variance for spectral classes. Consequently, classification algorithms lead to an overestimation of the contrast between categories (Campbell, 1981, p. 355). Thus, pixels are easily misclassified. Tubbs and Coverly (1978) suggest that classification algorithms need to be modified in order to take into account the autocorrelation structure of the homogenous training samples. Craig (1979) recommends avoiding spatial autocorrelation in training samples or in samples used to test accuracy; researchers should not sample any closer than every tenth pixel. Campbell (1981) concludes that much more research is needed in this area. He suggests improved sampling strategies to avoid spatial autocorrelation instead of revising classification procedures because of the unique characteristics of specific cover types.

Research relating to map patterns and eigenvectors is relatively new. Naturally, autocorrelation arises from particular spatial arrangements and these arrangements

can be quantified. Using a geographic connectivity matrix ($\mathbf{C}$), where each cell has a value of either one (two pixels are adjacent) or zero (no adjacency), a pseudo-correlation matrix can be derived by adding one to each diagonal element. Griffith (1996) states that the eigenvectors extracted from this correlation matrix summarize different, distinct map patterns present in georeferenced data. He stresses that each eigenvector "identif[ies] the possible mutually orthogonal geographic distributions of attribute values with given levels of spatial autocorrelation." (Griffith, 1996, p. 352) Since the geographic connectivity matrix is used to generate these eigenvectors, the resulting eigenvectors identify every possible distinct map pattern associated with a specific surface partitioning. Griffith (1996) adds that the second eigenvector essentially identifies that map pattern associated with the highest possible degree of spatial autocorrelation present in the attribute data, the third with the next highest, and so on. Of note is that these eigenvectors (second, third and so forth of the pseudo-correlation matrix) are close approximations to the first, second, and so forth, eigenvectors of the ensuing expression (1).

Griffith (1996) published an empirical analysis using three different datasets. Griffith (1996) found that in these three datasets, two of the first three eigenvectors illustrate two orthogonal linear trends while a third exhibits a concentric ring pattern. Griffith (1996) also theorizes that these eigenvectors and the map patterns they represent may be surrogates of significant environmental variables (Griffith, 1996, p. 366). Others, including Tiefelsdorf and Boots (1995), examined this topic, too.

Tiefelsdorf and Boots (1995) linked the small sample sampling distribution of the Moran Coefficient (MC) index of spatial autocorrelation to the eigenfunctions of

$$\left(I - \frac{\mathbf{1}\mathbf{1}^T}{n} \right) C \left(I - \frac{\mathbf{1}\mathbf{1}^T}{n} \right) \qquad (1)$$

Griffith (1997, p. 4) suggests that Tiefelsdorf and Boots (1995) and Griffith (1996) both show that eigenfunctions of expression (1) "describe the full range of all possible mutually orthogonal map patterns." Griffith (1998) suggests that those eigenvectors of expression (1), denoted by matrix $\mathbf{E}$, and the eigenvectors of matrix $\mathbf{C}$ deserve attention, especially since in practice these two sets of eigenvectors are essentially the same. In addition, Griffith (1998) shows

that the MC may be rewritten in terms of these eigenvectors as

$$\frac{n}{\mathbf{1}^T C \mathbf{1}} \; \frac{Y^T E \Lambda E^T Y}{Y^T E E^T Y} = \frac{n}{\mathbf{1}^T C \mathbf{1}} \; \frac{b^T \Lambda b}{b^T b} \qquad (2)$$

where $b = E^T Y$ comes from the regression equation
specification of $Y = E\beta + \varepsilon$.

The principle of parsimony dictates that those regression coefficients for which $E_j^T Y \approx 0$ can be ignored;
but since matrix $\mathbf{C}$ is n-by-n, there is a total of n of
these coefficients, of which at most half are associated with map patterns depicting nonnegative spatial
autocorrelation. In order to achieve parsimony, then,
an efficient procedure needs to be developed for identifying $E_j^T Y > 0$, where the Ys here are LANDSAT
spectral bands. The first step then is to compute MC
for a given Y; the second step is to restrict attention to
the set of eigenvalues

$$\lambda_j = 2 \left[\cos\left(\frac{\pi k}{P+1} \right) + \cos\left(\frac{l\pi}{Q+1} \right) \right]$$

—the analytical expressions (Ord, 1975) of the diagonal elements of matrix Λ in expression (2) for a P-by-
Q regular square tessellation (e.g., the geographic
configuration of pixels constituting a remotely sensed
image)—for which the corresponding MC values are
larger than some prespecified value (say, 0.5).

Relevant Theorems Pertaining to Remotely Sensed Image Surface Partitionings

Griffith (1998) proves a number of theorems about
eigenvectors of matrix $\mathbf{C}$ that furnish useful support
for the analysis of remotely sensed data. The first per-
tains to the aforementioned relationship between
eigenfunctions of matrix $\mathbf{C}$ and expression (1):

THEOREM 1. If the means of the nonprincipal
eigenvectors of matrix $\mathbf{C}$ are zero (i.e., $\mathbf{1}^T E_k = 0$, k = 1, 2, …, n), then the eigenvectors of
expression (2) asymptomatically are the cen-
tered eigenvectors of matrix $\mathbf{C}$, with the principal eigenvector E_1 being replaced with vector
$1/\sqrt{n}\,\mathbf{1}$.

The importance of this theorem lies in its allowing data
analytic attention to be restricted to matrix $\mathbf{C}$, rather

than its cumbersome expression (1) counterpart, which
becomes important when a researcher is dealing with
massively large georeferenced datasets.

A second theorem provides the analytical expression for these eigenvectors, when a regular square surface partitioning is being studied:

THEOREM 3. The eigenvectors, $\mathbf{E}$, of binary
matrix $\mathbf{C}$ depicting a regular square tessellation
surface partitioning forming a P-by-Q rectangular region are given by

$$E_{pq} = \frac{2}{\sqrt{(P+1)(Q+1)}} \left[\sin\left(\frac{\pi k}{P+1} \right) \times \sin\left(\frac{l\pi}{Q+1} \right) \right]$$

k = 1, 2, …, P, l = 1, 2, …, Q, p = 1, 2, …, P, and
q = 1, 2, …, Q.

The importance of this theorem lies in its allowing the
eigenvectors of matrix $\mathbf{C}$ for a remotely sensed image
to be calculated without having to numerically extract
them from an n-by-n binary connectivity matrix, regardless of the magnitude of P and/or Q.

Two additional theorems reveal that the means of
the nonprincipal eigenvectors of matrix $\mathbf{C}$ for a remotely sensed image are or converge upon zero:

THEOREM 4. For a finite surface partitioned
with a regular square tessellation, a number of
eigenvectors have exactly mean zero.

THEOREM 5. As both P and Q of a rectangular
remotely sensed image region, where n = PQ,
go to infinity, all of the means of the eigenvectors of matrix $\mathbf{C}$ converge on zero.

The importance of these two theorems lies in their
indicating that the analytical eigenvectors furnished
by Theorems 1 and 3 are not only orthogonal (matrix
$\mathbf{C}$ is symmetric), but also nearly or exactly uncorrelated
for a typical remotely sensed image. As such, these
eigenvectors can be used as predictor variables in regression without fear of encountering complications
due to the presence of multicollinearity.

Study Area and Data Characteristics

One LANDSAT TM image dated August 10, 1995
is analyzed in this chapter. The coordinates of the sub-

Table 37.1. Selected Summary Statistics for TM Subset.

Band	MC	K-S	R	DN Range	Ymax/Ymin
1	0.61305	0.13575***	0.80336	39–58	1.5
2	0.74656	0.22303***	0.88747	9–30	3.3
3	0.78745	0.19494***	0.91188	6–30	5.0
4	0.90988	0.07344***	0.97692	3–133	44.3
5	0.91063	0.07991***	0.97604	0–93	93+
6	0.96526	0.16748***	0.98663	127–147	1.2
7	0.88087	0.08762***	0.95967	0–29	29+

set image are as follows: upper left easting: 490232.50; upper left northing: 4886007.50; lower right easting: 502707.50; lower right northing: 4873532.50. The 1995 TM image was georeferenced to New York State Transverse Mercator projection, zone 18 (Clarke 1886 earth ellipsoid).

The TM scanner has a spatial resolution of 28.5 meters, except for band 6 which has a resolution of 120 meters. Each TM image pixel was resampled (nearest neighbor) when georeferenced in order to yield a 25-meter square pixel. The subset images used in this study have 500 (P) rows and 500 (Q) columns (of 25-meter square pixels). Each resampled pixel is 25 meters square so the area is approximately 12,500 meters square or seven and three-quarters miles on a side. Each of these pixels has a brightness value associated with it. Thus, there are 250,000 pixels or brightness values in one band of data. With all of the seven bands, the subset TM image contains 1,750,000 data values.

Exploratory data analysis was conducted on the 7 bands of data. Table 37.1 reports the Moran Coefficients (MC), the Kolmogorov-Smirnov (K-S) statistics, the correlation coefficients (r), the range of digital numbers (DN), and the Ymax/Ymin (an index of the effect transformations might have on a measure). The Moran scatterplots for each of the bands portray strong spatial autocorrelation. Band 6's MC scatterplot is a graphical illustration of a very high MC (0.96526). Band 6 was resampled from 120 meter pixels, to 25 meter pixels consequently inflating the spatial autocorrelation contained in the pixel values. The general distribution of the seven bands is quite variable. Bands 1, 2, 3, and 7 are symmetric but have a high spread. Bands 4 and 5 are essentially uniform, but frequencies do increase slightly toward the center of their respective distributions. Band 6, the thermal band, is highly skewed to the left.

METHODS

Two of the objectives of this research concern planning issues—fire management and recreation management. Consequently, this chapter uses a small-scale damage classification scheme that has four categories. The damage classification scheme quantifies percent of forest cover blown down. Dobson et al. (1990), Foster and Boose (1992), Cablk et al. (1994), and Foster (1988) also use this methods. The classification scheme presented in this chapter classifies *overstory* blowdown because, according to Cablk et al. (1994), and Gibbs (1997), understory blowdown is extremely difficult to classify. In this chapter, blowdown is defined as resulting from windthrow (uprooting or leaning extensively), wind snap (trunk snapping or splintering), and severe branch loss and leaf defoliation. This definition of blowdown corresponds to the definitions developed by Foster and Boose (1992) and Dobson et al. (1990).

Multivariate Image Analysis

As outlined above, the underlying methods for all classifications to be generated in this chapter is Multivariate Image Analysis (MIA). This procedure is considerably different from the well accepted and widely used supervised classification methods, such as minimum distance, maximum likelihood and its variant, Bayesian classification. "[T]he MIA approach is unequivocally distinct from the mainstream image processing tradition" (Esbensen and Geladi, 1989, p. 70) as it is conceptually the reverse of the spectral image training sample methods. Thus, the following research question naturally arises: Why is MIA a fruitful method in classification and what does it offer that the classical classification methods of remote sensing do not?

First, MIA relies heavily on principal components analysis (PCA), which is related to the factor analytic family of techniques. PCA usually is used to expedite variable reduction in complex datasets. It takes many highly correlated variables and through various linear transformations develops a new set of uncorrelated, orthogonal components. This lack of correlation lets the components define specific dimensions that explain the variance within the original data (Dillon and Goldstein, 1984, p. 24). Finally, unlike factor analysis, PCA defines the components sequentially and all components are orthogonal. Factor analysis may involve rotating the components, and can be implemented to construct nonorthogonal variables.

Since PCA components are independent, each principal component holds unique information. PCA enhances the accuracy of data classification because the nonparametric generation of components eliminates hidden relationships present in the original data. In multiband satellite data, PCA is especially useful because many redundancies and unusual patterns exist in the data. Once the data are converted through PCA into new independent synthetic variables, conclusions are drawn without fear of these hidden relationships or contextual patterns biasing them.

One of the most powerful aspects of computer-assisted PCA in remote sensing applications is its ability to aid in classifying an image through the analysis of scatterplots of pairs of principal components. This procedure is called Multivariate Image Analysis (MIA) (Esbensen and Geladi, 1989). Classification using scatterplots is user driven, and consequently classes are delineated that are relevant to the research problem (ERDAS®, Inc., 1990, p. 5). The "salient backbone" (Esbensen and Geladi, 1989, p. 74) of MIA is linking the principal component (PC) image to the PC scatterplots. This linkage functions with a pair of cursors in both a PC image and its corresponding PC scatterplot. One of the cursors is moved throughout the PC scatterplot while the corresponding cursor follows it in the PC image. Delineation of pixel classes in spectral images is conducted by overlaying the spectral image on top of a PC image. Using this technique, spectral classes are masked directly on the PC and spectral images from the pixel class delineations in the PC scatterplot. An appropriate pixel aggregation that includes training areas then is chosen through an iterative two-way process. The user constantly updates the classes by moving the cursor through the PC scatterplots. Through this procedure, a very precise class is generated that includes the training areas and other smaller regions with similar variance properties.

Throughout the MIA procedure, the user is interacting with both the PCA transformed variables and the original bands. In traditional image processing, a user only interacts with the scene variables (bands). As a consequence, a user thinks only in terms of the original scene space and its related spatial parameters. Training sites are assigned spatially in the scene and correspond to known cover type locations on the ground. Small areas of similar cover type with varying brightness values usually are not included in the training sample sites. After running classification algorithms such as minimum distance or maximum likelihood, these small areas may not be included in a particular class because they are outside of the probability region or beyond the distance vector created by the training sites. Additionally, the traditional training sample approach uses pixels in the spectral image to generate classes, unlike MIA, which uses pixel agglomerations in PC scatterplots to delineate classes. In MIA, training sites in the spectral image are used, but only as elements of an iterative process in which the analyst examines the variance structure of the PC image and PC scatterplots. Using this iterative procedure, highly variable and widely spatially distributed classes are slowly delineated. MIA is well suited to delineate outlying classes that are only obvious in the higher order scatterplots. In spectral images, these pixel agglomerations may be difficult to locate due to their scarcity, spatial irregularities, and lack of contrast. It may be nearly impossible to generate sufficient training samples, rendering the conventional methods obsolete.

The PC scatterplots consist of a linear combination of pixel brightness values in one principal component plotted against a second linear combination of corresponding pixel brightness values in a second component. These two-dimensional scatterplots give multimodal interpretations of the data now expressed in the two principal components. These scatterplots are valuable because (with the lower order components plotted) they yield much information about a data set (Figure 37.1). A PC1 versus PC2 scatterplot is generated by taking a location's pixel value from principal components one and two and plotting the paired data point on a set of axes corresponding to brightness values. Each point in the scatterplot shows the number of values with that specific pair of PC brightness values (Figure 37.1). These scatterplots are

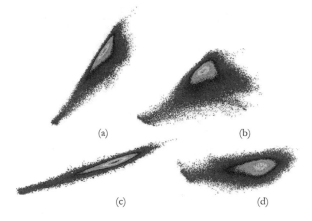

Figure 37.1. Principal components (PC) scatterplots derived from the spectral bands and the combined spectral bands and locational information. (a) PC1 versus PC2 (derived from spectral bands). (b) PC1 versus PC2 (derived from spectral bands combined with locational information). (c) PC2 versus PC4 (derived from spectral bands). (d) PC2 versus PC4 (derived from spectral bands combined with locational information). The components constructed with the locational information more clearly differentiate pixel agglomerations. The first component listed in the PC pairs corresponds to the horizontal axis.

coded for pixel modality with color intervals. The PC1 versus PC2 scatterplot often represents over 90% of the variability in an image (the seven spectral bands are highly correlated). Because of the high variability within the lower order scatterplots, unusual pixel associations are visible.

Eigenvectors as Variables in MIA

As illustrated above, the bands of multiband remotely sensed imagery are highly correlated. PCA accounts for these correlations between bands and generates a very useful summary of the data in the form of synthetic, uncorrelated factors that may be treated as synthetic variables per se. In addition to the correlation between the bands, there is also spatial autocorrelation (correlation inside each variable) within each band. This autocorrelation is not addressed by PCA because PCA only addresses the correlation between variables. Spatial autocorrelation may prove problematic because of the use of homogenous training areas in the iterative MIA classification process. It may be impossible to follow Craig's (1979) method and sample no closer than every tenth pixel for train-

ing sites due to monetary constraints or difficulties in navigating the terrain. If geographic continuity is sufficiently pronounced, spatial autocorrelation effects may extend beyond 10 pixels. In this study, both of these factors are present. If autocorrelation is not neutralized by the sampling procedure or addressed by modifying the classification method, users may generate a class that includes the training areas and other regions on the image, but is still not including highly variable pixels.

But redundant locational information represented by nonzero spatial autocorrelation can be accounted for, hence alleviating this problem. Eigenvectors illustrating significant map patterns present in the attributes of a specific data set will have $\mathbf{b}_j = \mathbf{E}_j^T \mathbf{Y} \neq 0$ and contribute to the value of MC as specified by expression (2). Because remotely sensed images routinely exhibit marked positive spatial autocorrelation, only those $\mathbf{E}_j$s characterizing relatively strong positive spatial autocorrelation map patterns need to be considered.

The vector $\mathbf{b} = \mathbf{E}^T \mathbf{Y}$ represents locational information associated with all prominent as well as trivial eigenvectors, and hence $\mathbf{C}(\mathbf{Y} - \overline{\mathbf{Y}})$ is the sum total of this locational information. Once significant eigenvectors are identified, these eigenvectors can replace $\mathbf{CY}$ in the MIA analysis with little or no loss of locational information. These variables summarize considerable, if not close to all, significant locational information present in the LANDSAT data. Using these eigenvectors in the MIA, Griffith (1998) suggests, will adjust for the spatial autocorrelation in the data converting the distribution of error in the data to one characterized by independence. This application treats the eigenvectors as missing variables accounting for spatial autocorrelation (Griffith, 1992). These eigenvectors will not combine with each other in the PCA since they are orthogonal and uncorrelated, but will combine with the spectral data to generate components derived from both attribute (spectral) data and locational data. These components will form the basis of MIA.

Griffith (1998) argues that the eigenvectors can be viewed as a spatial manifestation of an unknown environmental variable. In a regression context, researchers should examine the spatial pattern outlined by each prominent eigenvector and try to identify an environmental variable that displays the same spatial distribution. This chapter does not try to identify associations—rather, this chapter employs the locational information for classification purposes. In addition, with the MC scatterplots revealing marked levels of positive spatial autocorrelation, the number of candidate eigenvectors here ranges from 36,000 to more than 63,000. An efficient strategy for sifting through this massive number of distinct map patterns is required. One possibility is to draw a simple random sample, and explore it using stepwise regression of selected eigenvectors on specific spectral bands. This approach is feasible because for all practical purposes the n = PQ eigenvectors are mutually uncorrelated. A second possibility is to perform a cluster analysis on the set of eigenvalues, and then search within each group for that eigenvector displaying the largest correlation with a given spectral band. A third possibility is to sequentially estimate the (k,l) values using nonlinear least squares regression. This approach also can involve initially working with a random sample of pixels. And a fourth possibility is to implement the first approach, but with canonical correlation rather than stepwise regression. The first of these approaches was adopted for this analysis, with its results summarized in Table 37.2. The percent of variance accounted for by each of the eigenvectors listed in Table 37.2 was computed with the entire data set, making it more precise than that calculated with the simple random sample.

Of particular note in Table 37.2 is the prevalence of eigenvector (1,4). This is probably an eigenvector representing a distinct map pattern covering the complete study area. It relates to all bands including band 4, but (1,4) only accounts for 0.5% of the variance in band 4. Additionally, in band 6 the eigenvectors having large MCs account for much of the variance in this highly autocorrelated spectral measure, with just (1,2) and (1,4) accounting for approximately 25% of variance. This finding is not surprising, given that band 6 is based upon a much coarser geographic resolution.

Once the appropriate quantification of locational information has been selected it can be rescaled to the range [0,255], with each value being converted to an integer. Each integer is attached to a pixel in the remotely sensed image. These synthetic bands represent locational information and are a spatial manifestation of prominent map patterns within the LANDSAT TM subset. A PCA can be conducted on the synthetic locational bands combined with the spectral bands. While these locational bands are measured here with **CY**, research continues on how to sufficiently identify the prominent underlying E_js. This PCA generates a series of component scatterplots that can be used to classify the TM subset.

RESULTS AND DISCUSSION

Two classifications were generated for this chapter. One classification was conducted using just the 7 spectral bands, and a second classification was generated using the 7 spectral bands and latent locational information. The locational information utilized in this classification was the spatial lag for each pixel—the sum of the regression-coefficient weighted eigenvectors of matrix **C**. The spatial lag was defined as the rook-case adjacency and these pixel values then were averaged to get the spatial lag value. These spatial lags are the summary of all possible distinct map patterns underlying an image. Seven spatial lag variables were generated, one for each band. Selected eigenvectors themselves were not used directly because, as stated earlier, they were extremely difficult to identify across the entire data set. Exploratory results reported above, however, strongly suggest that such eigenvectors will further refine the methods being developed here.

There were several differences between the principal components resulting from spectral bands and the principal components resulting from the spectral bands and locational information. The final classifications themselves also differ. The first notable difference is the structure of the principal components scatterplots. Figure 37.1 shows four principal components scatterplots. The scatterplots for components 1 and 2 are clearly different. The scatterplot resulting from the spectral bands (a) is much more condensed while the scatterplot resulting from the spectral bands and the locational information (b) is clearly less condensed. The most striking difference between the PCA output from these two data sets is shown in Table 37.3. The components of both data sets have a similar variance structure. In each data set, components one and two account for approximately 55% and 24% variance each. Yet visual examination of the scatterplots of these components (Figure 37.1), reveals that they are conspicuously different. The locational information captured in (b) and (d) (Figure 37.1) clearly highlights several unusual patterns.

A set of 10 GPS points and corresponding field notes were used as reference data to generate the classes. These reference data were preliminary and future research will be supplemented by information available through aerial photographs. The classification generated from the spectral bands shows the scattered nature of the blowdown. Similar spectral signatures seem to be much more concentrated in the second classification that incorporates locational information. It was

Table 37.2. Prominent Eigenvectors According to % Variance Accounted For.

Band	MC	(k,l)	% Variance Accounted For
1	0.61305	(1,4), (2,1)	1.41%, 1.19%
2	0.74656	(1,4), (2,3)	1.35%, 1.23%
3	0.78745	(1,4), (2,3), (1,6)	2.42%, 1.94%, 0.87%
4	0.90988	(1,1), (2,3), (6,5), (5,3), (6,4)	2.00%, 1.64%, 0.89%, 0.82%, 0.80%
5	0.91063	(1,1), (4,6), (2,6), (4,2), (1,4), (6,4), (1,4)	2.35%, 2.29%, 1.30%, 1.14%, 0.96%, 0.93%, 0.85%
6	0.96526	(1,2), (1,4), (2,3), (1,6) (2,1), (5,3), (3,2), (1,14)	13.50%, 8.09%, 2.52%, 1.86%, 1.83%, 1.42%, 1.16%
7	0.88087	(4,6), (1,4), (1,1), (2,6), (4,2)	2.65%, 2.52%, 1.63%, 1.36%, 1.20%

Table 37.3. Eigenvalues and % Variance Explained for the Two PCA Data Sets.

Spectral and Locational Data		Spectral Data Alone	
7.87873	56.2766%	3.84549	54.9355%
3.46425	24.7446%	1.70227	24.3181%
0.84282	6.0202%	0.52467	7.4953%
0.67412	4.8152%	0.42037	6.0052%
0.48969	3.4978%	0.32501	4.643%
0.19615	1.4011%	0.15214	2.1734%
Remainder	1.5326%	0.03006	0.4295%

more difficult to generate large, rather ambiguous classes with the second set of components derived from the spectral and locational information. This phenomenon is visible if one compares the two classifications. The classification generated with the spectral bands has classes that cover a larger area—especially the moderate blowdown class. The second classification generated from the spectral bands combined with the locational information yields classes that are more concentrated.

During the iterative classification process, only scatterplots of PC1 versus PC2 were used to classify the image. It is very easy to generate large classes using the component scatterplots corresponding to the spectral data alone. The scatterplot associated with the combined spectral bands and the locational information is more discriminating during class delineation.

In conclusion, much of this research is still in a preliminary stage. Incorporating locational information into the MIA enhanced the classification process. One goal of future research will be to include only eigenvectors that represent distinct map patterns found within the image in question, hence eliminating both noise and correlation among the CY_js. Furthermore,

classifications need to be established using higher-order scatterplots of the principal components. In this analysis, an accuracy assessment was not conducted due to time and monetary constraints. A full accuracy assessment is planned for summer 1998.

Finally, this research represents a very modest attempt to pursue the application of statistical and particular spatial statistical methods to massively large data sets. Cressie et al. (1995) emphasize the importance of such research—especially concerning environmental data sets. They specifically mention the EPA EMAP study as well as that the DEM, LANDSAT TM, and AVHRR satellite systems merit attention (Cressie et al., 1995, p. 118). Further work on research problems associated with the one problem presented in this chapter relating to spatial autocorrelation in remotely sensed data can only improve our general knowledge of these data, thus enhancing our understanding of reality.

REFERENCES

Cablk, M.E., B. Kjerfve, W.K. Michener, and J.R. Jensen. Impacts of Hurricane Hugo on a Coastal Forest Assessment Using LANDSAT TM Data, *Geocarto International* 9, pp. 15–24, 1994.

Campbell, J.B. Spatial Correlation Effects upon Accuracy of Supervised Classification of Land Cover, *Photogrammetric Eng. Remote Sensing,* 47, pp. 355–363, 1981.

Cressie, N., A. Olsen, and D. Cook. Massive Data Sets: Problems and Possibilities, with Application to Environmental Monitoring, in *Massive Data Sets: Proceedings of a Workshop,* 1995, pp. 115–119.

Crosta, A.P. and J. McMoore. Enhancement of LANDSAT TM Imagery for Residual Soil Mapping in SW Minais Gerais State, Brazil, in *Proceedings of the Seventh (ERIM) Thematic Conference: Remote Sensing for Exploration Geology,* 1989, pp. 1173–1187.

Craig, R.G. Autocorrelation in LANDSAT data, in *Proceedings of the Thirteenth International Symposium on Remote Sensing of Environment,* 1979, pp. 1517–1524.

Dillon, W.R. and M. Goldstein. *Multivariate Analysis: Methods and Applications,* John Wiley & Sons, Inc., New York, 1984.

Dobson, J.E., R.M. Rush, and R.W. Peplies. Forest Blowdown and Lake Acidification, *Ann. Assoc. Am. Geogr.,* 80, pp. 343–361, 1990.

ERDAS® Inc., *Multivariate Image Analysis,* Atlanta, ERDAS® Inc., 1990.

Esbensen, K. and P. Geladi. Strategy of Multivariate Image Analysis, *Chemometrics Intelligent Lab. Syst.,* 7, pp. 67–86, 1989.

Foster, D.R. Disturbance History, Community Organization and Vegetation Dynamics of the Old-Growth Pisgah Forest, South-Western NH. *J. Ecol.,* 76, pp. 105–134, 1988.

Foster, D.R. and E.R. Boose. Patterns of Forest Damage Resulting from Catastrophic Wind in Central New England, USA. *J. Ecol.,* 80(1), pp. 79–98, 1992

Geladi, P., H. Isaksson, L. Lindqvist, S. Wold, and K. Esbensen. Principal Component Analysis of Multivariate Images, *Chemometrics Intelligent Lab. Syst.,* 5, pp. 209–220, 1989.

Gibbs, J. Supervising Forester, DEC Region Six. Interview by author. November 13, 1997. DEC office, Canton, NY, 1997

Griffith, D.A. What is Spatial Autocorrelation? Reflections on the Past 25 Years of Spatial Statistics, *L'Espace geographique,* 3, pp. 265–280, 1992.

Griffith, D.A. Spatial Autocorrelation and Eigenfunctions of the Geographic Weights Matrix Accompanying Geo-Referenced Data, *Can. Geogr.,* 40, pp. 351–367, 1996.

Griffith, D.A. Visualization and Spatial Autocorrelation, paper presented at the PRSCO/ANZRSA Conference. Wellington, NZ, 1997.

Griffith, D.A. Useful Eigenfunction Properties of Popular Geographic Weights Matrices employed in Spatial Autocorrelation Analyses, submitted to *Linear Algebra and Its Applications,* 1998.

Ord, J. Estimation Methods for Models of Spatial Interaction, *J. Am. Stat. Assoc.,* 70, pp. 120–126, 1975.

Tiefelsdorf, M. and B. Boots. The Exact Distribution of Moran's I, *Environ. Plann. A.,* 27, pp. 985–999, 1995.

Tubbs, J. D. and W.A. Coverly, Spatial Correlation and its Effect Upon Classification Results in LANDSAT, in *Proceedings of the Twelfth International Symposium on Remote Sensing of the Environment,* 1978, pp. 775–781.

Wang F. A Knowledge-Based Vision System for Detecting Land Changes at Urban Fringes, *IEEE Trans. Geosci. Remote Sensing,* 31, pp. 136–145, 1993.

Wienke, D., W. van der Broek, and L. Buydens. Identification of Plastics in Mixed Waste by Remote Sensing Near-Infrared Imaging Spectroscopy, *Anal. Chem.,* 67, pp. 3760–3766, 1995.

CHAPTER 38

High Order Uncertainty in Spatial Information: Estimating the Proportion of Cover Types Within a Pixel

M. Hughes, J. Bygrave, L. Bastin, and P. Fisher

INTRODUCTION

Land-cover classification of satellite imagery using traditional methods places each pixel in a single class. In medium resolution imagery, such as SPOT or Landsat, the structure of the landscape is such that few pixels are occupied by one cover type, especially in suburban areas and the majority of the image will contain mixed pixels (Fisher and Panthirana, 1990). Several methods have offered solutions to the mixed pixel problem by deriving partial memberships including fuzzy classification techniques (e.g., Foody, 1996a), linear mixture modeling (e.g., Shimabukuro and Smith, 1991) and artificial neural networks (Foody, 1996b). Fuzzy memberships allow a pixel to belong to more than one class, with a degree of membership in each class. These proportions are subject to errors such that for any estimate of fuzzy membership there is a higher order uncertainty, which is the probability of the estimate being correct.

As part of a project to study the viability of extracting fuzzy memberships from satellite imagery (FLIERS), a verification data set was compiled. This data set describes the detailed land cover of a suburban area and is derived from high-resolution aerial photography. In the project neural network classifiers are trained using Landsat image pixels and corresponding land-cover class proportions from the data set. We recognize that there are many sources of error in such a process and here we describe several methods for investigating the effects of these errors on proportion estimates.

One of the main sources of error is the locational uncertainty of the verification data within the satellite image. For example, the pixel value does not relate to ground cover in a pixel-shaped square on the ground, but an area which is controlled by the sensor sampling function and usually overlaps into neighboring pixels (Fisher, 1997). The actual pixel area will also be affected by atmospheric distortion and terrain relief, such that the exact location of a pixel can only be assumed to fall somewhere in a 3 by 3 window (Justice and Townsend, 1981). Imagery is also geometrically corrected to the verification; ground data to some RMSE. Furthermore, those verification data will have their own locational error which will add to the uncertainty of their position within the satellite image. If such data are being used to provide pixel by pixel cover proportions, then an understanding of the sensitivity of the proportion estimates to these error sources is crucial.

In this chapter we suggest a number of ways to sensitize the estimation process to these uncertainties.

STUDY AREA

A suburb of Leicester which contains many different land uses including medium and high density housing, light industrial and manufacturing complexes, parkland, and university buildings has been chosen for one test site. Verification data was derived from interpretation of aerial photography taken at 25 cm resolution of an area covering 1 km^2. The photography was corrected for aircraft distortion and was georeferenced to the British National Grid. The absolute accuracy of the aerial photograph was determined by ground control points taken from 1:10,000 mapping and found to have approximately 5 meters

RMSE. The interpretation, carried out by on-screen digitizing, separated the area into different land-cover types according to a detailed classification scheme, although for clarity only two classes, slate roof and asphalt, are shown here (Figure 38.1), and only one, asphalt, is discussed in detail.

SINGLE ESTIMATE OF LAND-COVER PROPORTION

In order to provide a single estimate of land-cover proportion per pixel, the verification coverage was co-registered with a subset of Landsat imagery (10 m RMS error). The resulting transform equation was inverted and used to transform the polygon coverage so that it could be overlaid on the satellite imagery. ArcInfo GIS was then used to derive a polygon coverage of pixel outlines with proportion values for each class attached as polygon attributes. These data can be viewed as proportion images which use gray scales to represent the extent of a class within each pixel. Figure 38.2a shows the proportion estimates for asphalt.

To model the uncertainty of the location of the satellite image pixels, 3×3 focal area filters were applied to the verification data to produce various statistical summaries of local dispersion (Justice and Townsend, 1981). Mean and standard deviation images are shown in Figure 38.2b,c.

HIGHER MOMENTS OF THE PROPORTION ESTIMATE FROM PIXEL PERTURBATIONS

To model further the effects of positional error, the pixel coverage was perturbed in a Monte Carlo simulation and new pixel proportions calculated for each realization. The pixel perturbation routine takes a polygon coverage and creates any number of equally probable versions of it based on user-input error grid size and standard deviation value (Hunter and Goodchild, 1996; Hunter et al., 1996). These realizations were then used to provide a set of equally probable pixel proportions. Summary statistics for each class were generated from the set of new proportions.

The statistics generated on a per-pixel basis are: mean, standard deviation, skewness, kurtosis, range and difference (Figure 38.2d–i). We call the mean proportion of all realizations the "fuzzy mean" since it accounts for some degree of error in the original data set. Again, only the results for asphalt are shown. The number of realizations was 20 with an error grid spacing of 15 m (Hunter et al., 1996), and an RMSE for

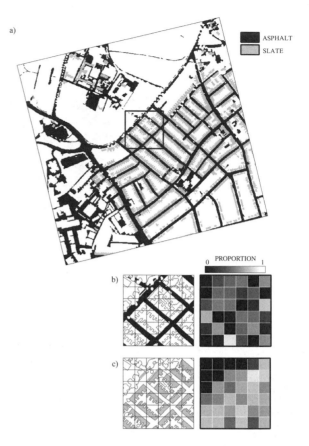

Figure 38.1. (a) Verification data for 1 km^2 derived from aerial photography and transformed to satellite image coordinates. A subset is enlarged to illustrate the process of pixel grid overlay and production of proportion data for: (b) asphalt and (c) slate.

the pixel locational error of 3 meters. These initial variables produced plausible versions of the original pixel grid, and simulate the effects of pixel area uncertainty discussed earlier.

DISCUSSION

We have produced several visualizations of mean proportion and higher order statistics for land-cover classes in the study area based on three methods. First, we use the single estimate proportion from an overlay of the pixel grid and the verification data. The second method is a simple filtering procedure that computes statistics in a 3 by 3 window. Finally, we use a Monte Carlo analysis to randomly sensitize the pixel outlines and produce a number of equally probable proportion images which are used to derive summary statistics.

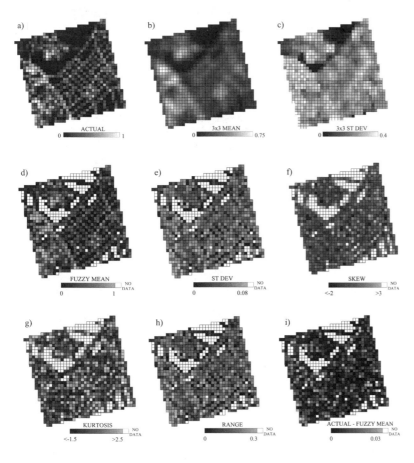

Figure 38.2. The actual proportions of asphalt derived from a single overlay of pixel grid in (a) are filtered to produce: (b) mean and (c) standard deviation from a 3 x 3 window. The set of 20 pixel grid perturbations were processed to produce the following statistics for asphalt on a per-pixel basis: (d) fuzzy mean, (e) standard deviation, (f) skewness, (g) kurtosis, (h) range, i.e., maximum proportion minus minimum proportion and (i) difference between actual proportion and fuzzy mean proportion.

The single estimate proportion image gives an easily interpreted output but conveys no information about the certainty of the proportion being correct, given that there are a number of potential error sources. The result of the 3 by 3 filtering gives an estimate of land-cover proportions that accounts for uncertainty due to poor knowledge of pixel location. The filtered mean may be a more appropriate training set for a neural network and the filtered standard deviation provides information on textural qualities of the cover-type as well as the error estimate of the filtered mean. Pixels with a high standard deviation for asphalt occur where, within the window, there is an isolated block or large strip surrounded by empty pixels. For homogenous areas larger than the window size the standard deviation would be zero, with intermediate values where the asphalt is spread fairly evenly across the window.

The 3 by 3 window is one of a number of possible neighborhoods that can be used in this way. In order to model a pixel location uncertainty that was not in whole pixel units a more complex filter could be used. This would entail resampling the satellite imagery to a pixel size compatible with the filter. Alternatively, a succession of possible pixel grids could be created with random shifts in the X and Y direction such that the RMS shift matched the RMS error in image registration. Such a procedure could be included in a Monte Carlo analysis which would result in summary statistics similar to the pixel perturbation method.

The pixel perturbation routines provide another way of simulating uncertainty in vector data. As expected, the fuzzy mean is well correlated with the actual mean for the parameters used (Figure 38.3a). However, if the RMS error in the perturbation routine is made larger, then the relationship can be expected to deteriorate. It is apparent that the uncertainty in land-cover proportion should be influenced by the scale of the land-cover objects in relation to pixel size and how the land-cover objects are distributed across the image.

Associated with these fuzzy means are higher order statistics. Figure 38.3b–d shows that these statistics are independent of the fuzzy mean except when proportions tend to 0 or 1, and therefore each displays a different property of the proportion estimate. The

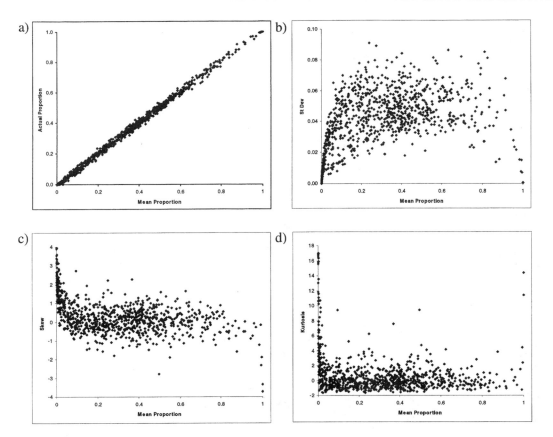

Figure 38.3. Scatter plots of mean proportion of asphalt per pixel from 20 realizations of the perturbed pixel grid against: (a) actual proportion, (b) standard deviation, (c) skewness, (d) kurtosis.

standard deviation indicates the spread of proportions in the set of realizations. A low value implies one of three things: first, that only a few of the realizations contained the class at all, second, that the proportion of the class in the pixel is very small, or third, that the pixel is dominated by the cover-type. The skewness and kurtosis describe further this set of realizations by giving measures of symmetry and peakedness of the distribution of proportion values.

Taken together, these statistical parameters could be used to provide a qualitative measure of the reliability of the mean proportion. For example, the mean proportion for a pixel with low standard deviation and a peaked (leptokurtic) nonskewed distribution is a more reliable estimate than one with the same mean but large standard deviation and a skewed distribution. Furthermore, skewness could be used to provide an adjustment to the mean, making it more reliable.

The evaluation of uncertainty of fuzzy membership proportions has become an area of major research.

Monte Carlo analysis, though computationally expensive, is still the most convenient method for experimenting with the effect of uncertainty in source data on the quality of GIS-derived products (Canters, 1997). We have used it here to randomly sensitize the pixel outlines used to calculate land-cover proportions. In the future, we will develop this approach for perturbing the verification coverage itself to model locational errors in the aerial photographs and digitizing error. However, the pixel perturbation routine is slow, and experimentation with the coverage used here has not been possible to date.

CONCLUSIONS

The methods described above have been used successfully to model the uncertainty that arises from locational error in both the satellite image and the verification data. They are applicable to data at any scale, and provided enough computer resources are available, exhaustive analyses can be carried out. The outputs from this process could be used to train a neural network classifier with mean proportions instead of the usual single proportion estimate. Alternatively, training could be carried out with the single proportion estimate and an a priori measure of certainty or confidence that the estimate is correct. This measure or index of certainty would be related to the higher order statistics, although its exact definition requires further work.

Clearly there are many factors contributing to error and uncertainty when trying to derive proportions of land-cover types within satellite image pixels. The results presented here are an indication of how we will develop insights as to the nature of these errors, and highlight the potential use of higher order statistics such as skewness and kurtosis.

ACKNOWLEDGMENTS

The authors wish to thank Gary Hunter for the use of his vector perturbation code. This work is supported by the European Union, DG XII, under the Environment and Climate Programme, Contract No: ENV4-CT96-0305.

REFERENCES

Canters, F. Evaluating the Uncertainty of Area Estimates Derived from Fuzzy Land-Cover Classification, *Photogrammetric Eng. Remote Sensing,* 63(4), pp. 403–414, 1997.

Fisher, P. The Pixel: A Snare and a Delusion. *Int. J. Remote Sensing,* 18(3), pp. 679–685, 1997.

Fisher, P.F. and S. Panthirana. The Evaluation of Fuzzy Membership of Land Cover Classes in the Suburban Zone, *Remote Sensing Environ.,* 34, pp. 121–132, 1990.

Foody, G.M. Approaches for the Production and Evaluation of Fuzzy Land-Cover Classifications from Remotely-Sensed Data, *Int. J. Remote Sensing,* 17(7), pp. 1317–1340, 1996a.

Foody, G.M. Relating the Land-Cover Composition of Mixed Pixels to Artificial Neural Network Classification Output, *Photogrammetric Eng. Remote Sensing,* 62(5), pp. 491–499, 1996b.

Hunter, G.J. and M.F. Goodchild. A New Model for Handling Vector Data Uncertainty in GIS, *J. Urban Reg. Inf. Syst. Assoc.,* 8(1), pp. 51–57, 1996.

Hunter, G.J., B. Hock, M. Robey, and M.F. Goodchild. Experimental Development of a Model of Vector Data Uncertainty, in *Spatial Accuracy Assessment in Natural Resource and Environmental Sciences: Second International Symposium, Fort Collins, Colorado,* 1996. USDA Forest Service, 1996, pp. 217–224.

Justice, C.O. and J.R.G. Townsend. Integrating Ground Data with Remote Sensing, in *Terrain Analysis and Remote Sensing,* J.R.G. Townsend, Ed., George Allen and Unwin, 1981.

Shimabukuro, Y.E. and J.A. Smith. The Least-Squares Mixing Models to Generate Fraction Images Derived from Remote Sensing Multispectral Data, *IEEE Trans. Geosci. Remote Sensing,* 29, pp. 16–20, 1991.

CHAPTER 39

Successive Approximation of Multiband Images Using Hyperclusters: The "PHASE" Approach

W.L. Myers

INTRODUCTION

New perspectives on old ways of doing things can have major benefits, which is proving to be the case for clustering in regard to image analysis. Clustering is central to computer-assisted thematic mapping in the so-called *unsupervised* mode, whereby the number of clusters extracted is commensurate with the number of thematic categories that are contemplated. In a break with tradition, Kelly and White (1993) showed that a much larger number of clusters could be advantageous if the task of labeling clusters could also be computer facilitated. Although their so-called *hyperclustering* did not really change the role of clustering for image analysis, their computer facilitation of cluster labeling involved colorizing clusters according to mean values of the respective image bands. It was noting that colorized clusters had considerable fidelity to conventional image renderings which inspired the current development for a broader purpose.

It seemed plausible that hyperclusters might serve as surrogates for entire image data sets if clustering could be customized to the special nature of landscapes instead of being naively numerical as conventionally conducted. Such a surrogate data form would constitute a hybrid between images and thematic map data. With a table of band means by cluster, the format would be essentially that of a multiattribute cellular GIS layer and it could be displayed in pseudocolor mode accordingly. Conversely, the clusters would effectively comprise distributed superpixels or spatial ensembles for which image analysis analogs could transpire very rapidly in the table domain. This idea was given the acronym PHASE for *Pixel Hyperclusters Approximating Spatial Ensembles*, and implemented as modular working prototype software (Myers, 1997).

A *PHASE* is a layer w/table(s) approximation to multiband digital (image) signals over an area which identifies uniformity based (patchwork) segments in the layer and gives segment statistics for band characteristics in table(s). The original image data is thus represented by segment averages, leaving intrasegment departures from average as residuals. Since residual variation precludes *exact* image restoration, the PHASE compression becomes a separate information product apart from copyright on the original data. Precedent in this respect has been set by EROS Data Center (Sioux Falls, South Dakota) redistribution of hyperclusters done in the manner of Kelly and White as introduced above.

Continued experimentation has shown this to be a very fertile approach that spawns new spatial capability for image analysis and encompasses more general forms of environmental signals (Myers et al., 1997a,b). The approach has evolved in sophistication of strategy and tactics, particularly with regard to compressive clustering. Clustering is progressive with relatively straightforward initial stages and provisions for subsequent adaptive refinement involving different degrees of complexity. Basic utility is not encumbered by advanced considerations.

FIRST PHASE COMPRESSION

The essence of the PHASE approach is image segmentation by clustering in which *each* segment or cluster forms a geospatial patchwork. Band averages over the respective segment are then used as image surrogates for *all* pixels belonging to the segment, wherever they may occur in the image. Clustering is done in (spectral) feature space without explicit concern for the nature of the patchwork that a cluster will form in geospace. It is, however, the patterning of signals in (two-dimensional) geospace that induces densification and rarefaction in feature space for which the signal (band) variables constitute axes.

Two things are of major importance for choice of clustering strategy. One is to have a way of exerting influence according to general knowledge of scene properties. The other is to have protocol that is amenable to parallelism in future generations of computers. These considerations led to the choice of the ISODATA clustering strategy (Tou and Gonzalez, 1974) as a basis for modification. ISODATA uses a predetermined number of cluster *seeds* as virtual condensation nuclei on which pixels coalesce to form cluster aggregates. The strategy involves a series of tentative aggregation cycles, each of which adjusts positions of seeds in feature space. In each cycle the pixels associate with the seed that is nearest in a euclidean distance sense. The seed is then moved to the centroid position among its associates. The number and positions of seeds provide a way of introducing general knowledge about the scene. Pixels can be considered in subsets that yield partial sums for purposes of parallel computation within a cycle.

The ISODATA strategy is too simplistic, however, to serve present purposes without modification. One issue concerns number and placement of initial seeds. ISODATA is most appropriate when the number of seeds is intrinsic to the application setting, which happens to be true for compressive image clustering. Capability for handling byte binary layers is almost universal among image analysis and raster GIS systems. Such layers have a capacity for 256 distinctions, with zero being a natural candidate to flag missing data. The compressive information capacity for the layer is thus 255 clusters or segments, with anything less being informationally wasteful.

Appropriate positions in feature space for the 255 initial seeds are much less obvious, with the usual ISODATA heuristics being questionable. Landscape understanding thus comes into play. For environmental signals generally, extreme or near extreme values tend to lend definition to landscape spatial structure. It thus becomes important for extremes to retain some separateness of identity in the compression. This can

be assured by having the initial seeds well dispersed over the occupied portions of feature space. The PHASE solution to this is a special preparatory algorithm which picks actual pixels as seeds while continually seeking to increase the nearest neighbor distance among the seed pixels. This is roughly analogous to placing a repulsive charge on the seeds. The first operation for PHASE compression is therefore to run the seed generator program. This program is fairly computationally intensive, but is not iterative.

There is also a counter concern regarding extremes for the iterative mainstage of ISODATA clustering. A *fringe* of outliers can effectively dominate the clustering so that most of the informational capacity for distinction is absorbed by a minority comprised of the more extreme values. This is the ultimate of informational inefficiency, since informational efficiency favors a balance in number of pixel associates among the seeds. It thus becomes important to have capability for controlling retention or dissolution of small clusters. The usual heuristics with ISODATA in this regard simply reduce the number of clusters by deleting seeds that have few associates, which would have the unfortunate consequence of underutilizing the byte binary layer capacity. This concern has been addressed by a major augmentation of ISODATA that provides for transplanting small cluster seeds in a manner that splits other clusters having large sums of squared distances from centroid. The latter is accomplished by tracking seed migration for each cluster, and using a previous seed position as a transplant site. Whereas ISODATA itself is agglomerative, the PHASE clustering facility is thereby dually agglomerative and divisive.

A final concern for initial or 1stPHASE clustering is that of computational feasibility for large data sets on PC-level computers. For a given number of bands, computational time is essentially linear in number of pixels. Whereas gridlets of a few hundred pixels on a side can be processed in a few minutes, large images on the order of 50 million pixels can take several days running in the background to accomplish a dozen iterations. It thus becomes desirable to run a few cycles, examine the state of compression, and then make a determination regarding need for additional cycles. This has been accommodated by having identical formats for the input file of seeds and output file of cluster centroids. Therefore the cluster centroids from a previous run can be used as seeds for a subsequent continuation run. Tactical logistics thus favor sequential runs of a few cycles each, as opposed to one very long run consisting of many more cycles.

Provisions are also made to facilitate rapid assessment of current compression. Clusters are renumbered before each cycle in order of increasing length for centroid vector. Despite minor perturbation that may occur in the ensuing cycle, this permits the grid of cluster numbers to be displayed as a gray-scale image. A facility is also available for subtracting the compressive approximation from the original image data to obtain residuals, and then making a byte binary layer of residual vector lengths for viewing. Pervasive spatial pattern in a gray-scale display of residual vectors is indicative that the approximation is subject to appreciable improvement. When the compressive approximation changes little in a further cycle, then 1stPHASE clustering has been consummated.

Postclustering operations for a 1stPHASE approximation are conducted as necessary to obtain two additional tables that supplement the table of cluster means (centroids). One of these is an intermediate lookup table of *enhancements* that is needed for setting up various displays. The other is a table of cluster variability statistics.

A 1stPHASE compression is intended to be generic in the sense that it should serve as an image surrogate for any purpose. PHASE software facilities provide for colorizing clusters to mimic any desired three-band view, as well as offering a variety of other innovative depictions. This includes ability to highlight ten clusters at a time in color on a gray-scale background of the remaining clusters. Clusters can be selectively collapsed, and clusterwise capability for both supervised and unsupervised classification is available as well. Clusterwise unsupervised classification involves further adaptation of ISODATA to allow for weighted clustering. This produces clusters of clusters wherein each cluster carries weight according to its number of pixels.

SUBPHASE EXPANSION

For particular purposes, some things in an image may be of more interest than others. Forests are of special interest to a forester, and forests happen to have relatively low contrast and relatively subdued spatial detail. In such circumstances, a customized subPHASE can be developed to provide additional detail regarding the focus of interest.

SubPHASE development begins with a preliminary supervised classification to put all things of interest in one class while leaving everything else unclassified. This serves to *mask out* things that are not of interest. A new clustering is then conducted on a limited basis

whereby 255 clusters are determined just for the preclassified portion of the scene, with the unclassified portion of the scene being designated as *offsite* by assigning such pixels a cluster number of zero which indicates missing data.

Taking full advantage of a subPHASE requires display software that can do thematic overlays or image backdrops. The generic 1stPHASE is selected as the bottom layer or backdrop with one color scheme. The subPHASE is selected as the top layer or overlay with a different color scheme. The zero or missing data portion of the top layer (overlay) is then made transparent for the viewer. The subPHASE detail thus stands in contrasting coloration to the remainder of the scene which "shows through" in the other color scheme.

SubPHASEs constitute an extension of the unsupervised strategy that image analysts call *cluster busting*. A more generic version of this idea involves doing a generalized preclassification by overlaying a residual image on the 1stPHASE and making a class that includes all sectors of the image having markedly strong residuals. A full complement of 255 clusters is then determined for this more deviant part of the scene as a subPHASE.

DUOPHASE ENHANCEMENT

The subPHASE strategy requires that a separate layer be maintained for a focal component of the image. A more modest version of the idea can be used to enhance a portion of an image that tends to have low relative contrast without requiring the maintenance of a second layer. The subPHASE is of a more limited nature, and gets *folded back* into the 1stPHASE, creating a hybrid that is called a duoPHASE.

DuoPHASE development must be anticipated, since it requires modification in the way the 1stPHASE is developed. The modification consists of having less than the full 255 clusters for the 1stPHASE in order to "leave room" for the enhancement. One might, for instance, develop the 1stPHASE with 200 clusters. This would leave 55 *additional* clusters available for doing enhancement. One then proceeds with preliminary classification for focal components of the scene in the same manner as for a subPHASE. Suppose that 20 of the original 200 clusters were included in the preliminary classification. A subPHASE would then be developed using only 20+55=75 clusters, which could more than double the level of detail for the enhanced components of the scene. PHASE program facilities are next used to integrate the 1stPHASE and subPHASE by causing the enhanced clusters to replace their original cluster counterparts while leaving other portions of the scene intact. The two sets of cluster numbers are also resequenced as a single series.

SUPERPHASE AUGMENTATION

SubPHASEs and duoPHASEs usually focus on particular components of a scene, although it is possible to do a more generic subPHASE that is motivated by strong residuals as suggested earlier. SuperPHASEs constitute a rather different strategy for dealing with pronounced residuals that have definite spatial pattern.

The superPHASE methodology consists of developing a PHASE for residuals about a previous PHASE, thus yielding a second order PHASE or 2ndPHASE. This method can be applied recursively to produce still higher order PHASEs if there still remains pronounced spatial pattern in the residuals about a superPHASE. Data volume obviously increases with each higher order PHASE since an additional layer is generated at each stage. A key point, however, is that a subPHASE can be generated from superPHASEs as well as from original data. This allows the recipient of PHASE compressed data to hold open the option for later exploration of subPHASEs as need might arise. A series of superPHASEs thus becomes a sort of image analog for series approximations to complicated mathematical functions. As each additional layer (term) in the series is generated, it should be examined for strength of spatial pattern in itself, as well as for any spatial pattern that might persist in the residuals that still remain. When the residuals assume the character of spatial white noise, then development of further PHASEs would be futile.

If sufficient media capacity is available, it may also be desirable to have a companion residual image for a PHASE distribution so that the recipient is capable of determining where approximation has particularly low fidelity with its parent image. Current residual image capability in the PHASE software uses lengths of residual vectors. Another conceptually appealing way to convey a sense of residuals via a single layer would be to map the first principal component of residual vectors.

INTERPHASE LINKAGE

Another intriguing line of PHASE investigations has apparent potential relative to issues of change de-

tection. Most past efforts regarding change detection have attempted the "apples and oranges" undertaking of directly comparing a time one image and a time two image. PHASE techniques hold prospects for transmuting time one apples into time two oranges on the basis of comparative spatial structure. Time two means can be compiled for time one clusters, and vice versa. If spatial structure is unchanged between dates, then the cross-compilation should contain equivalent information to the self-compilation. This should seemingly hold even when different sensors are used for the two dates. Furthermore, the distance between cross-compilation and self-compilation centroids should be large only where substantial reorganization of spatial structure has occurred. This is one of the topics for further PHASE research.

PERSPECTIVE

PHASE compression carries more than the obvious benefits of facilitating storage, transmission, sharing, and use of image data. It also has the more subtle effect of inducing spatially explicit structure where such structure was only implicit in the parent image. As a first expression of this, it is fairly straightforward to compute a matrix of edge apportionment by cluster and compare this to random expectation. Clusters exhibiting preferential juxtaposition must represent spatial complexes. There are several prospective ways in which this sort of information might be used to bolster classification. Interspersed occurrence is supporting evidence for common class membership in thematic mapping by supervised analysis. Landscape structure should likewise be expressed in the rate of spatial simplification with progressive collapse of the spectrally more similar clusters. Since PHASE compression opens new avenues of investigation for challenges like these, there are grounds to argue that the condensation infuses value as opposed to reducing information.

ACKNOWLEDGMENTS

Prepared with partial support from the NSF/EPA Water and Watersheds Program, National Science Foundation Cooperative Agreement Number DEB-9524722. The contents have not been subjected to Agency review and therefore do not necessarily reflect the views of the Agency, and no official endorsement should be inferred.

REFERENCES

Kelly, P. and J. White. Preprocessing Remotely-Sensed Data for Efficient Analysis and Classification, in *Applications of Artificial Intelligence 1993: Knowledge-Based Systems in Aerospace and Industry, Proceedings SPIE*, 1993, pp. 24–30.

Myers, W. *PHASE Approach to Remote Sensing and Quantitative Spatial Data*, Report ER9710, Environmental Resources Research Institute, Penn State University, University Park, PA, 1997.

Myers, W., G.P. Patil, and C. Taillie. *PHASE Formulation of Synoptic Multivarite Landscape Data*, Technical report 97-1102, Center for Statistical Ecology and Environmental Statistics, Dept. of Statistics, Penn State University, University Park, PA, 1997a.

Myers, W., G.P. Patil, and C. Taillie. *Adapting Quantitative Multivariate Geographic Information System Data for Purposes of Sample Design: The PHASE Approach*, Technical report 97-1201, Center for Statistical Ecology and Environmental Statistics, Dept. of Statistics, Penn State University, University Park, PA, 1997b.

Tou, J.T. and R.C. Gonzalez. *Pattern Recognition Principles*, Addison-Wesley, Reading, MA, 1974.

CHAPTER 40

Spatio-Temporal Prediction of Level 3 Data for NASA's Earth Observing System

J. Gabrosek, N. Cressie, and H.-C. Huang

INTRODUCTION

The Earth Observing System (EOS) involves a complex series of satellite measurements, data analysis, and statistical modeling. The goal of the program is to provide long-term, comprehensive observations of the chemical and physical processes that shape the Earth's environment (NASA, 1992).

There are three main components of the EOS program. The **EOS Observatories** consist of a series of remote-sensing platforms that carry instruments designed to measure key global parameters affecting the Earth's environment. The **EOS Scientific Research Program** provides support for interdisciplinary study of satellite data and modeling of planetary-scale processes. The **EOS Data and Information System** processes, validates, documents, archives, and distributes the massive amounts of data generated by EOS and other related scientific missions (Kahn and Wenkert, 1997).

The amount of data generated by the EOS satellites is truly massive. A single instrument on a polar-orbiting satellite, such as the Multi-Angle Imaging SpectroRadiometer (MISR), will generate roughly 80 gigabytes of basic data per day (Kahn, 1995). It is impossible for the scientific community to use this massive amount of data in its raw form. Thus the data are processed, analyzed, and summarized until distilled into a usable form. It is in this process of distillation that the statistical community can play a major role.

There are five classes of spacecraft data (Committee on Data Management and Computing, 1982):

- **Level 0**—The raw data stream from the spacecraft as received at Earth.
- **Level 1**—Measured radiances, geometrically and radiometrically calibrated.
- **Level 2**—Geophysical parameters at the highest resolution available.
- **Level 3**—Averaged data, providing spatially and temporally "uniform" coverage.
- **Level 4**—Data produced by a theoretical model, possibly with measurements as inputs.

Many instruments on spacecraft measure radiances and then use these values to infer the value of geophysical parameters. For instance, the Total Ozone Mapping Spectrometer (TOMS) instrument aboard the Nimbus-7 satellite measures backscattered radiation and then uses a set of climatological profiles to infer the total column ozone (McPeters et al., 1996). It is the set of inferred geophysical parameters that makes up the Level 2 data. Certainly, the method for inferring a given geophysical parameter is crucial but that is more the domain of the physicist than the statistician. The statistician's role in converting Level 2 data into Level 3 data should be substantial.

Level 2 data is spatially and temporally nonuniform. Specifically, for polar-orbiting satellites there is oversampling near the poles and undersampling near the equator. The scientific community has generally preferred to use spatially and temporally uniform Level 3 data. Typically, the production of Level 3 data from Level 2 data proceeds in two steps: (1) the Earth is divided into a grid, often based on longitude and latitude, and (2) an "average" value is computed for each grid cell. The "average" value may be a simple arithmetic average or the median or some type of weighted average, but rarely is the spatio-temporal dependence in the data explicitly used to improve the Level 3 data value. We shall focus on the production of Level 3 data from Level 2 data wherein we incorporate the spatio-temporal dependence in the data to generate statistically optimal predictors of Level 3 data values. Level 3 data are often the only data a scientist will use when trying to understand the earth's climate, general circulation models (GCMs). Using a sub-optimal technique for the production of Level 3 data is a problem that propagates into all subsequent uses of the data, such as in GCMs.

When considering an "optimal" approach to handling satellite-derived atmospheric data, one must consider four facts that make the processing of such data a challenge. First, atmospheric processes are multivariate in nature. Atmospheric chemistry involves a set of complex interactions, some still poorly understood (Asrar and Dozier, 1994). Second, atmospheric data exhibit spatiotemporal dependence. Nearby observations in space and time tend to be more alike than observations that are far apart. Also, there are often periodicities in atmospheric data that correspond roughly to seasonal changes. Third, polar-orbiting satellites introduce data artifacts owing to the nature of the data collection. Often, a banded structure appears with bandwidths equal to the length of the orbital swath (see, e.g., Zeng and Levy, 1995). Fourth, as discussed previously, satellites generate massive amounts of data.

We investigate the effectiveness of using tree-structured models in the production of Level 3 data. The particular data set we use for illustration is the TOMS Nimbus-7 total column ozone data for the latter part of 1988.

TREE-STRUCTURE MODELS

There are several approaches that have been or can be used to handle large volumes of polar-orbiting satellite data. Fang and Stein (1997) investigate variations in ozone levels through the use of a moving average (with seasonal dependence) for a fixed latitude. Zeng and Levy (1995) use a three-dimensional interpolation technique to fill in missing values for grid-cell locations at certain time points. Sampson et al. (1994) suggest multi-dimensional scaling to alleviate the problem of nonstationarity in the spatial covariance structure. And Barry and Ver Hoef (1996) detail a technique they call

"blackbox kriging" to fit flexible variogram models to an empirical spatial covariance structure.

We propose the use of a change-of-resolution Kalman filter for tree-structure models to obtain statistically optimal estimates of Level 3 data from Level 2 data. Tree-structure models have been used in the analysis of remote-sensing data (Fieguth et al., 1995; Daniel and Willsky, 1996; Fosgate et al., 1997) and the extension to polar-orbiting satellite data appears natural. Tree-structure models have advantages that are not enjoyed by some or all of the techniques mentioned above.

Tree-structure models are multi-resolutional, scale-recursive models. Thus, in theory, spatial statistically optimal estimators can be calculated for different scales of resolution. This is a particularly appealing feature for atmospheric data. The level of resolution needed often depends on whether the data are to be used in local, regional, or global calculations. Also, the models should be able to handle different data sources that are themselves at different levels of spatial resolution (i.e., the problem of combining satellite data, ground-station data, balloon data, and so forth).

Tree-structure models capture the spatial dependence in the data through parent-offspring relationships. Figure 40.1 shows a quadtree structure. Note that each cell (parent) is subdivided into four smaller cells (offspring). Notice that, conditional on the parent's value, any two offspring are independent. This simplifies the actual spatial dependence in the data.

The change-of-resolution Kalman filter allows one to calculate statistically optimal estimates with associated error bars at each level of spatial resolution. Thus, when using the results in GCMs, we can provide not only an estimated value but also an associated measure of reliability of the estimate.

The change-of-resolution Kalman filter can handle missing data values or asymmetric sampling patterns quite easily. Finally, and perhaps most importantly, the technique is computationally efficient. Since we do not have to calculate a variogram or in some other fashion estimate the actual spatial covariance matrix, we can perform the calculations in a less time-consuming fashion. When working with gigabytes of daily data, this is a crucial consideration.

TOTAL COLUMN OZONE APPLICATION

We used total column ozone, as measured by the Nimbus-7 TOMS instrument for the latter part of 1988, as our data set. The data were obtained from the NASA

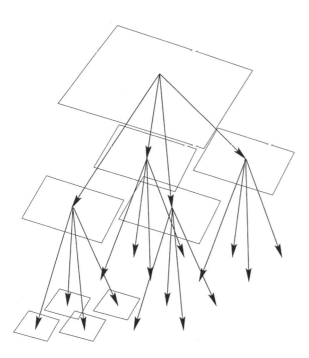

Figure 40.1. Tree-structured model with four offspring to each parent (quadtree model).

Goddard Distributed Active Archive Center and were stored in Hierarchical Data Format (HDF) as developed by the National Center for Supercomputing Applications (NCSA) at the University of Illinois. The TOMS ozone data has many of the same features as those we expect to see in the EOS data.

Nimbus-7 TOMS Instrument

Nimbus-7 was launched on October 24, 1978 with the TOMS instrument aboard. Nimbus-7 is a polar-orbiting satellite. The TOMS instrument scans in three-degree steps to an extreme of 51 degrees on each side of nadir, in a direction perpendicular to the orbital plane (McPeters et al., 1996). Each scan takes roughly eight seconds to complete, including one second for retrace (Madrid, 1978). The altitude of the satellite and scanning pattern of the TOMS instrument are such that consecutive orbits overlap, with the area of overlap depending on the latitude of the measurement. The TOMS instrument covers the entire globe in a 24-hour period.

Exploratory Analysis of Ozone Data

Stratospheric ozone has spatial and temporal dependencies that must be accounted for in any attempt

to predict ozone values optimally. Ozone dependence on latitude and season was first shown by Dobson and coworkers almost 70 years ago (Dobson et al., 1929). Periodic variations of an annual, semiannual, and quasibiennial frequency have also been noted (London, 1985). Furthermore, variations with the approximately 11-year sunspot cycle and with cosmic perturbations have been identified (see, e.g., Turco, 1985).

Initially, we shall consider the ozone field at a specific point in time, namely October 1, 1988. Thus, we shall model only the spatial nature of the ozone distribution. The temporal component will be discussed in a subsequent section.

Figure 40.2 shows the latitude dependence of ozone using the TOMS data for October 1, 1988. Level 2 data summaries were averaged over one-degree latitude bands. Negative latitudes correspond to the Southern Hemisphere and positive latitudes to the Northern Hemisphere (similarly, negative longitudes correspond to the Western Hemisphere and positive longitudes to the Eastern Hemisphere). Ozone values reach a minimum near the South Pole, increase to a maximum near 50°S latitude, and then are relatively stable in the Northern Hemisphere. This is to be expected, as the "ozone hole" was first discovered over Antarctica during the latter half of 1988 using the TOMS data. The hole corresponds to ozone values of less than 225 Dobson units (DU).

Our analysis proceeded on the mean-corrected (residual) ozone values:

$$\text{residual ozone} = \text{level 2} - \text{zonal mean} \qquad (1)$$

Similarly to Fang and Stein (1997), we treat these residual ozone values as meeting the intrinsic hypothesis. That is, we assume a constant mean for the residuals over the globe and we assume that the covariation in the residuals between any two locations depends only on the lag distance separating the locations and not on the actual locations themselves.

Application of Tree-Structure Models to Residual Ozone Values

Level 3 data products for the TOMS data are given on a 180×288 grid; that is, 1° latitude by 1.25° longitude grid cells (McPeters et al., 1996). Consequently, we based our statistical modeling on the same grid structure. NASA presently plans for the EOS Level 3 data product to be on a 180×360 grid; that is, 1° lati-

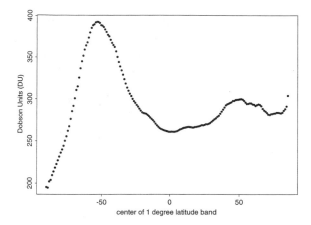

Figure 40.2. October 1, 1988 TOMS one-degree latitude band averages. There are 180 latitude bands beginning with (90°S, 89°S) and ending with (89°N, 90°N).

tude by 1° longitude grid cells (Kahn, 1997). The modification of our modeling to this new grid structure is straightforward. Our initial efforts are directed at modeling the residual ozone levels for a given day, which for illustration is October 1, 1988, over the entire globe. As such, our initial efforts are univariate and independent of time.

We begin with some notation:

Scale: 0 - entire globe,
  1 - 5 latitude bands and 8 longitude bands,
  2 - 15 latitude bands and 24 longitude bands,
  3 - 45 latitude bands and 72 longitude bands,
  4 - 90 latitude bands and 144 longitude bands,
  5 - 180 latitude bands and 288 longitude bands.

At scale 0, there is only one cell (i.e., node in the parlance of tree-structure models). This cell is subdivided into 40 smaller, equal-degree cells at scale 1 (the large cell is called the parent and the 40 smaller cells are called the offspring). Each of these 40 cells is then further subdivided into 9 equal-degree cells to obtain scale 2, and so on. At scale 0, we cannot estimate the variance since there is only one node; thus, the Kalman-filter algorithm considers scale 1 to be the coarsest scale and scale 5 to be the finest scale.

Location: $(i_1, i_2, ..., i_k)$; k = 1,2,3,4,5,
  $i_1 \in \{1,...,40\}$,
  $i_2 \in \{1,...,9\}$,
  $i_3 \in \{1,...,9\}$,
  $i_4 \in \{1,...,4\}$,
  $i_5 \in \{1,...,4\}$.

The location of a 1° latitude by 1.25° longitude grid cell (i.e., a grid cell at the finest resolution, which is scale 5) would be denoted by a five-number location label. The location of a node at scale 1 would be denoted by a one-number location label.

Noise-free residual ozone: $S(i_1, i_2, ..., i_k)$;
  k = 1,2,3,4,5.
Observed residual ozone: $Z(i_1, i_2, ..., i_5)$.

We observe data at scale 5. However, we can predict the actual residual ozone value at any scale we wish by defining a Gaussian tree-structure model. We assume that the Gaussian process evolves from parent to offspring in a Markov manner according to the following state-space equation:

$$Z(i_1, i_2, ..., i_5) = S(i_1, i_2, ..., i_5) + \varepsilon(i_1, i_2, ..., i_5) \qquad (2)$$

$$S(i_1, i_2, ..., i_k) = S(i_1, i_2, ..., i_{k-1}) + \eta(i_1, i_2, ..., i_k); \\ k = 1, ..., 5 \qquad (3)$$

where $\varepsilon(i_1, i_2, ..., i_5)$ and $\eta(i_1, i_2, ..., i_k)$ are independent, zero-mean, Gaussian, white-noise processes. Further, $S(i_1, i_2, ..., i_5)$ and $\varepsilon(i_1, i_2, ..., i_5)$ are independent, as are $S(i_1, i_2, ..., i_{k-1})$ and $\eta(i_1, i_2, ..., i_k)$. Equation 2 is called the measurement equation or observation equation, and Equation 3 is called the state equation or transition equation. Finally, we assume that the 40 cells at scale 1 are independently distributed Gaussian random variables with zero mean and variance σ_1^2.

The extension of the Kalman-filter algorithm to tree-structure models was developed by Chou et al. (1994). For the tree-structure state-space model defined by Equations 2 and 3, our goal is to obtain optimal predictors (i.e., conditional expectations) for all grid cells at all scales. The generalized Kalman filter has an uptree filtering step and a downtree smoothing step. The uptree filtering step begins at the finest level of resolution (scale 5) and proceeds upward by calculating optimal predictors at each node based on the data at that node and its descendents. Once we have reached the root of each tree (scale 1, since we are assuming independence at this level), we proceed to the downward smoothing step, finding the optimal predictor based on all of the data. Further details of the algorithm can be found in either Chou et al. (1994) or Huang and Cressie (1997).

We adopt the following notation:

$$\mathrm{var}\big(S(i_1)\big) = \sigma_1^2,$$

$$\mathrm{var}\big(\eta(i_1, i_2, ..., i_k)\big) = \sigma_k^2; \ k = 2,3,4,5,$$

$$\mathrm{var}\big(\varepsilon(i_1, i_2, ..., i_5)\big) = \sigma_\varepsilon^2.$$

We estimate the vector of parameters, $\theta \equiv (\sigma_1^2, \sigma_2^2, ..., \sigma_5^2, \sigma_\varepsilon^2)'$, by maximum likelihood based on the model (2) and (3) and a subset of the data $\{Z(i_1, i_2, ..., i_5)\}$ that allows estimates to be computed rapidly.

Since total column ozone are per unit volume data then, at any given resolution, we would like the area of the cells to be equal. Also, across resolutions, the cells of different levels should nest. Longitude-latitude based grids have the property that the cells nest at different levels of resolution; however, longitude-latitude based grids do not have equal area cells within a resolution. As one moves toward either pole, the area of a cell becomes very small. The Icosahedral Snyder Equal Area (ISEA) Grid can be used to solve the problem of unequal-area cells (Kimerling et al., 1997).

RESULTS AND DISCUSSION

Figure 40.3 shows the geographical locations of 80 ground stations that recorded total column ozone values on October 1, 1988. The ground-station data were obtained from the World Ozone Data Center, Downsview, Ontario. Notice the paucity of stations in the Southern Hemisphere. In fact, there are no stations in either South America or tropical and southern Africa. Since 1988, some new ozone monitoring ground stations have gone on-line but there is still a shortage of available ground-station data in these regions. Also, note that there are no stations at extremely high latitudes. The Amundsen-Scott station (latitude −89.983°) in the Antarctic is the closest station to the South Pole. (However, this station did not record data for October 1, 1988.) Since comparisons of ozone values are typically made to "ground truth," it is problematic that we have these huge spatial gaps in ground-station ozone data.

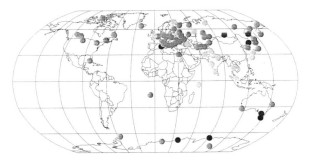

Figure 40.3. Location of 80 ground-based ozone monitoring stations that recorded total column ozone values on October 1, 1988. Values are color coded from least amount of ozone (light) to greatest amount of ozone (dark).

There are 51,840 grid cells on the entire globe, each of which will have a Level 3 datum associated with it. However, the spatial gaps in ground-station locations means that we can only use a small number of these Level 3 data to compare to the ground-station data. Nevertheless, in what is to follow, our method of comparison of various Level 3 algorithms will be based on the Level 3 data values in the cells within which ground stations fall geographically.

An estimated grid-cell value (i.e., Level 3 datum) for the Kalman-filter algorithm is calculated as:

$$\text{grid-cell value} = \\ \text{zonal mean} + \text{Kalman-filter residual} \quad (4)$$

The sum-of-squares error (SSE) for a Level 3 data product algorithm is calculated as:

$$SSE = \sum_{i=1}^{80} \big(GS(i) - L3(i)\big)^2 \quad (5)$$

where the sum is over the 80 ground stations. Further, GS(i) represents the ground-station value for station i on October 1, 1988 and L3(i) represents the Level 3 data value for the grid cell within which ground station i falls. Table 40.1 compares the NASA Level 3 data product to our Kalman-filter Level 3 data product (Equation 4). The Kalman-filter Level 3 data product enjoys a small reduction (5.1%) in the SSE but generally the two data products are very similar. Further enhancements we have planned for the Kalman-filter Level 3 data product will likely produce considerably bigger gains.

Table 40.1. Comparison of Methods Used to Produce a Level 3 Data Product.

Level 3 Data Product	SSE	Max. Absolute Diff.	Median Absolute Diff.
NASA	11,629	41	6.5
Kalman-filter	11,036	42	7.0

Spatiotemporal Tree-Structure Models

We know that atmospheric processes are dynamic with temporal dependence. We propose to model the temporal dependence at the daily time scale using an autoregressive process of order 1 [AR(1)]. The model is:

$$S_t = AS_{t-1} + \varepsilon_t \qquad (6)$$

where $S_t \equiv (S_t(1), S_t(2), ..., S_t(40))'$ and, in our special case, we assume $A = \alpha I$. Thus, the temporal dependence is captured at scale 1. For example, S_2 is an offspring of S_1. In this notation, S_1 is considered as a single node at scale 1. Thus, there is dependence within the 40 nodes that make up S_1 and dependence between S_2 and S_1 that is captured by the AR(1) process. Eventually, we will be adding temporally periodic components to the model.

CONCLUSION

While quite a bit more work remains in handling multivariate processes, incorporating multiple data sources, and carrying out model validation, the generalized Kalman filter for tree-structure models holds great promise for processing the massive amounts of data generated by polar-orbiting satellites. The most important advantage of the technique is that we obtain optimal estimators at different scales of resolution simultaneously. We also obtain measures of reliability in terms of estimation variances. Typically, such a measure is completely lacking in current Level 3 data. If a measure is given, it is usually the variance of all the observations used to calculate the sample mean of the observations that happen to fall in a given grid cell. But we know that this variance is inappropriate because it assumes independent observations, and ignores obvious spatial dependencies originating from combining nearby data.

ACKNOWLEDGMENTS

This research was supported by the U.S. Environmental Protection Agency under Co-operative Agreement Number CR822919-01-0. The authors are grateful to Ralph Kahn, Jon Kimerling, and Kevin Sahr for their help in various aspects of this research. TOMS data were obtained from the NASA Goddard Distributed Active Archive Center (DAAC). Ground-station data were obtained from the World Ozone Data Center, Downsview, Ontario.

REFERENCES

Asrar, G. and J. Dozier. *Science Strategy for the Earth Observing System,* AIP Press, NASA, Washington, DC, 1994.

Barry, R.P. and J.M. Ver Hoef. Blackbox Kriging: Spatial Prediction without Specifying Variogram Models, *J. Agric., Biol. Environ. Stat.,* 1, pp. 297–322, 1996.

Chou, K.C., A.S. Willsky, and R. Nikoukhah. Multiscale Systems, Kalman Filters, and Riccati Equations, *IEEE Trans. Autom. Control,* 39, pp. 479–492, 1994.

Committee on Data Management and Computing. *Data Management and Computation, Volume 1: Issues and Recommendations,* Space Science Board, National Academy of Sciences, Washington, DC, 1982.

Daniel, M. and A.S. Willsky. A Multiresolution Methodology for Signal-Level Fusion and Data Assimilation with Applications to Remote Sensing, *Technical Report LIDS-P-2373,* Laboratory for Information and Decision Systems, Massachusetts Institute of Technology, Cambridge, MA, 1996.

Dobson, G.M.B., D.C. Harrison, and J. Lawrence. Measurement of the Amount of Ozone in the Earth's Atmosphere and Its Relation to Other Geophysical Conditions, Part III, *R. Soc. (London) Proc., Series A,* 122, pp. 456–486, 1929.

Fang, D. and M. Stein. Some Statistical Models and Methods for Analyzing the TOMS Data, unpublished manuscript from the Department of Statistics at the University of Chicago, 1997.

Fieguth, P.W., W.C. Karl, A.S. Willsky, and C. Wunsch. Multiresolution Optimal Interpolation and Statistical Analysis of TOPEX/POSEIDON Satellite Altimetry, *IEEE Trans. Geosci. Remote Sensing,* 33, pp. 280–292, 1995.

Fosgate, C.H., H. Krim, W.W. Irving, W.C. Karl, and A.S. Willsky. Multiscale Segmentation and Anomaly Enhancement of SAR Imagery. *IEEE Trans. Image Process.,* 6, pp. 7–20, 1997.

Huang, H.-C. and N. Cressie. Multiscale Spatial Modeling, in *1997 Proceedings of the Section on Statistics and the Environment,* American Statistical Association, Alexandria, VA, forthcoming, 1997.

Kahn, R. "Why Do We Need Discrete Global Grids," from a presentation at the symposium, Global Grids: New Approaches to Global Data Analysis, held at Oregon State University, Corvallis, OR, May 19, 1997.

Kahn, R. What Shall We Do with the Data We Are Expecting in 1998?, *Proceedings of the Massive Data Sets Workshop,* July 7–8, 1995. National Academy of Sciences, Washington, DC, 1995.

Kahn, R. and D. Wenkert. Earth Observing System, in *The Encyclopedia of Planetary Sciences,* J.H. Shirley and R.W. Fairbridge, Eds., Chapman and Hall, New York, 1997.

Kimerling, A.J., K. Sahr, D. White, and M. Gregory. Terra Cognita Global Grid Research Team, Oregon State University, Corvallis, OR (web site: http://bufo.geo.orst.edu/tc/firma/gg), 1997.

London, J. The Observed Distribution of Atmospheric Ozone and its Variations, in *Ozone in the Free Atmosphere,* R.C. Whitten and S.S. Prasad, Eds., Van Nostrand Reinhold Company, New York, 1985.

Madrid, C.R. *The Nimbus-7 User's Guide,* Goddard Space Flight Center, NASA, MD, 1978.

McPeters, R.D., P.K. Bhartia, A.J. Krueger, J.R. Herman, B.M. Schelsinger, C.G. Wellemeyer, C.J. Seftor, G. Jaross, S.L. Taylor, T. Swissler, O. Torres, G. Labow, W. Byerly, and R.P. Cebula. *Nimbus-7 Total Ozone Mapping Spectrometer (TOMS) Data Product's User's Guide,* NASA Reference Publication 1384, NASA, Washington, DC, 1996.

National Aeronautics and Space Administration. *Report to Congress on the Restructuring of the Earth Observing System,* (March 9, 1992), NASA, Washington, DC, 1992.

Sampson, P.D., P. Guttorp, and W. Meiring. Spatio-Temporal Analysis of Regional Ozone Data for Operational Evaluation of an Air Quality Model, in *1994 Proceedings of the Section on Statistics and the Environment,* American Statistical Association, Alexandria, VA, 1994, pp. 46–55.

Turco, R.P. Stratospheric Ozone Perturbations, in *Ozone in the Free Atmosphere,* R.C. Whitten and S.S. Prasad, Eds., Van Nostrand Reinhold Company, New York, 1985.

Zeng, L. and G. Levy. Space and Time Aliasing Structure in Monthly Mean Polar-Orbiting Satellite Data, *J. Geophys. Res.,* 100, pp. 5133–5142, 1995.

Part IX

Characterizing and Obtaining Spatial Uncertainty Information for Specific Situations

Section A: Geographic Information Systems/Cartography

Section B: Remote Sensing

Section C: Points and Lines

Section D: Combinations of Data Types

In addition to a generalized advancement of the study of spatial uncertainty, there are practitioners who are interested in knowing the magnitude and characteristics of uncertainty in particular settings. Essentially, previous Parts have described general issues to be addressed relative to spatial uncertainty—types of uncertainty, modeling uncertainty, characterizing uncertainty, etc. Eventually, however, it is necessary to relate those generalized issues to specific situations. The chapters presented in this Part do exactly that.

This Part is divided into four sections, as it is apparent that practitioners are considering uncertainty for four different situations:

1. thematic information on conventional maps incorporated into geographic information systems,
2. information extracted from remotely sensed images as part of spatial data sets and used for decision-making,
3. point-based information, and
4. combinations of the preceding three types of information.

The breadth and diversity of chapters presented in this Part is truly encouraging. In addition to a generalized body of knowledge being developed for spatial uncertainty, it is apparent that the study of spatial uncertainty is not simply remaining in the realm of the theoretical. The chapters in this Part demonstrate that many individuals are working in a corner of the discipline in order to come to grips with, and overcome the problem of spatial uncertainty in natural resources databases.

CHAPTER 41

On Some Limitations of Square Raster Cell Structures for Digital Elevation Data Modeling

A.M. Shortridge and K.C. Clarke

INTRODUCTION

Digital elevation models (DEMs) are used in applications including hydrologic surface flow modeling, land use planning, vegetation mapping, terrain analysis and visualization, and flood and landslide hazard identification (Petrie and Kennie, 1990; Moore et al., 1991; Weibel and Heller, 1991). These applications frequently employ raster data models for visualizing and analyzing elevation data. A raster structure is typically logically based on a regularly spaced rectangular matrix of equal-sized cells. Cells may be square or rectangular; that is, the north-south dimension of each cell may be equivalent to the east-west dimension, or it may differ. Square-celled raster structures are commonly enforced in geographic information systems; Arc/Info, an industry-standard GIS, for example, supports only square raster cells (ESRI, 1996). This chapter examines some implications of the employment of square-celled structures to model elevation data from a nonsquare grid, when that grid must be resampled and interpolated due to reprojection.

The decision to employ any particular data model and any particular projection should involve not only identifying both the format and the projection of the available data, but also the impact data processing and spatial data structure will have on subsequent analysis. In one sense this is both a resolution and a sampling problem, and continues an existing line of research (Wolock and Price, 1994; Zhang and Montgomery, 1994; Guth, 1995; Monckton, 1994; Fisher, 1996; Gao, 1997). In another sense it is a data modeling problem. Discrepancies can exist between conceptual geographic

data models and the manner in which the data are structured in the GIS. When certain processes like projection are employed on the data structure, these discrepancies can result in fundamental changes to both locations and elevations in the output (Steinwand et al., 1995). Finally, decisions relating to data "preprocessing" can be fundamental to the outcome of an application and deserve special consideration.

The following section includes a discussion of a commonly available DEM format and how this format relates to the surface of the earth. Data modeling decisions introduce constraints on how the data may be transformed, and elevation values themselves can be affected by transformations of their locations. The third section introduces an actual data set and tests the magnitude of these effects on both global and neighborhood indicators. The chapter concludes with a discussion of the findings.

ELEVATION DATA MODELS, RASTER STRUCTURES, AND PROJECTION

U.S. Geological Survey (USGS) 1:250,000 digital elevation model series, also called "one-degree" DEMs after the areal coverage of each file, offer the highest resolution data for which complete coverage of the 48 conterminous United States is publicly available. For each 1 degree quadrangle, data points are stored on a 1201×1201 grid with 3 arc-seconds of latitude and longitude spacing in both east-west and north-south directions (USGS, 1993). A single integer elevation value is provided for each point. This is an example of a geographic data model for a continuous surface.

Samples of the elevation surface are taken at discrete point locations; the elevation calculated for each point is not an areal estimate for the region immediately surrounding the point, but a specific point elevation. Locations of objects in this model are defined by the model parameters. They are not separate occurrences of unique objects (Goodchild, 1992; Kemp, 1997).

The elevation data model may be usefully contrasted with that of a data set containing a collection of points marking the locations of maple trees and information about the trees' ages. In both cases the data structure is similar—the locations for a set of points along with an attribute about that location. However, the underlying data model is different. In the case of the trees, each point is defined because the object it represents exists uniquely at that point. Locations indicate the presence of that feature, and the absence of a point indicates that no object exists at that location. In the case of the elevation data model, the phenomenon is present in every part of the data set, whether a point measurement is at that location or not. If the USGS had sampled elevation at locations at the same density but offset by an arc-second, the data would be equally valid. This would not be the case with the hypothetical maple tree data.

An important point about the elevation data model is that no assumptions are made about the elevation of locations not falling on sample points. Although all locations, whether corresponding to a grid point or not, certainly possess an elevation, this model does not define what those intermediate values might be. In a raster data model of space, continuous surfaces are treated somewhat differently. Like the point model discussed previously, the raster model consists of a series of values of some phenomena, and the locations of these values are predefined by the parameters of the raster. However, locations are defined not as points but as areas. In GIS, these areas are frequently square or regular tesselations. A single value is stored in each cell; all parts of the cell's space possess the same value. A raster data set of this type consists of a rectangular matrix of cells which completely cover the study region. In the case of unprojected 3 arc-second data, the resulting structure from this raster model will appear identical to the point model. Data is spaced at 3 arc-second intervals, the number of columns and rows in the raster correspond with the number of profiles and observations per profile in the original point data, and all elevation values are identical. However, it is critical to note that the raster stores these values in identically shaped, regularly spaced cells.

Many applications require that this DEM data be projected into a standard coordinate system such as UTM, which stores coordinates as ground units in meters. Because of the convergence of the meridians, however, 3 arc-seconds in the east-west direction will be a shorter distance in ground units than 3 arc-seconds in the north-south direction for nonequatorial locations in many map projections including UTM. At the latitude of Santa Barbara, California, for example, a distance of 3 arc-seconds in the east-west direction is about 76 meters, while it is about 92 meters in the north-south direction. For a point-based structure, such differences are unimportant since each point location is projected independently of the others. Elevation values are unchanged, though the mapped grid spacing of points is no longer equal in both dimensions. However, the difference presents some problems for a raster-structured GIS when data must be projected. Figure 41.1 illustrates the problems. The gridded area in Figure 41.1b indicates the form of the raster cells in 41.1a following projection into UTM. Cells are narrower in the east-west direction, and cells at the top of the raster are narrower than cells at the bottom. Additionally, the grid has been rotated relative to its original orientation. The box which bounds this distorted grid is oriented to the UTM zone; the lines across top and bottom, for example, have constant northings, while the lines on either side maintain constant eastings.

To maintain equal area, square cells, the number of rows and columns can no longer match the original raster. Additionally, the fact that grid spacing in meters decreases from south to north conflicts with the requirement for a constant raster cell width. As a consequence, cells in any raster structure must be shifted and relocated and, in a square raster, rows need to be added or columns subtracted. This is represented in the transition from Figure 41.1b–41.1c. The number of columns in Figure 41.1c decreased to 7, while the overall resolution has actually coarsened since blank fill has been added in each corner of the new raster grid. No clear one-to-one relationship exists between the elevations of cells in the original arc-second raster and elevations of cells in the projected raster, so the new raster must be filled with interpolated or resampled values (Steinwand et al., 1995). In projecting to UTM, ARC/INFO (for example) resamples this data set to enforce its square-cell data structure (ESRI, 1996). The nonprojected raster for the full one degree DEM covering the Santa Barbara area is a 1201 × 1201 square coverage with cells measuring 3 arc-seconds on a side, as with any 3 arc-second DEM. Choosing the default,

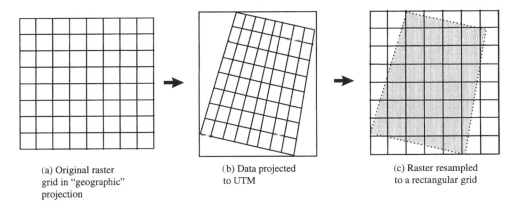

(a) Original raster
grid in "geographic"
projection

(b) Data projected
to UTM

(c) Raster resampled
to a rectangular grid

Figure 41.1. Projection of a square raster arc-second grid to Universal Transverse Mercator grid.

coarsest cell size results in a 1003×1201 grid of cells 94 meters on a side. Sampling density in the east-west direction therefore decreases by over 16%, as compared to a rectangular cell raster structure.

This reduction in sampling density may be of concern for many applications. For example, algorithms for calculating slope are sensitive to grid resolution. Guth (1995) found that slopes tend to decrease regardless of the particular algorithm employed as the elevation grid is coarsened. Hydrologic modeling applications use related transformations of elevation data like upstream contributing area which are also sensitive to grid resolution (Wolock and Price, 1994; Zhang and Montgomery, 1994). Environmental models which use these transformed grids are likely to be sensitive to coarsening of the original grid. Reduction in sampling density may also affect the spatial autocorrelation of the elevation surface. Spatial autocorrelation is the degree to which phenomena that are close together are more similar than phenomena that are farther apart. Elevation surfaces are autocorrelated to varying extents depending on the roughness of the terrain. Even if global measures are very similar, if the resampled surface possesses markedly different spatial autocorrelative structure then some applications will produce widely different results. For example, consider both a rugged terrain surface and the same surface following smoothing. The smoothed surface is similar in overall form to the original but it exhibits higher spatial autocorrelation. A least-cost path algorithm applied between two points on both surfaces might identify the same route, but the estimated cost for the smoother depiction would be lower.

The experiments in the following section include global measures like the mean and variance of elevation, and also employ measures of spatial dependence in the form of directional variograms, to examine the impact of projection on raster elevation data. Since the resolution is especially coarsened in the east-west direction, the degree to which the surfaces differ in particular directions is of interest.

TEST DATA AND METHODS

The Aztec-e, NM three arc-second DEM was used to examine grid resampling due to reprojection. Aztec-e covers the one degree block between 106°W and 107°W, and 36°N and 37°N, in northern New Mexico, USA. Like all three arc-second DEM files, the block consists of 1,201 profiles with 1,201 observations in each profile. To speed processing, only the central 600×600 region was used in the experiments discussed here. This subset of the entire DEM was imported to Arc/Info and saved as a "geographic" raster grid, with three arc-second spacing.

The three arc-second point file was exported and projected to UTM using the point projection algorithm in GRASS 4.1 to preserve the number of points and the elevation at each point. Conceptually, this file is similar to the original data. It consists of 600 north-south running profiles. The profiles are not parallel to each other as the columns in a raster are; at the southern end of the study area they are about 74.89 meters apart, while at the northern end they are about 74.41 meters apart. Each profile contains 600 points, spaced approximately 92.45 meters apart. This distance varies only slightly throughout the study area. Elevations for each point are identical to those in the original arc-second data set, since point locations have not been interpolated, only reprojected as points.

Twelve UTM-projected raster grids were also generated from the original geographic raster grid. As Table 41.1 indicates, four cell resolutions were used. The coarsest is the default setting in the Arc/Info projection command. This results in a UTM raster grid which maintains the same number of rows as the original raster grid. The number of columns is considerably truncated. The second coarsest resolution is 92.45 meters. At this cell size, the projected data has approximately the same grid spacing as the point data file. The number of rows is greater than 600 because the reprojected grid has blank cells on the top, bottom, and sides of the area (refer to Figure 41.1 for an illustration of the blank cells and their effect on cell resolution).

A third plausible resolution is 76 meters, which maintains the same number of columns as the original raster grid. However, this increases the number of rows to 738, a considerably finer resolution than the original data provided. The fourth and finest resolution employed is identical to the average distance between profiles at the southern end of the point data set. Again, because of the addition of blank cells to the projected UTM raster, the 76 meter cell resolution is coarser than the original data. At 74.4 meters, this grid is maintaining the data density of the finest portion of the study region, but this level is higher than the sampling density of much of the original data set. The increase in north-south resolution for the third and fourth cell size choices is artificial.

Each of the three interpolation methods Arc/Info supports were used to generate raster grids for each of the four cell sizes. The nearest neighbor method simply assigns the value of the single closest observation to each cell. Bilinear interpolation assigns weights to the nearest four surrounding points to evaluate the elevation of each cell. Kumler (1994) likened the resulting surface within the four points to a twisted ladder, with sides and rungs remaining straight but not resting in the same plane. The 16 nearest values are used for cubic convolution, which is the third method supported by Arc/Info. Weights based on distance are assigned to each of the values for the interpolation. ESRI (1996) notes that cubic convolution tends to smooth the resultant surface more than the other methods.

The comparison of square-celled raster data sets and the original point file took several forms. Since the raster dimensions are not all identical, and an entirely

Table 41.1. Test Raster Descriptions.

Production Method	Grid Dimensions col x row	Resolution (meters)
93 m, Nearest	488 x 600	93.6
93 m, Bilinear	488 x 600	93.6
93 m, Cubic	488 x 600	93.6
92 m, Nearest	494 x 608	92.5
92 m, Bilinear	494 x 608	92.5
92 m, Cubic	494 x 608	92.5
76 m, Nearest	600 x 738	76.1
76 m, Bilinear	600 x 738	76.1
76 m, Cubic	600 x 738	76.1
74 m, Nearest	614 x 755	74.4
74 m, Bilinear	614 x 755	74.4
74 m, Cubic	614 x 755	74.4

different model is employed for the nontransformed data, the grids were exported to a statistics package and examined individually to compare them further. Slope (measured in degrees of rise) raster maps were also generated from each of the UTM data sets; their global characteristics are of interest as well, since slope is a frequently employed attribute of DEMs and slope is known to be sensitive to changes in cell resolution. The Arc/Info slope algorithm was also implemented in the C programming language so that an identical measure could be made on the point data (this measure is presented in Burrough, 1986). Directional semivariograms (subsequently referred to as variograms) were generated in the north-south (or along-column) and east-west (or along-row) directions, respectively, for both elevation and slope raster maps and the elevation point data. Semivariance for any particular lag $\mathbf{h}$, where $\mathbf{h}$ is a vector indicating both distance and direction, is defined as:

$$\gamma(\mathbf{h}) = \frac{1}{2N(\mathbf{h})} * \sum_{(i,j)|\mathbf{h}_{ij}=\mathbf{h}} \left(v_i - v_j\right)^2 \qquad (1)$$

where v_i is the elevation at point i, v_j is the elevation at point j, and N is the number of (ij) pairs separated by the vector $\mathbf{h}$ (Isaaks and Srivastava, 1989).

The particular distance lags into which the data were partitioned varies depending on the cell size, but the overall shape and magnitude may be compared across data sets to determine how well different cell resolutions and interpolation methods capture spatial struc-

Table 41.2. General Comparisons (Slope Measured in Degrees).

Name	Mean	Std. Dev.	Range	Mean Slope	Std. Dev. Slope	Max. Slope	No. of Cells (% Change)
74mx92m Point	2410	314.6	1889–3308	5.30	5.70	56.67	360,000
93 m, Nearest	2409	314.5	1889–3308	5.27	5.60	57.35	292,800
93 m, Bilinear	2410	314.5	1889–3300	5.21	5.39	53.07	(81.3%)
93 m, Cubic	2410	314.6	1881–3305	5.31	5.62	57.68	(81.3%)
92 m, Nearest	2410	314.6	1889–3308	5.28	5.62	56.09	300,352
92 m, Bilinear	2409	314.5	1889–3301	5.21	5.40	55.42	(83.5%)
92 m, Cubic	2409	314.6	1881–3305	5.31	5.63	56.48	(83.5%)
76 m, Nearest	2409	314.5	1889–3308	5.35	5.89	59.37	442,800
76 m, Bilinear	2409	314.5	1889–3299	5.26	5.57	54.09	(123%)
76 m, Cubic	2409	314.6	1880–3305	5.38	5.86	58.23	(123%)
74 m, Nearest	2409	314.5	1889–3308	5.35	5.89	60.87	463,570
74 m, Bilinear	2409	314.5	1889–3300	5.27	5.59	56.40	(128.8%)
74 m, Cubic	2409	314.6	1880–3305	5.39	5.89	59.22	(128.8%)

ture. Of particular interest is the comparison of east-west (or along-row) variogram estimates for square and raster sets with the point data set to see if generalization in this direction has affected autocorrelative relationships.

RESULTS

The 12 test raster interpolation grids for Aztec-e were developed and compared following the methods outlined above. Table 41.2 indicates that the general statistics for each elevation data set are quite similar. Elevation means and standard deviations are practically identical for all permutations of the elevation data. Information on the elevation range of each interpolation suggests that the bilinear method reduces the peaks slightly, while the cubic convolution method raises the troughs. A comparison of the total number of cells (or, in the case of the first row, of data points) indicates substantial change in the number of data observations comprising these general statistics, however. Using the default cell resolution of 93 meters results in a reduction of 67,200 cells, over 18%, as a result of projecting the data to UTM. Employing the smallest resolution adds 103,570 cells, almost 29% of the original 360,000 data observations.

Statistics for the slope maps are somewhat more diverse. The mean slope of the highest resolution raster map ranges from 5.27 to 5.39 degrees, while the lowest resolution raster map ranges from 5.21 to 5.27 degrees. The actual data is near the upper limit of this range. Nearest neighbor and cubic convolution inter-

polation methods produce higher mean slopes than bilinear interpolation. Slope standard deviations follow a similar pattern—bilinear interpolation results in distributions with less spread than the other methods. Also, standard deviation increases with decreasing cell size. Maximum slope also increases as cell size decreases. All of these signs are in agreement with previous findings on the relationship between slope and grid resolution.

Variograms of the elevation data revealed that spatial autocorrelative structure varied only slightly across all permutations. Semivariance rose steadily with distance and showed no sign of reaching a range at 10,000 meters, the maximum distance for which semivariance was calculated. Row and column variograms were different in all cases; Figure 41.2 depicts the row and column variograms calculated on the gridded elevation data. The row variogram (that is, semivariance is calculated only in the east-west direction, along rows) rose more sharply, and semivariance was higher at all distances. This may be due to natural anisotropy in the terrain itself, but is more likely a factor of the DEM production method, which generates data along north-running profiles. Every data value is most heavily influenced by the one immediately beneath it in the profile (Cook, 1974), which would tend to reduce semivariance in the north-south direction.

Of particular interest is the comparison of cell resolution semivariances at short lags. Table 41.3 presents this information. The first row indicates the semivariance of the elevation values that were nontrans-

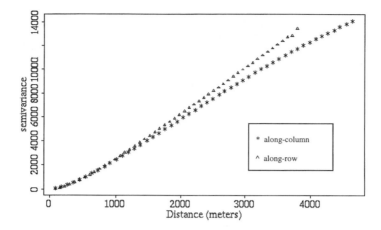

Figure 41.2. Semivariogram for gridded elevation data.

Table 41.3. Semivariance at Short Lags for Elevation Data at Different Resolutions.

Elevation File	Lag 1 Semivariance		Lag 2 Semivariance		Lag 3 Semivariance	
	Row	Col	Row	Col	Row	Col
Point DEM (74x92)	37	49	108	168	251	336
74 m, Nearest	38	40	127	123	254	239
74 m, Bilinear	32	29	117	106	240	221
74 m, Cubic	38	33	129	119	258	239
92 m, Nearest	60	50	188	169	366	338
92 m, Bilinear	49	43	173	158	349	323
92 m, Cubic	56	50	188	172	369	343

formed as raster data, but were instead projected as points. The remaining rows contain semivariances for rasters with two different cell resolutions and three interpolation methods. The ground distance of each lag (the columns in Table 41.3) depends on the cell resolution. For the point data, the lag was 75 meters for the along-row semivariance, and 93 meters for the along-column semivariance. One feature of this table is the increase of along-row semivariance in coarse resolution data. The 74-m resolution data maintains the original row semivariance at the cost of extrapolating the column semivariance. A second important feature of this table is that bilinear interpolation is especially smoothing. Both row and column semivariances are substantially underestimated at all lags. Nearest neighbor and cubic convolution interpolation methods appear to preserve spatial structure to a greater degree.

CONCLUSION

Problems with the employment of raster cell structures for modeling field elevation data sampled at points were examined in this chapter. The projection of raster data introduces problems that must result in resampling of the original data. This resampling is usually accompanied by a loss of sampling density when the raster model adheres to a square-celled structure. An assessment of the impact of such resampling on a sample elevation data set and its slope derivative was made. Twelve production methods within the same software produced twelve plausible UTM elevation surfaces of varying cell resolution and interpolation techniques. The number of elevation data observations dropped by over 18% for the default resampling resolution, regardless of interpolation method. Other plausible resolutions also resulted in substantial changes to the number and numerical value of data observations.

A variety of approaches was undertaken to detect the impact of these changes. Elevation means and standard deviations for the different surfaces were very similar, though maxima and minima were altered by interpolation due to resampling. Spatial autocorrelation of the elevation surface was also affected. A study of directional variograms indicated that elevations did vary more with distance in the east-west (along-row) direction than in the north-south (along-column) direction, which was probably at least partially due to the properties of the original DEM production method. Bilinear interpolation introduced considerably more smoothing of spatial variance at all distances than the other two resampling methods. Comparisons of slope revealed considerable differences between interpolation methods and between cell resolutions, indicating that local spatial structure is affected by raster reprojection.

Geographic information systems often impose particular data models for spatial data. For DEMs, square raster cell models are frequently the default. This chapter has demonstrated that such modeling decisions, when combined with transformation algorithms, result in fundamental changes to the data. Some of these changes may be difficult to detect through the use of global measures, but they can profoundly affect the character of the elevation surface, with implications for applications which use them. We suggest that GIS users give careful consideration to the potential impact of raster reprojection on their data and on the results of subsequent analysis upon that data.

REFERENCES

Burrough, P.A. *Principles of Geographical Information Systems for Land Resources Assessment,* Oxford, New York, 1986.

Cook, H.R. Numerical Mapping, *14th Congress, International Federation of Surveyors,* Washington, DC, 1974.

ESRI. *ARC/INFO Version 7 User's Guide,* Environmental Systems Research Institute, Redlands, CA, 1996.

Fisher, P. The Effect of Database Generalization on the Accuracy of the Viewshed, *Proceedings, 2nd International Symposium on Spatial Accuracy Assessment in Natural Resources and Environmental Sciences,* Fort Collins, CO, 1996, pp. 272–280.

Gao, J. Resolution and Accuracy of Terrain Representation by Grid DEMs at a Micro Scale, *Int. J. Geogr. Inf. Sci.,* 11(2), pp. 199–212, 1997.

Goodchild, M.F. Geographical Data Modeling, *Comput. Geosci.,* 18(4), pp. 401–408, 1992.

Guth, P.L. Slope and Aspect Calculations on Gridded Digital Elevation Models: Examples from a Geomorphometric Toolbox for Personal Computers, *Zeitschrift für Geomorphologie,* 101, pp. 31–52, 1995.

Isaaks, E.H. and R.M. Srivistava. *Applied Geostatistics,* Oxford, New York, 1989.

Kemp, K.K. Fields as a Framework for Integrating GIS and Environmental Process Models, Part 1: Representing Spatial Continuity, *Trans. GIS,* 1(3), pp. 219–234, 1997.

Kumler, M.P. An Intensive Comparison of Triangulated Irregular Networks (TINs) and Digital Elevation Models (DEMs), *Cartographica Monogr. 45,* 31(2), pp. 1–99, 1994.

Monckton, C.G. An Investigation into the Spatial Structure of Error in Digital Elevation Data, in *Innovations in GIS 1,* M.F. Worboys, Ed., Taylor & Francis, London. 1994, pp. 201–210.

Moore, I.D., R.B. Grayson, and A.R. Ladson. Digital Terrain Modelling: A Review of Hydrological, Geomorphological, and Biological Applications, *Hydrol. Process.,* 5, pp. 3–30, 1991.

Petrie, G. and T.J.M. Kennie, Eds. *Terrain Modelling in Surveying and Civil Engineering,* Whittles, London, 1990.

Steinwand, D.R., J.A. Hutchinson, and J.P. Snyder. Map Projections for Global and Continental Data Sets and an Analysis of Pixel Distortion Caused by Reprojection, *Photogrammetric Eng. Remote Sensing,* 61(12), pp. 1487–1497, 1995.

USGS, *National Mapping Program Technical Instructions. Standards for Digital Elevation Models,* U.S. Geological Survey, U.S. Dept. Interior, 1993.

Weibel, R. and M. Heller. Digital Terrain Modelling, in *Geographical Information Systems: Principles and Applications,* D.J. Maguire, M.F. Goodchild, and D.W. Rhind, Eds., Longman, London. 1991, pp. 269–297.

Wolock, D.M. and C.V. Price. Effects of Digital Elevation Model Map Scale and Data Resolution on a Topography-Based Watershed Model, *Water Resour. Res.,* 30(11), pp. 3041–3052, 1994.

Zhang, W. and D.R. Montgomery. Digital Elevation Model Grid Size, Landscape Representation, and Hydrologic Simulations, *Water Resour. Res.,* 30(4), pp. 1019–1028, 1994.

CHAPTER 42

Assessment of the Spatial Structure and Properties of Existing Ecoregionalization Systems of Ontario

D.J.B. Baldwin, L.E. Band, and A.H. Perera

INTRODUCTION

Ecoregionalization systems are widely recognized as critical tools in effective resource management (Hills, 1959; Burger, 1993; Bailey, 1988). Several existing systems contain many intuitive delineations that many experts would not dispute represent ecologically significant units. The problem lies in quantifying this significance. Many of these systems were developed prior to extensive, accurate spatial databases and include many subjective construction properties. Unfortunately, these construction methods are often not well documented and the specific properties of the units within a system are not known. To meet the growing need to include ecologically-based methods in resource management, reliance on these systems increases. Effective use of existing systems can only come from knowing what they mean quantitatively, including their strengths and weaknesses for stratifying the phenomena relevant to questions we want to answer.

This chapter summarizes the methodological aspects of work completed for Master's research to help provide this knowledge (Baldwin, 1997). The goal of the thesis was to quantify three existing systems in Ontario using spatial surfaces of geoclimatic variables relevant to ecosystem function and composition. The main objective of this quantification was to compare the strengths and weaknesses of each system in delineating these variables, illustrating which phenomena are delineated most effectively by each system, at each location. In this chapter, we present the techniques developed to quantify and compare ecoregionalization systems, rather than the detailed results relating to the three Ontario systems. The results obtained for one of the systems (Hills, 1959) are presented to illustrate these methods. We also outline conclusions relevant to the development of new ecoregionalization systems.

METHODOLOGY

Spatial Data Set Development

Variables relevant to basic ecosystem composition and function were determined from the literature. Table 42.1 outlines the series of geoclimatic variables included in the analyses.

Climate Variables

The climate variables were extracted from the Ontario Climate Model (OCM) outlined in detail by Graham (1995). Each surface was registered and maintained as a 1-km resolution, quadtree compressed raster using SPANS GIS software (TYDAC Technologies, 1994). These rasters were resampled to 10 km resolution for the core analyses outlined later.

Terrain Variables

The elevation surface was obtained from Mackey et al., 1994 and registered at 1 km resolution using SPANS. A surface was derived to quantify terrain complexity relationships. This was accomplished using a boundary modeling function in SPANS (TYDAC Technologies, 1994). The computational algorithm returns the number of cell boundaries that have differ-

Table 42.1. Summary of Variable Surfaces Included in the Analyses.

Variable	Source	Abbreviated Name
Mean annual min. temperature	OCM[a]	annmin
Mean annual max. temperature	OCM[a]	annmax
Mean monthly min. temperature-Jan.	OCM[a]	minjan
Mean monthly max. temperature-Jul.	OCM[a]	maxjuly
Mean monthly temperature - Jan.	OCM[a]	jantemp
Mean monthly temperature - Jul.	OCM[a]	julytemp
Mean annual temperature	OCM[a]	anntemp
Growing season length	OCM[a]	growlen
Annual growing degree days > 5°C	OCM[a]	anndd5
Mean annual precipitation	OCM[a]	annpm
Mean monthly precipitation - Jan.	OCM[a]	janpm
Mean monthly precipitation - Jul.	OCM[a]	julypm
Elevation	Mackey et al., 1994	km1dem
Elevation complexity	Baldwin, 1997	dembound
NDVI	Band, 1994	psnraw
Geology mean patch size	Baldwin, 1997[b]	ge_mps
Geology edge density	Baldwin, 1997[b]	ge_ed
Geology diversity	Baldwin, 1997[b]	ge_sim
Geology contagion	Baldwin, 1997[b]	ge_con
Land-cover mean patch size	Baldwin, 1997[c]	vg_mps
Land-cover edge density	Baldwin, 1997[c]	vg_ed
Land-cover diversity	Baldwin, 1997[c]	vg_sim
Land-cover contagion	Baldwin, 1997[c]	vg_con

[a] Derived from the Ontario Climate model maintained by the Genetics Program, Ontario Forest Research Institute, Ministry of Natural Resources, Sault Ste. Marie, Ontario.
[b] Derived using classified LANDSAT TM developed by Spectranalysis (1992).
[c] Derived using surficial geology database developed by Perera et al. (1996).

ent adjacent values within an N × M neighborhood surrounding each cell. Only nonzero values in the neighborhood are used. For example, the following 3 × 3 neighborhood returns a boundary function value of 6 (bars and dashes indicate interfaces with different adjacent nonzero values).

```
1  |  3  |  2
–
5  |  3  |  2
–
2     0     2
```

The raw elevation surface was preclassified into 25 m intervals to eliminate excessive noise. The boundary function was applied to this elevation surface, using a 5 × 5 neighborhood to delineate complex vs. gentle terrain. Three-dimensional representations of the elevation model were inspected to confirm correlation of this surface to complex terrain areas.

Land Cover

Surfaces of normalized difference vegetation index (NDVI) derived from Advanced Very High Resolution Radiometer (AVHRR) data were obtained from Band (1994). An annual aggregated surface was used in this study to represent a broad measure of land cover conditions. This surface was obtained at a resolution of 1 km. This layer was imported and registered in the SPANS project study area.

Land Cover and Geology Spatial Structure Variables

The nominal classifications of the surficial geology and LANDSAT land cover layers were not appropriate for the statistical tests used to analyze the ecoregionalization systems. A series of spatial indices was developed based on these layers to provide variables defining the structure of these attributes. A program was written in TCL to combine the spatial data manipulation functions of SPANS with Fragstats. The latter is an application developed to calculate a series of spatial metrics for landscape ecological applications (McGarigal and Marks, 1993).

This program was used to calculate localized measures of the spatial characteristics of the landcover/geology for 10 km × 10 km "landscapes" centered on each of the points used to sample the other variables (defined later). The land cover and surficial geology data were examined at 125 m resolution within each landscape. Four of the many spatial metrics derived by Fragstats were chosen for this study to quantify the land cover and geology layers: mean patch size, edge density, Simpson's diversity index, and a contagion index. These indices are detailed by McGarigal and Marks (1993).

Multivariate Data Set Preparation

The values of each variable and the region identifiers were appended to points generated at 1 km intervals for the boundary gradient analyses and 10 km intervals for the region core comparisons outlined later. The variable surfaces were resampled to 10 km resolution for the region core comparisons. The points were exported to create a multivariate data set for analysis in the SAS statistical package (SAS, 1993). All of the variables were standardized to have a mean of 0 and a standard deviation of 1 to eliminate discrepancies in measurement scales.

Compositional statistics

Standard univariate measures were generated to characterize each surface and each regional unit. These values were generated as a reference and a quantitative basis for users of the ecoregionalization systems for decision-making and analysis.

Region Core Analyses

Core Area Definition

The strength of the geoclimatic variable partitioning was examined. Inherent in the construction of each of these systems is a limited locational precision of the regional boundaries. Comparisons among the regions were confined to "core" areas. These core areas were modeled by imposing a buffer inward from the boundaries. A buffer distance of 40 km was selected to define robust region cores with viable region sizes for sampling.

The region lines were extended past the provincial extent to eliminate the effect of the arbitrary provincial limit. Ecologically, this administrative delineation has no true meaning except in cases where the boundary is defined by significant natural features such as the Great Lakes or Hudson Bay.

Variance Component Analysis

The within-unit to between-unit variance was examined. An effective classification should display greater between-unit variance than within-unit variance. A breakdown of the variance of each variable across the regional units of each ecoregionalization system into these two components was completed using the SAS Varcomp procedure.

Multivariate Distance

The internal structure of the system was examined by calculating the pairwise squared distance (Mahalanobis distance) between each of the regions:

$$D^2(i|j) = (\overline{X}_i - \overline{X}_j)' \, cov^{-1} (\overline{X}_i - \overline{X}_j)$$

The distances between each region and all others were calculated and sorted to examine the multivariate similarity between the regions. This provides a method to examine the relationships between multivariate similarity and spatial proximity and illustrates the relative strength of the boundaries between units. Choropleth maps were generated for the total of all multivariate distances for each region as well as maps showing the pairwise values between an individual region and all of the other regions.

Surficial Geology Similarity

The regional core units were compared in terms of their surficial geology compositional similarity using the Morisita-Horn index of similarity (Turner and Gardner, 1991). This index was calculated as:

$$MH = \frac{2 \sum p_{ij} p_{ik}}{\sum p_{ij}^2 + \sum p_{ik}^2}$$

where   MH = Simplified Morisita-Horn index
of overlap between region j and
k

p_{ij}, p_{ik} = Proportion geology type i is of
the total area in both regions
(i=1,2,3,…,n)

n = Total number of geology types

Choropleth maps were generated similar to the multi-variate distance comparisons outlined earlier.

Principal Components Analysis (PCA)

PCA was performed to examine the internal structure among regions. PCA was conducted for each regional unit, using all variables. The eigenvector loadings of each variable for the first few principal components were sorted for each regional unit to identify the key variables related to the region's structure. Maps were created showing a bar chart in each region with the eight largest positive loadings color-coded by the relevant variable. These maps were generated for the first three principal components which together typically explained more than 70% of the variance in each region

Region Boundary Analyses

Database Preparation

The boundaries between regional units were examined to identify the trends and key variables associated with each interface. A gradient was generated across each regional interface by creating 10-km wide bands, 50 km in each direction from the boundary. Points were generated at 1-km intervals; the region identifiers, buffer distance, and geoclimatic variable values were attached to the points. The spatial metric variables for vegetation and geology structure generated earlier for the core analyses were not included in the data set. Some interfaces were not examined, particularly northern units where climate interpolation was weak due to low climate station density.

Means, Multivariate Distance, and Geologic Similarity Comparisons

The means for each of variables were calculated for each band across each regional interface. The multivariate distance (Mahalanobis distance) and the geologic similarity (Morisita-Horn index) were calculated

between adjacent band pairs. These values were plotted against an x axis of the 10-km bands, centered on the regional boundary. This allowed examination of trends across the region interface and the sharpness of the gradient between regions.

Trend Comparison Statistics

A series of indices was developed to examine the slopes at different points over the gradient to quantify the trends across the region interface. The gradient was broken into three sections: one "core" area (30 km on each side of the boundary) and two "tail" areas (remaining 20 km in each region). The slopes of the lines were calculated for each variable between successive bands. Each section's slopes were compared as proportions of the entire trend, in order to standardize the measure for the many different scales of the various variables.

This relationship was calculated and interpreted differently depending on the type of analysis. For example, comparing means, a strong boundary trend is illustrated by a gradual slope in the first region tail, followed by a sharp increase or decrease across the boundary and then more gentle slopes away from the boundary into the second region. In contrast, strong multivariate distance trends are shown by a rapid increase in dissimilarity approaching the boundary from each region. Examples of strong trends for different comparison types are shown in Figure 42.1. Graph (a) shows a strong trend for the mean comparison (index value is 0.9024 calculated as the sum of slopes in the core area as a proportion of all slopes). Graph (b) illustrates a strong multivariate distance comparison trend (index = 0.7404 calculated as the sum of the absolute values of slopes in the core area as a proportion of all slopes). A strong trend for geologic similarity would be a vertical mirror image of Graph (b).

The shapes of all curves were examined in conjunction with the index values, to create a ranking scheme for the variables within each region. A ranking of strong, medium, weak, or no trend (scored 3, 2, 1, and 0, respectively), was determined for each variable, for each interface using threshold values of the index in combination with line shape. These scores were tabulated to provide a summary of the trends across each interface, for each ecoregionalization system. Total scores, standardized by the number of interfaces in each ecoregionalization system, were calculated to compare ecoregionalization systems. Summary maps were prepared, showing each bound-

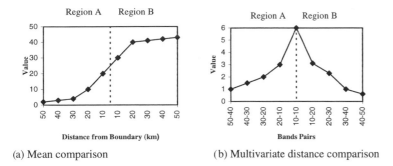

Figure 42.1. Examples of strong trend patterns for means comparison and multivariate distance comparison methods.

(a) Mean comparison

(b) Multivariate distance comparison

ary in a line width proportional to the strength of the boundary, based on the scores. The top five variables for each interface were graphed as bars showing their index values to illustrate which variables display the strongest trends for each interface.

RESULTS AND DISCUSSION

Region Core Analyses

The variance component analysis showed lower within-unit variance than between-unit variance for almost all variables except the spatial structure variables for geology and vegetation. This suggests that all of the system partition the geoclimatic variables well. This test, however, may be overstating this result. In Ontario, the variance of many of the climate variables, is linear and almost unidirectional, either latitudinal (temperature) or longitudinal (precipitation). This makes drawing 10 or 20 regions that partition most of the variance of this multivariate data set, a fairly rudimentary exercise. This does not diminish the merits of the systems studied; the authors of these systems could do this better than anyone, with knowledge of the subtleties of the ecological composition of the landscape well beyond the data. This does, however, limit the usefulness of this type of analysis for quantifying ecoregionalizations in Ontario or other locations where the variance patterns are similar. The differentiation of regions is better illustrated by the multivariate distance and geologic similarity analyses.

Figure 42.2 illustrates the maps used to examine the trends in multivariate distances among regions. Map (a) shows the total of all pairwise multivariate distances between a region and all others. Maps (b) and (c) show the pairwise multivariate distances between a focus region and the other regions. The southernmost and northernmost regions are the most strongly defined; the southern units were more distinct among themselves than the northern ones. This

is logical because they are separated by more isotherms and have no counterparts at the similar latitudes. The strength of the differentiation of southern units could also be related to stronger climate interpolations resulting from higher climate station frequency.

Trends in geologic similarity among regions were illustrated using maps like those in Figure 42.3. These proved very useful in illustrating that this geology data set is likely most useful for defining units at a spatial level above the region, such as ecozones or provinces. The regions approximating the Hudson Bay Lowland showed very low similarity values to other regions except among themselves. In addition, many of the central regions are not well differentiated among themselves. This suggests that the regions, as they are defined, may form complexes of geologic units which are broken down subsequently by other variables.

The regional PCAs were performed primarily as a reference for detailed study of individual regional units and are not presented here. In combination with the eigenvalues for each region, they proved very useful for examining the importance of each variable to each region.

Region Boundary Analyses

Figure 42.4 provides a summary of the boundary analyses results. A standardized score for all interfaces was used to rank the three systems in terms of their overall boundary strength; however, the boundary analysis results are most useful for examining specific interfaces. This information is critical to place the strength of the boundaries in a specific context. For example, if a system is needed to stratify a sampling program for a temperature-based question, information about boundary strength is needed in the context of temperature gradients.

One interesting trend these results show is that regions may stratify one type of variable well but their

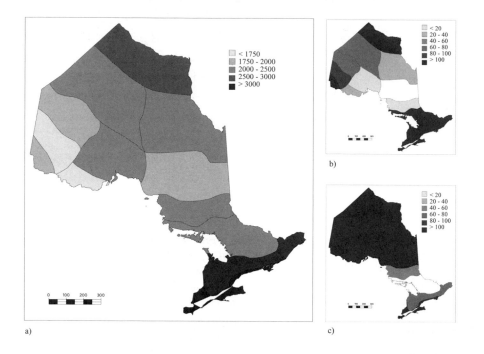

Figure 42.2. Multivariate distance values among regions. Map (a) shows the sum of pairwise multivariate distances for each region. Maps (b) and (c) show the pairwise values between a focus region (in white) and all other regions.

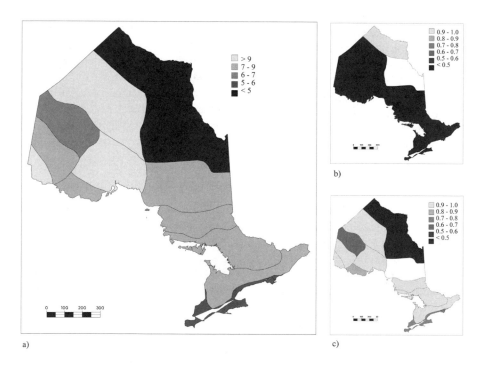

Figure 42.3. Geologic similarity values among regions. Map (a) shows the sum of Morisita-Horn similarity for each region. Maps (b) and (c) show the pairwise values between a focus region (in white) and all other regions.

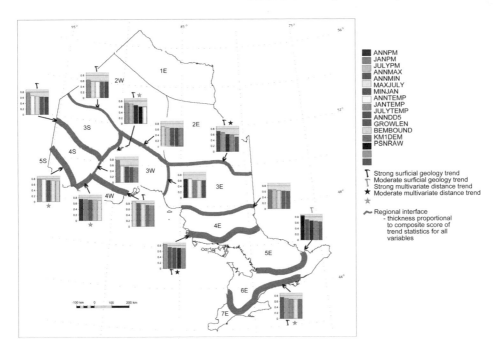

Figure 42.4. Regional interface characteristics. Region lines are drawn proportional to their total ranking scores. Bars indicate value for trend statistic of the top five variables for each interface.

exact boundary may be defined by another type. For example, we found in many cases that regions were most related to temperature variables (from the PCA results) but the exact boundaries were defined by a strong geology trend. This is consistent with the construction of many of these systems that defined regions based on landform subdivisions of climatically homogeneous areas.

CONCLUSIONS

This study has generated a number of tools to examine and compare differences in ecoregionalization systems at specific region and interface levels. In Ontario, and likely in other areas where climate variation patterns are similar, simple examinations of within-unit to between-unit variance is not very useful. On the other hand, mapping the multivariate distance and geologic similarity between regions provides an excellent visualization tool to examine the system's internal structure. The PCA results provide a useful reference to study the variables most relevant to specific regions. The methods developed in this study to quantify boundary strength, particularly in the context of variable type (i.e., climate, geology, terrain)

provide an excellent quantitative basis for using existing systems more intelligently.

The selection of a particular ecoregionalization systems or specific regions within an ecoregionalization systems must be made considering both the variables that define the region(s) as a whole as well as those that define the specific boundaries of the region(s). Choosing among many regions, however, amounts to a subjective combination of delineations to define a system for the specific phenomena of study. This results in the same limitation of using many of these existing systems—a subjective process that may not be reproducible.

This concept must be considered when new systems are developed. An effective ecoregionalization system should allow the user to interactively decide which variables are relevant to the phenomena, weight these variables according to their relative importance to this phenomena, and then generate a system by a reproducible, explicit methodology. Subjective improvements could be incorporated later, much like satellite classification. These changes, however, would be relevant to a specific analysis and would not limit stratification in subsequent analyses or for different phenomena. This rationale is being adopted in the development of the Hierarchical Ecoregional Framework (Perera et al., 1995).

REFERENCES

Bailey, R.G. *Ecogeographic Analysis: A Guide to the Ecological Division of Land for Resource Management,* USDA Forest Service Miscellaneous Publication 1465, 1988.

Baldwin, D.J.B. *Quantification of Existing Eco-Regionalizations of Ontario,* thesis presented to the University of Toronto in partial fulfillment of the requirements for the degree of Master of Science, 1997.

Band, L.E. *Development of a Landscape Ecological Model for Management on Ontario Forests,* Rep. No. 7, Forest Fragmentation and Biodiversity Project, Ont. Min. Nat. Resour., Ont. For. Res. Inst., Sault Ste. Marie, Ont., 1993.

Band, L.E. *A Pilot Landscape Ecological Model for Forests in Central Ontario,* Rep. No. 17, Forest Fragmentation and Biodiversity Project, Ont. Min. Nat. Resour., Ont. For. Res. Inst., Sault Ste. Marie, Ont., 1994.

Burger, D. *Revised Site Regions of Ontario: Concepts, Methodology and Utility,* For. Res. Rep. No. 129, Ont. Min. Nat. Resour., Ont. For. Res. Inst., Sault Ste. Marie, Ont., 1993.

Graham, B.J. *Modelling Distributions of Rare Carolinian Tree Species in Southwestern Ontario,* thesis presented to the University of Western Ontario, London, in partial fulfillment of the requirements for the degree of Master of Science, 1995.

Hills, G.A. *A Ready Reference to the Description of the Land of Ontario and Its Productivity,* Prelim. Res. Report, Ont. Dep. of Lands and Forests, Maple, Ont., 1959.

Hills, G.A. *The Ecological Basis for Land-Use Planning,* Research Report No. 46, Ont. Dep. of Lands and Forests, Maple, Ont., 1961.

Mackey, B.G., D.W. McKenney, C.A. Widdifield, R.A. Sims, K. Lawrence, and N. Szcyrek. *A New Digital Elevation Model of Ontario,* NODA/NFP Tech. Rep. TR-6, Dep. of Nat. Resour. Can., CFS—Ont., Sault Ste. Marie, Ont., 1994.

McGarigal, K. and B.J. Marks. *Fragstats: Spatial Pattern Analysis Program for Quantifying Landscape Structure,* Oregon State University, Corvallis, 1993.

Perera, A.H., J.A. Baker, L.E. Band, and D.J.B. Baldwin. A Strategic Framework to Eco-Regionalize Ontario, *Environ. Assess. Monitor.,* 39, pp 85–96, 1995.

Perera, A.H., J. Laguna, D.J.B. Baldwin, R. Bae, H. Godschalk, and M. Ouellette. *Surficial Geology of Ontario: A Digital Database,* Rep. No. 22, Forest Fragmentation and Biodiversity Project, Ont. Min. Nat. Resour., Ont. For. Res. Inst., Sault Ste. Marie, Ont., 1996.

SAS Institute. *Procedures Manual for the SAS System,* 1993.

Spectranalysis Inc., *Development of a Digital Database of Red and White Old-growth Pine Forests in Ontario— East,* Rep. No. 22, Forest Fragmentation and Biodiversity Project, Ont. Min. Nat. Resour., Ont. For. Res. Inst., Sault Ste. Marie, Ont., 1992.

Turner, M.G. and R.H. Gardner. *Quantitative Methods in Landscape Ecology,* Springer, New York, 1991.

TYDAC Technologies. *SPANS Reference 5.3 Reference Manual,* Ottawa, 1994.

CHAPTER 43

Toward Real Image Quality Improvement: The Development of a Decision Model for Controlling Image Fusion Processes

L. Li and B.C. Forster

INTRODUCTION

Information fusion deals with the integration of information from several different sources, aimed at an improved quality of results. With the availability of a range of data in fields such as remote sensing, GIS, medical imaging, machine vision, as well as military applications, information or data fusion has emerged as a new and promising research area.

Information fusion can be classified as signal-level fusion, pixel-level fusion, feature-level fusion, and symbol-level fusion (Li et al., 1995; Pohl, 1996; Costantini et al., 1997). Signal-level fusion refers to the combination of a group of sensors with the objective of producing a single signal of greater quality and reliability. It is sometimes referred to as sensor fusion. Feature-level fusion enables the detection of useful features with higher confidence and requires the extraction of objects recognized in various data sources. Symbol-level fusion allows the information from multiple sources to be effectively used at the highest level of abstraction. Pixel-level fusion, generally referred to as image fusion, by combining two or more different images to form a new image (Van Genderen and Pohl, 1994), serves to increase the useful information content or to improve the quality of an image such that the performance of image processing tasks such as segmentation and feature extraction can be improved.

For image fusion, there are a number of algorithms or approaches that have been developed over the years. It can be performed by simply overlaying the images, by using the intensity-hue-saturation (IHS) transform merge (Haydn et al., 1982; Carper et al., 1990), component substitution (Shettigara, 1992), wavelet transformation (Yocky, 1995), fuzzy sets (Russo and Ramponi, 1997), and numerous other approaches. More detailed description of image fusion algorithms can be found in the publications of Filiberti et al. (1994), Vrabel (1996), Hall and Llinas (1997), Wilson et al. (1995), and Pohl's Ph.D. dissertation (1996). Research on evaluating the fusing algorithms can also be found in the current literature; for example, Garguet-Duport et al. (1996), Vrabel (1996), and Yocky (1996). However, the fusion results are mostly evaluated visually.

There is no quantitative performance measurement for evaluating image fusion algorithms. The problem lies in the difficulty of defining an "ideal" composite image based on multisensor images or images taken at different times. Li et al. (1995) used a performance measure, which is the standard deviation of the difference between the test (ideal) image and the fused image, to evaluate the algorithms. The experimental results are generally consistent with visual evaluation, but this method is not applicable to real situations where it is not possible to obtain the ideal fusion manually. This chapter, however, discusses a quantitative solution.

Figure 43.1 depicts the general processes involved in image fusion. Image registration and resampling are the important preprocessing steps in image fusion. It is the process of matching two images so that corresponding coordinate points in the two images correspond to the same physical region of the scene being imaged. A wide range of registration techniques has been developed over the years for different types of applications and data. Given the diversity of the data, it is unlikely that a single registration technique will

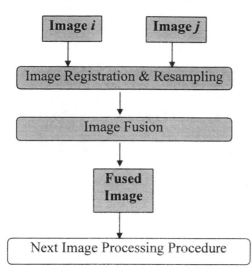

Figure 43.1. Image fusion processes.

satisfy all of the different applications. Fonseca and Manjunath (1996) summarized issues involved in registering multisensor remote sensing images. The primary issue is to identify suitable ground control points and to choose the "right" registration and resampling method for the registration. The robustness of the registration process determines the quality of the fusion output at the end of the process, thus influencing the decision-making process.

The motivation for image fusion is that the fusion result has improved quality for better decision-making. In remote sensing, for certain areas, there are many different sources of data at different spatial and spectral resolution. There are two important properties of these remotely-sensed image data sets. These are:

- there is a high spectral/spatial redundancy
- a relationship exists between the image spatial continuity and the spatial resolution

In addition, all the data sets are of different quality. Thus, it is essential to determine if the data sets are "good enough" for image improvement. It is not necessary to spend time to find suitable and sufficient control points to coregister images and go through all the processes for fusion if there is no or limited improvement to be expected.

In this chapter, the authors are presenting a quantitative assessment model, based on geostatistical analysis and information theory, that can be used to evaluate the spatial structure and spatial correlation of the in-

put data sets and to present the relationship between the data sets and the quality improvement of the fused outputs. The model enables users to make decision on whether to perform the fusion task at the commencement stage and to assess the improvement of the fusion process.

THE MODEL

Shannon and Wiener's entropy approach to express the information content of a message has been extended by researchers in the image processing field for measuring the information content of the image sets; e.g., spatial and radiometric entropy developed by Maitre et al. (1994), mutual information or redundancy method by Le Hegarat-Mascle et al. (1997), and the quantitative method by Atkinson (1995). Although there are different ways of measuring the information content of the image data, the information provided by the image is strongly related to spatial structure or spatial variation of the data set. In this chapter, we introduce a quantitative model which uses a set of simple statistical indicators to measure spatial data structure

There are three methods for describing spatial structure or spatial continuity of a data set; namely, correlation function, covariance function, and experimental variogram, and any of the three serve as well as the others. The experimental variogram is a convenient tool for the analysis of spatial data as it is based on a simple measure of spatial dissimilarity or continuity (Wackernagel, 1995; Isaaks and Srivastava, 1989). Recently, it has been widely used in remote sensing applications for estimating information content, spatial resolution selection, and error assessment (Curran, 1988; Curran and Williamson, 1988; Atkinson, 1993, 1997a, 1997b; Atkinson and Curran, 1997; Van Der Meer, 1997). The behavior at very small scales, near the origin of the variogram, is of importance, as it indicates spatial structure of the data. If the variogram value is constant for all lags, there is no spatial structure in the data. Conversely, a nonzero slope of the variogram near the origin indicates structure (Wackernagel, 1995, p. 33).

The experimental variogram is fitted to a mathematical model curve. The fitted models are described by three parameters: range, sill, and nugget variance. The sill approximates the variance of the data. The range is the distance at which the variogram reaches its sill; it is a measure of the spatial continuity (or structure) of the data. The nugget variance may be contributed by the spatial variability as well as the measurement error (Van Der Meer, 1997; Wacker-nagel, 1995). In the model, we use these three statistical values as the indicators to measure the spatial structure of the data sets.

With the development of sensor systems, more data sets with similar spatial and spectral properties will be available to users, and in many cases they will provide the same or similar information; for example, adjacent bands of hyperspectral imagery. It is necessary to measure the correlation between data sets, so that users can choose the most appropriate set for their application. Spatial correlation between the image data sets can be measured by using a cross correlogram.

Image quality is usually measured with respect to two aspects (Frey and Susstrunk, 1996):

- Objective image quality, evaluated through physical measurements of image properties (e.g., resolution, bit depth, reflectance, density, modulation transfer function, etc.)
- Subjective image quality, evaluated through judgment by human observers of variables such as darkness, sharpness, colorfulness.

So far, most of the literature that reports the "successful application" of image fusion has presented the "improvement" of quality in subjective terms. Some researchers used objective measurements, such as GSD (effective ground-sample distance) (Vrabel, 1996), to compare the relative effectiveness of different algorithms; however, the GSD values were subjectively determined by the experts (image analysts and image scientists) in the field. In the model, we propose to use modulation transfer function to measure the improvement. However, this approach is not dealt with in the current chapter.

Figure 43.2 illustrates the concept of the quantitative data assessment model. It is comprised of two models: prefusion assessment model and postfusion assessment model. Existing data sets are identified based on application areas. Data sets will be initially selected by the users' expert knowledge on spatial and spectral requirements of the data sets based on the nature of the application. The "goodness" of selected data sets are then tested by using the prefusion assessment model. If the decision is to undertake the fusion, then the fused result will be evaluated by using the postfusion model to measure the improvement of the fusion process. The prefusion assessment model is comprised of a set of programs that can be used to calculate indicators for measuring spatial structure of

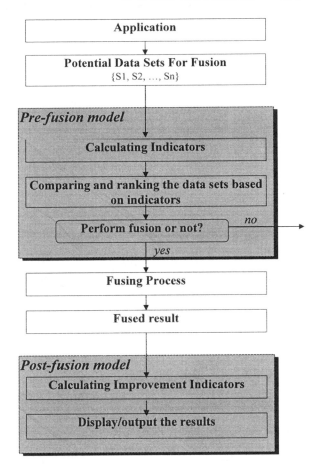

Figure 43.2. The quantitative assessment model.

the data sets and the spatial correlation between the data sets and a decision support model to display the indicators that can help the users to make decisions on whether to pursue the fusion process. If data sets are not coregistered, only variograms that present the spatial structure are calculated and fitted with mathematical models. If they are coregistered sets; for example, hyperspectral data sets, then correlation between data sets are calculated and evaluated. The postfusion model consists of programs that can generate quality indicators such as classification accuracy, point spread function, etc., to show the improvement of the fused result.

EVALUATION OF THE MODEL

In order to test the effectiveness of the model, we use an urban case study from the Sydney metropolitan area, Australia. The main reason for selecting an urban area is because it is the most difficult type of land-

Table 43.1. Statistical Indicators of the Test Data Sets.

Image Type	Blue Band			Green Band			Red Band		
	N[a]	R[a]	S[a]	N	R	S	N	R	S
1m x 1m	340	13.77	1428	720	13.26	2250	1150	16.83	4400
2m x 2m	330	17.85	1305	450	15.3	2232.7	817	18.36	4042
4m x 4m	127.4	18.87	921.2	272	20.4	1547	450	21.93	3000

* N - nugget variance; R - range; S - sill.

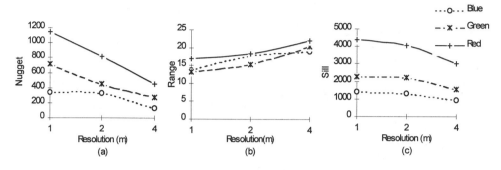

Figure 43.3. Geostatistical indicators of the data sets (a) nugget variance, (b) range value, and (c) the sill of fitted models.

scape to deal with, due to high variability of information in space and spectral band, and the diversity of features, both in size and nature (Wald et al., 1997), and it is also the potential application area of high resolution imagery systems. According to Welch (1982), IFOV or resolution equivalent of 5 meters or better is most effective for visual interpretation for urban studies, particularly in dense Asian environments. Jensen et al. (1989) have noted that there are no data of 1 m to 5 m spatial resolution from space sufficient to meet the urban mapping requirements which are becoming so important for use in Geographic Information Systems (GIS).

The original test data are in the red, green, and blue bands with approximated 1 meter by 1 meter ground resolution (R_1, G_1, B_1) obtained by scanning the airphoto taken in October 1994. The data sets were then degraded to 2 meters (R_2, G_2, B_1) and 4 meters (R_4, G_4, B_4) resolution, respectively. A sample data set was selected from all the image data sets and variograms were calculated and fitted with Gaussian models (Pannatier, 1996, p. 83; Wackernagel, 1995, p. 40). Figure 43.3 shows the statistical values of the models and Table 43.1 shows the statistics. It can be seen from the blue band that the statistics are very similar for 1 meter and 2 meter data sets (ratio value of

nugget variance is close to 1). The nugget value drops significantly for the red and green bands when the resolution decreases. The decision can then be made, based on the statistics, that there is no need to replace 2-meter resolution blue band data with 1-meter resolution blue band since they have similar information content.

Classification was carried out by using the nine combinations of data shown in Table 43.2 by using the maximum likelihood classification method. Eleven classes were predefined; they are: red roof houses, brown roof houses, grey roof houses, white concrete roof building, grass land, tree/shrub, bare soil surface, road (asphalt), concrete tile surface, swimming pool, and shadow areas.

Figure 43.4 shows the classification results, and the total number of pixels classified for each class are shown in Table 43.3. By visual observation, we can see that the result of 1-meter resolution data (conventional combination) is adequate for urban application; most of the features can be clearly and accurately classified. Details are lost in the 4-meter combination. There is a significant improvement if fusing 1-meter data with 4-meters, and little with 1- and 2-meter combination except the R(1)G(2)B(2) combination [Figure 43.4(h)]. The statistics of classification are shown in Figure 43.5. Figure 43.5 plots the accumulative

Table 43.2. Band Combination for Classification.

Type	No.	Red Band	Green Band	Blue Band
Conventional	1	1 m	1 m	1 m
Conventional	2	2 m	2 m	2 m
Conventional	3	4 m	4 m	4 m
Fusion	4	1 m	2 m	2 m
Fusion	5	2 m	1 m	2 m
Fusion	6	2 m	2 m	1 m
Fusion	7	1 m	4 m	4 m
Fusion	8	4 m	1 m	4 m
Fusion	9	4 m	4 m	1 m

Legend
- red roof
- white roof
- concrete tile
- road
- barren land
- grass land
- tree
- shadow
- swimming pool
- brown roof
- grey roof

Figure 43.4. Original image and the classification results of different combinations. (a) $R_4G_4B_4$, (b) $R_2G_2B_2$, (c) $R_1G_1B_1$, (d) $R_4G_4B_1$, (e) $R_2G_2B_1$, (f) original image, (g) $R_1G_4B_4$, (h) $R_1G_2B_2$, (i) $R_4G_1B_4$, (j) $R_2G_1B_2$.

number of pixels per class, and it can be seen that there is a significant difference between the results of 1- and 4-meter data, and the improvement of replacing band data of 4 meter resolution with 1 meter is

very significant, which is not the case with a replacement by 2-meter data. Considering Table 43.3, it can be seen that different combinations also have different impacts on individual classes when compared to the classification resulting from the original image. This is a function of both the spatial resolution and spectral signature of the class in question. Further research is required in this area before an optimum fusion decision can be made for all classes. In the program to be developed for this model, criteria for replacing one data set by another will be the ratio of their nugget values. A value close to one will assume that replacement will not lead to a significantly improved result. The current study suggests that a ratio of 1.5 to 2 or greater would be an appropriate value to initiate the fusion process.

Future Research

The work presented above indicates some intermediate results of the research. For assessing the improvement of the fusion of the case study, more tasks need to be performed in the area of collecting in situ data to assess the classification accuracy in terms of commission and omission errors. However, the method used for assessing the improvement in the case study is a "subjective" method and was only employed to show the effectiveness of using the geostatistical indicators for controlling fusion process. It is not practical since there wouldn't be a "true" value available in reality. Our next task is to input a set of "objective" measurements such as point spread function or MTF, etc., into the postfusion model.

The most common fusion technique is to fuse or merge higher spatial resolution panchromatic data with lower resolution multispectral image. This technique is often called band sharpening. In band sharpening, the product has the spatial resolution of the panchromatic image and the spectral characteristics of the multispectral image. The spectral characteristics are useful for identifying features such as trees, water, soil, etc. With increased spatial resolution, the features can be more accurately delineated, thus making the resulting product more useful for various applications, even more useful if there is no change in the spectral content of the sharpened product. More importantly, band sharpening with a single high-resolution panchromatic image allows the multispectral band data to be acquired at a lower spatial resolution. This permits systems to be designed that have lower bandwidth and storage requirements. Lower multispectral spatial resolution

Table 43.3. Comparison of the Classification Results.

Classes	$R_1G_1B_1$	$R_2G_2B_2$	$R_2G_1B_2$	$R_1G_2B_2$	$R_2G_2B_1$	$R_4G_4B_4$	$R_1G_4B_4$	$R_4G_1B_4$	$R_4G_4B_1$
Red roof	3600	5080	5783	6254	5688	5648	8688	6765	3984
White roof	9720	7124	6878	7149	7463	2824	4073	3899	3861
Concrete tile	2065	4492	2408	4185	2048	4249	5020	6124	6654
Road	1824	1916	1007	1097	978	12008	3111	2078	2804
Barren land	2957	2512	2574	1997	3896	1416	1643	2171	3094
Grass land	1749	1696	711	867	2131	608	636	538	1500
Tree	831	1704	4051	1354	1335	3976	2978	2825	3226
Shadow	6786	5169	5183	5822	5412	2080	4685	3374	3563
Swimming pool	23	1616	1166	393	860	552	210	456	273
Brown roof	2973	2616	5026	3153	5220	4800	4527	8191	9783
Grey roof	7873	6476	5614	8130	5370	2240	4830	3980	1659

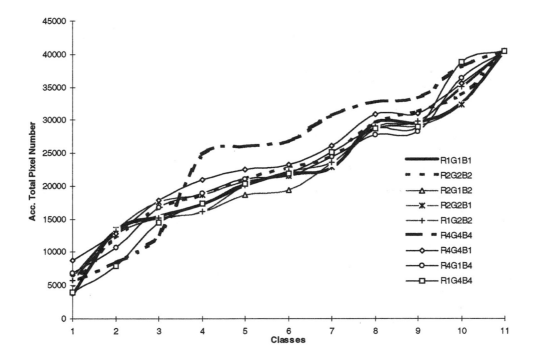

Figure 43.5. Classification results (accumulative number of pixels for each class).

can also lead to the implementation of increased spectral resolution on future sensors (Vrabel, 1996). Investigation in this area will be followed up, once suitable data sets become available.

SUMMARY AND CONCLUSIONS

In this chapter, we have presented a quantitative assessment model to control image fusion process and have used a case study to test the effectiveness of such a model. The case study shows that the ratio of nugget variance of two data sets can be used to present the similarity of information content provided by the data since information content is determined by the spatial structure. It suggests that there will be a significant improvement when replacing one data set with "less" information content with the "better" set if the ratio is 1.5 to 2 or greater. However, more research is needed to test the effectiveness of the proposed model, especially when choosing the data from data sets of the same or similar spatial and spectral accuracy, such as hyperspectral data.

REFERENCES

Atkinson, P. The Effect of Spatial Resolution on the Experimental Variogram of Airborne MSS Imagery, *Int. J. Remote Sensing,* 14(5), pp. 1005–1011, 1993.

Atkinson, P. A Method for Describing Quantitatively the Information Content, Redundancy and Error in Digital Spatial Data, in *Innovations in GIS II,* P. Fisher, Ed., Taylor & Francis, Washington, DC, 1995.

Atkinson, P.M. Selecting the Spatial Resolution for Airborne MSS Imagery for Small-Scale Agricultural Mapping, *Int. J. Remote Sensing,* 18(9), pp. 1903–1917, 1997a.

Atkinson, P. On Measurement Error in Remotely-Sensed Images with the Variogram, *Int. J. Remote Sensing,* 18(14), pp. 3075–3084, 1997b.

Atkinson, P. and P.J. Curran. Choosing an Appropriate Spatial Resolution for Remote Sensing Investigations, *Photogrammetric Eng. Remote Sensing,* 63(12), pp. 1345–1351, 1997.

Carper, W.J., T.M. Lillesand, and R.W. Kiefer. The Use of Intensity-Hue-Saturation Transformation for Merging SPOT Panchromatic and Multispectral Image Data, *Photogrammetric Eng. Remote Sensing,* 56(4), pp. 459–467, 1990.

Costantini, M., A. Farina, and F. Zivilli. The Fusion of Different Resolution SAR Images, *Proceedings of the IEEE,* 85(1), pp. 139–146, January 1997.

Curran, P.J. The Semi-Variogram in Remote Sensing: An Introduction, *Remote Sensing Environ.,* 3, pp. 493–507, 1988.

Curran, P.J. and H.D. Williamson. Selecting a Spatial Resolution for Estimation of Pre-Field Green Leaf Area Index, *Int. J. Remote Sensing,* 9(7), pp. 1243–1250, 1988.

Filiberti, D.P., S.E. Marsh, and R.A. Schowengerdt. Synthesis of Imagery with High Spatial and Spectral Resolution from Multiple Image Sources, *Optical Eng.,* 33(8), pp. 2520–2528, 1994.

Fonseca, L. and B. Manjunath. Registration Techniques for Multisensor Remotely Sensed Image, *Photogrammetric Eng. Remote Sensing,* 62(9), pp. 1049–1056, 1996.

Frey, F. and S. Susstrunk. Image Quality Requirements for the Digitization of Photographic Collections, in SPIE Symposium on Very High Resolution and Quality Imaging: *Proceedings of SPIE: Very High Resolution and Quality Imaging, San Jose, California, 1996,* SPIE Proceeding Series Vol. 2663, SPIE-The International Society for Optical Engineering, 1996, pp. 2–7.

Garguet-Duport, B., J. Girel, J.M. Chassery, and G. Pautou. The Use of Multiresolution Analysis and Wavelet Transformation for Merging SPOT Panchromatic and Multispectral Image Data, *Photogrammetric Eng. Remote Sensing,* 62(9), pp. 1057–1066, 1996.

Hall, D. and J. Llinas. An Introduction to Multisensor Data Fusion, *Proceedings of IEEE,* 85(1), pp. 1–8, Jan. 1997.

Haydn, R., G.W. Dalke, J. Henkel, and J.E. Bare. Application of the IHS Color Transform to the Processing of Multisensor Data and Image Enhancement, *Proceedings of the International Symposium on Remote Sensing of Arid and Semi-Arid Lands,* Cairo, Egypt, 1982, pp. 599–616.

Isaaks, E.H. and R.M. Srivastava. *Applied Geostatistics,* Oxford University Press, New York, 1989.

Jensen J.R., J. Campbell, J. Dozier, J. Estes, M. Hodgson, C.P. Lo, K. Lulla, J. Merchant, R. Smith, D. Stow, A. Strahler, and R. Welch. Remote Sensing, in *Geography in America,* G. Gaile and C.J. Williams, Eds., Merrill Publishing Co., Columbus, OH, 1989, pp. 746–775.

Le Hegarat-Mascle, S., D. Vidal-Madjar, O. Taconet, and M. Zribi. Application of Shannon Information Theory to a Comparison Between L- and C-Band SIR-C Polarimetric Data Versus Incidence angle, *Remote Sens. Environ.,* 60, pp. 121–130, 1997.

Li, H., B.S. Manjunath, and S.K. Mitra. Multisensor Image Fusion Using Wavelet Transform, *Graphical Models & Image Processing,* 57(3), pp. 235–245, 1995.

Maitre, H., I. Bloch, and M. Sigelle. Spatial Entropy: A Tool for Controlling Contextual Classification Convergence, in IEEE International Conference Image Processing, *IEEE ICIP Proceeding,* Vol. 2, Austin, TX, 1994, pp. 212–216.

Pannatier, Y. *Variowin—Software for Spatial Data Analysis in 2D,* Springer, New York, 1996.

Pohl, C. Geometric Aspects of Multisensor Image Fusion for Topographic Map Updating in the Humid Tropic, thesis presented to the International Institute for Aerospace Survey and Earth Sciences in partial fulfillment of the requirements for the Degree of Doctor of Philosophy, ITC Publication No. 39, ITC, 1996.

Russo, F. and G. Ramponi. Fusion Methods for Multisensor Data Fusion, *IEEE Transaction Instrum. Meas.,* 43(2), pp. 288–294, 1997.

Shettigara, V.K. A Generalized Component Substitution Technique for Spatial Enhancement of Multispectral Images Using a Higher Resolution Data Set, *Photogrammetric Eng. Remote Sensing,* 58(5), pp. 561–567, 1992.

Van Der Meer, F. What Does Multisensor Image Fusion Add in Terms of Information Content for Visual Interpretation?, *Int. J. Remote Sensing,* 18(2), pp. 445–452, 1997.

Van Genderen, J.L. and C. Pohl. Image Fusion: Issues, Techniques and Applications, in *EARSel Workshop,* Proceedings EARSel Workshop on Intelligent Image Fusion, J.L. Van Genderen and V. Cappellini, Eds., Strasbourg, France, 1994.

Vrabel, J. Multispectral Imagery Band Sharpening Study, *Photogrammetric Eng. Remote Sensing,* 62(9), pp. 1075–1083, 1996.

Wackernagel, H. *Multivariate Geostatistics: An Introduction with Applications,* Springer-Verlag, Berlin/Heidelberg, 1995.

Wald, L., T. Ranchin, and M. Mangollni. Fusion of Satellite Images of Different Spatial Resolutions: Assessing the Quality of Resulting Images, *Photogrammetric Eng. Remote Sensing,* 63(6), pp. 691–699, 1997.

Welch, R. Spatial Resolution Requirements for Urban Studies, *Int. J. Remote Sensing,* 3(2), pp. 139–146, 1982.

Wilson, T., S.K. Rogers, and L.R. Myers. Perceptual-Based Hyperspectral Image Fusion Using Multiresolution Analysis, *Optical Eng.,* 34(11), pp. 3154–3164, 1995.

Yocky, D. Image Merging and Data Fusion by Means of the Discrete Two-Dimensional Wavelet Transform, *J. Optical Soc. Am. A—Optics & Image Science,* 12(9), pp. 1834–1841, 1995.

Yocky, D. Multiresolution Wavelet Decomposition Image Merger of Landsat Thematic Mapper and SPOT Panchromatic Data, *Photogrammetric Eng. Remote Sensing,* 62(9), pp. 1067–1074, 1996.

CHAPTER 44

Monitoring Defoliation of Forest Trees by Means of Large-Scale Digital Image Processing

A. Floris and G. Scrinzi

INTRODUCTION

Situation of Forests in Europe

Since the late 1970s the phenomenon known as "forest decline" or *Waldsterben* (citaz.) has increased to such an extent in Europe to induce many forest research and assessment institutions in several countries to show a strong demand for interventions for the systematic observation of the phenomenon.

In 1986 the European Community (EC), passing Regulation No. 3528/86, started an organically coordinated action aimed at protecting forests from air pollution and, more generally, from "new type damage." This action has three fundamental goals:

1. Research into the causes of these diseases (air pollution, temperature increase, etc. ...);
2. Research into more suitable observation and measurement methods of these phenomena;
3. Research into damaged forest preservation and restoration methods.

Our Institute's interest has been addressed to some aspects of the second topic, for which the EC action has developed (EC Regulation No. 1696/87) the guidelines of a periodical inventory that is carried out by the membership countries on the basis of a common methodology.

Current Monitoring Methods

The EC monitoring protocol provides for a systematic observation points network (16 km × 16 km), in each of which a sample area is set up including from 20 to 30 trees, which are monitored once a year (generally in the same period in summertime).

After careful visual observation, done also with the aid of binoculars, a well-trained estimator gives each tree an evaluation concerning with various parameters, the most significative of which is the level of tree-crown transparence. This index can assume values from 0% (full-density crown, healthy tree) up to 100% (completely absent crown, no leaves, dead tree). Each class of estimation is 5% wide.

Almost 120,000 trees distributed in more than 5,300 inventory points of 30 countries are yearly monitored with this method (Vel et al., 1997).

Demand for a More Objective and Sensitive Monitoring Procedure

Concordance of estimate criteria among different surveyors (whose data should be comparable also at a transnational level) should be guaranteed by several forms of "synchronization," such as periodic collective training stages, field handbooks containing pictures with examples of different transparence levels (Conedera, 1989; Ferretti et al., 1994a) and a sort of combined analysis of some case studies.

During the data processing, too, some mathematical normalization and adjustment methods aimed at improving data comparability can be used (Ferretti et al., 1994; Tesi et al., 1997).

Despite these expedients, a number of studies have shown the visual estimate limitations (Innes, 1988a, 1988b; Landmann, 1989; Scrinzi and Confalonieri,

1991), especially in terms of subjectivity and poor sensitivity, thus emphasizing the necessity developing alternative monitoring procedures, with more objectivity and sensitivity and, at the same time, able to maintain the same promptness typical of visual estimates.

MATERIALS AND METHODS

Procedure Overview

The main indicator of a tree's health is the foliation level (or, better, the leaf weight) of its crown, related to a physiologically optimal (but very often merely theoretical) level.

Leaf biomass, in terms of weight, can be measured only by means of destructive techniques. This procedure, moreover being very laborious, could have some justification only when applied to very small samples (a few trees) and in special circumstances, using if possible only trees already intended for felling.

The level of tree crown transparence is currently regarded as the best indicator of leaf biomass.

Remote sensing carried out by means of satellite images is useful to supply multitemporal data about whole forests, but it is absolutely unsuitable to give information on single or small groups of trees, due to its spatial resolution limitations (in the best case, each pixel represents a 100 square meter area).

The aerial photograph analyses, especially those on large and macro scale (1:2,000 to 1:500; Howard, 1991), can surely supply more detailed information for a single tree, but are unfortunately not yet sufficient to carry out automatic measurements on individual trees. Moreover, not all European countries currently have organizations which are able to prepare a monitoring system for the whole area they are responsible for (province, region, or state) based on large-scale aerial photography with the desired time range (1–3 years). This is one of the reasons the EC has chosen to prepare a monitoring protocol based on ground estimations.

Starting from these assumptions, the method shown in this study consists in taking tree images (in particular, tree crown silhouettes against the sky background) from the ground using a camera, on a very large scale (1:50–1:300).

Afterward, these images are digitalized by a scanner with the aim of measuring the tree crown area, this parameter being strictly related to the tree foliation level, and therefore to its health condition, as many studies have demonstrated (Honer and Collins, 1974;

Lee et al., 1983; Benincasa et al., 1987; Scrinzi and Floris, 1992).

Considering the features of forest environments, this kind of procedure has not only to be suitable from a scientific point of view, but also simple enough to be adopted in a permanent inventory with samples distributed over the whole territory, which means:

- field surveys should be carried out by the personnel of regional bodies, using portable and not too expensive or complex equipment;
- the subsequent image-processing should be carried out by means of simple hardware and software (therefore, PC);
- survey criteria should be adaptable to most of forest scenery and to as many tree species as possible, to obtain a good sample distribution covering the entire range of real situations.

Providing that the proposed procedure respects these requirements, it could supplement the traditional evaluation method in a short time, and it could even replay it in the future.

Data Collection

The research is being carried out in the Friuli-Venezia Giulia region (northeastern Italy). Most of the forests are located in the mountainous and hilly part of the region, which covers about 62% of the total area (784,600 ha). The main species are beech (*Fagus sylvatica*), red spruce (*Picea excelsa*), white fir (*Abies alba*) and larch (*Larix decidua*); Scots pine and black pine are also represented and, to a small extent, some species of oak and the ash tree.

Although the research design provided for a total sample of about 100 subjects, 168 trees were selected, in case some trees had fallen down or been felled, or some images were not suitable for the elaboration (Table 44.1).

Surveys were carried out in the summertime (from July to September) in 1995, 1996, and 1997, in order to obtain a multitemporal set of three images for each subject; we took care to photograph the same tree in the same period each year.

Pictures were taken from fixed station points, using a ranging pole to position the camera on exactly the same point and at exactly the same height. We consider a tripod unnecessary and it could increase the equipment volume and weight.

Table 44.1. The Observed Sample with Its Distribution by Species.

Species	Number of Trees
Beech	50
Red spruce	42
White fir	27
Larch	19
Oaks	15
Black pine	8
Scots pine	5
Ash	2
TOTAL	168

The camera was a normal reflex 35 mm camera with interchangeable optics. In this particular application a camera is preferable to a videocamera for its cost-effectiveness, simplicity of use, and easier digitalization.

The distance between the station point and the subject was generally within a range of 20–30 m; therefore, focal length values have been changed in relation to tree dimensions, taking care to include in the frame most of the tree crown and external reference points (other trees, rocks, etc.) suitable for referencing. The most frequent focal lengths went from 28 to 70 mm.

Although it would be right to use the same photographic parameters in each survey for the same subject, some little differences in scale or shot can be corrected during the image processing phase, as will be explained subsequently.

Normal panchromatic color film with medium sensitivity (ISO 100/21°) was used. It gave the expected performance in terms of spatial resolution and chromatic resolution and fidelity. Some trials were carried out using black-and-white panchromatic film with a blue-absorbing filter (Kodak Wratten 47) with the aim of centering the collected information in the blue region (the most interesting for our purposes), increasing at the same time spatial resolution. This option was rejected for three main reasons:

1. Digital resolution of the image cannot exceed certain values (300-max 600 dpi), in order to avoid slow and complicated processing (a 13 × 18 cm format picture, digitalized at 300 dpi and 24 bit, requires about 7–8 mb of ram);
2. Difficulties in processing this kind of film using a standard procedure in most laborato-

ries (sometimes we obtained very poor results);
3. In view of further development in the analysis method, it could be important to preserve any information concerning the whole visible spectrum.

Film for color slides could also be used, but we have observed it doesn't bring many advantages and it makes digitalization more complicated (you need a more sophisticated scanner and a larger amount of memory).

In spite of the well-known usefulness the near IR could have, its employment has been rejected because of difficulties in obtaining a stable film and processing it correctly.

Digitalization Parameters

Prints in 13 × 18 cm format were used (edges of standard 10 × 15 cm prints are often cut in relation to the contents of the negative). Digitalization parameters are: RGB color (24 bit), 300 dpi. Some A/D conversion (acquisition, transformation) modules apply filters and enhancement functions automatically with the aim of improving the visual quality of images. The use of these functions should be avoided (deactivating them before the digitalization) because they modify and reduce the quality of the distribution of frequencies in the spectrum (histogram) and make the following processing steps more difficult.

Dimensioning and Referencing of "Analysis Windows"

The measure of the tree crown area is not an indicator of the health conditions of the tree by itself. In fact, this value depends on several factors as species, genotype and fenotype, age, climate, site conditions, and so on.

Better information could be furnished by the "changes" of the tree crown area during the time. They can be measured accurately only if carried out each time on the same portion of the tree-crown. To do this we have to "cut" for each subject an analysis window geometrically referenced and scaled on "conspicuous points" (the position of which doesn't change during the time) that can be recognized at the following sessions of the survey.

Only in this way can the changes of the tree crown area be measured with a suitable degree of sensitivity and accuracy to produce a real improvement with respect to the visual estimate.

The procedure for dimensioning and referencing the analysis window is carried out using some common functions of most image processing software packages. We used Photoshop[1] 3.0. It can be described in the following steps:

1. display of the images set concerning the same tree;
2. stem slope measurement (it can be different because of little changes in the camera positioning[2]);
3. slope correction (do not take all the stems at 90°, but only rotate the minimum value to obtain the same slope for all the images);
4. image scale measurement;
5. choice of the "basic" image (it is usually the latest of the sequence);
6. selection and cutting of the analysis window on the basic image;
7. cutting of a homologous analysis window on the other images of the set.

More details about this procedure are shown in Figures 44.1 and 44.2.

Discriminating the Tree-Crown from the Background

To measure an object area (expressed as number of pixels) on a digital image, we have to produce a *binary* image, containing only 0 and 1 reflectance values (Di Zenzo, 1984). The *thresholding* requires that different components of the image (in our case, the tree crown and the sky) have different reflectance distributions, better if not overlapping.

The choice of the threshold level has to be carried out on the basis of objective criteria so that different surveyors reach the same results, and avoiding the image components dilation/erosion.

In our images, a high contrast of brilliance between the tree and the sky can be noted (the tree crown is dark, the sky is bright). Observing the histogram of such an image (Figure 44.3), we can recognize two

1996   1995

Figure 44.1. Measurement and correction of image slope and scale. α is the stem slope; d is the "homologous" distance between the conspicuous points x,y and x1,y1. The frames drawn into the original pictures are the final analysis windows.

populations of pixels, which have a not-normal (not-Gaussian, not-Gauss) distribution and different variance. A more or less wide zone of low frequencies of pixels is interposed between these two populations.

The threshold level has to be found necessarily within this low-frequencies zone called "transition" or "uncertainty" zone. Within this portion of the histogram, both pixel-tree and pixel-sky cohabit, and they have very similar or even overlapped brilliance values. This happens for two reasons:

1. presence of nonresolved pixels (when a pixel samples both a tree and a sky portion);
2. presence, in the same image, of very bright crown's portions (on the edges or in the case of wide and strongly enlightened leaves) and very dark sky's portions (clouds, subexposition of the picture and so on).

The segmentation procedure has the aim of reducing this uncertainty zone as much as possible, and distributing its pixels between tree and sky by means of an objective criterium which considers their respective pixel distributions.

The procedure has been carried out in the blue region of the image, for the following reasons:

(a) blue radiation is prevalent in the background (clear or little cloudy sky). Therefore the sky shows very high brilliance values (in this region);

---

[1] Registered trademark of Adobe Systems Incorporated, San Jose, California.

[2] Differences of more than 3–4 degrees have to be considered anomalous. In this case the images should be rejected.

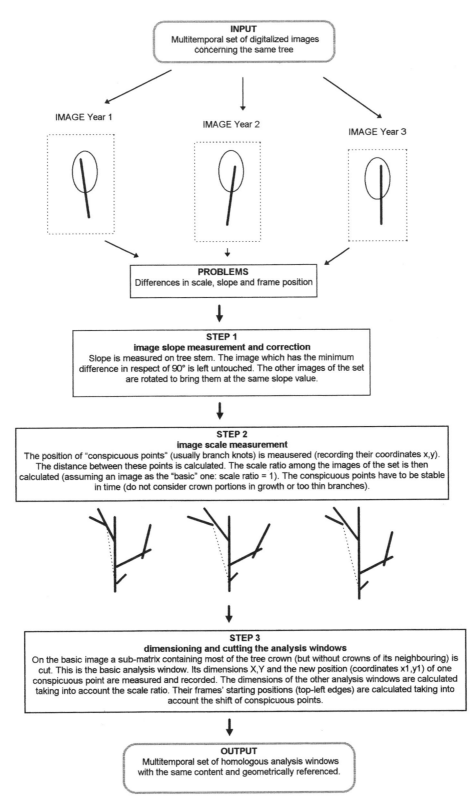

Figure 44.2. Flowchart of the procedure for dimensioning and cutting the analysis windows.

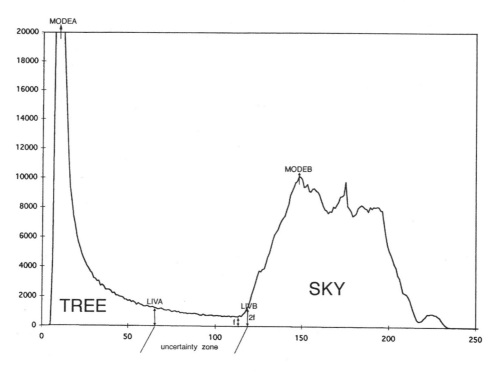

Figure 44.3. An example of image histogram with the conspicuous points.

(b) the vegetation has a very low reflectance in the blue region (4–7% of the whole energy amount); therefore it is dark and dull;

(c) the high diffusion of the light in the blue region, due to motes and steam, reduces the contrast between lights and shadows to the minimum.

The other components of the visible spectrum show some problems: in the green region the light diffusion in the background decreases while both transparence and reflectance of leaves grow up in sensible measure. A good dullness and a little reflectance of the tree crown could be obtained In the red region, but the background should be less brilliant and strongly contrasted within itself (especially in the presence of clouds); furthermore, the light diffusion is very limited in this region and therefore high contrasts between parts of the image enlightened differently should exist.

The adoption of the blue channel has reduced the difficulties due to the high reflectance of some parts of the tree crown caused by the direct illumination of the leaf, typical in the broad-leaved.

When taking pictures in the field, however, one should absolutely avoid having the sun at the back, because if the leaf and the incident light make an angle

less than the "limit angle" (about 40°) we have total reflection and therefore it's impossible to distinguish the three components of color.

Minimizing the "Uncertainty Zone"

A number of observations on the image histograms show that the uncertainty zone covers the continuous range of the lower and higher levels around the level with the minimum frequency, up to a double value of frequency. In other words, if LIVMIN (Figure 44.3) has frequency f, the lower and upper limits of uncertainty zone are LIVA and LIVB, which have frequency 2f. Out of these limits each pixel can be assigned with certainty to tree-population or to sky-population.

In the uncertainty zone the neighboring pixels are characterized by high brilliance gradients, while both in tree-zone and in sky-zone, gradients are absent or very low.

The following procedure is aimed to put in evidence high gradients and to reduce as much as possible the uncertainty zone. Its steps are:

1. Make two copies of BLUE image and threshold them at the above-described LIVA and LIVB, respectively; call them A and B;

2. Calculate the image INT = A – B (difference of modules);

3. Calculate the image STR = INT * BLUE (stretching LIVA and LIVB at 0 and 255, respectively[3]);

4. Make two copies of STR. Assign the minimum and the maximum values on submatrix 3×3 (minimum and maximum filters, provided by the software, assign the same value to all 9 pixels of the submatrix). Call them MIN and MAX;

5. Calculate the image GRA = MIN – MAX (difference of modules). The maximum gradient within each 3×3 submatrix is obtained in this way;

6. Make a copy of STR. Apply to a "mosaic" filter on submatrix 3×3. Call it MOS;

7. Calculate the image REG = STR – MOS (difference of modules). Apply to it a "mosaic" filter on submatrix 3×3. The "regional" variation of the brilliance is obtained in this way;

8. Calculate the image MSK = (GRA + REG)/2. Threshold it at 128. A mask to extract only the "uncertain pixels" is obtained in this way;

9. Calculate the image END = MSK * BLUE. The uncertain pixels are extracted in this way.

At the end of this procedure the threshold value T is chosen, within the range LIVA÷LIVB on the histogram of END image, as the weighted mean of L_i levels which have n_i presence on total number of pixels N.

$$T = (\Sigma_i \, L_i \times n_i)/N$$

The transfer of the histogram data into a spreadsheet or other calculation environment is necessary to perform this operation, but not all image processing softwares allow this transfer. We used Optilab[®4].

The last steps of the procedure are thresholding the BLUE image and counting the number of pixels that have value 0, which form the tree crown area. It can

also be expressed as ratio $A = n_{tree}/N_{win}$, where n_{tree} and N_{win} are the number of pixel-tree and the total number of pixels of the analysis window, respectively. Obtaining these data for a multitemporal set of homologous windows concerning the same subject, the changes of crown area δA during the time-range δt_{f-i} (where f and i are the final and the initial instants of the monitoring period) can be easily calculated as percentage:

$$\delta A_{f-i} \% = ((A_f/A_i) - 1) * 100$$

The Speedy Procedure

The analytical procedure described above is quite slow and exposes the operator to the risk of mistakes when he runs complex but repetitive steps, even if computer-aided and automatized where possible (some softwares allow one to build a macro-function which contains more steps of the processing).

During several processing sessions we observed that the determined threshold level was positioned along the image histogram in a congruent manner in respect to the position of some characteristic points of the histogram easily recognizable. For example, when the mode of the pixel sky population was positioned at a higher level of the considered spectrum (0–255), the threshold "moved" to a higher level as well.

For this reason we thought it possible to study and carry out a *speedy procedure* based on the relation between the threshold level and one or more characteristic points of the histogram.

On this hypothesis we carried out a specific experimentation with the aim of elaborating a regression model able to estimate in a speditive manner a suitable threshold level.

Images of two years (1995–1996) concerning 32 trees (16 conifers and 16 broad-leaved) were processed by the analytical procedure; the threshold level so determined is our observed dependent variable (64 observations). Therefore, data of tree crown area are also available for these trees.

Five characteristic points were then found on the histogram of each analysis window of the sample (see Figure 44.3). Level and respective frequency of pixels were recorded for each point. The positions of these points are the independent variables:

1. MODEA: the level where the mode of tree-pixel population is positioned;

---

[3] It is possible to apply a linear stretching of the frequencies to 0÷255 at any step of the procedure each time the algorithms used tend to reduce the frequencies range.

[4] Registered trademark of Graftek, 78960 Voisins-Le-Bretonneux, France.

2. MODEB: the level where the mode of sky-pixel population is positioned;
3. LIVMIN: the level which has the minimum frequency, and comprised between A and B;
4. LIVA: the level which has the double of the minimum frequency, positioned near the mode of tree-pixel population and higher than it;
5. LIVB: the level which has the double of the minimum frequency, positioned near the mode of sky-pixel population and lower than it.

After recording this data on a spreadsheet (Table 44.2), charts concerning the distribution of the dependent variable in the function of each independent one, considered singularly, were plotted (Figure 44.4). They show that two variables appear more strictly related to the observed threshold level: LIVMIN and LIVB.

Their distributions suggested the adoption of a rectilinear function: models have been elaborated by Systat[5]. Main statistical parameters, as well as the observation of charts, show that the model which uses LIVB as an independent variable is the best.

As the model which uses LIVMIN as an independent variable had also good statistical parameters (Figure 44.5), we thought that a multiple regression model using both variables could improve the estimate quality. At the same time, the introduction of a new dummy variable at two levels (conifers/broad-leaved) called GRUPPO was decided upon.

The statistical analysis on this model (Figure 44.6) has shown that:

(a) the variability explained by the estimate (adjusted smr) doesn't increase in an important measure;
(b) the variable GRUPPO is not significant (high value of P): in fact, it's eliminated by the *stepwise* procedure;
(c) LIVMIN and LIVB are strongly auto-correlated: value 0.892 in the matrix of regression coefficients;
(d) this model shows a certain degree of heteroscedasticity.

All these considerations suggest adopting a singular regression model using LIVB as an independent variable (Figure 44.7): it supplies a very good estimate (adjusted smr = 0.951) and requires only one variable. The selected regression function is the following:

$$T = -1.143 + 0.766*LIVB$$

Differences between observed and expected values (Figure 44.8) are within a range of 10–15 levels, with mean value of 6.3 levels (without sign) and 0.05 levels (with sign).

Since the threshold level serves to measure the tree crown area, it's interesting to verify the differences of area measured by means of the two procedures (the analytical one and the speedy one).

The mean difference of area is 0.48%, with maximum values of about 1.2%. In the same sample the mean variation of tree crown area between 1996 and 1995 (giving value 100 to the initial area) were of 6.03%, with maximum values of 15–20%. The error of estimate is therefore quite small in respect to the phenomenon dimensions. Moreover, changes in the crown area probably increase during the time, while the error of estimate tends to be constant.

RESULTS AND DISCUSSION

Results concerning the sample so far processed (the work is still in progress) are presented in Table 44.3. Since it is impossible, even if very useful, to show images of a significant number of sample trees, two case studies are presented (Figure 44.9). Times to process each image are, on average, the following:

1. digitalization, extraction of blue channel and recording: 5 minutes.
2. cutting of the analysis window: 15 minutes.
3. thresholding with analytical procedure: 30 minutes.
3b. thresholding with speedy procedure: 10 minutes.

Naturally, all procedures concerning data recording and processing can be carried out implementing the algorithms in a spreadsheet; we used Excel[®] 4.0.[6]

Changes of the tree crown area can be very small in one or two years, and it should be very difficult to mark them in a correct way by means of visual estimate.

---

[5] Registered trademark of Systat Inc., Evanston, Illinois.

[6] Registered trademark of Microsoft Corporation.

Table 44.2. Data Concerning the Regression Model to Estimate the Threshold Level by Means of a Speedy Procedure.

No.	Year	LIVMIN	LIVA	LIVB	MODEA	MODEB	Observed T	Estimated T
1	1995	154	64	193	6	211	142	147
2	1996	158	65	189	6	241	144	144
3	1995	122	53	162	6	217	114	123
4	1996	113	50	149	8	185	112	113
5	1995	83	49	90	6	100	72	68
6	1996	187	110	207	11	222	176	157
7	1995	150	76	202	16	230	157	154
8	1996	94	45	120	7	219	89	91
9	1995	54	33	86	6	125	63	65
10	1996	94	43	110	9	124	84	83
11	1995	94	50	109	8	205	83	82
12	1996	134	58	166	6	191	123	126
13	1995	162	97	185	8	202	151	141
14	1996	182	102	220	13	243	178	167
15	1995	140	54	206	9	249	147	157
16	1996	128	51	181	10	229	130	138
17	1995	139	99	146	6	170	125	111
18	1996	195	107	233	11	249	166	177
19	1995	149	108	163	11	178	141	124
20	1996	157	84	206	11	246	160	157
21	1995	91	58	97	7	111	80	73
22	1996	142	72	165	10	231	133	125
23	1995	165	76	235	9	249	178	179
24	1996	158	64	186	9	243	136	141
25	1995	94	39	116	6	132	84	88
26	1996	68	31	77	8	91	57	58
27	1995	151	79	182	8	209	140	138
28	1996	120	58	133	9	220	99	101
29	1995	140	74	153	11	179	121	116
30	1996	128	69	141	9	167	112	107
31	1995	118	44	146	11	191	103	111
32	1996	64	25	77	10	98	55	58
33	1995	137	73	184	11	199	144	140
34	1996	142	79	178	10	199	149	135
35	1995	154	47	220	8	249	169	167
36	1996	87	40	96	7	109	70	72
37	1995	90	40	113	7	198	83	85
38	1996	127	32	204	8	242	151	155
39	1995	112	43	132	6	151	96	100
40	1996	161	47	186	6	204	138	141
41	1995	155	51	222	16	249	168	169
42	1996	128	46	203	16	249	165	154
43	1995	112	61	146	6	164	121	111
44	1996	174	94	205	11	240	175	156
45	1995	144	42	186	6	249	129	141
46	1996	118	46	137	8	223	99	104
47	1995	144	54	185	7	229	130	141
48	1996	114	44	128	7	200	90	97
49	1995	113	39	151	6	199	112	115
50	1996	128	47	197	15	218	152	150
51	1995	142	50	211	7	233	150	160

Table 44.2. Data Concerning the Regression Model to Estimate the Threshold Level by Means of a Speedy Procedure (continued).

No.	Year	LIVMIN	LIVA	LIVB	MODEA	MODEB	Observed T	Estimated T
52	1996	122	45	148	6	240	106	112
53	1995	183	80	229	12	249	184	174
54	1996	168	76	198	9	249	160	151
55	1995	143	62	154	21	187	116	117
56	1996	157	52	182	15	203	134	138
57	1995	154	53	213	13	250	152	162
58	1996	141	47	219	8	249	175	167
59	1995	169	56	226	9	249	165	172
60	1996	131	44	210	6	249	148	160
61	1995	157	65	226	10	249	165	172
62	1996	128	45	221	8	249	157	168
63	1995	121	42	138	7	195	96	105
64	1996	46	26	51	6	64	39	38

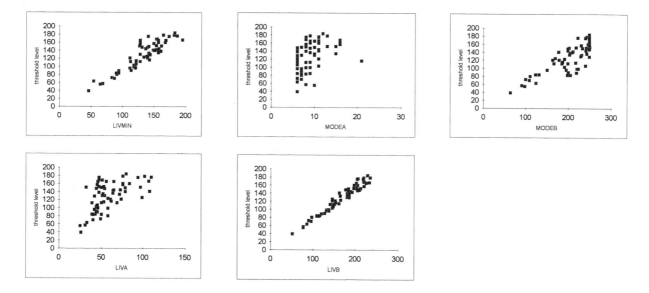

Figure 44.4. Distributions of the independent variables considered for estimating the threshold level by means of a speedy procedure.

Nevertheless, at the moment our method has some limitations and difficulties of employment that it is correct to point out:

1. A completely randomized choice of each sample tree is not possible, because the shot requires the crown to be visible against the sky. The randomization can be applied during the choice of sample sites. This problem could be resolved taking hemispherical pictures of the crowns along a vertical axis

(Bruciamacchie and Jover, 1992), but in this way it could be difficult to have data referred to a single tree. Moreover, significant changes of scale in the same image could occur because of the differences of distance between the camera and the different portions of the crown.

2. Referencing and cutting the analysis window is not a very simple procedure, because it requires that some conspicuous points (absolutely stable during the time) be found on

DEP VAR: SOGLIA   N: 64 MULTIPLE R: 0.909 SQUARED MULTIPLE R: 0.827
ADJUSTED SQUARED MULTIPLE R: .824   STANDARD ERROR OF ESTIMATE: 15.012

VARIABLE	COEFFICIENT	STD ERROR	STD COEF	TOLERANCE	T	P(2 TAIL)
CONSTANT	−6.156	7.969	0.000	.	−0.773	0.443
LIVMIN	1.013	0.059	0.909	1.000	17.223	0.000

ANALYSIS OF VARIANCE

SOURCE	SUM-OF-SQUARES	DF	MEAN-SQUARE	F-RATIO	P
REGRESSION	66848.000	1	66848.000	296.645	0.000
RESIDUAL	13971.485	62	225.347		

DEP VAR: SOGLIA   N: 64 MULTIPLE R: 0.975   SQUARED MULTIPLE R: 0.951
ADJUSTED SQUARED MULTIPLE R: .951   STANDARD ERROR OF ESTIMATE: 7.968

VARIABLE	COEFFICIENT	STD ERROR	STD COEF	TOLERANCE	T	P(2 TAIL)
CONSTANT	−1.143	3.821	0.000	.	−0.299	0.766
LIVB	0.766	0.022	0.975	1.000	34.798	0.000

ANALYSIS OF VARIANCE

SOURCE	SUM-OF-SQUARES	DF	MEAN-SQUARE	F-RATIO	P
REGRESSION	76883.018	1	76883.018	1210.920	0.000
RESIDUAL	3936.467	62	63.491		

Figure 44.5. Regression analysis of estimated threshold level as a function of LIVMIN (above) and LIVB (below).

the subject or near it. On the conifers this operation is not difficult (usually they have a lot of knots or other stem features easily recognizable), but to find suitable points on some broad-leaved trees is sometimes impossible, due to their typical crown morphology. This problem could perhaps be resolved using artificial targets positioned on the same point in each session.

Keeping in mind these limitations, the proposed monitoring method seems to be able to exceed the operative and methodological drawbacks of visual estimation, in terms of objectivity, accuracy, and sensitivity. Moreover, it can be used as a useful training method for the estimators, and allows the creation of a permanent file of the sample with the possibility

of reexamining the overall monitoring period and making analytical comparisons among data collected from different sources.

In the near future, some specific topics of this research concerning the geometric parameters of tree crown shape will be developed in depth: perimeter/area ratio, fractal dimension, analysis of alternances; they could be more suitable indicators of tree health than the biomass change. They could perform the field survey in an easier and speedier way , because it seems that their application doesn't need precise referenced shots. Their relationship with the tree crown density has been demonstrated by several studies (Mandelbrot, 1983; Pfeifer, 1988; Zeide, 1990a,b; Zeide et al., 1991a,b; Strand, 1990), but always analyzing too few samples.

On the other hand, our trials suggest considering the results of these kinds of analysis with caution, be-

DEPENDENT VARIABLE: SOGLIA

MINIMUM TOLERANCE FOR ENTRY INTO MODEL= .010000

fORWARD STEPWISE WITH ALPHA-TO-ENTER= .150 AND ALPHA-TO-REMOTE= .150

STEP # 1 R= .909 RSQUARE= .827
TERM ENTERED: LIVMIN

VARIABLE	COEFFICIENT	STD ERROR	STD COEF	TOLERANCE	T	'P'
IN						
---						
1 CONSTANT						
2 LIVMIN	1.013	0.059	0.909	.1E+01	296.645	0.000
OUT	PART. CORR					
---						
3 LIVB	0.873	.	.	0.20440	195.733	0.000
4 GRUPPO	−0.273	.	.	0.99274	4.895	0.031

STEP #2 R= .979 RSQUARE= .959
TERM ENTERED: LIVB

VARIABLE	COEFFICIENT	STD ERROR	STD COEF	TOLERANCE	F	'P'
IN						
---						
1 CONSTANT						
2 LIVMIN	0.215	0.064	0.193	0.20440	11.335	0.001
3 LIVB	0.630	0.045	0.803	0.20440	195.733	0.000
OUT	PART. CORR					
---						
4 GRUPPO	0.059	.	.	0.87581	0.211	0.647

THE SUBSET MODEL INCLUDES THE FOLLOWING PREDICTORS:

CONSTANT
LIVMIN
LIVB

DEP VAR: SOGLIA   N: 64 MULTIPLE R: 0.909 SQUARED MULTIPLE R: 0.827
ADJUSTED SQUARED MULTIPLE R: .824   STANDARD ERROR OF ESTIMATE: 15.012

VARIABLE	COEFFICIENT	STD ERROR	STD COEF	TOLERANCE	T	P(2 TAIL)
CONSTANT	−6.799	3.916	0.000	0	−1.736	0.088
LIVMIN	0.215	0.064	0.193	0.204	3.367	0.001
LIVB	0.630	0.045	0.803	0.204	13.990	0.000

ANALYSIS OF VARIANCE

SOURCE	SUM-OF-SQUARES	DF	MEAN-SQUARE	F-RATIO	P
REGRESSION	77499.845	2	38749.923	712.049	0.000
RESIDUAL	3319.639	61	54.420		

Figure 44.6. Multiple regression analysis of estimated threshold level as a function of LIVMIN, LIVB, and GRUPPO.

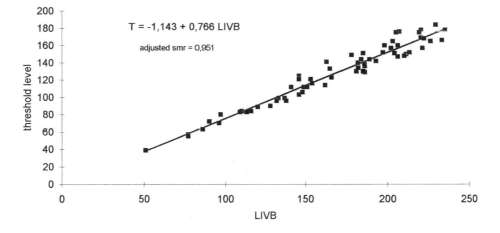

Figure 44.7. The selected regression model to estimate the threshold level.

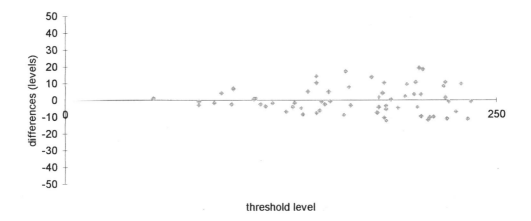

Figure 44.8. The regression model to estimate the threshold level. Differences between observed (analytical procedure) and expected (speedy procedure) values.

cause often the phenomenon magnitude is so small as to be confused with its bias.

Our purpose is to improve the use of these geometric parameters and to apply them to our whole sample, in order to obtain affordable results about their role in the study of defoliation.

ACKNOWLEDGMENTS

The authors are grateful to the following colleagues: Daniele Bini and Emilio Gottardo of the Servizio della Selvicoltura of the Regione Autonoma Friuli-Venezia Giulia for their demanding field survey. Arnaldo Tonelli, our "image processing Master." Sandro Gecele, who carried out the image processing of the whole sample. Luigia Maltoni and Flora De Natale for their valuable suggestions on preparing figures and tables. Giuseppe Cappalonga for his valuable help in translating the article.

This research has been promoted and financed by Direzione Regionale delle Foreste e die Parchi of the Regione Autonoma Friuli-Venezia Giulia, within the EC Regional Project "PACID-FVG 2°," Subproject "DIGIDEP."

Table 44.3. Results of the Three-Year Survey on 80 Trees of the Sample, with Differences of Tree Crown Area Between 1997 and 1995.

No.	1995		1996		1997		1997–1995
	T	A%	T	A%	T	A%	dA%
1	172	56.60	183	53.55	105	56.84	0.42
2	98	37.77	114	39.13	72	43.55	15.30
3	184	42.84	160	42.13	142	42.58	−0.61
4	59	45.69	65	42.48	128	39.80	−12.89
5	114	46.06	112	42.37	93	35.34	−23.27
6	83	49.11	123	53.48	123	56.11	14.25
7	141	57.52	160	59.40	103	56.47	−1.83
8	80	56.94	133	57.12	113	53.47	−6.09
9	142	51.94	144	52.67	174	52.65	1.37
10	72	54.57	176	61.72	93	49.86	−8.63
11	97	59.84	160	66.40	154	67.87	13.42
12	160	45.29	127	43.07	128	41.66	−8.02
14	103	26.91	55	20.44	118	27.96	3.90
15	88	41.93	58	36.31	147	43.90	4.70
16	183	54.48	130	52.61	188	54.60	0.22
17	157	48.43	89	50.79	136	52.42	8.24
18	140	50.41	99	46.46	176	48.71	−3.37
19	96	30.91	39	30.17	55	31.56	2.10
20	169	30.11	70	27.87	81	31.84	5.75
21	121	53.95	112	55.22	189	57.32	6.25
22	116	41.42	134	37.85	118	37.76	−8.84
23	165	41.58	148	42.56	131	45.49	9.40
24	129	45.24	99	47.33	143	53.64	18.57
25	165	50.85	157	50.79	172	55.45	9.05
26	150	43.50	106	43.66	185	47.24	8.60
27	162	34.99	113	35.96	107	37.34	6.72
28	120	60.13	156	59.13	121	56.87	−5.42
29	118	47.90	154	45.09	128	44.35	−7.41
30	83	36.48	151	32.67	65	22.86	−37.34
31	152	40.03	175	40.19	89	40.74	1.77
32	168	37.02	165	33.02	62	28.87	−22.02
33	172	58.00	79	58.87	180	61.80	6.55
34	98	44.75	50	48.69	179	50.75	13.41
35	188	60.90	139	61.11	189	65.52	7.59
36	149	50.63	178	52.23	139	57.90	14.36
37	141	54.59	158	55.30	159	58.41	7.00
38	131	48.46	138	48.17	153	50.77	4.77
39	162	54.25	158	59.84	185	60.30	11.15
40	134	71.67	177	67.96	184	69.53	−2.99
41	180	38.60	138	46.82	179	52.73	36.61
42	176	52.24	127	55.12	164	61.45	17.63
43	138	52.17	85	53.46	128	51.18	−1.90
44	172	39.24	177	41.80	82	43.03	9.66
45	174	47.07	169	47.86	96	50.60	7.50
46	160	38.24	144	33.94	129	43.26	13.13
47	174	46.28	168	45.73	177	44.64	−3.54
48	182	45.86	177	42.38	140	51.03	11.27
49	139	46.41	173	46.79	144	46.32	−0.19
50	184	51.92	187	52.83	76	46.96	−9.55
51	173	33.17	173	33.87	127	37.21	12.18

Table 44.3. Results of the Three-Year Survey on 80 Trees of the Sample, with Differences of Tree Crown Area Between 1997 and 1995 (continued).

No.	1995 T	1995 A%	1996 T	1996 A%	1997 T	1997 A%	1997–1995 dA%
52	97	50.32	189	55.50	115	61.45	22.12
53	110	54.98	189	54.58	115	61.35	11.59
54	179	48.40	177	47.71	135	52.94	9.38
55	183	39.18	144	41.09	144	47.04	20.06
56	158	53.69	110	47.71	160	55.63	3.61
57	102	29.65	124	32.68	95	31.01	4.59
58	128	30.79	107	29.22	95	31.01	0.71
59	127	37.57	48	31.66	79	42.16	12.22
60	104	35.02	38	27.07	52	30.56	−12.74
61	164	46.37	159	47.76	174	51.08	10.16
62	120	39.90	65	30.82	79	44.57	11.70
63	126	35.90	93	35.09	95	38.43	7.05
64	111	34.25	182	36.11	144	39.41	15.07
65	128	33.96	65	31.43	80	35.36	4.12
66	118	47.88	105	48.01	130	52.66	9.98
67	146	51.05	172	61.64	188	60.79	19.08
68	119	60.07	183	66.90	134	70.37	17.15
69	164	34.21	163	41.02	124	41.35	20.87
70	175	28.24	173	32.93	54	33.32	17.99
71	133	32.71	172	34.49	173	38.38	17.33
72	135	45.22	180	47.13	173	47.65	5.37
73	180	51.24	170	49.97	180	52.57	2.60
74	178	46.30	160	46.71	180	50.90	9.94
75	189	46.61	101	48.41	174	54.23	16.35
76	189	46.06	171	50.15	183	52.23	13.40
77	81	50.29	81	51.46	146	54.48	8.33
78	161	43.85	127	44.53	150	47.15	7.53
79	155	41.66	92	40.66	170	42.60	2.26
80	180	54.36	127	55.45	124	55.81	2.67
81	146	48.10	125	46.75	137	49.59	3.10

Figure 44.9. Two case studies of the sample, a beech (above) and a white fir (below).

REFERENCES

Benincasa, F., I. Bernetti, and A. Pierguidi. Un nuovo metodo fotografico per la valutazione della defogliazione degli alberi forestali, *Monti e Boschi,* 38(6), pp. 55–58, Edagricole (Bologna - I), 1987.

Bruciamacchie, M. and L. Jover. Utilisation du traitement d'images en matière forestière, *Revue Forestière Française,* XLIV, 1-1992, pp. 43–53, 1992.

Conedera, M. *Danni boschivi di nuovo tipo ed inquinamento atmosferico,* corso di aggiornamento per il Servizio Forestale del Sud delle Alpi, SANASILVA coord., Sud delle Alpi, Informazione, CH 6500 Bellinzona/TI, 1989.

Di Zenzo, S. Segmentazione di immagini digitali, *Note di Informatica* 6, IBM Italia, pp. 3–20, 1984.

Ferretti, M., E. Cenni, and A. Cozzi. Indagini sulle condizioni dei boschi. Coerenza e confrontabilità dei dati sulla trasparenza delle chiome degli alberi in Italia, *Monti e Boschi,* 45(2), pp. 5–13, Edagricole (Bologna - I), 1994a.

Ferretti, M., A. Economou, E. Beccu, G. Canu, S. Cocco, F. Bussotti, E. Cenni, A. Cozzi, M. Andrada de Conceiçào, and G. Sanchez Peña. *Alberi della Regione Mediterranea. Guida per la Valutazione delle Chiome,* CEC - UN/ECE, Brussels, Geneva, 1994.

Honer, T.G. and K. Collins. Ground Photography for the Measurement of Open Growth Tree Crowns, *Proceeding of the International Union of Forest Research Organizations,* S4.01-4 *Royal College of Forestry,* 1974, pp. 91–100.

Howard, J. A. *Remote Sensing of Forest Resources: Theory and Application,* Chapman & Hall, London, 1991.

Innes, J.L. Forest Health Surveys—A Critique, *Environ. Pollut.,* 54, pp. 1–15, 1988a.

Innes, J.L. Forest Health Survey: Problems in Assessing Observer Objectivity, *Can. J. For. Res.,* 18, pp. 560–565, 1988b.

Landmann, G. *La surveillance au sol de l'état sanitaire des forêts: première évaluation rétrospective des résultats français (période 1983–1988) et comparaison avec les résultats de spays voisins,* La Santé des forêts, Ministre de l'agriculture et de la forêt, 1988, pp. 33–51.

Lee, Y.J., R.I. Alfaro, and G.A. Van Sickle. Tree-Crown Defoliation Measurement from Digitized Photographs, *Can. J. For. Res.,* 13(5), pp. 956–961, 1983.

Mandelbrot, B. *The Fractal Geometry of Nature,* W.N. Freeman, New York, 1983.

Pfeifer, P. Fractal in Surface Science: Scattering and Thermodynamics of Absorbed Films, in *Chemistry and Physics of Solid Surfaces,* Vol. VII, R. Vanselow and R.F. Howe, Eds., 1988, pp. 283–305.

Scrinzi, G. and M. Confalonieri. Monitoraggio dello stato di deperimento forestale: analisi di concordanza tra classificazioni soggettive dello stato di defogliazione, *Indagine conoscitiva sullo stato sanitario delle foreste in provincia di Trento,* 1991, pp. 43–55.

Scrinzi, G. and A. Floris. *Un metodo di misura degli stati di defogliazione delle chiome arboree mediante analisi quantitativa di immagini digitali,* Comunicazioni di Ricerca ISAFA n. 92/2 (Trento, I), 1992.

Strand, L. Crown Density and Fractal Dimension. *Communications of the Norwegian Forest Research Institute,* 43.6. ISBN 82-7169-471-5, ISSN 0332-5709, 1990.

Tesi, G., E. Cenni, and M. Ferretti. Normalizzazione dei dati di trasparenza delle chiome per l'analisi statistica dei risultati dei rilevamenti delle condizioni degli alberi, *Monti e Boschi,* 48(3), pp. 54–58, Edagricole (Bologna - I), 1997.

Vel, E., Ed. *Protezione delle foreste nell'Unione Europea contro l'inquinamento atmosferico, 1987–1996,* CE, DG VI/DG X, Bruxelles, ISSN 1012-2133, 1997.

Zeide, B. *Fractal Analysis of Crown Structure,* Publication no. FWS-2-90, School of Forestry and Wildlife Resources, Virginia Polytechnic Institute and State University, Blacksburg, VA, 1990a, pp. 232–241.

Zeide, B. *Fractal Geometry and Forest Measurements,* in State of the Art Methodology of Forest Inventory, USDA Forest Service, General Technical Report PNW 263, 1990, pp. 260–266.

Zeide, B. and C.A. Gresham. Fractal Dimensions of Tree Crowns in Three Loblolly Pine Plantations of Coastal South Carolina, *Can. J. For. Res.,* 21, pp. 1208–1212, 1991a.

Zeide, B. and P. Pfeifer. A Method for Estimation of Fractal Dimension of Tree Crowns, *For. Sci.,* 37(5), pp. 1253–1265, 1991b.

CHAPTER 45

Spatial Monitoring Protocol to Optimize the Monitoring of Forest Entities with Remote Sensing Digital Images and GIS

R.A. Fournier, P.E. Bonhomme, and C.-H. Ung

INTRODUCTION

Monitoring status and changes in the forest landscape is an essential activity for effective management of the natural resources, especially in the context of sustainable development. Monitoring programs rely, first, on the ability to map precisely targeted entities, and second, on the implementation of an efficient method tailored for timely interventions/questioning on the forest landscape. A common approach for spatial representation in forestry involves the interpretation of aerial photographs and the layout of the targeted entities. This approach is still largely dominant in current forestry practices (Avery and Berlin, 1992; Howard, 1991; Roscoe et al., 1955; Spurr, 1960). Photo-interpretation meets a wide array of activities such as forest inventory (Paine, 1981; van Zuidam, 1986), the assessment of forest damage (Murtha, 1972), and sustainable development of forest resources (Kimmins, 1997). Unfortunately, the monitoring of complex structures such as the forest canopy is fraught with severe limitations. For example, the ability to assign boundaries to forest stand is highly variable from one interpreter to another (Lowell et al., 1996). The heterogeneity of forest entities is intrinsic to most monitoring exercise in forestry. Thus, new monitoring approaches and techniques must take this variability into account.

Emerging geographic information systems (GIS) and digital remote sensing tools provide new opportunities in monitoring the forest landscape. More specifically, satellite remote sensing images provide a low-cost monitoring alternative to aerial photographs.

Unfortunately, presently available spatial resolutions for satellite digital images are much coarser than those used in aerial photography. Nevertheless, the launch of new satellite systems (Nieke et al., 1997) and the increasingly competitive digital airborne alternatives (King, 1995) raise the issue of the appropriate spatial resolution of digital images to monitor forest landscapes. Moreover, the field of digital image analysis provides a wide array of analysis tools which offers many efficient algorithms to assist in the monitoring procedure. In addition to digital images, GIS accommodate many of the requirements in forest management or research. GIS are consequently implemented in most management units of forestry offices. The possibilities now available with GIS and digital remote sensing thus put the question of spatial accuracy for forest resources in a new light: the crossroads of tools and techniques forces those monitoring the forest landscape to follow strict guidelines while selecting a spatial resolution to best meet their objectives.

Digital remote sensing brings a new perspective into spatial monitoring in forestry; the digital image can be seen as a regular sampling grid, the image pixels, over a given spatial extent. In addition, each pixel can be seen as an integrative measure of the spectral reflectance over the pixel footprint, thus summing the radiometric contribution of all forest entities within that viewshed. Airborne sensors offer the possibility of collecting digital remote sensing images at resolutions comparable to those used for photo-interpretation (King, 1995). However, economic reasons favor the use of satellite systems, such as Landsat-MSS (Goetz and Prince, 1996) and Landsat-TM (e.g., Bauer et al.,

1997) and AVHRR (e.g., Liu et al., 1997), for applications and modeling studies. While the spatial resolutions available from digital sensors is limited, it is relevant to raise the question of which spatial resolution is adequate to monitor forest landscapes. This chapter addresses two objectives related to the optimal spatial resolution for monitoring forest resources using digital images: (1) to define the major considerations for the optimal spatial resolution with digital imagery, and (2) to suggest basic guidelines and limitations to establish a useful strategy to select spatial resolution in forestry.

DEFINITION OF OPTIMAL SPATIAL RESOLUTION

An optimal spatial resolution is rarely defined as such; rather, its quest should be understood as the search for windows of spatial resolutions that maximize information retrieval. Elaborate mathematical frameworks may emerge while considering the problem in general terms. However, such a framework becomes weakly applicable when the scenes for which it applies include very heterogeneous entities, as in forestry. Consequently, rules on optimal spatial resolution are increasingly effective if they address directly: the work objective, practical considerations tied to monitoring consistency, the structural and organizational aspects of the entities in the concerned field.

Too often the optimal spatial resolution of digital images is treated as a purely mathematical problem, a pragmatic perspective to this problem also carries important practical and economic considerations that should not be ignored. Four considerations must be addressed to fulfill the requirement of a useful window of spatial resolution: (1) the availability of sensors and images with the targeted resolution, (2) the economic aspects attached to the solution, (3) the radiometric and geometric integrity of the image, and (4) the ability to distinguish the entities and organizational structure. Thus a practical/tractable strategy can only be developed with these four considerations in mind. Furthermore, the chance of adopting a spatial strategy relies greatly on answering satisfactorily these four considerations from the perspective of those actively involved in a field, in the case here, in forestry.

First, the availability of sensors and images is critical for a monitoring program. The longer the monitoring is required, the more critical is the sensor availability for the targeted spatial resolution. Landsat, SPOT, and AVHRR satellites have steadily been supported in the past years and show an excellent record of archived images. These satellites, as well as many others (see Nieke et al., 1997, for a list of available and forthcoming sensors), are well suited for continuous monitoring. The availability also refers to a suitable set of spectral bands and the repeatable coverage over a given time period. Although not always exact, it is generally the case that a finer resolution satellite may provide consecutive images at shorter intervals. For example, a given region in Canada may only have one or two suitable Landsat-TM images (e.g., with less than 15% cloud cover) a year during the growing season. The time period for consecutive coverage over an area is 18 days for Landsat. In contrast, one image a day can be collected by AVHRR if good weather persists. Supplemental capabilities of several new sensors, such as variable viewing optics, may facilitate timely acquisition in the near future. Meanwhile, it is largely the case that satellite images are much more easily available than those from an airborne mission. Airborne imagery, although often more flexible for customizing image specifications, requires much more effort in mission planning, and is often flawed by many unpredictable malfunctions.

Secondly, an economically profitable sampling strategy is likely to be adopted, pursued for a long period, and even supplant older procedures. This consideration is often reduced as the simple balance between the finest spatial resolution and the lowest acquisition cost to cover a given territory. Given the economic and availability constraints, the effort of selecting a spatial resolution is worthwhile only if it results in a practical strategy. The search for optimal spatial resolution is too often oversimplified to a single incentive: *select the coarser spatial resolution that suits the purpose at hand*. In reality, the economic factors are much broader if the issue of imaging a territory is brought in a multiuser, multipurpose, and multiplatform context. Digital imagery has shown reliability for data availability and will improve this record by providing a larger choice of sensors. Consequently the choice of spatial resolutions will increase and the time period between successive coverages will be reduced. In addition, the price ranges for satellite images are advantageous compared to airborne missions. Such a context is therefore suitable to the increased use of digital images in operational forestry. This progression is, however, highly dependent on the ability of digital interpretation techniques to provide the information required by the current methods; more specifically, interpretation of aerial photography can only be supplanted or complemented by techniques involving digital images if the information extraction capabilities are perceived to be economically and technically profitable by the people currently involved in operational forestry.

Thirdly, a rigorous radiometric and geometric integrity of the acquired images ensures that interpretation from image to image, and image to map be consistent (Toutin, 1995). Surprisingly, most image suppliers do not provide full radiometric calibration of images. This creates an important schism between trained users in research and operational users (Teillet et al., 1997). Successful use of digital remote sensing image in operational fields depends largely on the progressive removal of that schism. The radiometric correction of a digital image also involves the removal of atmospheric effects. The resulting image has pixel values in radiance or surface reflectance. In contrast to radiometric calibration, reliable softwares for geometric correction are more commonly available. Most GIS and image analysis systems provide modules designed to ensure a good image to map registration. The end-user requires time- and space-consistent image products; therefore, planning images acquisition also requires a plan for rigorous radiometric and geometric correction.

Fourth and finally, a specific answer on the adequate spatial resolution is highly application-dependent and therefore is defined from the hierarchy of entities inherent to the selected application. Applications such as the modeling of global change, biodiversity in ecozones, and inventory of forest condition, all involve scales from national, regional, to the landscape and local levels. Consequently, a monitoring or a mapping effort must be linked with specific forest entities. This link defines the admissible spatial resolution windows. We suggest, as a definition, that a forest entity is a physical structure, composing the forest canopy, delimited and identifiable in space; this also including the grouping of forest elements in an organized fashion. Among the forest entities exists a spatial hierarchy. A spatial sampling strategy is therefore qualified as entity-driven first, but with an explicit link to the application.

The rest of the text will focus on the dependency of spatial resolution on application and entity. Although no more attention will be placed on the first three considerations, they remain crucial in the acceptance of the selected strategy. Up to now, the choice satellite digital images applications were severely restrained by the limited selection of satellite sensors. Therefore the issue of

optimal spatial resolution was primarily relevant for airborne imagery. Now that a large selection of sensors will rapidly become available, the question of optimal spatial resolution can be addressed more strategically; i.e., find the best window of spatial resolution to suit the application, establish the image availability according to desirable technical specifications, ensure that geometric and radiometric calibration are possible, and estimate the economic viability of the approach.

FOREST ENTITIES

The search for windows of relevant spatial resolutions requires organizing the entities related to the forest. The management and monitoring of forest land with GIS and digital remote sensing also require entity-relationship models (Tokola et al., 1997). The fields of landscape and forest ecology are useful sources for descriptive lists of entities and quantitative methods to study them (Turner and Gardner, 1990). For instance, the hierarchical view of the landscape proposed by Allen and Starr (1982), O'Neill et al. (1986), and Walker and Walker (1991) provide a strong theoretical background. From another perspective, photo-interpreters developed from years of practical experience while observing forest entities for a wide range of forestry applications. Manuals on aerial photography, such as those from Sayn-Wittgenstein (1960), Zsilinszky (1964), and Murtha (1972), established relationships between the scale of the photograph and the interpretation of specific forest entities. Although these relationships are strongly ecosystem-dependent, their generalization must be approached with relative care. The experience from landscape ecology and photo-interpretation in forestry both provide a theoretical and practical basis for a hierarchical table of forest entities.

Monitoring and modeling forest landscape with GIS require a strategic approach to spatial resolution, particularly when juxtaposing layers of digital maps and remote sensing images. A table of forest entities is given in Table 45.1. The link between the entities and the approximate scale is based on photo-interpreters' experience (e.g., Sayn-Wittgenstein, 1960). The translation of photograph scale into a pixel size of a digital image assumes that a photograph digitized at 600 dots per inch (dpi) corresponds to what an interpreter can identify (Figure 45.1). This resolution integrates finer grains of the emulsion in a 42.3 micron viewing footprint. Experience from mapping (Gagnon et al., 1991; Cowen et al., 1995) suggests this resolution best

Table 45.1. An Approximate Relationship between Forest Entities and the Required Scale for Monitoring. (Pixel size is equivalent to the human discrimination abilities of the corresponding scale.)

Forest Entities	Approx. Scale	Pixel Size
Leaf arrangement: Softwood (in tufts) Hardwood (individual leaf) Branch structure	1:600	0.025 m
Understory: Herbaceous plants, shrubs, soil, detritus	1:3,000	0.127 m
Individual crown	1:8,000	0.338 m
Group of crowns	1:50,000	2.11 m
Stand	1:200,000	8.46 m
Ecosystem-forest	1:1 000,000	42.3 m
Watershed, habitats, ecozones	1:8,000,000	338 m

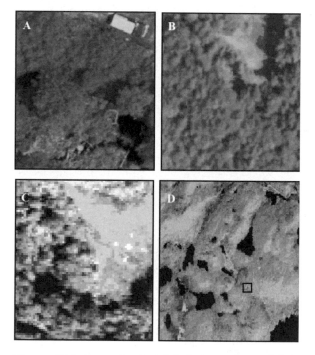

Figure 45.1. Example of aerial photographs and digital remote sensing images over the same mixed forest at Duchesnay, Quebec, Canada. The aerial photographs of scale (A) 1:2,000 and (B) 1:150,00 are scanned at 600 dpi. Digital images have an approximate pixel footprint of (C) 1 m and (D) 30 m.

is equivalent to the pixel size for the digital image. We therefore selected 600 dpi to best simulate this elemental point on a photographic emulsion based on the human perception into an equivalent pixel of the digital image. Multipurpose and multiscale studies in forest ecophysiology (Fournier et al., 1997) concur with the interpretative link given in Table 45.1 between entity and pixel size. The associated pixel size with the entity must, however, be treated only as a useful marker for planning. Forest entities are so variable in size and texture that any generalization oversimplifies the problem. The first element of a strategic approach to spatial resolution nevertheless starts by targeting a specific entity and investigating its dimension using the most precise available means (e.g., field measurements, aerial photographs, satellite images).

Once the approximate size of the targeted entity is known, supplemental information must be acquired to select the best approach to measure its spatial distribution. Conventions exist for each forest entity of Table 45.1 to subdivide them in classes. These classes can be itemized, like a list of tree species, or stratified, like stands (Perron and Morin, 1997). The visual attributes used to describe an entity provide the tools to assess the differences between the classes. The differentiation between entity classes depends primarily on the physical attributes of the targeted entity such as the variability in size, the spatial organization of the smaller entities composing it, the spatial heterogeneity, and the optical properties. Thus the ability of finding the borderline dividing two adjacent entity classes must be considered while identifying the admissible windows of spatial resolution. Therefore the relevance of

matches the discriminating potential from a photograph translated into a digital image. A photo-interpreter cannot discriminate anything at the grain level but rather requires a visual integration of the grain structure to construct an elemental image point. This point a spatial resolution is also based on the ability to distinguish between the classes of a given forest entity.

As the arrangement of units composing the entity becomes more complex, the problem of spectral mixing is increasingly important for the analysis of digital remote sensing images. If the spatial resolution of the sensor is too coarse, it will combine signatures from independent entities on the ground into aggregated spectral responses for a particular response. More spatial resolution is not necessarily better. If spatial resolution is too fine, objects may appear more heterogeneous than they really are; this misrepresentation masks their inherent homogeneity. The spectral mixing problem thus becomes increasingly important when scenes are heterogeneous, as is often the case in forestry.

The ability to delimit the boundary of an entity class with the neighboring classes depends first on a clear definition of the entity, and second on the available visual attributes describing the entity. While structural units like a leaf or a tree have a tangible physical boundary, stratified entities such as stands, ecosystems, or habitats require much interpretative knowledge for successful discrimination. The boundaries between strata are fuzzy. Better results for strata discrimination are usually obtained when interpretation is made by an experienced forester/ecologist, by using a complex algorithm, or by a combination of both. A clear understanding of the classes and an encyclopedic knowledge of the visual texture in relation to the stratification of the entity must therefore support an efficient approach. Often, equally acceptable interpretations of an image lead to different positions of the entity classes' boundaries. For example, significant differences are always noticed between stand maps made by several photo-interpreters. In the context of the optimal spatial resolution for digital remote sensing, the problem of consistence for mapping the entity classes may be dealt with in two different ways: by adopting a pixel size of the same dimension of the entity or by oversampling the entity with a smaller pixel size to best discriminate the visual textures. The importance of discriminating between entity units or classes therefore imposes an interpretation strategy that leads to two different windows of relevant spatial resolution: a reductionist and a descriptive strategy.

TWO STRATEGIES FOR RELEVANT SPATIAL RESOLUTIONS

A strategy to find the relevant ranges of spatial resolutions requires one to clearly identify the following considerations listed in chronological order of attention: the purpose of the application, the targeted entity, a description of the structural units or the entity strata, and the available tools for the spectral/texture analysis. Two complementary strategies can be used while selecting suitable spatial resolutions for digital images: the reductionist and the descriptive (Figure 45.2). The reductionist strategy searches for the lowest spatial resolution that best allows the identification of an entity. In contrast, the descriptive strategy involves a spatial resolution significantly finer than the targeted entity in order to increase the ability to extract descriptive attributes. The finest spatial resolution dictated by the descriptive approach is, in theory, limited by the size of the next organizational level composing the entity and by the available analysis tools to deal with the increased radiometric texture. For instance, an investigator interested in trees that adopts a reductionist strategy must select a pixel size in the order of the crown dimension. In contrast, the descriptive strategy requires a spatial resolution such that several pixels are required to cover a crown, but pixels must be large enough not to reveal clearly the branches' structure. When applied to forest stands, the finer spatial resolution of the descriptive approach would match the size of a typical tree grouping in the stand; this could encompass a dominant tree and the surrounding codominant and intermediate crowns. In addition to size considerations, the descriptive strategy also requires that spectral and texture analysis tools be identified to deal with the increased complexity of scene elements. Both considerations on entity size and analysis tools must be compatible for a successful application of the descriptive strategy. The theoretical advantage of the descriptive strategy is often flawed by the inability of the analysis algorithms to deal with the emerging complexity of the radiometric texture with finer resolution. In addition to the extra efforts for the analysis of a complex texture, increased spatial resolution also carries an economic disadvantage. Consequently, most theoretical studies on optimal spatial resolution understandably usually focus on a reductionist approach. Nevertheless, approaches like Hay et al. (1997) combine both strategies by first oversampling and then merging pixels until the entities are integrated. Both strategies are worth more discussion in the context of forestry applications. For conciseness, the next section discusses the impact of the proposed strategies on the two most common entities in forest operation: the tree and the stand.

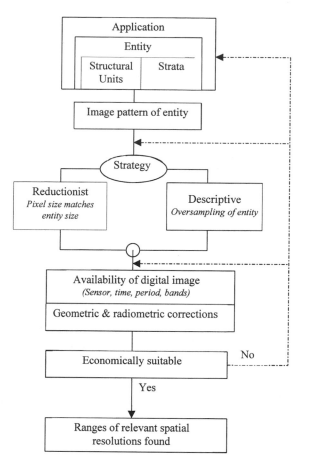

Figure 45.2. Algorithm proposed for a strategic approach to the selection of the relevant spatial resolutions.

SPATIAL STRATEGIES APPLIED TO TREES AND FOREST STANDS

Most studies on optimal spatial resolution refer to the reductionist approach. This strategy when applied to digital images places the impetus on pixel radiometric variance and spatial texture. For example, Fournier et al. (1995) and Gougeon (1995) developed tree recognition algorithms based on the radiometric texture. Meanwhile, most approaches use local variance as a probe to assess optimal spatial resolution. For instance, Woodcock and Strahler (1987) suggest that the highest local variance peak of a histogram based on the radiometric texture of a digital image corresponds to the spatial resolution that maximizes the information content. They found that this maximum variance peak generally occurs at a spatial resolution of about one-half to three-fourths of the size of the entity. In the case of the forest scene tested, the peak corresponded to 6 m pixel size. Other mathematical analyses of forest scenes by Woodcock et al. (1988), and Jupp et al. (1988, 1989) led to similar results. For example, Atkinson and Curran (1997) suggest 6 m as the upper limit to take into account the local variance of the radiometric texture at the tree level. However, when considering more complex forest scenes, ranges of spatial resolution corresponding with the maximum variance peak are not as clearly defined. Marceau et al. (1994) showed that optimal spatial resolutions in the range of 5 to 10 m occur in a variety of forest scene configurations. The pixel size is correlated with the average size of the pattern formed by a tree crown and its shadow. These results show that a reductionist approach applied to the tree requires a translation of the structural unit into an image unit. The window of spatial resolution therefore ranges from the size of the image unit to about half its area.

All entities defined by strata are complex to delimit. The forest stand is a good example of a forest entity that is widely used in operational forestry but is inherently variable in space. While building a stand map, the interpreter must place boundaries between strata. The variability is spatially continuous and the thresholds between adjacent strata vary substantially from one interpreter to another (Lowell et al., 1996; Biging et al., 1991). For example, stand strata are made for various portions of stem density. An interpretation of stand strata in percent of crown closure using classes like 0–25%, 25–50%, 50–75%, and 75–100% can only be approximated in practice. Information on the distribution of stand area is given by a histogram. When applied to a mixed deciduous-coniferous forest, the histogram of the stand area in Figure 45.3 shows a peak at 3 hectares. However, the histogram also shows a very rapid drop in number of stands smaller than 3 hectares. This lack of small stands reflects a restriction in the interpretation to limit the number of small stands. In this context, overpopulating a map with small strata is avoided by merging small strata into the neighboring strata. The resulting polygon is given the best suitable description with the understanding that merging a polygon with close characteristics is better than interpreting several small polygons. Exceptions to this rule are frequent, but the principle of merging small strata is applied when possible. Despite these approximations on the boundaries, the stand remains a widely used forest entity for management, operation, and research. Such an impetus on the stand comes from the scale

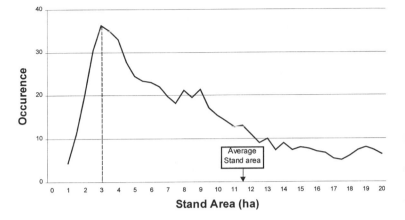

Figure 45.3. The histogram of the occurrence of stand by area for a mixed forest deciduous-coniferous at the Duchesnay Forest, Quebec, Canada, on latitude 46° 55′ and longitude −71° 40′.

of the unit, which corresponds to the scale of the direct human intervention. Therefore, even if the boundaries between strata are fuzzy, the entity remains well suited to forestry applications.

The spatial variability intrinsic to forest stands forces the question of optimal spatial resolution of digital images to be addressed under a different perspective than for aerial photographs. Digital remote sensing has an advantage over the visual layout of strata boundaries from aerial photography: it provides a continuous spatial sampling of a scene with pixel values that are physically meaningful (radiance or reflectance in a specific spectral bandpass). Therefore, the digital medium benefits from a wide range of spectral/texture analysis tools in supplement to the encyclopedic knowledge of an interpreter, if available. The digital image therefore provides a medium that is well suited to mathematical modeling, statistical analysis, and continuous mapping of the forest landscape (e.g., Franklin et al., 1997), given that the ranges of suitable spatial resolution are well identified.

SUMMARY

The increasing use of digital imagery in forestry and the scale dependence on the inference of information require a rigorous approach to selecting the appropriate spatial resolution. The optimal pixel dimensions of digital remote sensing images depend on the objectives of the application or study and also on the inherent characteristics of the landscape. Landscape is divided into a number of information levels. The approach proposed in this chapter to assess the relevant windows of spatial resolution for digital im-

ages requires first identifying the entity that best suits the analysis.

The success of linking spatial resolution with information extraction on an entity is affected by the ability to detect the boundaries. The interpretability of boundaries for a given entity is dependent on its composition. More specifically, it depends on the fact that an entity is defined as a structural unit or is made of strata. An entity composed of a structural unit like a twig or a tree has a clear physical boundary. Difficulties in delimiting the boundary are more dependent on the optical properties and the arrangement of the composite material. Meanwhile, an entity composed of strata is more subject to misinterpretation. Thus, the radiometric and textural differences between the composite element of the entity (i.e., structural units or strata) determine the spatial discrimination potential.

The intrinsic variance of the forest landscape combined with the discrete nature of digital images requires a compromise: a digital image can be used either as an integrator or as a descriptor of the selected entity. The choice of a spatial strategy must be defined by taking into consideration the discrimination potential and the analysis tools available. The strategy is either reductionist or descriptive. For both strategies, the physical entity dimensions must be translated into the visual dimensions perceptible in an image. The distinction between the physical and image dimensions is important in particular when visual features, like shadows, bidirectional reflectance, and atmospheric mask, become significant. The reductionist strategy requires the larger pixel dimensions necessary to identify, differentiate, describe, and map the targeted entity. It involves a pixel footprint that integrates the entity components but should not exceed the entity's dimensions. Many analysis tools exist, in particular those based on semivariograms, to assess the most suited pixel size for a given scene. In contrast, the descriptive strategy assumes an oversampling of the entity. The potential pixel size associated with the descriptive strategy spans from the dimension of the next finest entity up to the size found for the reductionist strategy. The availability of analysis tools able to handle the increase in radiometric texture is crucial to finalize a practical range of spatial resolutions associated with a descriptive strategy. The two strategies provide a flexible approach that takes into account the radiometric texture of a digital image and the available analysis tools. The digital image thus becomes a continuous radiometric sampling from an imaging sensor adapted to a specific entity in the context of a given application.

Once a range or a series of potential spatial resolutions are defined, supplemental considerations must be addressed. More specifically, three considerations dominate: image availability, calibration, and economics. The availability of the digital image is crucial. This availability must be determined in terms of the spatial resolution, the spectral bands, and the time frame for data acquisition (date of data acquired and repeatability, if required). The limited choice of sensors often forces a decision on the closest available spatial resolution. Great care must be taken that the selected resolution still falls in the range of admissible spatial resolutions. For example, a possible scenario could be that a pixel size for a reductionist strategy is defined as 10 m, but only sensors with 20 m pixel size are available. Usefulness of the 20 m image may be jeopardized if information on the entity cannot be extracted as expected from an analysis tool. The next consideration, calibration, ensures image compatibility with other images or georeferenced products. For instance geometric calibration is only important if the image must correspond to a cartographic reference. Similarly, radiometric calibration matters only in cases where several sensors or multitemporal images are used or if the pixel values are required in radiance or in reflectance. Lastly, the economic considerations are crucial to selecting a spatial sampling strategy. The benefits in information content of finer spatial resolution must compensate for the supplemental costs. In the eventuality of an approach not meeting one of the considerations favorably, the process must be reassessed from the first levels: the objectives, the targeted entity, and the analysis tools.

REFERENCES

Allen, T.F.H. and T.B. Starr. *Hierarchy: Perspectives for Ecological Complexity,* University of Chicago Press, Chicago, 1982.

Atkinson, P.M. and P.J. Curran. Choosing an Appropriate Spatial Resolution for Remote Sensing Investigations, *Photogrammetric Eng. Remote Sensing,* 63(12), pp. 1345–1351, 1997.

Avery, T.E. and G.L. Berlin. *Fundamentals of Remote Sensing and Airphoto Interpretation,* Macmillan Publishing Co., New York, 1992.

Bauer, M.E., T.E. Burk, A.R. Ek, P.R. Coppin, S.D. Lime, T.A. Walsh, D.K. Walters, W. Befort, and D.F. Heinzen. Satellite Inventory of Minnesota Forest Resources, *Photogrammetric Eng. Remote Sensing,* 60(3), pp. 287–298, 1997.

Biging, G.S., R.G. Congalton, and E.C. Murphy. A Comparison of Photo-Interpretation and Ground Measurements of Forest Structures, *Technical Papers at the 1991 ACM-ASPRS Annual Convention,* Baltimore, MD, 3, pp. 6–15, 1991.

Cowen, D.J., J.R. Jensen, P.J. Bresnahan, G.B. Ehler, D. Graves, X. Huang, C. Wiesner, and H.E. Mackey, Jr. The Design and Implementation of an Integrated Geographic Information System for Environmental Applications, *Photogrammetric Eng. Remote Sensing,* 61(11), pp. 1393–1404, 1995.

Fournier, R.A., P.M. Rich, and R. Landry. Hierarchical Characterization of Canopy Architecture for Boreal Forest, *J. Geophys. Res.,* 102(D24), pp. 29,445–29,454, 1997.

Fournier, R.A., G. Edwards, and N.R. Eldridge. A Catalogue of Potential Spatial Discriminator for High Spatial Resolution Digital Images of Individual Crowns, *Can. J. Remote Sensing,* 21(3), pp. 285–298, 1995.

Franklin, S.E., M.B. Lavigne, M.J. Wulder, and E.R. Hunt, Jr. Landsat TM Derived Forest Covertypes for Modelling Net Primary Production, *Can. J. Remote Sensing,* 23(3), pp. 243–251, 1997.

Gagnon, P.A., C. Nolette, J.P. Agnard, and C. Larouche. Extending the Field of the Surveyor's Practice with PC-Videometry, *Proceedings of the 15th Conference of Commonwealth Surveyors,* 5–15 August, Cambridge, England, Vol. 1, 1991.

Goetz, S.J. and S.D. Prince. Remote Sensing of Net Primary Production in Boreal Forest Stands, *Agric. For. Meteorol.,* 78, pp. 149–179, 1996.

Gougeon, F.A. A Crown-Following Approach to the Automatic Delineation of Individual Tree Crowns in High Spatial Resolution Aerial Images, *Can. J. Remote Sensing,* 21(3), pp. 274–284, 1995.

Hay, G.J., K.O. Niemann, and D.G. Goodenough. Spatial Thresholds, Image-Objects, and Upscaling: A Multiscale Evaluation, *Remote Sensing Environ.*, 62, pp. 1–19, 1997.

Howard, J.A. *Remote Sensing of Forest Sesources: Theory and Application,* Chapman & Hall, London, 1991, p. 420.

Jupp, D.L.B., A.H. Strahler, and C.E. Woodcock. Autocorrelation and Regularization in Digital Images, I, Basic Theory, *IEEE Trans. Geosci. Remote Sensing,* 26, pp. 463–473, 1988.

Jupp, D.L.B., A.H. Strahler, and C.E. Woodcock. Autocorrelation and Regularization in Digital Images, I, Basic Theory, *IEEE Trans. Geosci. Remote Sensing,* 27, pp. 247–258, 1989.

Kimmins, H. *Balancing Act: Environmental Issues in Forestry,* Second Edition, UBC Press, Vancouver, Canada, 1997, p. 305.

King, D.J. Airborne Multispectral Digital Camera and Video Sensors: A Critical Review of System Designs and Applications, *Can. J. Remote Sensing,* 21(3), pp. 245–273, 1995.

Liu, J., J.M. Chen, J. Cihlar, and W.M. Park. A Process-Based Boreal Ecosystem Productivity Simulator Using Remote Sensing Inputs, *Remote Sensing Environ.,* 62(2), pp. 158–175, 1997.

Lowell, K.E., G. Edwards, and G.L. Kucera. Modelling Heterogeneity and Change in Natural Forests, *Geomatica,* 50(4), pp. 425–440, 1996.

Marceau, D.J., D.J. Gratton, R.A. Fournier, and J.-P. Fortin. Remote Sensing and the Measurement of Geographical Entities in a Forested Environment, 2, The Optimal Spatial Resolution, *Remote Sensing Environ.,* 49, pp. 105–117, 1994.

Murtha, P.A. *A Guide to Aerial Photographic Interpretation of Forest Damage in Canada,* Canadian Forestry Service, Dept. of the Environment, Publication No. 1292, 1972.

Nieke, J., H. Schwarzer, A. Neumann, and G. Zimmermann. Imaging Spaceborne and Airborne Sensor Systems in the Beginning of the Next Century, *Proceedings of the European Symposium on Aerospace Remote Sensing (Europto Series): Sensors, Systems, and Next-Generation Satellites,* 22–25 September, London, UK, 1997.

O'Neill, R.V., D.L. DeAngelis, J.B. Waide, and T.F.H. Allen. *A Hierarchical Concept of Ecosystems,* Princeton University Press, Princeton, NJ, 1986.

Paine, D.P. *Aerial Photography and Image Interpretation for Resource Management,* Forest Management Department School of Forestry, Oregon State University, John Wiley & Sons, 1981, p. 571.

Perron J.-Y. and P. Morin. *Normes d'Inventaire Forestier: Les Placettes-échantillons Permanentes,* Dircction de la Gestion des Stocks Forestiers, Service des Inventaires Forestiers, May 1997, p. 248.

Roscoe, J.H., L.D. Black, H. Weiner, H.D. Young, P. Maynard, F.C. Whitmore, and R.N. Colwell. Photo Interpretation Keys, *Photogrammetric Eng.,* 21, pp. 703–724, 1955.

Sayn-Wittgenstein, L. *Recognition of Tree Species on Air Photographs by Crown Characteristics,* Technical Note No. 95, Forest Research Div., Department of Forestry, Canada, 1960, p. 56.

Spurr, S.H. *Photogrammetry and Photo-Interpretation,* Ronald Press Co., New York, 1960, p. 472.

Teillet, P.M., D.N.H. Horler, and N.T. O'Neill. A Quality Assurance and Stability Reference (QUASAR) Monitoring Concept for Calibration/Validation, *Proceedings of the European Symposium on Aerospace Remote Sensing (Europto Series): Sensors, Systems, and Next-Generation Satellites,* 22–25 September, London, UK, 1997, pp. 262–270.

Tokola, T., A. Turkia, J. Sarkeala, and J. Soimasuo. An Entity-Relationship Model for Forest Inventory, *Can. J. For. Res.,* 27, pp. 1586–1594, 1997.

Toutin, T. Intégration de Données Multisources: Comparaison de Méthodes Géométriques et Radiométriques, *Int. J. Remote Sensing,* 16(15), pp. 2795–2811, 1995.

Turner, M.G. and R.H. Gardner, Eds., *Quantitative Methods in Landscape Ecology,* Ecological Studies, Vol. 82, Springer-Verlag, 1990, p. 536.

van Zuidam, R.A. *Aerial Photo-Interpretation in Terrain Analysis and Geomorphologic Mapping,* International Institute for Aerospace and Earth Sciences, 1986, p. 442.

Walker, D.A. and M.D. Walker. History and Pattern of Disturbance in Alaskan Arctic Terrestrial Ecosystems: A Hierarchical Approach to Analysing Landscape Change, *J. Appl. Ecol.,* 28, pp. 244–276, 1991.

Woodcock, C.E. and A.H. Strahler. The Factor of Scale in Remote Sensing, *Remote Sensing Environ.,* 21, pp. 311–322, 1987.

Woodcock, C.E., A.H. Strahler, and D.L.B. Jupp. The Use of Variograms in Remote Sensing: I. Scene Models and Simulated Images, *Remote Sensing Environ.,* 25, pp. 323–348, 1988.

Zsilinszky, V.G. The Practice of Photo Interpretation for Forest Inventory, *Photogrammetria,* 5, p. 17, 1964.

CHAPTER 46

Designing an Accuracy Assessment for a USGS Regional Land Cover Mapping Program

Z. Zhu, L. Yang, S.V. Stehman, and R.L. Czaplewski

INTRODUCTION

The U.S. Geological Survey (USGS), in cooperation with the U.S. Environmental Protection Agency (EPA), is conducting a conterminous U.S. land-cover mapping project using Landsat Thematic Mapper (TM) 30-meter data acquired in the early 1990s. The objective of this large mapping project is to produce a generalized, consistent, and application-driven land-cover product with reasonable accuracies (Vogelmann et al., 1998). The project is being carried out on the basis of the 10 EPA federal regions which make up the conterminous U.S.; each region is a study area composed of multiple states. As the land-cover classification is completed for a federal region, the thematic accuracy of a modified 23-class Anderson classification system (Anderson et al., 1976) will be assessed relative to objectives of the mapping effort and to provide users of the data with a quality measurement.

Federal Region II, consisting of New York and New Jersey, is one of the first federal regions completed under the mapping project. This region servers as a pilot study to develop a suitable accuracy assessment plan for this and the rest of the federal regions based on a statistically valid sampling design. Of the 23 land-cover classes, 15 exist in the region.

Assessing accuracy for the mapping project is a complex task, largely due to the fact that study areas (federal regions) are very large relative to TM's 30-meter spatial resolution. Few reference data sets exist that are consistent and suitable for all federal regions. And for a federal region of multiple states, not all land-cover classes are distributed commonly.

Treatment of rare classes is an important issue as it is difficult to find corresponding reference data for these rare map classes.

Given these constraints, it is clear that a statistically sound and yet practical sampling protocol is the centerpiece of the accuracy assessment. Stehman and Czaplewski (in press) stressed the importance of a valid probability sampling design to an accuracy assessment, and noted that the choices of sampling unit and sampling method are two major decisions required of the assessment. Additionally, as the two elements are dependent on the reference data set to be used, the choice of a reference data set is a fundamental decision in a real-world accuracy assessment.

When the map area of the study is large and field observations are impractical, choosing reference data is often limited by existing ancillary data sets. These data sets should be carefully evaluated to determine if any of them can be used as the source of reference data. In this study, we reviewed several existing large programs or data sets, including the National Resource Inventory (NRI) of the U.S. Department of Agriculture (USDA) Natural Resource Conservation Service, Forest Inventory and Analysis (FIA) of USDA Forest Service, U.S. Department of Commerce agricultural statistics, U.S. Department of Interior GAP program, and the National Aerial Photography Program (NAPP).

The usefulness of these data sets, except for NAPP, is limited in two aspects: incomplete coverage and different land-cover classification systems. For example, NRI does not cover federal lands and FIA data are limited to forest land only. The differences in classification systems also hinder a direct comparison using these data sets. While it is not yet clear that these existing data sets would satisfy the objectives of the current assessment, the supplemental utilities of the data sets are still being explored.

Because NAPP covers all lands in the U.S., it provides a satisfactory source of reference data from which to design a suitable sampling plan. NAPP photo frames serve as the first-stage sampling unit in a probabilistic sampling design such as multistage or cluster sampling. Another important consideration is that the date of the current NAPP photography (1992) in this region is in close approximation of the date of the Landsat data used for the classifications. The disadvantages of using NAPP aerial photos include the potentially high cost of labor for photo interpretation and photo interpretation error. However, these limitations are not related to the statistical validity of using NAPP. Given these considerations, we selected NAPP aerial photos as the sampling frame and the basic sampling unit.

OBJECTIVES

The central objective of the pilot study is to estimate and analyze misclassification errors in the regional land-cover map. Errors will be estimated and described using the traditional error matrix (Congalton, 1991) as well as a number of other important measures, including the overall proportion of pixels correctly classified, user's and producer's accuracies, and omission and commission error probabilities. Because the statistical foundation for estimating these parameters requires a probability sample of reference data, we limited consideration to probability sampling design. In this chapter, we describe the process of selecting the reference data, including defining both the sampling unit and the sampling design to be used for this pilot study. We also discuss how various decisions were made concerning various choices necessary to selection of the sampling protocol.

SAMPLING PROTOCOL

With NAPP aerial photos selected as the source of reference data, several probabilistic samplings were considered and compared with each other using criteria adopted for the study. These criteria are: (1) a probability sampling design to ensure objectivity and a valid statistical basis (Stehman and Czaplewski, in press); (2) small variances of the estimated accuracy parameters; (3) a good spatial distribution of sample elements to ensure adequate precision within regional

subpopulations; (4) an adequate representation of all classes including rare classes; (5) low cost, and (6) simplicity.

For Federal Region II, various probabilistic sampling methods were tested. Results showed that designs such as simple random, simple systematic, or stratified by class allowed little control over the number of NAPP photos and did not guarantee that the sample would be spatially well distributed. Instead, repeated tests showed that a two-stage, cluster sampling with an additional treatment of rare classes achieved these desirable criteria and at the same time retained the desired simplicity feature. We elaborate characteristics of this design as follows.

Sampling Frame and Sampling Units

The sampling frame for the pilot study is defined as the entire Federal Region II represented by NAPP photo coverage (scale 1:40,000). A NAPP photo is defined as a primary sampling unit (PSU) in the two-stage cluster sampling design, whereas a sampled pixel within each PSU is treated as a secondary sampling unit (SSU).

First-Stage Sampling

To select PSUs to achieve a spatially well-distributed sample, the entire sampling frame was partitioned in 333 grids based on NAPP flight-line and frame numbers, with each grid measuring 15' × 15' and consisting of 32 NAPP photos. Some gaps in the NAPP coverage result in the actual population assessed being smaller than the full region. Approximately 3% of the target region is not covered by NAPP photos, and thus the accuracy estimates apply to the remaining 97% of the region.

Next, a stratified random sample was selected employing the 333 grid cells as geographic strata (equal area for all strata). One photo was selected from each stratum with all photos having an equal probability of being selected, thus satisfying the probability sampling criterium as well as providing a spatially well distributed sample of first-stage sampling units (photos).

Strata on the boundary of Region II were treated as if they were the complete strata except that if the sampled PSUs fell outside the regional boundary they were not used. This restriction is necessary to maintain the equal-probability characteristic of the design. A total of 278 NAPP photos were selected as PSUs for the first-stage sample (Figure 46.1).

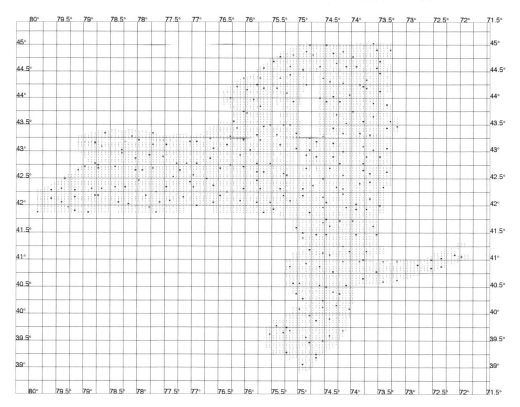

Figure 46.1. The sampling frame of New York and New Jersey consisting of NAPP photos (represented in the figure by small dots), and 278 sampled PSUs (represented by large, bold dots). Note that there are a few gaps in the state of New York where there are no NAPP photos.

Second-Stage Sampling

Second stage sampling was accomplished by selecting four SSUs (pixels) within each PSU to provide the actual locations for obtaining the reference land-cover classification. Each photo (PSU) was divided into four equal-area quadrants, and one pixel was selected at random with equal probability from within each quadrant. If the PSU was a boundary photo and the selected pixel was outside the target region, no sample pixel was obtained for that quadrant of the photo. This protocol maintained the equal probability feature at the second stage of the design. A total of 1,112 SSUs were selected (Table 46.1).

Additional Sample for Rare Classes

Seven land-cover classes were treated as rare classes, as the area of each was less than 2% of the total map area and received few SSUs from the extensive, general sampling design (Table 46.1). Increasing sample sizes for the rare classes was not a desirable option as it would lead to unequal probabilities between the rare and common classes. An additional, separate design for rare classes was implemented for this study to augment the first, general design. The desire to exercise some control over the spatial distribution of the SSUs continued to be one of the key criteria influencing the rare-class treatment. Consequently, all NAPP photos selected during the first stage were used as the starting point for the rare class design. Sixty SSUs were selected randomly from each of the rare map classes. Within each rare class stratum, pixels have equal inclusion probabilities, but these inclusion probabilities differ from those resulting from the general sampling design. This needs to be accounted for in the data analysis and parameter estimation stage.

The sampling design described above produced the spatial distribution of sampled NAPP photos shown in Figure 46.1, and two separate but related samples (Table 46.1). The first sample encompasses all mapped land-cover classes, whereas the second sample con-

Table 46.1. Output of the Two-Stage, Cluster Sampling Design for All of the 15 Classes and the Second, Additional Design for Rare Classes. (Background class designates pixels outside the study area.)

Class Name	Class Number	No. of Pixels	% of Pixels	No. of First Sample	% of First Sample	No. of Rare Sample
Background	0	352354300	NA	92	NA	
Water	1	24829781	13.67	81	7.70	
Low intensity residential	2	7645892	4.21	39	3.71	
High intensity residential	3	2236720	1.23	20	1.90	60
Commercial/industrial	4	1967410	1.08	16	1.52	60
Hay-pasture	5	16704716	9.20	110	10.46	
Row crop	6	22224280	12.24	137	13.02	
Urban grass	7	1357851	0.75	8	0.76	60
Conifer	8	9924031	5.46	59	5.61	
Mixed forest	9	29794345	16.41	163	15.49	
Deciduous	10	57903977	31.88	371	35.27	
Forest wetland	11	5277962	2.91	33	3.14	
Nonforest wetland	12	1301204	0.72	12	1.14	60
Barren1-quarry/strip mine	13	234341	0.13	2	0.19	60
Barren2-sand beach	14	86994	0.05	0	0.00	60
Barren3-transitional	15	127422	0.07	1	0.10	60

tains only additional SSUs for the rare classes. These two samples can be either combined for accuracy estimation or be treated separately; we discuss this in more detail in the next section.

DISCUSSION

This study showed that, when dealing with accuracy assessment of a large map area, exercising spatial control over the distribution of SSUs is one of the important considerations for sampling design. In our case, it strongly influenced the design planning because it limited the number of NAPP photos that could be selected for the reference sample. This constraint virtually eliminated from consideration options such as simple random or systematic sampling, because these options allow little control over the number of photos in which the reference sample pixels will fall. Similarly, stratifying by map class and selecting a stratified random sample does not permit strong control over the number of photos required.

An additional advantage of not stratifying on the map classes is that the sample is not influenced by the accuracy of the map itself. Specifically, stratifying by map classes implicitly places a priority on estimating user's accuracy and commission error probabilities (Stehman, 1995). It is possible that such a stratified sample will perform poorly for estimating producer's accuracy and omission error probabilities because too few reference pixels may occur in some of the less common land-cover classes, particularly those having poor user's accuracy, to allow useful accuracy estimates.

During the design process, some technical details were strongly influenced by two considerations, namely the criteria of good spatial distribution, and equal probability sampling. The geographically stratified design of the first stage created the desired spatial distribution of the sample, much like a systematic sample, but the stratified design is less susceptible to the potential variance inflation effects a systematic sampling may have if periodicity is present in spatial patterns of classification error. Similarly, the procedure of selecting SSUs from divided quarters of a PSU is motivated by the desire to achieve good spatial distribution for SSUs within each PSU.

The technicalities of treating boundary PSUs and SSUs are important to this design because they maintain the equal probability feature of the overall design. The equal probability feature allows for estimating the error matrix and accompanying accuracy measures using formulas appropriate for simple random sampling. That is, no weighting is necessary to account for pixels being sampled with unequal probability. This feature satisfies part of the simplicity criterium, ease of analysis.

A two-stage sampling design such as the one described here involves a compromise decision balancing the better precision of sampling more PSUs with few SSUs per PSU, with the cost savings of sampling many SSUs in each PSU. The combined extensive, general sample and the additional rare sample results in over 1,500 locations requiring photo interpretation. More detailed sample size planning would require high quality information on the within-cluster (photo) correlation of classification error, but such information is itself difficult and expensive to obtain, and was unavailable to guide our choice. Because SSUs are selected with equal probability, it would be a simple matter to supplement the sampling within each PSU, should we later find that the sample size is too small to provide adequate precision of estimates. Supplementing the sample within each PSU by an additional simple random sample of pixels results in a design that is still equal probability.

The use of an additional, separate sample for rare classes is often a useful treatment for rare classes (Edwards et al., 1998). In this study, two options for selecting the additional sample within each rare map class were considered. The first option was to select a simple random sample of all pixels identified in that class in the first-stage sample—a stratified random sample where the stratification is applied to all pixels in the first-stage PSUs. This option, which results in an equal probability sample of pixels, was the one chosen. However, the concern with this design was that if most of the rare class pixels were concentrated in one or a few of the PSUs with many other PSUs having just one or two pixels in the class, the precision of the estimates for this class could be high because of the potential for a high within-cluster correlation. An alternative design option would be to select at least one pixel for each first-stage PSU in which the rare class was found. This would ensure that the sampled pixels would be from spatially dispersed PSUs (assuming the rare class was found in more than a few PSUs). A problem with this design is that SSUs would then be sampled with unequal probability. For example, a PSU with just two pixels of the rare class would have a much higher probability than a PSU with, say, 300 pixels of the rare class. In addition to making the analysis more complex, this discrepancy in inclusion probabilities could also result in higher variance. The actual sample selected following the first option did not result in the sampled pixels being concentrated in just a few PSUs, so the concern with the first option did not materialize in

this region. However, further analysis is needed to determine which of these options is more efficient under different scenarios of the spatial distribution of rare classes. This information will be important to assessment for other regions.

The data from the additional rare class sampling design may be used to estimate user's accuracy and commission error probabilities for each rare class separately. Ideally, we would like to combine this additional rare class sampling with the extensive, general sample (for common and rare classes) to obtain more precise estimates of accuracy measures resulting from the larger sample size of the combined data. General estimation formulas to combine data from two probability samples exist (Cochran, 1977), but translating these results to the estimated parameters used in accuracy assessment has not yet been done. The issue is that the two sampling designs result in different probabilities of sampling a particular pixel, so the data from the two samples combined must account for these different probabilities. It is important to recognize that estimating results from the two sampling designs separately is always a viable option.

Photo interpretation errors, either due to time lapse between date of the TM data and that of the NAPP photography or due to interpreter's misclassification, should be addressed during the analysis and estimation phase. Besides ensuring that each SSU will be read by multiple interpreters, a statistical device can be designed to measure the accuracy of photo interpretations. The current design of PSUs is advantageous if field visits to validate photo interpretation are required.

SUMMARY

Sampling for accuracy assessment of large-area land-cover classification is best achieved with a probability sampling design and considerations to cost, spatial distribution, and simplicity. The two-stage, geographically stratified cluster sampling design described in this chapter allows an equal-probability sample spatially well distributed over the map region and over each sampled aerial photo. The sampling design is also flexible for additional sample of SSUs if the sample size needs to be increased.

Because this study in Federal Region II is a pilot for the other regions, it is possible that the design may be modified based on what we have learned from this region.

ACKNOWLEDGMENT

The work in this chapter was performed under U.S. Geological Survey Contract No. 1434-CR-97-CN-40274.

REFERENCES

Anderson, J.R., E.E. Hardy, J.T. Roach, and R.E. Witmer. *A Land Use and Land Cover Classification System for Use with Remote Sensor Data,* U.S. Geol. Survey Prof. Paper 964, 1976, p. 28.

Cochran, W.G. *Sampling Techniques,* 3rd ed., John Wiley & Sons, New York, 1977.

Congalton, R.G. A Review of Assessing the Accuracy of Classification of Remote Sensed Data, *Remote Sensing Environ.,* 37, pp. 35–46, 1991.

Edwards, T.C. Jr., G.G. Moisen, and D.R. Cutler. Assessing Map Accuracy in a Remotely Sensed Ecoregion-Scale Cover Map, *Remote Sensing Environ.,* 63, pp. 73–83, 1998.

Stehman, S.V. Thematic Map Accuracy Assessment from the Perspective of Finite Population Sampling, *Int. J. Remote Sensing,* 16 pp. 589–593, 1995.

Stehman, S.V. and R.L. Czaplewski. Design and Analysis for Thematic Map Accuracy Assessment: Fundamental Principles, *Remote Sensing Environ.,* in press.

Vogelmann, J.E., T. Sohl, and S.M. Howard. Regional Characterization of Land Cover Using Multiple Sources of Data, *Photogrammetric Eng. Remote Sensing,* 64(1), pp. 45–57, 1998.

CHAPTER 47

Uncertainty in Automatically Sampled Digital Elevation Models

M.J.P.M. Lemmens

INTRODUCTION

Because of the high burden human beings put on nature both quantitatively and as a consequence of their life style, the need for geoinformation has increased dramatically in the last decades. This trend will only increase for the coming millennium. Quality has become an important issue in the field of geographical information. Quality is often defined in generic terms as "fitness for use." With respect to Digital Elevation Models (DEMs), fitness for use means: how well do the measurements in the data set represent the actual terrain surface given the application domain? Within this context a high degree of uncertainty may be associated with low quality.

DEMs are a basic data source for many applications in the earth and the engineering sciences. Their potential for solving a broad variety of problems is well known. They include extraction of drainage networks, determination and delineation of ecozones, determination of beach and dune erosion, determination of earthwork volumes, and hydrological modeling. Within the mapping and remote sensing community an important application of DEMs is rectification of aerial and satellite images in order to reduce the effects of relief distortions when georeferencing the data.

The terms Digital Elevation Model (DEM) and Digital Terrain Model (DTM) are often considered to be synonyms. We look at a DEM as a set of elevation data in numerical format stored in a computer accessible format. The sampling points can be captured and stored in grid or in irregular format. It has to be noted

that direct acquisition in grid format is not a common option. Due to operational limitations, elevation data is usually collected as an irregularly distributed set of points. Next, by interpolation the data are transferred to grid format for purposes of ease of storage and processing. Generally, the grid representation obtained from interpolation has a higher resolution than the original irregular data set. A DTM is a DEM extended with structural features such as drainage channels, ridges, hilltops, depressions, and other terrain discontinuities. These structural features are usually stored as strings of points.

DTM data can be acquired from existing maps, from photogrammetric stereomodels, from ground surveys, or from other systems. At present, operational techniques are available that enable one to acquire elevation data by highly automated means. Digital photogrammetry and airborne laser-altimetry are at present the most important operational automatic data capturing methods for DEM creation. SAR interferometry (InSAR) is rapidly becoming another fully operational automatic DEM capturing technique.

In conceptual terms a DEM can be defined as the boundary between lithosphere and atmosphere. The problem of DEM generation concerns now the determination of the location of this boundary with respect to a certain reference system. It is becoming more and more realized that the uncertainty problem lies not only in the measuring process itself, but is also due to the fact that real-world phenomena have to be abstracted and mapped into numerical formats.

Data capturing, representation, and use of DEMS are changing dramatically presently. Until one decade ago DEMs have usually been considered to be a tool for deriving contour lines for map production (e.g., Ackermann, 1978; Balce, 1987). We observe now that the numerical terrain representation is processed and combined with other data sources by computational means, in particular by GIS-systems, to determine a wide range of earth-related natural phenomena. Furthermore, due to the highly automated data sampling process, we face now fundamental problems which force us to reconsider the concepts of DEMs. The basic aim of this chapter is to provide an inventory of these problems, in particular related to uncertainty. We carry out this task in order to improve our insight into these problems, which are far from trivial. Within the present context we focus on automatic DEM sampling techniques.

The nature of measuring is such that all observations are inevitably contaminated by error; errors are an empirical reality. Errors in data may be induced by the variations (stochastics) in the measuring process yielding random errors and by blunders resulting in gross errors (outliers). Since uncertainty in the measuring processes is the best investigated part of the entire geodata capturing process, including the (geometric) disturbing effects of atmosphere and gravitational forces, we will leave this subject undiscussed here. The uncertainty in the information derived from the DEM, possibly combined with other data sources, depends not only on the quality of the data but also on the quality of the models that link the data with the information. Although very important, we also leave this subject out of consideration within this chapter.

Uncertainty in DEMs, besides measurement inaccuracies and errors, is also introduced by choice of interpolation technique, point sampling interval (resolution), characteristics of the terrain surface, representativeness of the sampled points, and fuzziness of the boundaries. We consider these aspects more thoroughly in the following sections. The single point accuracy of an interpolated DEM surface point is an appropriate indicator for expressing DEM uncertainty. Therefore, we pay special attention to deriving an expression for this accuracy measure.

INTERPOLATION

Interpolation aims at providing elevation data in regions where no data exist or to transfer an irregular distributed height set into grid format. Many interpolation techniques have been developed in the course

of time. Tests have shown that these interpolation techniques have similar performance, provided that the data behaves well with respect to point distribution and fluctuations in relief (Ackermann, 1996).

An important issue is the robustness of a certain interpolation technique against the changes in the geomorphological structure within the vicinity of the point of which the elevation has to be determined. It is important that the points used in the interpolation and the point to be interpolated all belong to the same landscape type. For example, when we determine the height value of a point which is located in a valley from two valley points and one slope point, the interpolated height value may be too large; that means erroneous. So the goodness of an interpolation can be judged by considering how well the method is insensible to the changes in geomorphological characteristics of the terrain.

To perform well prior to interpolation, a classification of the points has to be carried out in order to determine to which type of landscape these points belong; e.g., flat area, watershed, and so on. The interpolation method used should be advanced enough to prevent points situated on two or more landscapes type being combined in the interpolation process.

The single point accuracy of an interpolated point is an important measure to describe the quality of a DEM. The single point accuracy of an interpolated height will depend on the interpolation method, the accuracy of the points from which the point is interpolated, and disturbing factors caused by the terrain slope, the microrelief, and other effects. When we model the influence of these distortions for the time being as one integrated variance factor (VAR_X) which is independent of the measurement precision of the individual observations (VAR_{MSP}), and supposing that a linear interpolation function INT is applied involving n vicinity points, the accuracy of the interpolated height can be obtained by error propagation, resulting in the variance factor:

$$VAR_{ISP} = INT_{i=1}^{n}\left(VAR_{MSP_i} + VAR_{X_i}\right) + VAR_{X_{ISP}} \qquad (1)$$

Now we have derived with Equation 1 an expression to describe the height accuracy of a DEM surface, it becomes the task to obtain more detailed insight into the individual terms of the joined variance factor VAR_X. We carry out this task in the next sections.

RESOLUTION

When the terrain surface in an area is perfectly horizontal, one height point will suffice to describe the elevation of the entire area. However, when relief fluctuations are present in the terrain, more height points are necessary in order to obtain a representation which is sufficiently precise. The more rapid the fluctuations occur in the terrain, the more points will be necessary. In case of a high point accuracy though a low resolution, interpolated points will show a low point accuracy; the terrain is undersampled.

It has been early recognized that single point accuracy alone is not sufficient to describe the precision of a DEM. Also, the resolution in relation to the terrain fluctuations plays an important role. This theme has been subject of several studies. The relation between accuracy and resolution has been investigated for photogrammetric data acquisition by Ackermann (1978), Ayeni (1982), Balce (1987), and Li (1993). For DEMs derived from contour lines this relation has been studied by Carrara et al. (1997) and Gao (1997). Both Li and Gao derived numerical expressions for the relationship. For photogrammetric data Li (1993) argued, based on theoretical analysis and some experimental results, that terrain fluctuations (roughness) can be expressed by two main descriptors: average slope and wavelength. For surfaces linearly constructed from grid data, the following relationship among variance of the interpolated surface point (VAR_{ISP}), the variance factor of the measurement points (VAR_{MSP}), the mean slope angle (α), the resolution of the measured points expressed as the point sampling interval (Δx) and the average wavelength of the fluctuations of the relief W were found:

$$VAR_{ISP} =$$
$$K_1 VAR_{MSP} + K_2 \left(1 + \frac{4\Delta x}{W}\right)(\Delta x \tan\alpha)^2 \qquad (2)$$

where K_1 is a constant between 0 and 1 of which the value depends on the chosen interpolation technique; K_2 is a constant which depends in addition to the chosen interpolation technique on the characteristics of the terrain surface and the resulting DEM surface. Values for K_1 and K_2 can be obtained by linear regression. It is assumed that the measured points in the data set all have the same variance (VAR_{MSP}). For points on the surface linearly interpolated from composite data Equation 2 reduces to:

$$VAR_{ISP} = K_1 VAR_{MSP} + K_2 (\Delta x \tan\alpha)^2 \qquad (3)$$

By using regression techniques, Gao (1997) found for DEMs constructed from map sheet contour lines the following expression:

$$\sqrt{VAR_{ISP}} = (7.274 + 1.666\Delta x)D \qquad (4)$$

With Δx the DEM resolution; that means the distance between two sampling points on the contour lines, and D the contour density expressed as meter contour lines per square meter. It can be argued from theoretical considerations that slope and wavelength do not form the optimal parameters to describe the influence of terrain roughness on the precision of the interpolated points. We consider this issue now in greater detail.

- First, the definition of mean slope of an area is resolution or scale dependent. Features at the surface of the earth can be described at a multitude of scales. For a certain application domain, only a limited range of scales is adequate to describe that feature. The description one obtains does not only depend on the characteristics of the surface but also on the scale of measurement. An obvious example is a valley. On top of the slope of the valley many details may be present that are not of interest when one wants to compute; for example, the erosion. However, when gradients are computed that do not represent the trend of the slope but the small-scale fluctuations on top of the valley slope, erroneous results will be derived. So, at too fine a scale, one is swamped with extraneous detail. On the contrary, at too coarse a scale, important features may be missed. From a signal theoretical viewpoint the surface of the earth may be considered as a two-dimensional signal. It may be sufficiently well reconstructed from sampling values when the sampling interval obeys the theorem of Shannon. However, even perceptually simple landscapes may contain an overwhelming amount of detail, which can manifest themselves as height discontinuities in the DHM. Therefore, in general it can not be stated that a higher resolution accomplishes a better terrain description. Whether a better terrain description is achieved depends on the application domain.

- Second, the influence of the trend surface (slope) on the measuring precision depends, besides the gradient of the slope, on the applied measurement technique. For example, for airborne laser-altimeter data, we found from theoretical considerations that the pointing accuracy of the scanning mirror affects the accuracy of points in hilly terrain, according to:

$$\sqrt{\text{VAR}_{\text{slope}}} = \sqrt{2} \cdot z \cdot \tan \alpha \cdot \sqrt{\text{VAR}_{\Delta\phi}} \qquad (5)$$

with $\text{VAR}_{\Delta\phi}$ a variance factor describing the random pointing jitter error, α the slope of the terrain, and z the flying height. A realistic value of the pointing jitter error for a flying height of 1000 meter and a slope of 50 degrees is one meter.

- Third, the question of how well a measured point represents the local elevation depends largely on the microrelief present within the region of representation (RoR); that means the region for which the height measurement is supposed to be representative for the elevation. The influence of these local terrain fluctuations within a RoR can be expressed by a random error, in particular the height variance $\text{VAR}_{\text{mircorelief}}$ of the quasi random terrain variations:

$$\text{VAR}_{\text{microrelief}} = \int_{x \in \bigcirc} \left[h(x) - \overline{h} \right]^2 w(x) dx \qquad (6)$$

where $\bigcirc$ is the region of representation (RoR), $h(x)$ a height value within the RoR, $\overline{h}$ the mean height within the RoR, $w(x)$ the weight of each height value with $\int w(x) dx = 1$. The above equation shows that terrain fluctuations and resolution should be taken together to describe the microrelief; i.e., the terrain fluctuations at local level.

The above considerations demonstrate that the relationship among the accuracy of any arbitrary point of the DEM, the accuracy of the measured points, the terrain slope, the resolution and the terrain fluctuations is highly nonlinear. So, it is rather questionable whether

a direct relationship between these parameters can be established by linear regression. We propose to model the uncertainties introduced by slope, resolution, and microrelief by two variance factors of the type expressed by Equations 5 and 6, where it should be noted that the expression for $\text{VAR}_{\text{slope}}$ depends on the measuring device used. Both $\text{VAR}_{\text{slope}}$ and $\text{VAR}_{\text{microrelief}}$ form two individual terms of the joined variance factor VAR_{X} in Equation 1.

REPRESENTATIVENESS

The measuring accuracy of individual points may be rather high due to the use of advanced equipment. However, it is an experimental reality that when the same terrain surface is sampled a second time with the same resolution and the same measuring equipment/method, and sampling is carried out independently of the first sampling, the variation in the interpolated height values is much larger than one would expect from error propagation of the measuring accuracy. For example, with leveling, an accuracy of two centimeters is achievable. However, between two representations of the same relatively flat terrain surface, covered by grass, using the same equipment and sampling interval, differences up to 20 centimeters are observed. With respect to photogrammetrical DHMs, captured manually, Li (1992) observed: "the accuracy results obtained from two data sets of the same area may be quite different, even if the sampling interval is the same for both of them."

Sampling of a signal can be carried out in two modes. In the first mode only the signal value exactly located at the sampling location is quantized. In the second mode, the signal values that are within the area represented by the sampling location are averaged. This way of sampling is, for example, carried out by satellite remote sensors for imaging the earth surface. When applying the first method, which we call for convenience point sampling, the sampled value may not be representative of the signal values in the vicinity. The probability that the sampled value will not be representative of its vicinity will be high when the signal fluctuations within that vicinity is high. When carrying out sampling in the second mode (area sampling), the integration process causes us to obtain a value which is more appropriate for terrain description.

The land surveyor and photogrammetric operator carry out *a point sampling* of the terrain surface. During selection of the points, they perform an interpretation process in which they choose locations of which the height is close to the average terrain height within the vicinity of the point. So, an inherent integration process is carried out by the land surveyor and the photogrammetric operator by choosing representative points. The choice of representative points is inevitable associated with errors. These errors may be much larger than the measuring accuracy of the equipment, including atmospheric and gravitational effects. So, different samplings of the same terrain surface carried out by human operators may produce quite different result.

When using the automatic DEM capturing techniques mentioned earlier, the interpretation part of the sampling process is lost. The terrain locations recorded during measuring are arbitrarily chosen. Measurements may hit any detail. Next, these points are assumed to be representative for the entire vicinity. As a consequence of the automatic sampling, points will in general not be representative for their vicinity. The effect is that repetitive collection of the terrain surface by automatic techniques show higher random errors than when applying manual sampling methods, assuming that the measuring accuracy and the sampling interval are the same for both. So, manual techniques are associated with a notion of representativeness of the sampled points. This representativeness is introduced during an interpretation process carried out by human beings. Since automatic techniques sample the terrain by blind capturing, no interpretation is carried out, causing the collected height points not to be representative for their vicinity. Consequently, breaklines and other characteristic terrain features remain usually unrecorded. When assuming that the selection of representative points corresponds to an integration process, the loss of representativeness of automatic methods can be resolved by measuring more points than strictly required for obtaining an appropriate terrain description. By selecting—using an averaging process—one point from the set of points covering a neighborhood, the notion of representativeness is reestablished. Measuring more points than actually necessary for describing the terrain surface properly is not only necessary for representativeness purposes, but also for removal of unwanted objects such as buildings and trees. The redundancy enables one to check the height values on the presence of discontinuities by using filtering techniques. Finally, a dense point coverage of the terrain enables one to trace breaklines from the DEM data set itself, using automatic filtering techniques.

TERRAIN CHARACTERISTICS

By terrain characteristics we mean slope, terrain roughness, coverage, and reflectivity. We now briefly consider these issues, with the exception of terrain roughness, since this subject is already treated under the heading Representativeness.

- **Slope** causes two main types of errors. The first type refers to errors in the horizontal position, causing height errors of which the value depends on the slope angle. Pointing jitter, discussed in the section on Resolution, is one of the sources that cause this type of error. The second error is also related to the measurement technique. When using airborne laser-altimetry terrain, slopes cause a part of the footprint to hit the ground earlier than other parts. The error depends, besides slope, on the footprint size and the signal-to-noise ratio. In digital photogrammetry, errors may occur when the slope is not adequately incorporated in the matching model. Both types of sources cause random errors which are mutually independent. So, the variance factors resulting from these error sources can be combined by addition, yielding the variance factor VAR_{slope}.

- **Coverage** may introduce significant errors when using automatic capturing techniques. Since automatic elevation-capturing techniques are nondiscriminating, unwanted objects, such as houses, trees, hedges, and cars will be recorded. Vegetation (e.g., forest) may cause the foliage to be captured instead of the bare ground. This introduces the need for removal of these unwanted objects. Since manual removal is often too time-consuming, a solution by computational means (filtering) is preferred. However, filtering techniques alone cannot solve the entire problem because they also suffer from nondiscriminating characteristics. Terrain coverage results in gaps in the observed points and systematic errors which may occupy a wide range of values. Consequently, the errors introduced by terrain coverage are difficult to model.

- **Reflectivity.** Automatic capturing methods rely heavily on sufficient signal reflection on the terrain surface. Water bodies are a problem for both digital photogrammetry and laser-altimetry. For laser-pulses, water bodies cause for some portions', absorption of the signal, and for a larger portion, specular reflection yielding reflection of the signal in the direction away from the recording platform. Using digital photogrammetry, water bodies show less texture, too, to successfully match corresponding points. Sand bodies such as beaches and dunes also show less texture, too, for matching purposes. Specular reflection of laser pulses may cause corner reflection by multipath effects when buildings or other discontinuous objects are in the neighborhood of the specularly reflected signal, causing elevation dips. All these sources produce systematic errors or data gaps in the DEM. In addition, reflectivity influences the signal-to-noise ratio of the measuring system producing random height errors on the data. For laser-altimetry, for example, the less the reflected signal of the laser pulse reaches the receiving sensor, the larger the random error on the height data will be. This type of random error can be expressed by the variance factor: $VAR_{reflection}$.

FUZZY BOUNDARIES

What constitutes the boundary between the lithosphere and the atmosphere? This question is often difficult to answer, especially when the boundary refers to natural phenomena. Well-defined boundaries are usually related to man-made objects such as houses and roads. Most geographic natural features have boundaries which are unsharp and fuzzy (Frank, 1996). The problem is now how to describe the degree of fuzziness of a terrain surface. To describe fuzzy features, fuzzy measures have been developed. Although fuzzy measures result in an appropriate description of the fuzzy terrain surface, it introduces the problem of combining these measures with measures which make use of a probabilistic framework for describing uncertainty. The fuzziness of a terrain surface boundary results in uncertainty in the height value. This uncertainty or variation can be described by a pseudovariance factor $VAR_{fuzziness}$. The introduction of a pseudovariance factor supports the ease of processing since it enables one to combine this fuzziness value with other probabilistic uncertainty measures by error propagation. The value of the pseudovariance factor $VAR_{fuzziness}$ depends on the type of boundary. For man-made objects, such

as roads covered by concrete, and for beaches this value will be, in general, low. For natural terrain surfaces a wide range of values will be covered.

ACCURACY DESCRIPTION

An important parameter to describe the quality of a DEM is the single point accuracy of interpolated points. With Equation 1 we found an expression to describe the accuracy of a DEM surface. The major part of this chapter was devoted to finding the individual terms that build up the joined variance factor VAR_X. The individual variance terms we identified are:

- local-relief within the region of representation (RoR): $VAR_{microrelief}$;
- terrain slope: VAR_{slope};
- reflectivity of the terrain surface: $VAR_{reflection}$;
- fuzziness of the terrain surface: $VAR_{fuzziness}$.

In addition to the above terms, the time dimension has to be taken into account when considering DEM uncertainty. Usually there is a time delay between capturing of terrain and use of the data. This time difference can occupy a wide range of intervals and may even be on the order of several decades. In the time interval between capturing and use, the landscape may have changed, sometimes even drastically. When the height changes are a consequence of human action, they are usually of a systematic nature. However, when natural phenomena are involved the changes are often a result of random processes, which can be readily assumed to cause random variations: say VAR_{time}. It is reasonable to suppose that the individual variance terms are mutually independent. Accordingly, the joined variance factor VAR_X may be decomposed as follows:

$$VAR_X = VAR_{microrelief} + VAR_{slope} + \\ VAR_{fuzziness} + VAR_{time} \qquad (7)$$

With Equation 7 we have derived a basic expression describing the random errors that are due to influences other than those introduced by the measurement process.

EXAMPLES

In order to confront the theoretical model we developed above with real measurement data, we use figures published by the Survey Department of

Rijkswaterstaat, the Netherlands (Huising, 1996; Vaessen, 1997). For a variety of terrain types in the Netherlands, such as flat terrain (bare soil and covered with grass or reed), dunes and beaches, laser-altimeter data and also height data captured by digital and analytical photogrammetry were collected. A control data set was obtained by land surveying using GPS stop-and-go method. For all captured landscape types analytical photogrammetry performed the best. For example, in hilly terrain sparsely covered with vegetation, the RMS (root mean square) error of analytical photogrammetry was 0.12 m, for digital photogrammetry 0.22 m, and for laser altimetry 0.24 m. This result indicates that the interpretation process of human operator is essential and that terrain representativiness should be recovered for automatic capturing methods by using an averaging process while simultaneously reducing the number of height points. For flat terrain (bare soil or covered with grass) with laser-altimetry, a RMS error beneath the decimeter level could be achieved. For flat terrain covered by reed, laser altimetry shows an offset error of more than half a meter, indicating a large systematic error, while the RMS error lies around two decimeter. When comparing the last figure with the figure of flat terrain (bare soil or covered with grass), which appeared to be less than one decimeter, we may conclude that vegetation covers of reed type introduce a fuzziness of the boundary of approximately one decimeter. For hilly terrain covered with shrubs, analytical photogrammetry yielded a RMS error of 0.15 m, while digital photogrammetry yielded a figure of 0.31 m, and laser altimetry a figure of 0.40 m, again indicating that the fuzziness of the boundary introduces a random error of about one decimeter for both digital photogrammetry and laser altimetry. When comparing the figures of hilly uncovered terrain with the figures of flat terrain we observe that the random error introduced by slope and microrelief lies around one decimeter for the test areas. The above are examples to demonstrate the feasibility of the theoretical model introduced in this chapter. However, a more systematic analysis of test areas classified according landscape type is necessary to arrive at good insight into the values of the diverse uncertainty factors. This has to be part of future research.

DISCUSSION

The identification of uncertainty factors in DEM data is a complex endeavor, since uncertainty can be approached from at least three viewing points: (1) measuring process, (2) nature of the objects, and (3) the user's view of the world. As stated earlier, the measuring process is usually the best-understood uncertainty aspect of DEM generation.

- The inherent accuracy limitations of the different types of natural phenomena should be the subject of future research. Furthermore, features at the surface of the earth can be described at a multitude of scales. For a certain application domain, only a limited range of scales is adequate to describe that feature. Frequently there is no good basis for choosing the scale of measurement a priori. In addition, it may often be desirable to describe the same terrain surface at more than one scale in the course of interpreting it. Therefore we argue that there should be a difference between the resolution of data acquisition and resolution of data use. Because of the loss of the representativeness aspect, when moving from a manually to an automatically captured DEM, which subject we discussed earlier extensively, the resolution of DEM data use should be coarser than the sampling resolution. Now, we face the problem of finding an appropriate function to integrate neighboring height values into one representative value to arrive at a coarser resolution. This can be established by using a scale-space concept where data are integrated by using a Gaussian weighted averaging technique. The scale reduction is defined by just one parameter: the "standard deviation" of the Gaussian function. It is recommended to carry out experiments to test the suitability of the Gaussian scale-space for DEM representation at coarser resolutions. One of the problems one will face is undoubtedly the multitude of landscape types often present within the area covered by the Gaussian scale-space function.
- The real world contains phenomena or entities that we wish to describe in some format in order to arrive at a better understanding of these phenomena and their relationships. Because of the characteristics of information technology, we are forced to represent these entities as objects in the form of collections of numerical values; the concept of geographic objects is a product of GIS-technology

(Couclelis, 1996). The objects represented in a geoinformation system are representations of conceptual entities that play a role in some descriptive model of the surface of the earth (Molenaar, 1996). To arrive from a real world phenomenon to a description that is suited to be stored in a GIS, some mapping of the real-world to a conceptual world is necessary. The conceptual world is called by CEN the *nominal ground,* sometimes also called *abstract view of the universe* (David et al., 1996). The mapping of the real world onto a conceptual world is called abstraction. The way the abstraction is carried out is user and task domain dependent. A better understanding of how the real world is mapped onto concepts that play a role in the variety of geodisciplines is necessary.

- Due to the numerical representation of real-world phenomena, the models that relate the observable quantities to the required descriptive parameters need to be mathematized; i.e., transferred to expressions, algorithms, and processes that can be handled by computer. The mathematization of models introduces the problem how well these models describe real-world phenomena and how errors in these models propagate through information derived from data sources using these data.

CONCLUSIONS

When the terrain surface is recorded by automatic capturing techniques, such as digital photogrammetry, laser-altimetry, and SAR-interferometry, we lose some important parameters, compared with manual data capturing, which affect the uncertainty of the terrain surface description in a negative sense. On the other hand, automatic capturing techniques enable us to win back some aspects that we lost by collecting more data than strictly necessary for describing the terrain surface. In particular, the possibilities concern improvement of the representativeness of the descriptive points, removal of undesired objects, and probably also the ability to trace breaklines. Furthermore, the numerical storage of elevation data using mass storage devices, allows us to add to the actual height values quality parameters which are important when the data are used in subsequent processes to derive information. The quality parameters may concern signal parameters (strength of texture in digital photogram-

metry or strength of the reflected laser signal received by the sensor), terrain roughness, terrain slope, and terrain coverage. The uncertainty of height capturing of natural phenomena by automatic methods is not well understood yet. More research is necessary to arrive at a better description, both qualitatively and quantitatively, of these uncertainty factors.

REFERENCES

Ackermann, F. Experimental Investigation into the Accuracy of Contouring from DTM, *Photogrammetric Eng. Remote Sensing,* 44(12), pp. 1537–1548, 1978.

Ackermann, F. Techniques and Strategies for DEM Generation, in *Digital Photogrammetry: An Addendum to the Manual of Photogrammetry,* C. Greve, Ed., American Society for Photogrammetry and Remote Sensing, 1996, pp. 135–141.

Ayeni, O.O. Optimum Sampling for Digital Terrain Models: A Trend Towards Automation, *Photogrammetric Eng. Remote Sensing,* 48, pp. 1687–1694, 1982.

Balce, A.E. Determination of Optimum Sampling Interval in Grid Digital Elevation Models (DEM) Data Acquisition, *Photogrammetric Eng. Remote Sensing,* 53(3), pp. 323–330, 1987.

Carrara, A., G. Bitelli, and R. Carla. Comparison of Techniques for Generating Digital Terrain Models from Contour Lines, *Int. J. Geogr. Inf. Sci.,* 11(5), pp. 451–473, 1997.

Couclelis, H. Towards an Operational Typology of Geographic Entities with Ill-Defined Boundaries, in *Geographic Objects with Indeterminate Boundaries,* P.A. Burrough and A.U. Frank, Eds., Taylor & Francis, London, Bristol, 1996, pp. 45–55.

David, B., M. van den Herrewegen, and F. Salgé. Conceptual Models for Geometry and Quality of Geographic Information, in *Geographic Objects with Indeterminate Boundaries,* P.A. Burrough and A.U. Frank, Eds., Taylor & Francis, London, Bristol, 1996, pp. 193–206.

Frank, A.U. The Prevalence of Objects with Sharp Boundaries in GIS, in *Geographic Objects with Indeterminate Boundaries,* P.A. Burrough and A.U. Frank, Eds., Taylor & Francis, London, Bristol, 1996, pp. 29–40.

Gao, J. Resolution and Accuracy of Terrain Representation by Grid DEMs at a Micro-Scale, *Int. J. Geogr. Inf. Sci.,* 11(2), pp. 199–212, 1997.

Huising, E.J. Application of Laser Altimetry to Coastal Zone Management, in *Airborne Laser Scanning—Ein neues Verfahren zur berührungslosen Erfassung der Erdoberfläche,* Hansa Luftbild Symposium Münster, 1996.

Lemmens, M.J.P.M. Accurate Height Information from Airborne Laser-Altimetry, in *Remote Sensing: A Sci-*

entific Vision for Sustainable Development (Proc. IGARSS '97), T.I. Stein, Ed., Piscataway, NJ, 1997, pp. 423–426.

Li, Z. Variation of the Accuracy of Digital Terrain Models with Sampling Interval, *Photogrammetric Rec.,* 14(79), pp. 113–128, 1992.

Li, Z. Theoretical Models of the Accuracy of Digital Terrain Models: An Evaluation and Some Observations, *Photogrammetric Rec.,* 14(82), pp. 651–660, 1993.

Molenaar, M. A Syntactic Approach for Handling the Semantics of Fuzzy Spatial Objects, in *Geographic Objects with Indeterminate Boundaries,* P.A. Burrough and A.U. Frank, Eds., Taylor & Francis, London, Bristol, 1996, pp. 207–224.

Vaessen, E.M.J. A Qualitative Comparison of DEM Data Capturing Techniques (in Dutch), *Geodesia,* 39(11), pp. 483–490, 1997.

CHAPTER 48

Quality Control and Validation of Point-Sourced Environmental Resource Data

A.D. Chapman

INTRODUCTION

Databases of plant and animal specimens have been in existence for centuries in the form of dried plant specimens in herbaria or preserved specimens in museums. It is estimated that there are over three billion records worth around $135–150 billion held in this form around the world (unpublished OECD report). The costs of replacing these data with new surveys would be prohibitive. It is not unusual for a single survey to exceed $1 million (Burbidge, 1991). They are an essential resource in any effort to conserve the environment, as they provide the only fully documented historical record of occurrence of species in areas that may have undergone extensive vegetation change. The habitat from which these specimens were collected may no longer exist, having been cleared for urbanization, agriculture, or been modified in some other way. Even though the collections have rarely been made in a systematic fashion, they are unique in that they cannot be recollected or obtained from any other source. The vastness of the resource may be indicated by the fact that 200 years of biological collecting across Australia has created some 35 million records, which translates to just over 4 collections per square kilometer. We can't afford to not make this resource available in electronic form and to make it available to environmental decision makers and managers wherever they are.

It has only been the introduction of computer processing and, more recently, the ability to store large amounts of data relatively cheaply, that these databases have become more readily available for uses other than just taxonomy. Many of these records, however, carry little geographic information other than a general description of the location where they were collected (Chapman and Milne, 1998). Geographic codes (e.g., longitude and latitude) that define data records for use in distributional studies are seldom given on the original specimen labels. These codes are usually added at a later date, generally at the time of entering the data into an electronic database (Chapman, 1992). It essential that the geocodes be included in the final databases for use in environmental planning and decision making.

Types of Error and Causes

Adding locality codes to the data long after the collecting event can produce various kinds of error. For example, many of the place names used in historical collections no longer exist. Many collections only include very broad locality information (e.g., "Tasmania," "Nova Scotia"). In setting up its species databases at the Australian Environmental Resources Information Network (ERIN), an extra field was added to record location accuracy. Another source of error is that one place-name may refer to several different localities and is thus easily misapplied. For example, there are hundreds of "Stony Creeks" in Australia. It is also easy for recorders to misread latitude and longitude from a map or to confuse these with UTM coordinates. Another common source of error is the accidental swapping or transposition of characters; for example, 23 degrees South entered as 32 South or even more commonly as 34 degrees. All of the above errors have occurred in most data sets received at ERIN and have needed correction before analysis proceeded.

DATA QUALITY

Checks carried out at ERIN on data received from a range of biological institutions indicate considerable error in supplied geographic information (up to 18% in some cases) (Chapman and Busby, 1994). If one is going to base important environmental decisions on the available data; for example, the placement of a reserve to protect particular endangered species, then there is a requirement for the data to be accurate. Many of the collections involved have not been collected with databasing in mind, and many of them were collected so long ago that one cannot go back to the collector and check the facts.

It is also the case that we are dealing with large amounts of data. If one is loading a million specimen records it is not possible to check each record individually against a map or secondary resource. The cost of databasing a collection can be substantial (Armstrong, 1992). Few institutions can afford to carry out further detailed checking of data once they are in a database and thus the accuracy of the geographic information is often not checked. Because of this, automated validation tests have been developed to identify possible "suspect" records. Other data may be in error through misidentification and thus refer to a different species than what is expected.

Data Validation Tests

Apart from the general validation tests that are carried out as data are entered into a database—things such as "are the data in the right format?," "is the date of collection a realistic date?" etc., checks are needed on the "what" and "where" of the specimen. These tests look at the names used for the record and the geographic location of the record.

Taxon Validation

Australia is in a unique position as a continent in that it has a detailed bibliographic index of vascular plant names, the *Australian Plant Name Index* (Chapman, 1991a–d), maintained as an on-line database by the custodian, the Centre for Plant Biodiversity Research, and a census of vertebrate animals, the *Census of Australian Vertebrate Species*—also maintained as a database (ABRS, 1992; Just, 1997).

The ERIN database has been designed to include protocols to test incoming specimen "taxon name" fields against the current names in the bibliography and census. Where a mismatch occurs, the specimen is flagged and then the user can determine if he or she wishes to use the record in their analysis or not. For example, the species may be an as-yet undescribed species and thus still be a valid record for use in an analysis. Error reports for any mismatched records are returned to the data custodian for checking.

These tests identify a number of errors that relate to the name used for the specimens. It particularly identifies names that are spelled incorrectly, names that have changed due to a recent revision, specimens that have been given the wrong name due to a misidentification, etc.

Geographic Validation

ERIN has developed several procedures for checking geographic information (Chapman and Busby, 1994). As mentioned above, herbarium and museum specimens carry very little geocode information other than a free text location of collection. Latitude and longitude information, however, is essential if the data are to be used in a Geographic Information System (GIS).

The validation procedures developed at ERIN involve several steps. The first one checks that the latitude and longitude are not null, that they fall within the range of values for latitude and longitude, and that the direction is valid for the data set. Other checks are described in detail below.

Offshore Testing Procedures

The latitude and longitude values in the data are checked against a 1/40th of a degree Digital Elevation Model (DEM) (using the Oracle®[1] database) and the coastline to determine, for each record, if the record is offshore, on the Australian mainland, or on an offshore island. At the same time, an SQL script uses the DEM to place an altitude value in the altitude field if that field is null; the altitude is an essential element in the next step. This is a rapid process and thousands of records can be checked in just a few minutes. If the record is determined to be offshore it gets tagged with an "O" in the Data Quality Indicator field in the data-

---

[1] Registered trademark of the Oracle Corporation.

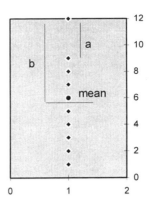

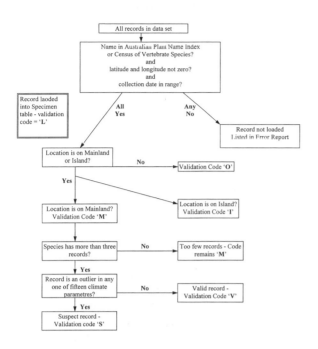

Figure 48.1. Validation diagram showing order of validation events and allocation of codes.

Figure 48.2. a = distance between record and its nearest neighbor, b = distance between record and the mean.

base. Records determined to fall on an offshore island get tagged with an "I" and the remainder—mainland records are tagged with an "M" (Figure 48.1).

Because the resolution of the data is such that many records have an accuracy of only about 5 or 10 minutes of latitude or longitude, such a precise delineation can sometimes be misleading. A record can be tagged as an offshore record, when in reality, the record is only offshore because its locality information is recorded only to those nearest 5 or 10 minutes. Refinements to the program, not introduced yet, use a 9 second DEM (resolution about 280 meters) and the Euclidean distance and Euclidean direction in ArcInfo™[2] GRID to determine how far a record is offshore in a number of possible categories—0.0025° (ca 280 meters), 0.005° (ca. 500 meters), 0.01° (ca.1,000 meters) and 0.017° (ca. 1.8 km or 1 minute of latitude), and to determine if a record is nearly equidistant from two islands, or the mainland and an island, to which that record should be moved within the limits of its accuracy. Different tags would be used for each category.

The next step looks for offshore records as opposed to records that are on the mainland or an island (Figure 48.2). The third step checks that the altitude is in the correct range. A final step uses species climate profiles generated by a bioclimatic modeling system (BIOCLIM) to look for outliers in any of 15 climate parameters.

Climate Outlier Detection

It is relatively easy to map the records for a particular species and to recognize that a number of records that should be on the mainland, are in fact out to sea and vice versa. This merely indicates, however, that a number of other records occurring on the mainland may also be in error. These records are far more difficult to identify, leading to the development in ERIN of techniques to identify and flag them. Thus records that are tagged with an "M" (see above) undergo a second step in the validation process. Of course, not all the spatial errors will be detected in this way as some will fall within the acceptable range of the test (i.e., the climate on both the east and west coasts at the same latitude may be quite similar). These records are identified at a later stage following mapping of the records.

BIOCLIM is a bioclimatic analysis and prediction system that has been successfully used in Australia and a number of other countries for modeling the potential distribution of plants and animals using climate profiles (Busby, 1986; Nix, 1986; Busby, 1991). A new module has been added at ERIN which utilizes the climate profiles generated by the modeling pro-

---

[2] Registered trademark of the Environmental Systems Research Institute, Inc.

gram to detect outliers. BIOCLIM has also been used in earlier studies for detecting possible outliers by excluding records that fall outside the 90 percentile range of the climate profile for the taxon (Busby, 1991), or by using cumulative frequency curves (Lindenmeyer et al., 1991). These techniques, however, do not allow for taxa which may not include any genuine outliers, or that include many outliers. They also are suspect for very small sample sizes.

BIOCLIM uses estimates of monthly mean maximum and minimum temperatures based on actual values from nearly 1,000 stations and estimates of monthly mean precipitation based on 15,000 stations across Australia (Hutchinson and Bischoff, 1983; Hutchinson, 1995; Nix, 1986). From each of these, BIOCLIM generates 36 primary climatic attributes, i.e., 12 monthly values for each of minimum and maximum temperatures and precipitation. These are then used in conjunction with a one-fortieth of a degree Digital Elevation Model (DEM) to produce a climate surface for Australia and to produce 16 climatic indices considered to have biological significance. These are:

- annual mean temperature
- minimum temperature for the coolest month
- maximum temperature for the warmest month
- annual temperature range
- mean temperature for the coolest quarter
- mean temperature for the warmest quarter
- mean temperature for the wettest quarter
- mean temperature for the driest quarter
- annual mean rainfall
- rainfall of the wettest month
- rainfall for the driest month
- coefficient of variation of monthly rainfall
- rainfall for the wettest quarter
- rainfall for the driest quarter
- rainfall for the coolest quarter
- rainfall for the warmest quarter

Recent versions of BIOCLIM—marketed as ANUCLIM (McMahon et al., 1995; Hutchinson et al., 1997), include many more indices (27 rainfall, temperature, and radiation related and 8 associated with soil moisture retention) and can use finer scale DEMs (Hutchinson and Dowling, 1991). Climate surfaces for use in BIOCLIM have been set up for Southern Africa (Hutchinson et al., 1996), Madagascar, China, Indonesia, Papua New Guinea, North America, and a number of other countries (M.F. Hutchinson, personal communication).

Using these 16 parameters (with the exception of the coefficient of variation of monthly rainfall) along with the latitude, longitude, and altitude or elevation for all available specimen records for the taxon, a climate profile is built for the taxon (Tables 48.1 and 48.2).

A reverse of a statistical jackknifing procedure (Barnett and Lewis, 1978) is applied to emphasize (rather than reduce) the effect of marginal records in climate space, leading to critical values being obtained for each specimen for each climate index.

The difference between each record and its neighbor is calculated (Figure 48.3). This is then multiplied by the distance between the mean and the outer record (i.e., for records less than the mean, the lower of the two records is used and for records larger than the mean, the higher of the two records is used). The result is divided by the standard deviation to give the critical value (C). If C is greater than the Threshold Value (T) for that number of records (Figure 48.2), then the record is regarded as an outlier and flagged as "suspect." The method is used to identify an unknown number of outliers at both the top and bottom of the array. Other methods exist for identifying either a known number of outliers at the top and bottom of an array, or an unknown number at either the top or the bottom of an array, but not both. Thus:

$$x < \bar{x}$$

if

$$y_{(i)} = \left(x_{(i+1)} - x_{(i)}\right)\left(\bar{x} - x_{(i)}\right)$$

else

$$y_{(i)} = \left(x_{(i+1)} - x_{(i)}\right)\left(x_{(i+1)} - \bar{x}\right)$$

then

$$C = \frac{y_{(i)}}{\sqrt{\frac{\sum_{i=1}^{n}\left(y_{(i)} - \bar{y}\right)^2}{n-1}}}$$

where C = Critical Value.

Once the critical value is obtained, it is examined in relation to a threshold value curve. The formula

Table 48.1. Temperature Profile for *Eucalyptus whitei* in Degrees Celsius.

	Annual Mean Temp.	Min. Temp. Coolest Month	Max. Temp. Warmest Month	Annual Temp. Range	Mean Temp. Coolest Quarter	Mean Temp. Warmest Quarter	Mean Temp. Wettest Quarter	Mean Temp. Driest Quarter
Mean	22.8	8.3	34.6	26.3	17.0	27.4	27.1	18.5
Min	1.5	1.5	2.5	3.0	1.5	1.7	1.8	1.5
5 Percentile	16.6	3.4	28.2	18.6	10.4	22.2	22.2	11.5
25 Percentile	20.0	5.4	28.7	19.4	14.6	23.6	23.1	16.2
50 Percentile	21.8	7.6	34.1	25.6	16.5	27.2	26.8	18.0
75 Percentile	23.6	9.0	36.1	27.8	17.8	28.3	28.2	19.3
95 Percentile	24.4	10.6	36.8	29.8	19.5	29.0	28.9	20.6
Max	25.4	12.8	38.2	30.3	20.6	30.9	30.1	21.8

Table 48.2. Rainfall Profile for *Eucalyptus whitei* in millimeters.

	Annual Mean Rainfall	Rainfall Driest Month	Rainfall Wettest Month	Rainfall Wettest Quarter	Rainfall Driest Quarter	Rainfall Coolest Quarter	Rainfall Warmest Quarter
Mean	658	142	88.3	381	39	51	337
Min	326	68	15.6	203	26	32	111
5 Percentile	390	78	46.2	204	0	4	154
25 Percentile	439	87	56.4	230	15	20	228
50 Percentile	479	103	78.9	273	26	37	270
75 Percentile	705	151	93.4	394	42	53	348
95 Percentile	1402	319	119.0	896	83	94	589
Max	1982	392	126.5	1154	141	181	902

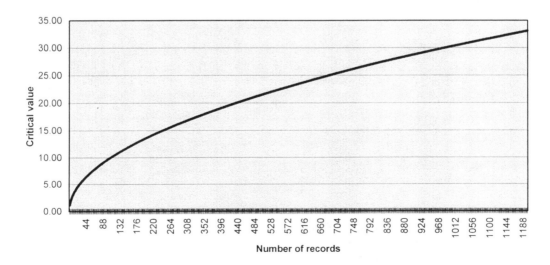

Figure 48.3. Threshold Value Curve. Values above the curve are regarded as "suspect," values below the line as "valid."

for this curve was determined after extensive experimentation with species with varying numbers of records.

The threshold value curve is based on the formula

$$T = 0.95\left(\sqrt{n}\right) + 0.2$$

where n = number of records.

The value is multiplied by 0.95 to extract a few more records than otherwise would be the case. This is because it is important not to miss possible "suspect" records, and we would rather identify a few more good records as suspect, than miss suspect records by their being identified as "valid." The 0.2 is added to allow identification of suspect records for species with few records.

From the graph—any record that falls above the line for any parameter is regarded as "suspect"—below the line as "valid" (Figure 48.2).

If the critical value is outside the appropriate range for that sample size, the record is regarded as "suspect" and is flagged with an "S" in the data quality field and copies returned to the custodian for further checking (Figure 48.1). This procedure not only identifies possible suspect records that may have an aberrant latitude or longitude, it also identifies errors in altitude, and occasionally, misidentifications of the name of the taxon. Tests have been successfully carried out on sample sizes ranging from 3 to over 5,000.

In some cases, records identified as "suspect" may, of course, not be in error. This is usually the case when dealing with species with small sample sizes, e.g., endangered and vulnerable species, or species with large climatic distributions.

One area where this technique does fall down is for species that occur fairly evenly on either side of the summer/winter rainfall isohyet where all records may be identified as "suspect," even for large sample sizes. This is because parameters such as rainfall of the warmest and coolest quarters and temperature of the wettest and driest quarters will have opposing effects. Those records in the summer rainfall zone will have high values for rainfall of the warmest quarter and temperature of the wettest quarter and low in the others, while those records in the winter rainfall zone will have high records for rainfall of the coolest quarter and temperature of the driest quarter and low in the others.

An example of the application of the technique can be shown using *Eucalyptus whitei*, an Australian gum tree. A number of records from several Australian her-

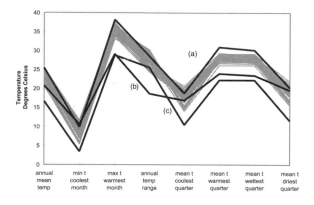

Figure 48.4. Temperature chart for *Eucalyptus whitei*. Dark lines show three suspect records—see text for explanation, light lines show '98 valid records.

baria were run through BIOCLIM to determine a climate profile (Tables 48.1 and 48.2). The records obtained have been graphed against various temperature (Figure 48.4) and rainfall (Figure 48.5) attributes using a standard spreadsheet package.

From this analysis, it can be seen that three records are considerably out of step with the others for several temperature or rainfall attributes. These are graphed in the darker color here for emphasis. All three fall above the line (Figure 48.2) and are thus flagged as "suspect." When the locality information supplied with the records is examined and mapped using ArcInfo™ GIS (Figure 48.6), it shows that the suspect records (a–c) have the following information:

(a) The latitude and longitude are –20°51′, 140°44′, respectively, and the locality—"56 km E of Hughenden toward Pentland."
(b) The altitude value is "1,000" meters.
(c) The latitude and longitude are –16°17′ and 145°05′, respectively, and the altitude 600 meters. It was collected in a State Forest.

Further examination of the records shows that for record:

(a) the longitude for that locality should be 144°44′,
(b) the altitude for that locality should be less than 450 meters and has possibly been recorded in feet rather than meters,
(c) may be a valid outlier record, but taxonomists are examining the record to check if it

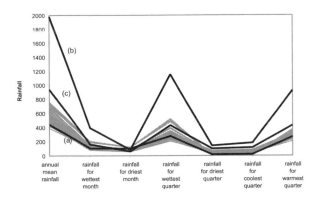

Figure 48.5. Rainfall chart for *Eucalyptus whitei*. Dark lines show three suspect records—see text for explanation, light lines show '98 valid records.

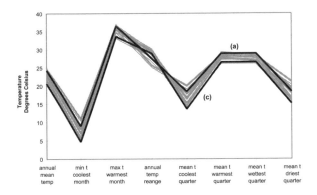

Figure 48.7. Temperature chart for *Eucalyptus whitei* after correction of suspect records. Dark lines show previously suspect records—see text for explanation, light lines show '98 valid records. Previously suspect record (b) removed due to possible misidentification (cf. Figure 48.4).

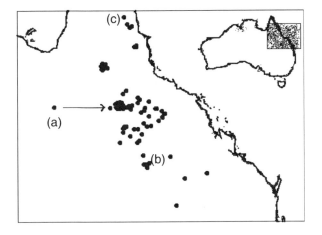

Figure 48.6. Localities for *Eucalyptus whitei*. (a), (b), and (c) show localities of suspect records, arrow indicates the correct locality for specimen (a).

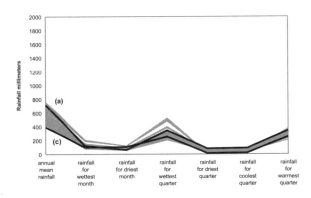

Figure 48.8. Rainfall chart for *Eucalyptus whitei*. Dark lines show previously suspect records—see text for explanation, light lines show 98 valid records. Previously suspect record (b) removed due to possible misidentification (cf. Figure 48.5).

may be a possible misidentification or new species.

Once these records have had their locality information corrected, and the species rerun through the program, it can be seen (Figures 48.7 and 48.8) that there are now no longer any species identified as suspect. Record (c) was removed from this stage as possibly belonging to another species.

This technique is used extensively on data sets as they are loaded into the ERIN Specimen Module. Reports are generated for return to the custodian (Sanford-Readhead, 1992), and maps prepared showing where

the suspect records occur (Sanford-Readhead, 1992; Chapman and Busby, 1994).

Assumptions and Limitations

Although about 85% of records identified as "suspect" have later proved to include actual errors, there are a number of limitations to these procedures. The whole procedure assumes that the locations of plant and animal species are closely tied to the climate parameters included in BIOCLIM. This is a reasonable assumption, based on a large number of experiments conducted since the program was first introduced in 1964 (see Chapman and Busby, 1996). Limitations include lack of consistency when dealing with small sample sizes, problems arising with species occurring across the summer/winter rainfall isohyet, and occasionally species with very disjunct distributions. Plant and animal collections in Australia (as elsewhere in the world) include significant collection bias—traditionally collections are located within the vicinity of roads, or near areas of high population. This effect is probably less in Australia than elsewhere, however, due to the flatness of the land and lack of steep climatic gradients within the country, meaning that roads often act as good transects. The unevenness of climate stations, particularly in the tropical north of the continent, causes some inconsistency in those areas, particularly the high rainfall areas of the Cape York Peninsula.

USE OF VALIDATION CODES IN ANALYSIS

Once the data have been validated, it is important to ensure that users of the data use only validated records and, unless there is a specific reason for not doing so, only those records identified as "valid." In ERIN, tools are being developed that extract the data straight from the Oracle™ database to the ArcView™[3] GIS. These are done through the use of standard SQL (structured query language) scripts.

The data are being used extensively in analysis within the Australian Environment Department, Environment Australia, for example, to help determine the adequacy of reserve systems (Bull et al., 1993; Thackway, 1991), in environmental impact assessment, in comprehensive regional assessment to develop forest agreements (RFA, 1997), in the conservation assessment of endangered and vulnerable species (Briggs and Leigh, 1995), in the assessment of migratory patterns for birds and other animals, in wildlife trade monitoring, and for studies on the impact of climate change (Chapman and Milne, 1998).

ERIN has also developed a number of data presentation tools, including interfaces to the database using the World Wide Web. The scripts for extracting this information again only allow for extraction of "valid" records. These include a species mapping and modeling system—Species Mapper (ERIN, 1993) and a pro-

totype map presentation tool—Enviromaps—also using species data (ERIN, 1997).

CONCLUSION

The procedures outlined above have several advantages where data are compiled into a single database such as at ERIN. For example, with endangered species, the total number of records is extremely low and the number in any one institution is generally so small as to make any statistical procedure unreliable (Chapman and Milne, 1998). Also, the geographic extent of a species' range in any one institution is unlikely to reflect the total range for the species. For example, it is likely that a state-based institution has obtained the majority of their records from within that state. The resultant climate profile generated from BIOCLIM would thus be skewed and identify otherwise valid records as suspect and vice versa. Through the process of supplying suspect record reports back to the data custodians, the quality of the nation's species data has been improved considerably.

One of the modifications being investigated is a requirement to allow for data custodians to carry out analysis on their own data, perhaps through the use of distributed database technology (Chapman and Croft, in press). The ERIN Unit is examining possible ways of doing this using the resources of the World Wide Web and database searching using the Z39.50 Protocol (NISO, 1992).

ACKNOWLEDGMENTS

Many of the staff of the Environmental Resources Information Network have contributed ideas on data validation techniques since its inception in 1989. Particular thanks must be given to Kate Sanford-Readhead, who turned many of my ideas into workable programs, and Gaston Rozenbilds and John Busby, who contributed ideas for modification and improvements to the process. Data for the projects for which the program was developed came from a large number of Commonwealth, State, and Territory museums, herbaria, and wildlife agencies as well as a number of nongovernment agencies and individuals. Their support and contribution is gratefully acknowledged. Gaston Rozenbilds, Richard Thackway, and Lynne Woodcock of the ERIN Unit reviewed drafts of the manuscript and provided valuable comments.

---

[3] Registered trademark of the Environmental Systems Research Institute, Inc.

REFERENCES

Armstrong, J.A. The Funding Base for Australian Biological Collections, *Aust. Biol.,* 5(1), pp. 80–88, 1992.

Australian Biological Resources Study. *Census of Australian Vertebrate Species (CAVS) Version 8,* unpublished list and database, ABRS, Canberra, 1992.

Barnett, V. and T. Lewis. *Outliers in Statistical Data,* John Wiley & Sons, Chichester, 1978, p. 365.

Briggs, J.D. and J.H. Leigh. *Rare or Threatened Australian Plants,* CSIRO Publishing, Melbourne, 1995, p. 446.

Bull, A.L., R. Thackway, and I.D. Cresswell. *Assessing Conservation of the Major Murray-Darling Basin Ecosystems, Towards a Conservation Strategy for the Conservation of Ecosystems and Endangered and Vulnerable Species: Phase 3,* Draft final report to the Murray-Darling Basin Commission, Environmental Resources Information Network, Australian National Parks and Wildlife Service, Canberra, Report Number 1, 1993.

Burbidge, A.A. Cost Constraints on Surveys for Nature Conservation, in *Nature Conservation: Cost Effective Biological Surveys and Data Analysis,* C.R. Margules and M.P. Austin, Eds., CSIRO, Canberra, 1991.

Busby, J.R. A Biogeoclimatic Analysis of *Nothofagus cunninghamii,* (Hook.) Oerst, in Southeastern Australia, *Aust. J. Ecol.,* 11, pp. 1–7, 1986.

Busby, J.R. BIOCLIM—A Bioclimatic Analysis and Prediction System, in *Nature Conservation: Cost Effective Biological Surveys and Data Analysis,* C.R. Margules and M.P. Austin, Eds., CSIRO, Canberra, 1991, pp. 64–68.

Chapman, A.D. Australian Plant Name Index A-C, *Austral. Fl. and Fauna Ser.,* No. 12, Australian Government Publishing Service, Canberra, 1991(a), pp. 1–898.

Chapman, A.D. Australian Plant Name Index D-J, *Austral. Fl. and Fauna Ser.,* No. 13, Australian Government Publishing Service, Canberra, 1991(b), pp. 899–1710.

Chapman, A.D. Australian Plant Name Index K-P, *Austral. Fl. and Fauna Ser.,* No. 14, Australian Government Publishing Service, Canberra, 1991(c), pp. 1711–2476.

Chapman, A.D. Australian Plant Name Index Q-Z, *Austral. Fl. and Fauna Ser.,* No. 15, Australian Government Publishing Service, Canberra, 1991(d), pp. 2477–3055.

Chapman, A.D. Quality Control and Validation of Environmental Resource Data, in *Data Quality and Standards: Proceedings of a Seminar Organised by the Commonwealth Land Information Forum, Canberra, 5 December 1991*, Commonwealth Land Information Forum, Canberra, 1992, p. 20.

Chapman, A.D. and J.R. Busby. Linking Plant Species Information to Continental Biodiversity Inventory, Climate Modeling and Environmental Monitoring, in *Mapping the Diversity of Nature,* R.I. Miller, Ed., Chapman and Hall, London, 1994, pp. 179–195.

Chapman, A.D. and J.R. Croft. Linking Species Diversity Data into the Clearing-House Mechanism Under the Convention on Biological Diversity, in *Species Diversity: Global Data Systems,* K.L. Wilson and F.A. Bisby, Eds., Springer-Verlag, Berlin, in press.

Chapman, A.D. and D.J. Milne. *The Impact of Global Warming on the Distribution of Selected Australian Plant and Animal Species in relation to Soils and Vegetation,* Environmental Australia, Canberra, 1998, p. 146.

Environmental Resources Information Network (ERIN). Species Mapper, 1993 [published electronically at http://www.environment.gov.au/search/mapper.html].

Environmental Resources Information Network (ERIN). Enviromaps, 1997 [published electronically at http://www.environment.gov.au/cgi-bin/enviromaps/enviromaps.pl].

Hutchinson, M.F. Interpolating Mean Rainfall Using Thin Plate Smoothing Splines, *Int. J. GIS,* 9, pp. 385–403, 1995.

Hutchinson, M.F. and R.J. Bischoff. A New Method for Estimating the Spatial Distribution of Mean Seasonal and Annual Rainfall Applied to the Hunter Valley, New South Wales, *Austral. Met. Mag.,* 31, pp. 179–184, 1983.

Hutchinson, M.F. and T.I. Dowling. A Continental Hydrological Assessment of a New Grid-Based Digital Elevation Model of Australia, *Hydrol. Process.,* 5, pp. 45–58, 1991.

Hutchinson, M.F., D. Houlder, H.A. Nix, and J.P. McMahon. *ANUCLIM Version 1.5 User's Guide,* Centre for Resource and Environmental Science, Australian National University, Canberra, 1997, [also published electronically at http://cres.anu.edu.au/software/anuclim.html].

Hutchinson, M.F., H.A. Nix, J.P. McMahon, and K.D. Ord. The Development of a Climate and Topographic Database for Africa, *Third International Conference/Workshop on Integrating GIS and Environmental Modelling*, NCGIA, University of California, Santa Barbara, 1996.

Just, J. *Census of Australian Vertebrate Species (CAVS) Version 8.1.,* Unpublished list and database, ABRS, Canberra, 1997, [also published electronically at gopher://www.anbg.gov.au:8080/00/publprog/abcavs].

Lindenmeyer, D.B., H.A. Nix, J.P. McMahon, M.F. Hutchinson, and M.T. Tanton. The Conservation of Leadbeater's Possum, *Gymnobelideus leadbeateri* (McCoy): A Case Study of the Use of Bioclimatic Modelling, *J. Biogeog.,* 18, pp. 371–383, 1991.

McMahon, J.P., M.F. Hutchinson, H.A. Nix, and K.D. Ord. *ANUCLIM User's Guide Version 1*, Centre for Resource and Environmental Studies, Australian National University, Canberra, 1995.

National Information Standards Organization (NISO). *Information Retrieval Service Protocol for Open Sys-*

tems Interconnection (ANSI Z39.50–1992), NISO, Bethesda, MD, 1992.

Nix, H.A. A Biogeographic Analysis of Australian Elapid snakes, in *Atlas of Australian Elapid Snakes,* R. Longmore, Ed., *Austr. Fl. Fauna Ser.,* 8, pp. 4–15, 1986.

Commonwealth and Victorian Government Regional Forest Agreements (RFA) Steering Committee. *Central Highlands Comprehensive Regional Assessment. Biodiversity Technical Report,* Canberra: Commonwealth of Australia, 1997, [also published electronically at http://www.environment.gov.au/land/forests/cra/vic/cenhigh/biodivers/contents.html].

Sanford-Readhead, K. Automated Production of Specimen Validation Maps Using ARC/INFO Version 6 and ORACLE, *OZRI6, Adelaide '92 Conference Program,* Session 3, 1992.

Thackway, R. *Assessing Conservation of the Major Murray-Darling Basin Rcosystems. Development of a Data Base of the Distribution of Endangered and Vulnerable Species: Phase 1,* Final report to the Murray-Darling Basin Commission, Environmental Resources Information Network, Australian National Parks and Wildlife Service, Canberra, Report Number 1, 1991.

CHAPTER 49

Preserving Spatial and Attribute Correlation in the Interpolation of Forest Inventory Data

M. Moeur and R. Riemann Hershey

INTRODUCTION

On-the-ground inventories can be used to collect information on forest composition. But an exhaustive sample is unrealistic. An alternative approach is to "map" forest composition by interpolation from sample data. We know that environmental factors, topographic attributes, and land-use patterns at different spatial scales have influenced present-day forest vegetation—its distribution in space, amounts, and species co-occurrences. In this study we emphasize both the spatial structure as well as attribute structure inherent in broad-scale patterns of forest species composition. We define spatial structure as the presence of nonzero spatial autocorrelation computed as semivariance or correlogram functions. We define attribute structure by the relationship between species and related covariates measured at a location. We argue that more accurate models of forest conditions across landscapes can be provided by making use of both spatial correlation and attribute correlation inherent in data describing environmental systems. To this end we demonstrate combined approaches to modeling that preserve to varying degrees both spatial and attribute structure of ecosystem data in a single analysis.

We use two very different modeling approaches in this study: geostatistical simulation (GS), and Most Similar Neighbor analysis (MSN). Each was developed independently to build interpolated data layers, or "maps" of environmental data (here, forest species composition). GS and MSN differ radically in many respects. However, for practical purposes their main distinction is in the way spatial and structural infor-

mation are processed. Geostatistical procedures estimate the distribution of species across the landscape as a function of observed spatial distribution, but do not consider species' relationships with other species or other covariates. Conversely, MSN analysis explicitly incorporates the correlations between multiple species and related data layers, but ignores their spatial interactions. As a result, interpolations developed from GS preserve spatial structure inherent in the sample data, and interpolations developed from MSN preserve attribute structure.

Sampling requirements for GS are less intensive than for MSN. For GS, minimum data requirements are the species values (here, occurrence and amount) measured on ground inventoried plots at known locations. For MSN, a two-phase sample is required, combining a source of complete coverage data (here, satellite imagery) with the ground inventory sample data. Spatial information may be used to condition MSN estimates, and attribute information may be used to condition GS estimates, but neither are required.

The goal of this study is to use the procedures to estimate forest species composition at locations where the properties have not been measured. Interpolated species maps and data sets are useful for landscape-level analyses and planning purposes, such as (1) translating inventory data into a more easily available spatial context for analysis with other spatially referenced data, (2) helping analysts better understand the pattern and distribution of the resource, and (3) providing data sets which can be queried to suit the individual management or research question at hand. In order to do this effectively, we require interpolation techniques that output a modeled estimate, a measure of the uncertainty associated with that estimate, and some indication of the local variability.

In this chapter we compare interpolated maps and data sets derived from GS and MSN techniques, and then propose a partial solution to the problem of simultaneously maintaining both spatial and attribute structure when interpolating species composition. The main objectives are to:

1. Demonstrate the application of independent GS and MSN analyses to a common data set, and compare how well each maintains spatial and attribute structure of the sample data.
2. Demonstrate approaches that combine some elements of the GS and MSN techniques, and measure how the results change.

In assessing the accuracy of the resulting data sets, we look at how well the procedures maintain insofar as possible the univariate, joint distribution, spatial, and local characteristics of the sample data. The better we maintain spatial and attribute structure, the more realistic our interpolations. More reliable interpolations can be used to better address questions of interest to managers and researchers. Examples include questions of location (e.g., where is the highest potential for managing for a species), co-occurrence (e.g., what effect do management practices have on multiple species occurrence), and amount (e.g., in how much land area does a species dominate).

DATA AND METHODS

Our interpolation frame is a 222×204 km area in the Finger Lakes region of New York State covered by Landsat Scene 15/30 (May 2, 1992) (Figure 49.1). The species values are obtained from sample data collected on 1,250 inventoried plots by the USDA Forest Service Forest Inventory and Analysis (FIA) unit (Hansen et al., 1992). They consist of the basal area of individual species (summed cross-sectional area of trees at breast height in ft^2/acre) and relative "importance value," defined as species basal area per acre as a percent of total basal area per acre (% ba/acre). The FIA plot locations are recorded to an accuracy of 100 m. The average distance between FIA plots in the New York scene is about 6 km. Sampling probabilities differ for forested areas (in New York, 1 plot for every 20 km^2) and nonforested areas (1 plot for every 32 km^2).

Because we are interested in the characteristics of forested areas, combined with the fact that nonforest information is more accurately obtained from other sources of information, the GS routines used only forested plots (n=713) in the modeling and estimation. Because we used the entire TM scene in the MSN analyses, including both forested and nonforested areas, all plots (n=1250) were used in the MSN runs.

In addition to the inventoried species data, MSN requires an independent data source spatially referenced to the FIA data. The global data source used in the MSN interpolations is obtained from Landsat Thematic Mapper satellite imagery, and consists of raw spectral band data measured on 30m $\times$ 30m pixels, and tasseled cap (Crist and Cicone, 1984) and Normalized Difference Vegetation Index (Reed et al., 1994) transformations. Because of the locational uncertainty associated with the plots, the TM band values and transforms were averaged using a 7×7 pixel window (.044 km) surrounding the plot location before performing the overlay.

We interpolated the occurrence and relative importance (% ba/acre) of five major tree species on a 2-km grid over this scene ($111 \times 102 = 11322$ interpolation points). The species are eastern white pine, eastern hemlock, red maple, sugar maple, and American beech. This list captures three of the most common species in the area, and two rarer ones, including several species commonly found together, and representing a range of abundances and distributions.

We compared the GS, MSN, and combined approach results by assessing how well each maintains spatial structure and attribute structure inherent in the sample data. The former is measured by how well the variogram of interpolated results matches the sample variogram. The latter is measured by how well the univariate (cumulative frequencies) and joint distributions (species pair-wise scatterplots and correlations) of the sample data are maintained. We also visually compared interpolated maps for resulting patterns of predicted species occurrence and amounts.

Geostatistical Estimates

The geostatistical simulation (GS) techniques studied were sequential Gaussian conditional simulation (sgCS) with indicator kriging (IK), and sgCS with sequential indicator conditional simulation (siCS). Briefly, the GS techniques make use of the spatial structure present in the data, defined by a model of the indicator variogram or the correlogram, respectively, two

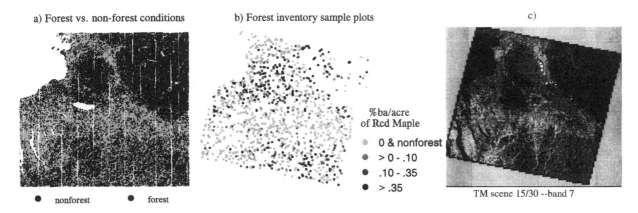

a) Forest vs. non-forest conditions

b) Forest inventory sample plots

c)

%ba/acre
of Red Maple

• 0 & nonforest

• > 0 - .10

• .10 - .35

• > .35

• nonforest   • forest

TM scene 15/30 --band 7

Figure 49.1. (a) Forest vs. nonforest conditions in the 222 x 204 km Finger Lakes scene; (b) locations of 1250 FIA sample plots (n=713 forested, n=537 nonforested) showing relative importance (% ba/acre) of red maple; (c) an example of the global data: Landsat TM band 7 (wavelengths in the mid IR ranges, 2.08 to 2.35 μm).

functions for summarizing spatial continuity which depict the variation between sample data values at increasing distances from each other (Isaaks and Srivastava, 1989). Kriging estimates are essentially weighted moving averages of the original data values, taking the distance, direction, and redundancy of neighboring points into account using the model defined from the variogram, and honoring the overall sample mean. The geostatistical simulation techniques are related to this, but rather than determining a single "best" estimate, conditional simulation generates, via Monte Carlo techniques, many equally probable realizations, resulting in a distribution of values for each cell. Each of the realizations (n=100 in all runs in this study) retains the univariate characteristics and honors the variogram of the original sample data. A summary statistic such as the median of this distribution is then chosen and used as the modeled "estimate" of % ba/ acre for that cell, and another value, such as the interquartile range is chosen as the expression of uncertainty about that estimate (Deutsch and Journel, 1992; Rossi et al., 1993). Previous application of these techniques to the FIA species data is detailed in Riemann Hershey et al. (1997).

The indicator routines make no assumptions about the distribution of the data, and are thus very reliable under many data conditions. Sequential Gaussian conditional simulation, on the other hand, assumes that the data are normally distributed. Species %ba/acre values present a highly skewed distribution. Although a 1–1, invertible normal-score transform (Deutsch and Journel, 1992) is applied to the species importance data prior to the sgCS, the highly skewed character of the

data can still affect results because we cannot ensure that the data are also bivariate normal and higher-moment normal. Separating the data such that the IK or siCS used all forested plots and sgCS used only those plots where a species occurs, improved this multivariate normality of the sample data used by sgCS (Riemann Hershey and Reese, unpublished manuscript).

Sequential Gaussian Conditional Simulation with Sequential Indicator Conditional Simulation (siCS/sgCS)

This is a two-step process, in which sequential indicator conditional simulation (siCS) is used to estimate species occurrence, and sequential Gaussian conditional simulation is used to estimate species % ba/acre for each point in a 2×2-km grid using models of the spatial structure in the indicator variogram and the correlogram, respectively (Deutsch and Journel, 1992; Riemann Hershey et al., 1997). In the first step an indicator transform divides the data into two classes (here, presence or absence), and the resulting 100 realizations thus predict either a "1" (presence) or "0" (absence) for every cell. This information is then fed directly into the sgCS routine. Where the results of siCS indicated species presence at a grid cell, the sgCS then estimates the % ba/acre that would be found there. Where siCS indicated species absence, then a value of 0 %ba/acre is recorded for that cell in that realization. In practical terms, this combines the uncertainties of the two simulation procedures relatively seamlessly. (This approach has been applied for mapping species composition in Maine: see Riemann Hershey and Reese, unpublished manuscript, for a more detailed description of these procedures).

Sequential Gaussian Conditional Simulation with Indicator Kriging (IK/sgCS)

Here, indicator kriging (IK) is used in place of indicator simulation (siCS) in a similar two-step process. Instead of producing 100 realizations that are fed individually into the sgCS routine, IK outputs a single data set containing a probability of species occurrence at each grid cell. A probability cutoff value (of, say, 0.4) is then used simply as a mask on the output of sgCS. Because IK is essentially a kriging vs. a simulation procedure, this approach provides a more smoothed data set of species occurrence, creating a less gradual transition into the "nonoccurrence" areas vs. use of siCS. The IK masking effect does have some influence on the characteristics of the output as a result, but its use can be warranted because it significantly decreases computation time. The IK/sgCS approach has been applied for mapping species composition in Pennsylvania (Riemann Hershey, 1996). In contrast to the next approach, it uses only the inventory plot data to perform the interpolation of species occurrence.

IK/sgCS Using Forest Cover Type "Soft" Information (IKfct/sgCS)

Both IK and siCS can easily incorporate "soft" information as well as the known inventory data into the estimation of unknown areas (i.e., make use of ancillary information that has a known relationship to the variable being estimated). As a first attempt to incorporate attribute structure into the GS routines, we incorporated a common data layer, a map of forest cover type groups, as soft information into the IK routine for each species. The forest cover type data set (northern hardwoods, pines, spruce/fir, aspen/birch, elm/ash, and nonforest) was translated into a grid of species' probability values, based on sampled species' occurrences within cover types in the study area. This additional information is incorporated into the IK as added sample points whose values are the conditional species' probability of occurrence. The intention was that this common data layer would incorporate species cooccurrence information, thus providing attribute structure information to the GS routines. The forest type group data used here was generated via sequential indicator simulation

from the FIA data, and was almost completely dominated by two cover-type groups, northern hardwoods and nonforest. Except for the incorporation of soft forest cover-type information in the IK step, this procedure is analogous to the IK/sgCS method.

MSN Estimates

The MSN approach uses the relationship between the species values at sampled locations, and the global variables measured at the same locations, to interpolate unsampled cells. Briefly, this is accomplished by finding the "most similar" sample plot to a cell based upon its satellite signature, and imputing the species % ba/acre values to the cell. "Similarity" is defined by a multivariate difference function incorporating the relationship among the species values and the satellite data. The MSN difference function between an unsampled cell i and an observation j from the FIA sample (Moeur and Stage, 1995), is given by $D_{ij} = (X_i - X_j) W (X_i - X_j)'$. The inner weight matrix, W, developed by canonical correlation analysis, summarizes the best (i.e., maximally correlated) multivariate linear relationship between the species values ($Y = [\%$ ba/ acre for 5 species]) and the global variables ($X = $ [TM values]). Unsampled cell i's most similar neighbor is found by minimizing D_{ij} over all j FIA plots. Following the most similar neighbor FIA plot selection, the sampled % ba/acre species values are imputed to the grid cell location. The salient result of the MSN procedure, for this study, is that the covariances of the species values and satellite data drive the selection of neighbors.

In previous applications, MSN formulations have excluded location information (Moeur et al., 1995), and thus, we speculate, have not taken full advantage of the information contained in spatially referenced data sets. However, our goal in this study was to design modifications to the interpolations that would make use of both sample data attribute **and** spatial structure. Therefore, we compared three different MSN formulations in this chapter. The first ("spatially naive") develops the MSN estimates from the inventory and global data using no location information. The second ("partially spatially informed") adds to the first variable list the X and Y coordinates of the sample plot and grid cell locations. The third is a combined MSN-GS approach ("fully spatially informed"). It adds to the previous variable list the spatial variogram function (γ) computed as a function of distance (or "lag") between pairs of grid points and sample plots.

Table 49.1. Three MSN Models Compared in the Study.

Name	Inventory Variables	Global Variables
Level 1		
Spatially naive	WP[a] % ba/acre[b]	TM Band 1[c]
	EH % ba/acre	TM Band 2
	RM % ba/acre	TM Band 3
	SM % ba/acre	TM Band 4
	AB % ba/acre	TM Band 5
	Forest/Nonforest (0,1)	TM Band 7
	WP Present? (0,1)	Greenness
	EH Present? (0,1)	Wetness
	RM Present? (0,1)	Brightness
	SM Present? (0,1)	NDVI
	AB Present? (0,1)	
Level 2		
Spatially partially informed	Above variables	Above variables, plus Utm-X[d] Utm-Y
Level 3		
Fully spatially informed (combined MSN-GS)	Above variables	Above variables, plus $\gamma(h)$[e] for WP $\gamma(h)$ for EH $\gamma(h)$ for RM $\gamma(h)$ for SM $\gamma(h)$ for AB

[a] Species codes: WP=eastern white pine, EH=eastern hemlock, RM=red maple, SM=sugar maple, AB=american beech.

[b] Species importance data are transformed using a 1-1, invertible normal-score transform (Deutsch and Journel, 1992).

[c] Landsat Thematic Mapper spectral band data and transformations (Crist and Cicone, 1984; Reed et al., 1994).

[d] X and Y coordinates in Universal Transverse Mercator projection (Zone N18).

[e] Variogram function computed at lag distance h (m).

MSN Level 1: Spatially Naive Model

In all three MSN levels, species importance values (% ba/acre by species), plus indicator variables for species presence/absence and forest/nonforest are used as the inventory variables in the MSN model (Table 49.1). In the level 1 model, the Landsat TM spectral band data and band transformations alone are used as the global data source. As a result, MSN neighbor selections are made to match inventoried data to each interpolated cell location based upon the values of its TM spectral signature alone, without including any location information.

MSN Level 2: Spatially Partially Informed Model

In the level 2 MSN analysis, geographic coordinates (X and Y) are added as additional variables in the global data layer (Table 49.1). In this model, the MSN selection now includes location information in the form of geographical proximity. Thus, distance between an interpolation cell and the FIA plots in both the latitudinal and longitudinal directions is included in the weighting function that determines the neighbor choice. But note that the simple inclusion of geographic proximity does not explicitly incorporate a model of spatial "structure" (i.e., the variogram of species importance values as a function of distance).

MSN Level 3: Fully Spatially Informed Model: Combined MSN-GS

In the combined analysis, variogram information is incorporated explicitly into the MSN model (Table 49.1). This is done by fitting variogram models γ(lag distance) to the sample data for each species, and solving the function $\gamma(h_{ij})$ for each grid point, where h_{ij} is the lag distance in meters between cell i and the jth FIA inventory plot. The effect of including the variograms in the MSN formulation is to weight the neighbor selection according to the distance and direction of the sample plots, in addition to their satellite signatures. The variogram models used for the five individual species were the same models used in the GS approaches.

Uncertainty Estimates

For the interpolated data sets to be useful, they need an uncertainty estimate, or prediction error, to associate with predicted species' values at a cell. Knowledge of this error is important in both the development and use of these output data sets. For example, it affects how much weight to give to different sources of information in both decisions and analyses. Both the GS and the MSN routines produce a measure of the model error associated with the estimate at each location. For the GS simulations, uncertainties are drawn directly from the realizations by specifying an "acceptable" measure of uncertainty. Here, this was defined as the interquartile range of the distribution (25th to 75th percentiles). MSN prediction errors are estimated empirically using FIA observed sample values and their predictions (obtained by choosing the second most similar neighbor to the FIA data, the most similar being the plot itself). The map of MSN prediction error is created using a two-step model of error as a function of calculated MSN normalized distance (Moeur and Stage, 1995). First, a point is calculated along the cumulative distribution of MSN distances over all the grid cells using a logistic function. Then, the prediction error is modeled as a point along the cumulative distribution of distances. The minimum prediction error for a species (% ba/acre) is constrained as the larger of the modeled error and the empirical root mean square error for the FIA observations. The resulting

functions are applied to the predictions for cells on the map, given their calculated MSN distance.

RESULTS AND DISCUSSION

We have chosen two of the five species analyzed to illustrate comparisons between the interpolation methods. Figure 49.1 shows the spatial distribution of forested and nonforested area in the Finger Lakes scene (estimated from photointerpretation), FIA sample plots with red maple importance values, and one view of the Landsat TM data. Features include Lake Ontario (NW quadrant), and the predominantly nonforested urban corridor surrounding Interstate 90 from Syracuse to Albany, and running generally E-W through the middle of the scene. The region is most heavily forested in the NE quadrant (encompassing a portion of Adirondack Park), NE of Oneida Lake (the large lake located SE of Lake Ontario), and in a scattered band running W to E through the southern third of the scene. Red maple is the most commonly occurring species, present on 34% of sample plots, scattered throughout the forested areas. Eastern hemlock occurs on a lower proportion of plots (17%), concentrated in the SE quadrant and in a horizontal band in the upper third. There is little hemlock present on FIA plots in the Adirondack Park area. Red maple and hemlock co-occur on 14% of FIA plots.

Figures 49.2 and 49.3 compare GS and MSN interpolation results on a 2-km grid over the Finger Lakes scene for red maple and hemlock. Examples are shown for some, but not all of the interpolation methods; where not shown, they are discussed in the text.

The IK/sgCS interpolations for red maple and hemlock are notable for two characteristics (Figures 49.2a,b and 49.3a,b). First, the species are predicted to occur in large, contiguous blocks, and outside of these blocks there are almost no isolated pixels with predicted occurrence. This is largely a result of using a kriging routine instead of a simulation routine for estimating species occurrence. Second, the percentiles chosen for summarization of the individual realizations (50th and 70th percentiles for red maple and hemlock, respectively) necessarily smooth the final estimates. This is particularly true in areas with a high degree of local spatial variability at the current scale of sampling. We realize this is a spatially summarized, but still useful view of species composition, as long as the limitations of the approach and data set are understood. Occurrence and amounts predicted by IK/sgCS overlay nicely with patterns of forested areas and FIA samples (com-

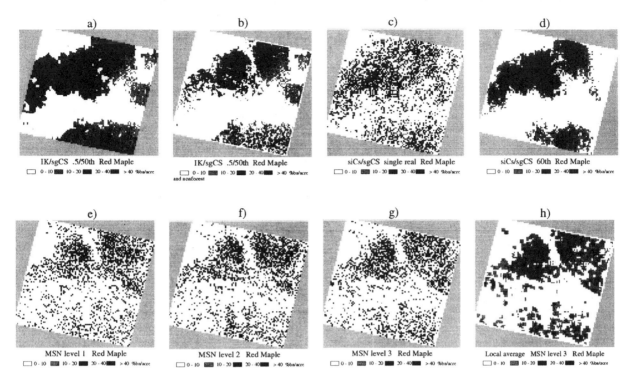

Figure 49.2. Interpolated maps for red maple. Top row is results of the GS interpolations, and bottom row the results of MSN interpolations. (a) Indicator Kriging/sequential Gaussian Conditional Simulation (IK/sgCS); (b) Forest/nonforest mask from Figure 49.1a overlayed on the IK/sgCS; (c) a single realization of sequential indicator simulation/sequential Gaussian Conditional Simulation (siCS/sgCS); (d) siCS/sgCS summary (60th percentile for red maple); (e) Level 1 MSN (spatially naive); (f) Level 2 MSN (partially spatially informed, with geographical coordinates); (g) Level 3 MSN (fully spatially informed with variogram models); (h) Local average (3 x 3 cell moving window), MSN Level 3.

pare Figure 49.1). The effect of the indicator simulation (siCS/sgCS—Figures 49.2d and 49.3d) is to create a larger masking effect (because more zeros enter the interpolation than in IK/sgCS), but the relative patterns and amounts are comparable. For comparison, a single realization of the indicator simulation approach is shown (Figures 49.2c and 49.3c), reminding us that the "summary" views are simply averages of a process that does honor the fine-scale spatial structure of the data. Single realizations, however, do not provide any uncertainty information—that is only provided by the range of values predicted for each cell after the calculation of many individual realizations.

The MSN results resemble the highly spatially variable map generated by a single GS realization (Figures 49.2e,f,g and 49.3e,f,g). This effect results from the selection of individual most similar neighbors, cell by cell. Ignoring this effect for a moment, there appears to be excellent general correspondence between the MSN patterns and amounts compared with the forest vs. nonforested scene, the FIA samples, and the GS-interpolated maps. This is more obvious for the abundant species (red maple) than for the rarer species (hemlock). To enhance these patterns, we computed local averages for the MSN interpolated maps by applying a moving average (using a 36 km^2 window) (Figures 49.2h and 49.3h). These views show more closely the quality of the correspondence between MSN and GS estimates for both species (at least in predicted locations; the MSN local averages are unrealistically low because of the preponderance of zeros included).

Differences in the MSN formulations are subtle. There appears to be an effect that might be called "cleaning up" of isolated pixels in those areas where the species did not occur on FIA plots, or that are known to be largely nonforested regions (compare Figures 49.2h and 49.3h with Figures 49.2b and

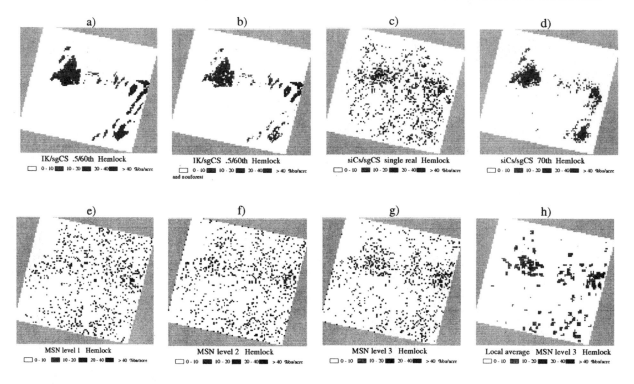

Figure 49.3. Interpolated maps for eastern hemlock. (a) IK/sgCS; (b) forest/nonforest overlayed on IK/sgCS; (c) single realization, siCS/sgCS; (d) siCS/sgCS summarization (70th percentile for hemlock); (e) Level 1 MSN; (f) Level 2 MSN; (g) Level 3 MSN; (h) Local average, MSN Level 3. See Figure 49.2 for explanation of codes.

49.3b—IK/sgCS results overlayed by forest/nonforest). The addition of spatial information yields at least a qualitative improvement in the level 2 and level 3 formulations over level 1. For red maple there are subtle differences between the MSN levels, but substantial improvements for hemlock. We judged this by comparing moving window views for the three formulations (not shown for levels 1 and 2).

These results also show that the MSN results are able to make use of both forest/nonforest and individual species information contained in the TM data. If there were only a forest/nonforest relationship, we would expect the different species to be predicted to occur in roughly the same general locations but vary primarily in magnitude. Instead, the patterns of occurrence and amounts are strongly species-specific. We interpret this to mean that the satellite signature for individual cells was of sufficient resolution to provide some discrimination between species, that this discrimination is captured in the MSN difference function, and ultimately in the interpolations resulting from selection of most similar neighbors.

We have not shown results for the IKfct/sgCS runs that incorporate soft information into the IK step in the form of forest cover type. Those results were disappointing, at best. Predicted species patterns were very different from those obtained with the other methods, and were not in concert with our picture of reality for species' distributions. For now, we have learned how dramatically this layer can affect the GS results, and that quality of the soft information is critical.

Measures of uncertainty for both the siCS/sgCS and MSN level 3 results for hemlock are presented in Figure 49.4. These represent the model error associated with each interpolation method. In the MSN runs, the largest prediction errors correspond to those categories that were undersampled, most notably water and some types of nonforest areas. Measures of uncertainty in the GS interpolations correspond to those areas of high local spatial variability. In both cases, the information is useful, for example, as a direct indication of what areas (in the case of GS) and what types of areas (in the case of MSN) should be most targeted for additional sampling if the final map is to be improved.

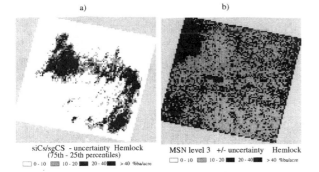

siCS/sgCS - uncertainty Hemlock
(75th - 25th percentiles)
☐ 0 - 10 ▨ 10 - 20 ▧ 20 - 40■ > 40 %ba/acre

MSN level 3 +/- uncertainty   Hemlock
☐ 0 - 10 ▨ 10 - 20 ▧ 20 - 40■ > 40 %ba/acre

Figure 49.4. Uncertainty estimates for hemlock interpolations. (a) siCS/sgCS interquartile ranges; (b) MSN Level 3 (see text for explanation).

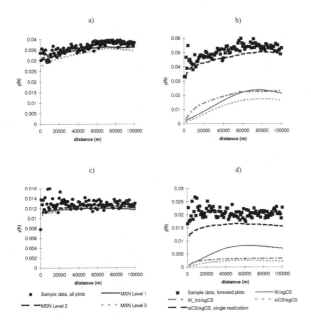

● Sample data, all plots — MSN Level 1

— MSN Level 2   · · · MSN Level 3

■ Sample data, forested plots — IK/sgCS

· · IK_tct/sgCS   · · siCS/sgCS

— siCS/sgCS, single realization

Figure 49.5. Semivariances γ(h) of predicted species importance (% ba/acre) resulting from the different interpolation methods, compared with sample data semivariances. (a) red maple MSN results; (b) red maple GS simulation results; (c) hemlock MSN results; (d) hemlock GS simulation results. See Figure 49.2 for explanation of interpolation method legends.

Spatial Structure—Species Variograms

Semivariograms calculated on the output from the different interpolation methods are shown in Figure 49.5. (Note that because both forested and nonforested FIA plots were included in the current MSN analysis, but nonforested plots only in the GS interpolations,

the MSN and GS results are compared against different sample plot sets.)

All three MSN methods reproduced the sample variograms remarkably well (Figures 49.5a and c). The shape is always closely followed. For both species, the level 2 estimates (using geographic coordinates) produced the closest match to the sill (global variance) of the original sample data. Clearly, use of the TM-derived global data sets to drive the MSN selections results in interpolations that reproduce the spatial structure inherent in the sample data. Because the MSN variograms honor the sample variograms at small and large distances, we conclude that both fine-scale and broad-scale spatial structure of the species composition is captured in the MSN interpolated data sets.

In the GS simulations, single realizations honor the sample variogram fairly well (Figures 49.5b and d). Shape is followed, and sill (global variance) is close to, but lower than, the sample sill. (The random choice of realization for display influences this; single realizations, overall, do honor the sample variogram). The GS summary estimates, because they are smoothed, produce sills much lower than the sample data. This is an expected result.

Attribute Structure—Univariate Frequencies and Joint Species Scatterplots

Figure 49.6 shows the simple cumulative frequency distributions for the species sample data and interpolation results. The MSN results always follow closely the original sample distribution (Figures 49.6a and c). This is an expected result, because the MSN predictions select from the original sample. As long as the selection is made uniformly, then the predicted cumulative distributions will reproduce the sample cumulative distributions. There is little difference in the three MSN formulations in retaining the univariate characteristics of the sample data.

The siCS/sgCS single realization also closely reproduces the original frequency distribution (Figures 49.6b and d), validating the GS simulation model. The simulation summaries (for both siCS/sgCS and IK/sgCS) show how highs and lows in the sample data have been truncated, smoothing the summarized estimates. Still, the smoothed summaries overlap the sample distribution reasonably well. The IK cutoff (0.5 used for both red maple and hemlock) influences the intercept (percent of cells where the species is not predicted to occur) on the cumulative frequency charts. Comparison of the univariate statistics can actually be

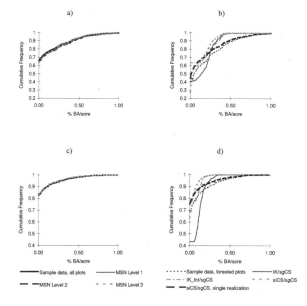

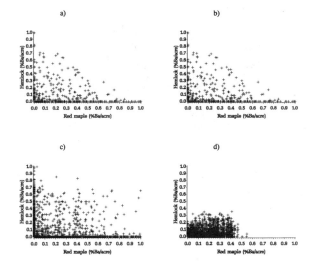

Figure 49.6. Cumulative frequency distributions for the species sample data and interpolation results. (a) red maple MSN results; (b) red maple GS simulation results; (c) hemlock MSN results; (d) hemlock GS simulation results. See Figure 49.2 for explanation of interpolation method legends.

Figure 49.7. Co-occurrence importance values of hemlock and red maple in the sample data, and in the interpolation results. (a) sample data; (b) MSN Level 3 (fully spatially informed); (c) single realization, siCS/sgCS; (d) siCS/sgCS summary.

used in the IK/sgCS run to determine the probability cutoff and percentiles to use in the summarization. This choice depends largely on the relative abundance of the species, and usually it is appropriate to choose lower values for more abundant species and higher values for rarer species.

Figure 49.7 shows the joint occurrence values of hemlock and red maple in the sample data, and in the interpolation results. All three MSN formulations recreated the joint distributions essentially exactly (example, Figure 49.7b). A siCS/sgCS single realization resulted in cells that had higher joint occurrences for red maple and hemlock than observed in the sample data (Figure 49.7c). The summarized GS interpolations (Figure 49.7d) severely truncated the sample joint distributions—no cell was predicted to contain more than about 55% red maple, or 40% hemlock, whereas the sample distribution showed much higher amounts individually for the species, and higher co-occurrences.

CONCLUSIONS

It is clear that continuing to develop methods that incorporate both spatial and attribute structure into interpolated map layers and data sets will result in more accurate information for planning and research objectives (Rossi et al., 1992; Biondi et al., 1994). However, the usefulness of data products and maps that can be obtained from interpolation procedures depends a great deal on the objectives of the user. The basic products can be thought of as a map of predicted species values (where and how much), and associated uncertainty estimates. The user may have additional requirements, such as a minimum mapping unit, so that the view created by one or another interpolation method is more appropriate (e.g., the smoothed view resulting from the GS summary vs. the fine-scale view resulting from GS single realizations or from MSN).

We have seen how the nature and scale of the maps depends on the interpolation method, as well as the data sources used to condition the estimates. For example, factors affecting the distribution of species can occur at a spatial scale smaller than that resolved by the FIA plots. For this reason, the geostatistical interpolations that use only the sample plot data cannot be expected to accurately estimate fine-scale distribution patterns. On the other hand we assume that the broad-scale species variation will be adequately represented using geostatistical techniques. Conversely, MSN relies entirely on the satellite data signature at an interpolated point (in the level 1 model) to estimate species importance, thus the (spatially naive) MSN predictions take advantage of the fine scale data. Bringing in species' variogram information in the level 3 model extends the predictions to include coarse-scale influences, and this is likely the reason we see a decrease in the fragmented pattern of the MSN interpolated maps. Without systematic ground-truthing, we don't actually know which view is more accurate—the smoothed GS estimate, or the fragmented, fine-scale MSN estimate.

Because they have different data requirements, the applicability of the GS and MSN approaches depends on the availability of appropriate data. When spatial structure is present in the data such that it can be modeled with variogram techniques, then geostatistical analysis can be applied to it. The MSN procedure will only yield useful results when a global data source is available that relates well to the characteristics of the attributes on the ground. Fortunately, in our study, the TM data proved to be useful apparently because of the strength of the relationship between the satellite signature at a location on the ground, and the species composition of the forest at the same point. Without that relationship, the MSN estimates would have been mere noise. How much spatial structure the MSN maintains is also highly dependent on the global data layer. The TM data itself provided good results, but these results seemed to be improved by the explicit addition of spatial information to the MSN procedure. These results have enormous implications for the potential utility for the use of MSN with TM-derived data to identify individual species in large landscapes. Up until now the identification of individual species via standard image classification methods has been unreliable. The results here suggest that the combination of remotely sensed imagery and the MSN-GS interpolation method may result in a dramatic improvement in species modeling.

Assessment of input data and interpolation results for attribute and spatial structure will help the user choose the most appropriate interpolation method. This assessment of differences between the interpolated data set and the sample data can be used to qualify what questions the modeled data set can be used to address. Example questions include, "In how much land area does red maple dominate?" vs. "Can red maple be found at a specific location?."

The two approaches studied here, GS and MSN, are different but complementary techniques. While we will continue to work on combining the best features of the approaches to bring the information together into a single output map, we also recognize that used independently, but in concert, they can give us differ-

ent, but equally useful pictures of the resource. For example, "where" MSN estimates hemlock is strictly driven by the satellite signature at a location; "where" GS estimates hemlock is strictly driven by the location of the FIA plots. When these two estimates coincide, the likelihood of hemlock really being there is increased.

Finally, we offer a few ideas that in the near future will be tested and incorporated to improve upon the combined MSN-GS approach. One is the seemingly considerable potential for incorporating results of one procedure, treated as an independent data layer, into the other. For example, a GS-interpolated data set predicting species occurrence and amounts might be included as a global data source in addition to the TM-derived data in an MSN model. Conversely, MSN results could be used to condition an IK layer in the GS simulation procedure.

Although our trial with using forest cover type as soft information in IK/sgCS was unsuccessful, the approach is sound. However, we need better soft information layers that could be provided by remotely sensed data sets that are correlated with species composition (examples are TM-derived data, data from digital elevation models, other species data sets, and of course, data layers estimated from other procedures, such as MSN).

Further work is required to understand the implications of estimated uncertainties to attach to predicted species' values. Systematic comparison between the GS and MSN, and combined procedures is needed. Reliable uncertainty estimates can yield very useful information; for example, in guiding new inventories in undersampled conditions.

REFERENCES

Biondi, F., D.E. Myers, and C.C. Avery. Geostatistically Modelling Stem Size and Increment in an Old-Growth Forest, *Can. J. For. Res.,* 24, pp. 1354–1368, 1994.

Crist, E.P. and R.C. Cicone. A Physically-Based Transformation of Thematic Mapper Data—The TM Tasseled Cap, *IEEE Trans. Geosci. Remote Sensing,* GE-22(3), pp. 256–263, 1984.

Deutsch, C.V. and A.G. Journel. *GSLIB: Geostatistical Software Library and User's Guide,* Oxford University Press, New York, 1992.

Fortin, M., P. Drapeau, and P. Legendre. Spatial Autocorrelation and Sampling Design in Plant Ecology, *Vegetatio,* 83, pp. 209–222, 1989.

Hansen M.H., T. Frieswyk, J.F. Glover, and J.F. Kelly. The Eastwide Forest Inventory Data Base: Users Manual,

USDA Forest Service, North Central Experiment Station, St. Paul, MN, General Technical Report NC-151, 1992.

Isaaks, E.H. and R.M. Srivastava. *An Introduction to Applied Geostatistics,* Oxford University Press, New York, 1989.

Moeur, M., N.L. Crookston, and A.R. Stage. Most Similar Neighbor Analysis: A Tool to Support Ecosystem Management, in *Analysis in support of ecosystem management. Analysis Workshop III, Fort Collins, CO, April 10–13, 1995,* J.E. Thompson (compiler), USDA Forest Service, Ecosystem Management Analysis Center, Washington, DC, 1995, pp. 31–44.

Moeur, M. and A.R. Stage. Most Similar Neighbor: An Improved Sampling Inference Procedure for Natural Resource Planning, *For. Sci.,* 41, pp. 337–359, 1995.

Reed, B., J.F. Brown, D. VanderZee, T.R. Loveland, and J.W. Merchant. Measuring Phenological Variability from Satellite, *J. Vegetation Sci.,* 5, pp. 703–714, 1994.

Riemann Hershey, R. Understanding the Spatial Distribution of Tree Species in Pennsylvania, in *Spatial Accuracy Assessment in Natural Resources and Environmental Sciences, Second International Symposium, May 21–23, 1996,* H.T. Mowrer, R.L. Czaplewski, and R.H. Hamre (Tech. Coords.), USDA Forest Service, Rocky Mountain Forest and Range Experiment Station, Fort Collins, CO. General Technical Report, RM-GTR-277, 1996, pp. 73–82.

Riemann Hershey, R., M.A. Ramirez, and D.A. Drake. Using Geostatistical Techniques to Map the Distribution of Tree Species from Ground Inventory Data, in *Modelling Longitudinal and Spatially Correlated Data: Methods, Applications, and Future Directions,* T.C. Gregoire et al., Eds., Lecture Notes in Statistics 122, Springer-Verlag, New York, 1997, pp. 187–198.

Riemann Hershey, R. and G.C. Reese. Modeling the Spatial Distribution of Tree Species in Maine Using Both Indicator and Sequential Gaussian Conditional Simulation Techniques, Presented at the Thirteenth Annual Meeting of the International Association for Landscape Ecology: Applications of Landscape Ecology in Natural Resource Management, March 17–21, 1998.

Rossi, R.E., P.W. Borth, and J.J. Tollefson. Stochastic Simulation for Characterizing Ecological Spatial Patterns and Appraising Risk, *Ecol. Appl.,* 3, pp. 719–735, 1993.

Rossi, R.E., D.J. Mulla, A.G. Journel, and E.H. Franz. Geostatistical Tools for Modeling and Interpreting Ecological Spatial Dependence, *Ecol. Monogr.,* 62, pp. 277–314, 1992.

CHAPTER **50**

Mapping Forest Site Potential at the Local and Landscape Levels

C.-H. Ung, P.Y. Bernier, R.A. Fournier, and J. Régnière

INTRODUCTION

In 1996, the Canadian Forest Service initiated the ECOLEAP project, a research initiative whose objective is to develop a spatialized model for predicting the net primary productivity of forest ecosystems both at the regional and the local scales for applications in forest management. The spatialization is based on the numerical description of elevation, slope, aspect, slope shape, slope position, and climate, as well as on the use of satellite images to capture data defining the state of the vegetation cover. The present work is an intermediate step in the derivation of this spatialized, process-based simulator of forest productivity, and is aimed at the development of a biophysically sound empirical methodology for mapping forest site potential. Biophysical variables retained are related to climate and topography. In this work, site potential is defined as the maximum height reached by a tree species on a site. In grid-based geographic information systems, the topographic, climatic, and edaphic variables may be available for every grid cell of a forest stand, whereas in current forest inventories this information is available from only a small sample of cells.

MATERIAL AND METHODS

Plot Data

The data used to develop the forest site potential model come from 46,000 temporary plots measured by the Quebec Ministry of Natural Resources as part of its ongoing forest inventory program (Figure 50.1). The plots are arranged along 5-plot 1 km transects.

Transect orientation is selected to maximize along-transect variability in forest types. The transect themselves are distributed in most of the forest types of the province according to a sampling scheme designed to characterize all forest strata. Within each plot, collected information includes variables related to the site, the trees, and the stand. Site-related variables measured are topography (elevation, slope, aspect, topographic type), drainage class, and soil characteristics, particularly texture. Drainage classes vary from 0 (excessive or dry) to 6 (poor or wet). Tree-related variables are the diameter at 1.3 m (dbh) of each stem with a dbh greater than 9 cm within a 400 m²-circular plot, dbh of all stems smaller than 9 cm within a 40 m² plot, and age of three dominant trees per plot as determined by ring count at stump height. Data on the forest stand relate to its canopy cover and its developmental stage. Measurements of those plots was carried out between 1992 and 1997.

Biophysical Variables

Independent or explanatory variables were chosen to explain maximum tree height (H) so that we could capture local site and climate characteristics. The variables were selected based on their functional relationship to tree growth, and on the possibility of computing their value for any point using numerical topographical and soils information, and regional climatic data. The variables had to be as independent from one another as possible.

Soil fertility potential was expressed as the sum of the percent silt and clay fractions (SC). Drainage

431

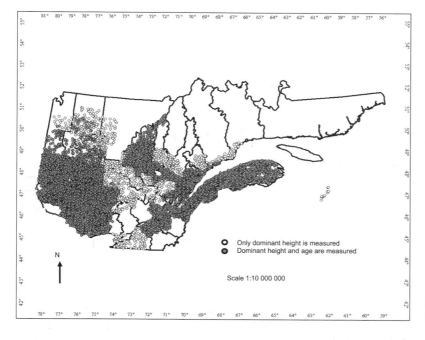

Figure 50.1. Spatial distribution of the temporary sample plots.

classes determined on site were also retained (DR), but the six initial classes were compressed into four classes by grouping classes 1 and 2 (dry) and classes 5 and 6 (wet). Climate-related variables were computed using the algorithms of BioSIM (Régnière and Bolstad, 1994; Régnière, 1996). Given the longitude, latitude, and elevation of a point, this model performs spatial interpolation from 30-year records of 120 permanent meteorological stations (Figure 50.2), as well as adjustments for elevation, slope and aspect, and maritime influences. Variables retained were degree-days above 5°C (DD), the sum of June, July, and August precipitation (PP), an aridity index (AI), and cumulative atmospheric vapor pressure deficit (VPD). All three temperature-related variables (DD, AI, and VPD) were computed from daily maximum and minimum temperatures. Finally, forest type (FT) was also included in the analysis as a class variable. In this first attempt at modeling site potential, we focused on three species, and plots were retained if they contained either sugar maple (*Acer saccharum* Marsh), balsam fir [*Abies balsamea* (L) Mill.], or black spruce [*Picea mariana* (Mill) BSP].

Analysis

Boundary-line analysis was used to determine the relationship between site potential and explanatory variables (Webb, 1972; Chambers et al., 1985). The

rationale for boundary-line analysis is that, in a scatter diagram of dominant tree height against an explanatory variable for all 46,000 plots, the upper limit of the cloud of points represents cases in which that particular variable actually limits growth (Jarvis, 1976). A function adjusted to this upper limit thus represents the relationship between this explanatory variable and dominant tree height.

Model fitting was performed on a subset of the 46,000 plots selected as follows. In each scatter diagram relating dominant height to one of the independent or explanatory variables, only the 5% uppermost points were retained for further analysis. The selected observations for each of the independent variables were combined into a final data set used for model fitting. All five explanatory variables were initially considered as fully independent so that their respective effect on the dependent variable would be multiplicative. When transformed logarithmically, the full model can be written as:

$$LnH = b_0 + b_{FT1}FT1 + b_{FT2}FT2 + b_{DR1}DR1 + \\ b_{DR2}DR2 + b_{DR3}DR3 + b_{DD}LnDD + \\ b_{FT1.DD}FT1*LnDD + b_{FT2.DD}FT2*LnDD + \\ b_{VPD}LnVPD + b_{PP}LnPP + b_{AI}LnAI + \\ b_{SC}LnSC + e_{LnH} \qquad (1)$$

where FT1 and FT2 are dummy variables which, in combination, represent the three possible species re-

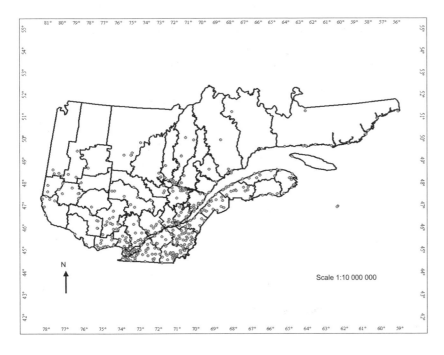

Figure 50.2. Spatial distribution of the long-term climatic stations used in the BIOSIM model.

tained, DR1 to DR3 are also dummy variables which, in combination, represent the four possible classes of drainage. All b's are estimated parameters, and other variables are as described above. The parameters of Equation 1 were estimated by ordinary least squares. When adjusted to the provincewide data set, this model was called the Provincial model. Descriptive statistics for the data used in fitting the Provincial model are given in Table 50.1.

In a second step, the same model was fitted to a much more geographically-constrained data set incorporating only temporary plots that were inside the boundary of a 3,000 km² pilot area. This pilot area is a long band of forest that stretches from the sugar maple stands just west of Quebec City, to the balsam fir stands of the *Réserve Faunique des Laurentides* about 130 km to the north. From north to south, this region is therefore composed essentially of three climatic domains: (1) balsam fir–white birch domain, ecological region 8f of Thibault (1985), with degree-day above 5°C of 890 to 1,000, (2) sugar maple–yellow birch domain, ecological region 3g with degree-day above 5°C of 1,220 to 1,550, and (3) sugar maple–American basswood domain, ecological region 2c with degree-day above 5°C of 1,660 to 1,780. Because black spruce is rare in the pilot region, only balsam fir and sugar maple were considered in model estimation. Descriptive statistics for the data used in fitting the Pilot Region model are given in Table 50.2.

Finally, the regional applicability of the Provincial model was tested by comparing predicted maximum tree heights with those obtained using the Pilot Region model.

RESULTS AND DISCUSSION

Ideally, all explanatory variables should have been independent. However, all four climatic variables proved to be highly correlated. Model fitting therefore required a few iterations to deal with these co-linearities. After fitting the full Provincial model, parameters for the LnPP, LnAI, and LnSC variables were found to be nonsignificant. Removing these variables yielded a reduced model which was fitted to the provincial data set. In this reduced model, the parameter associated with the LnVPD variable also proved to be nonsignificant. This variable was therefore removed and a new reduced model was fitted to the data. Finally, the parameter associated with DR3 (drainage class 4) also turned out to be nonsignificant. However, DR3 was retained because the parameters associated with the other two drainage-related dummy variables (DR1 and DR2) were significant. The final Provincial model was therefore:

$$LnH = 0.966587 + 4.254747*FT1 + 2.683395*FT2$$
$$- 0.013622*DR1 - 0.009983*DR2 +$$

Table 50.1. Descriptive Statistics of the Data Used in the Adjustment of the Provincial Model.

Variable	Mean	Standard Deviation	Minimum	Maximum
Black spruce: 732 observations				
Maximum height (dm)	193.2	14.1	169.0	307.5
Degree-day (°C)	1208.5	189.1	654.9	1739.8
Deficit of vapor pressure (mbar)	1273.4	162.7	457.0	1609.4
Precipitation (mm)	308.0	25.9	244.9	435.0
Aridity index (no unit)	0.10	0.33	0.00	2.05
Sum of clay and silt percentage	41.4	22.9	8.0	93.0
Balsam fir: 874 observations				
Maximum height (dm)	188.7	14.2	166.3	255.0
Degree-day (°C)	1301.3	239.9	642.7	1816.3
Deficit of vapor pressure (mbar)	1253.2	214.2	438.4	1666.9
Precipitation (mm)	306.0	40.8	241.0	435.0
Aridity index (no unit)	0.14	0.41	0.00	2.70
Sum of clay and silt percentage	53.7	22.6	8.0	93.0
Sugar maple: 356 observations				
Maximum height (dm)	240.6	21.7	150.0	327.7
Degree-day (°C)	1474.8	167.4	760.4	1861.7
Deficit of vapor pressure (mbar)	1361.5	164.2	469.4	1671.7
Precipitation (mm)	297.9	39.5	241.0	429.7
Aridity index (no unit)	0.36	0.61	0.00	2.51
Sum of clay and silt percentage	56.5	20.8	8.0	93.0

Table 50.2. Descriptive Statistics of the Data Used in the Adjustment of the Pilot Region Model.

Variable	Mean	Standard Deviation	Minimum	Maximum
Balsam fir: 29 observations				
Maximum height (dm)	181.2	22.5	151.7	241.0
Degree-day (°C)	1027.1	198.8	816.2	1494.8
Deficit of vapor pressure (mbar)	1159.2	129.0	913.3	1528.8
Precipitation (mm)	406.4	37.5	285.4	429.7
Aridity index (no unit)	0.00	0.00	0.00	0.00
Sum of clay and silt percentage	47.1	22.0	18.0	80.0
Sugar maple: 7 observations				
Maximum height (dm)	227.0	24.2	174.0	245.0
Degree-day (°C)	1371.6	141.2	1109.1	1522.9
Deficit of vapor pressure (mbar)	1370.4	141.1	1204.6	1533.3
Precipitation (mm)	400.4	20.4	380.3	429.7
Aridity index (no unit)	0.00	0.00	0.00	0.00
Sum of clay and silt percentage	49.0	11.2	37.0	58.0

$$0.000936*DR3 + 0.624831*LnDD - 0.615162*FT1*LnDD - 0.397782*FT2*lnDD \quad (2)$$

The final model presented in Equation 2 was also fitted to the Pilot Region data set. Parameter values and some underlying statistics for both models are presented in Table 50.3.

Equation 2 emphasizes the key role of degree-days in explaining maximum tree height, both as a main effect, and in interaction with tree species. This vari-

Table 50.3. Values and Statistics of the Parameters of the Provincial and the Pilot Region Models.

	Parameter Estimates	Standard Error	T Test Probability
Provincial model: 1356 observations; mean square error = 0.00447; adjusted R-square = 0.6698			
Intercept	0.966587	0.235075	0.0001
FT1	4.254747	0.277540	0.0001
FT2	2.683395	0.254913	0.0001
DR1	−0.013622	0.006100	0.0256
DR2	−0.009983	0.005039	0.0478
DR3	0.0000939	0.011533	0.9353
LnDD	0.624831	0.032384	0.0001
FT1*LnDD	−0.615162	0.038540	0.0001
FT2*LnDD	−0.397782	0.035254	0.0001
Pilot Region model: 36 observations; mean square error = 0.00968; adjusted R-square = 0.7313			
Intercept	−1.375271	2.701010	0.6139
FT1	11.525321	3.534985	0.0025
FT2	4.463306	2.788726	0.1187
DR1	−0.021194	0.054774	0.7012
DR2	0.006463	0.042860	0.8810
LnDD	0.940461	0.373445	0.0167
FT1*LnDD	−1.700434	0.496878	0.0016
FT2*LnDD	−0.635874	0.387268	0.1098

[a] DR3 is absent in the pilot area.

able represents large-scale variability in tree height across the data set domain. Its interaction with species indicates that the height of each of the three species is influenced differently by differences in degree-days. The drainage effect is also important by locally modulating the effect of degree-days. These inferences drawn from the model confirm much of the ecophysiological work on the forest productivity (Pastor et al., 1984) in which effects of drainage and interactions between energy inputs (degree-days) and tree species are taken as basic premises. Vapor pressure deficit and aridity index were not significant because of their colinearity with degree-days. Precipitation was not significant simply because it is not a limiting factor to height growth in Eastern Canada. The sum of clay and silt percentages was not significant, indicating that textural contribution to soil fertility is not a major factor in explaining height growth.

Results of the comparison of predictions made with the Provincial model for the sites in the Pilot Region with maximum tree heights predicted by the Pilot Region model are shown in Figure 50.3. Fit around the 1:1 line was good for both balsam fir and sugar maple, indicating that provincewide relationships derived us-

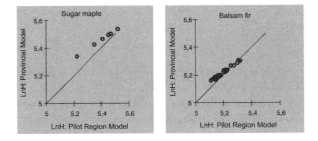

Figure 50.3. Maximum heights predicted by the Provincial model against maximum heights predicted by the Pilot Region model for sugar maple and balsam fir.

ing this method produce valid local estimates. The small bias for sugar maple possibly stems from the fact that drainage classes are less sampled than for balsam fir in the available data.

The procedure presented above was our first attempt at the derivation of a site productivity model based on biophysical variables. Its derivation revealed flaws that were not apparent at the outset, and convinced us to carry on using a different approach. In the next iteration of this procedure, the dependent variable, maximum height, will be replaced by the classical logistic function relating dominant height and age. This change will enable us to use all plots in our analysis, and not the upper 5% as is the case in the present analysis. For each species, the relationship between dominant height H and age A can be described by the following function:

$$H = c_1 \exp(c_2/(1 - c_3)A) + e_H \qquad (3)$$

Where c_1 is the maximum height achievable by that species, and c_2 and c_3 are adjusted coefficients. Unpublished work on that relationship has shown that c_2 is related primarily to broad ecological regions, whereas c_3 is related to local factors such as drainage. Consequently, these two parameters will be related to our biophysical variables in the following manner:

$$c_2 = c_{20} + c_{DD}DD + c_{VPD}VPD + c_{PP}PP + c_{AI}AI + c_{SC}SC \qquad (4)$$

and

$$c_3 = c_{30} + c_{DR1}DR1 + c_{DR2}DR2 + c_{DR3}DR3 \qquad (5)$$

The other development is to improve representation of the drainage class. Drainage is an important factor because it controls dominant height at the local scale. However, the use of the interpreted drainage class binds us to mapped values of drainage classes derived from the subjective interpretation of aerial photographs. Ideally, drainage should be derived from fixed terrain attributes, possibly in interaction with climatic variables such as the aridity index. Work is currently under way to derive such a potential drainage index.

CONCLUSION

By using data from the Quebec forest inventory, we quantified the significant effects of degree-days and drainage on maximum height reached by balsam fir, sugar maple and black spruce. After being fitted to all of the Quebec forest stands in our database, the maximum height prediction model was applied without large bias to a restricted Pilot Region. With boundary-line analysis, only 5% of the available data could be used. It is possible to improve site potential predic-

tions by incorporating in our analysis the functional relationship between dominant height and tree age, a relationship better known as the site index. This would enable the use of all the available data. Ongoing development of the procedure should therefore yield a site index model based on biophysical variables. With further refinements, we hope to replace drainage class by a quantitative drainage index derived from topographical and climate information. Finally, an accurate estimate of model error is a prerequisite for its eventual use in forest inventory or management. This problem will be addressed by quantifying both errors linked to site-specific predictions, and errors linked to spatial interpolation of underlying variables.

ACKNOWLEDGMENTS

We thank the Quebec Ministry of Natural Resources for free access to the forest inventory. The invaluable support of Mrs. Stéphane Camiré and Claude Lapointe in data analysis is gratefully acknowledged.

REFERENCES

Chambers, J.L., T.M. Hinckley, G.S. Cox, and R.G. Aslin. Boundary-Line Analysis and Models of Leaf Conductance for Four Oak-Hickory Forest Species, *For. Sci.*, 31(2), pp. 437–450, 1985.

Jarvis, P.G. The Interpretation of the Variations in Leaf Water Potential and Stomatal Conductance Found in Canopies in the Field, *Philos. Trans. R. Soc. London, Ser. B*, 273, pp. 593–610, 1976.

Pastor, J., J.D. Aber, and J.M. Melillo. Biomass Prediction Using Generalized Allometric Regressions for Some Northeastern Tree Species. *For. Ecol. Manage.*, 7, pp. 265–274, 1984.

Régnière, J. A Generalized Approach to Landscape-Wide Seasonal Forecasting with Temperature-Driven Simulation Models, *Environ. Entomol.*, 25, pp. 869–881, 1996.

Régnière, J. and P. Bolstad. Statistical Simulation of Daily Air Temperature Patterns in Eastern North America to Forecast Seasonal Events in Insect Pest Management, *Environ. Entomol.*, 23, pp. 1368–1380, 1994.

Thibault, M. Les Régions Écologiques du Québec Méridional, *Ministère de l'Énergie et des Ressources*, Québec, 1 map, 1985.

Webb, R.A. Use of Boundary Line in the Analysis of Biological Data, *J. Hortic. Sci.*, 97, pp. 309–319, 1972.

CHAPTER 51

The Role of Error Propagation for Integrating Multisource Data within Spatial Models: The Case of the DRASTIC Groundwater Vulnerability Model

M. Fortin, G. Edwards, and K.P.B. Thomson

INTRODUCTION

The need to produce maps of environmentally sensitive parameters has increased dramatically over the past decade, due to changes in public perception and the new social policies to which these gave rise. Hence Canadian government goals of sustainable development are currently driving interest in the quantification of a variety of indicators which can be used to measure progress in this area. Furthermore, local governments, at the provincial and municipal levels, increasingly desire access to model outputs which predict vulnerability of the environment to a variety of dangers. These outputs are used to determine when and how to protect vulnerable regions. Such an area of concern concerns the potential hazard of pollutants such as agricultural fertilizers to the water table. This is a complex problem, requiring understanding of the geological properties of aquifers below the level of the soil, the vegetation conditions above the soil, the topography of the region and its partitioning into watersheds, and other properties of a region.

Since the early 1980s a model developed in the United States (Aller et al., 1987) has found widespread use throughout North America, in a variety of adapted forms, as a basis for policy making on the vulnerability of the water table. This is the DRASTIC model, named after the seven parameters on which it is based (Depth to water, Recharge, Aquifer medium, Slope, Texture of soil, Impact of the vadose zone, and Conductivity).

However, the DRASTIC model, like many others which form the basis of policy decisions, may be used in situations where the data are scant or where the model has not been fully tested. Hence there are concerns over the reliability of its predictions, and the means to characterize and measure this reliability (Merchant, 1994). Furthermore, the model as it was originally designed does not exploit the spatial structure of the landscape, except in a very crude way. There is increasing interest in adapting the model to recognize the spatial characteristics of the landscape explicitly, or, alternatively, to develop new models which are better suited to such explicit spatial representation. In either case, there is a need to study how error is distributed spatially, how this can be characterized, and how it can be propagated.

Indeed, the need to characterize errors in spatial data for a variety of problems and to propagate these errors through the various transformations that data undergo has become chronic (Lunetta et al., 1991). Error propagation is, however, a complex problem. Data are usually correlated in various ways, and methods for predicting output errors must either incorporate understanding of these correlations or find means of sidestepping them. The DRASTIC model uses a range of parameters which in many cases are known to be correlated. Furthermore, the model uses data which are obtained at different scales and hence are characterized by different regimes of spatial autocorrelation.

Because of the complexity of correlations among different data, in general, analytical techniques have largely been abandoned in favor of Monte Carlo or stochastic simulation techniques. These latter consist of varying input data within the known range of error and under different correlation scenarios, and characterizing the range of variation this produces in the output data. Evaluating correlations, and especially spatial autocorrelation, is difficult at best. Hence many researchers advocate the use of simulations of different correlation scenarios, and a comparison of these scenarios with the data of interest in order to understand the correlation structure (Goodchild et al., 1992).

Monte Carlo techniques have been successfully applied to data sets obtained under similar circumstances, which were then used as input to a variety of modeling procedures. However, Monte Carlo techniques typically require studying several hundred, if not thousands of generated scenarios in order to ensure stability of the resulting distributions. For large regions with large numbers of input variables, the computing times involved may be significant. Although computer speeds are still increasing, the requirements of computer processing associated with Monte Carlo methods are also increasing and there is some debate over the ultimate practicality of Monte Carlo techniques. Furthermore, Monte Carlo simulations do not always lead to increased understanding of the data, as do analytical methods. Often the details of the correlation structure are not revealed by Monte Carlo methods, only their impact on the predicted maps. While the latter is often what is sought, these approaches do not lead easily to a capability to generalize beyond a given situation.

Similarly, there is a real need to determine the relative importance of remotely sensed data as a data source for models used in decision support. Remotely sensed data is often costly to obtain and, unfortunately, is often oversold as a product. The development of evaluation criteria for spatiotemporal data can be used to determine the real gains obtained when remotely sensed data are introduced into a data set and hence allow one to assess the real benefits of this data source. This should both help clients determine whether remotely sensed data is appropriate for a given problem and also help characterize data from new sensors as they come on line.

Hence, when input data sets are complex or are characterized by different error models, there is some interest in using analytical methods in order to better understand the way error propagates from one data set to the next and how different error models interact, even when such methods are incomplete. In this chapter, we present the results of using such an analytical model for propagating error in a complex prediction model which relies on three different types of source data, field samples, photo-interpretations, and remotely

sensed data. The analytical model used, although incomplete, allows us to identify both areas where uncertainty is high and source data sets where accuracy could be improved. The relative importance of using remotely sensed satellite data with respect to interpreted airborne data may also be evaluated using this approach.

THE DRASTIC MODEL

The DRASTIC model consists of the linear combination of seven indices (Depth to water, Recharge, Aquifer medium, Slope, Texture of soil, Impact of the vadose zone, and Conductivity) to form a final index of vulnerability:

$$\text{Index} = D_w\,D_i + R_w\,R_i + A_w\,A_i + S_w\,S_i + T_w\,T_i + I_w\,I_i + C_w\,C_i \qquad (1)$$

where X_w are the weights and X_i the indices associated with each DRASTIC variable. The source data used to produce the intermediate indices include meteorological data from stations, drill sites, soil maps, geological maps, classified Landsat imagery, and a digital elevation model.

The DRASTIC model is not a dynamic process model, but it is a fairly complex model nonetheless. Twenty-seven different data sources serve as input to the model. For the recharge variable alone, sixteen different sources of data are required. In order to characterize the uncertainty in these different kinds of data, we developed and used three different error models. Because all three models are pertinent in the determination of the Recharge, we limited our focus to this variable alone within the larger model.

The DRASTIC model was developed in order to produce a single value for the vulnerability of a fairly large region (roughly 40 hectares), which was treated as a homogeneous hydrogeological unit (Aller et al., 1987). Information was obtained from transparent map overlays and aggregated to this scale. In our study, we proceed by producing indices for cells of 25 meters on a side (0.06 hectares), hence at a scale far below that for which the model was designed. However, even in its original use, data at such microscales were aggregated or averaged over the hydrogeological unit. Such aggregation or averaging occurred without attempting to solve the correlation structure of the variables under consideration; in fact, it was generally carried out by adopting a subjective interpretation of the set of maps. Furthermore, the source data used to drive the model were produced at many different scales and aggregated to the target scale, again without due consideration for the correlation structure. Hence although our adaptation of the model is at a much finer scale than it was designed to address, we believe that exploring the characterization of the data and its error structures at the scales at which they are produced is necessary in order to understand the error at the scales at which the model outputs are produced and used.

The recharge consists of the water which infiltrates through the soil into the water table underneath. We used the following equation to characterize the recharge:

$$R = (1 - E/P) * (P - R_u) \qquad (2)$$

where E is the evapotranspiration, P is the precipitation, and R_u is the runoff. The equation is an adaptation of the usual simpler form:

$$R = P - E - R_u \qquad (3)$$

The adaptation was required in order to handle two different spatial scales present in the data. The evapotranspiration and precipitation estimates are both obtained at the locations of the meteorological stations, whereas the runoff is calculated according to the local conditions of a given site. As a result, the use of Equation 3 led to inconsistent results (e.g., the runoff was stronger than the difference between precipitation and evapotranspiration, leading to a negative value of the recharge). The modification shown in Equation 2 was obtained by reasoning that the regional average fraction of water precipated which remains behind after evapotranspiration [i.e., (1–E/P)], could be used to convert whatever remains of the precipitation after runoff ($P–R_u$) into recharge.

Both the precipitation and evapotranspiration, as mentioned above, are estimated at the locations of meteorological stations. In order to obtain spatially local estimates of these parameters, it is necessary to interpolate them between the meteorological stations. Such interpolation is rife with difficulties, since precipitation is characterized by a great deal of local behavior. However, neglecting such local structure, we used straightforward universal kriging to obtain both an interpolated surface and an "error" surface. Following the principles of kriging, this error surface is based on the assumption that precipitation and evapotranspiration are regionalized variables — that is, that they are the result of processes operating at a regional scale. The "error" then consists of the expected dispersion that measurements would take around the estimated value, given the observed variation in the variable itself across several meteorological stations and the estimated scale of spatial autocorrelation characteristic of the underlying process. However, the error produced by this means is more a prediction on the repeatability of measurements than it is a characterization of the deviation of the estimated values from some "true" value.

For evapotranspiration, the error estimate derived from kriging is likely to be reasonable for the most part, since evapotranspiration is controlled by temperature, which is a regionalized variable. For precipitation, as noted above, this is not necessarily the case. Hence the true uncertainty is very probably larger than the one that kriging would indicate.

The runoff is determined according to the following relation from Monfet (1979):

$$R_u = [P - 0.2 * (1000/CN - 10)]^2 / $$
$$[P + 0.8 * (1000/CN - 10)] \qquad (4)$$

where CN is the hydrological curve number. The latter is a concept in widespread use in the hydrological community—it consists of a series of curves which characterize the relationship between precipitation and runoff according to a variety of conditions, including soil type, slope, land use, and drainage conditions. Soil type is a type of interpreted map, slope an interpolated value, drainage is associated with the soil texture map and hence is likewise interpreted, while land use is usually the result of classification of remotely sensed imagery or from photointerpreted maps.

THE THREE ERROR MODELS AND THEIR INTEGRATION

The different sources of input data can be classified into three broad spatial categories: photo-interpreted boundaries maps (lineaments; soil types, textures and drainage conditions; aquifer locations and types of aquifer), remote sensing classifications (land use), and interpolated field or point data (meteorological parameters; water levels; flow rates; soil permeability; elevations and slopes; captivity status of the water table; etc.). For each of these kinds of data, we developed a separate error model:

1. A boundary error model based on the model developed by Aubert et al. (1994) for photo-

interpreted forest strata. Two types of interpreted data were used as input to the spatialized version of the DRASTIC model, soil maps and geological maps. Although these interpretation products have not been as carefully evaluated as have vegetation maps, we assumed that the same principles discovered for the latter apply to the former. Because our study was focused on the error propagation issues rather than the error characterization, we replaced the careful study of multiple interpretations of the same data by different interpreters, with interviews with interpreters. During interviews, interpreters estimated the spatial error of geological boundaries to be of the order of 200 meters (i.e., between 100 and 300 meters) for soils, and 500 meters for aquifers. The boundaries were simulated using fuzzy membership coefficients ranging from 0 to 1 across the width of the boundary.

2. A contextual pixel measure of local confusion in remotely sensed imagery. We desired to obtain a spatially distributed measure of class confusion, rather than the global measures usually adopted in remote sensing. Hence we assumed a pixel-based classification was globally correct, and examined the number of pixels classified in each category within a mobile window centered on each pixel. This measure of local confusion clearly underestimates the error in remotely sensed classification, in that any systematic misclassifications will not be modeled. However, the accuracy of the land-use classification used in the study was estimated to be at least 80%, and hence any systematic errors must be lower than 20%. Given the "salt-and-pepper" effect of the pixel classifier, we estimate that systematic errors are no greater than 10%. Furthermore, systematic effects, once known, might be applied to modify the probability estimates obtained from the frequency counts in local windows and hence obtain a better representation of local classification error.

3. A kriging error model was developed for the interpolated variables. We made no attempt to optimize or fine-tune the kriging model beyond getting a functional interpolation. In any case, the error model used is based on the variability of the sample values under the assumption of spatial autocorrelation expressed by the semivariogram. For many of the parameters of interest, this measure of error is likely to be too low. For example, precipitation is known to be characterized by microscale behaviors that are not well characterized by sparsely distributed meteorological stations, and hence interpolation of this variable will need to yield a very useful measure of distributed error, except for long-term average precipitations.

Because three quite different models were used to model uncertainty, a method or methods for combining them into a single propagation model is required. Furthermore, each model uses a different principle of error characterization: the boundary model uses fuzzy membership values, the contextual pixel model uses frequency counts which can be interpreted as probabilities, and the kriging error model produces spatially distributed variance estimates. If populations for each variable are assumed to be distributed normally, then it is possible to convert from variance estimates to probability and back again. Furthermore, the boundary model has been tested for normality of the distributions of interpretation boundary locations around a median interpretation, albeit in another context (Aubert et al., 1998). Hence it is possible to reinterpret the membership values used in the fuzzy boundaries as probability values, because the underlying assumptions have been tested previously. Hence we performed integration of the three models by working with both probabilities and variances, and converting from one to the other as required.

Hence, for example, the determination of the curve number CN requires the use of a "lookup table" which specifies curve number as a function of values in each of four controlling variables (land use, drainage class, soil type, and slope). The propagation of errors through the lookup table was handled in the following manner:

- for each location, we estimated the values of drainage, slope, land use, and texture;
- variance estimates on interpolated values (slope) were converted to probabilities of confusion;
- boundary error was likewise converted to a pairwise confusion probability (texture, drainage and soil type);

- contextual confusion obtained for the land-use classification was expressed as a probability;
- these probabilities were then used to estimate the mix of possible output values of the CN;
- the distribution of possible output values was then evaluated and converted back to a variance estimate.

Error propagation beyond this point was handled via standard analytical error propagation (Bevington, 1969); that is, using a first-order Taylor expansion. If $y = f(u,v)$, then

$$Var(y) = (df/du)^2 * Var(u) + (df/dv)^2 * Var(v) + 2(df/du)(df/dv) * Covar(u,v) \qquad (5)$$

Expression 5 can be generalized to an arbitrary number of variables. For the purposes of this (preliminary) study, we neglected the covariance terms. All manipulations involved in the determination of the recharge and its uncertainty were carried out within ArcInfo™ using Grid (hence a raster environment). The number of manipulations was already fairly large—including the covariance terms would have required an order of magnitude increase in the number of operations performed and the extra effort required to evaluate the covariance between the coverages. The final product consists of a variance surface (Figure 51.1) matched to the recharge index (Figure 51.2). Two variance surfaces were created, one for a conservative choice of uncertainty values, and one for a more liberal choice (e.g., the width of the boundary zones and the size of the contextual window). These surfaces may be interpreted as a measure of the repeatability of the results. Furthermore, we also varied those input variables which exhibit nonlinear behavior in the model. The most important of these was the precipitation.

RESULTS OF THE ANALYSIS OF THE INTEGRATED ERROR MODEL

The recharge index (like the final vulnerability index) ranges in value from zero (no change in the water table) to nine (large change in the water table). In Figure 51.2, the large gray surface corresponds to a saturated recharge index, and this includes more than 70% of the region studied. Furthermore, the recharge index never drops below a value of 6, even in the dark regions. The uncertainty associated with this index is of the order of one index value (1.1, actually) over the

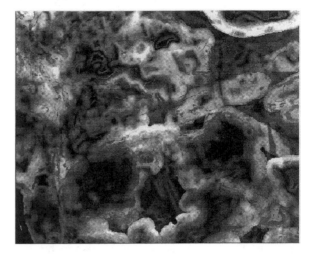

Figure 51.1. The uncertainty map in units of standard deviation (white means high standard deviation, black means low standard deviation).

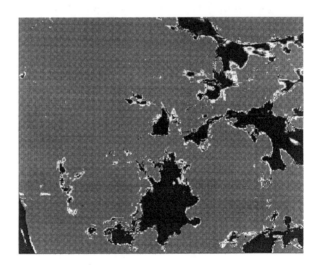

Figure 51.2. The recharge index (gray means high recharge, white means moderate recharge, black means low recharge).

majority of the region studied. Hence the model suggests that the recharge rate throughout this region is relatively high (which will lead to a high vulnerability of the water table to pollution), and that the uncertainty concerning these values is moderately low.

Sensitivity studies indicate that the mean uncertainty drops to about 0.9 index values when boundary error in the photo-interpreted maps is lower (e.g., 100 meters for soils), but that it changes much less (± 0.1 index values) when the contextual window size is varied from

3×3 to 19×19. However, lowering the precipitation beyond a critical value changes the behavior of the model, leading to quite different values of the recharge uncertainty surface (much higher uncertainty). Hence the uncertainty model can be strongly affected by the choice of certain parameters. However, precipitation values in this lower range lead to model behavior which is not physically reasonable, and hence these results should be treated with some caution.

Additional tests were carried out in order to identify the relative contributions of the different types of input data to the final error results. This was achieved by setting errors to zero in all but the model of interest, and then recomputing the output error surface. The resulting partial error surface corresponds to the contribution from a particular error model. The partial error surface for each model type was then subsampled and correlations between this and the full error surface were evaluated. These tests were performed for both values of precipitation. The results are quite revealing.

The boundary model appears to contribute the most strongly to the final error surface. A correlation value (R^2) of 0.72 between the partial boundary error surface and the full error surface was obtained, with an average standard deviation of 0.78. This corresponds to 59% of the variance. The correlation between the partial error surface for the contextual classification model of the remotely sensed data and the full error surface was found to be 0.36, corresponding to an average standard deviation of 0.61 and to 36% of the variance. Finally, the correlation of the interpolated partial error surface with the full error surface was negligable, with an average standard deviation of 0.28 and less than 5% contribution to the total variance.

This is interesting, because it leads one to the conclusion that increased use of remotely sensed image analysis may result in reduced errors. If appropriate image analysis techniques can be developed to replace the use of photo-interpretation, up to 60% of the output variance may be removed. This is a significant gain. Furthermore, cost/benefit analyses may be performed to determine how much precision can be achieved for how much cost, and hence it may become possible to quantify the impact of the use of remotely sensed data on models used for policy making.

DISCUSSION AND CONCLUSIONS

The uncertainty model presented in the previous section constitutes a first attempt to evaluate the spatial distribution of uncertainty in the context of policy making concerning water table vulnerability. The model presented is characterized by a number of limitations which are of a fairly general nature in this kind of model. First of all, the focus throughout this project was on characterizing the uncertainty rather than on obtaining the most reliable vulnerability index (to some extent, however, the latter had been done in previous work—see Pouliot et al., 1996). On several occasions, fairly arbitrary choices were made as to the way input data were used in order to facilitate the error propagation. Hence, for example, no attempt was made to improve the precipitation interpolation estimates using additional variables known to correlate with precipitation, such as elevation, or to use more recent and more sophisticated kriging methods. Secondly, as stated, we did not attempt to account for covariance between variables during the propagation. There are likely to be significant levels of covariance between several of the input coverages—soil type and drainage, for example, are partially correlated. The inclusion of the covariance terms is likely to increase the uncertainty estimates. An evaluation of covariance is clearly a critical next step.

Furthermore, it is important to attempt to determine the reliability of the coefficients used in the model itself. The reliability of the DRASTIC model in a nonspatial context has been examined previously, however (although there is some controversy over the applicability of the model under different conditions)—hence it is the reliability of the spatial distribution of the DRASTIC index which must be tested (Merchant, 1994).

The geometric fidelity of the base maps used, the topography, and the land-use classification are expected to be reasonably good compared to the uncertainties present in the data—there is no need to question this aspect of the problem. The final consideration which we have not addressed is the question of spatial autocorrelation. Clearly, the uncertainty or error component associated with each of the input data coverages will be characterized by its own level of spatial autocorrelation. Furthermore, the average user of the results of a model such as DRASTIC is not so much interested in the local uncertainty associated with each cell, but in the combined uncertainty associated with a region. Because of sampling properties, the relative uncertainty of a variable evaluation across a region will be smaller than that found locally, but the extent to which it is smaller will depend on the spatial autocorrelation structure of the error component. The uncertainty map shown in Figure 51.1 is useful for determining sources of uncertainty and for determining where uncertainty is lower or higher, but not yet fully functional for determining the absolute level of uncertainty across arbitrary regions. The result of combining several data coverages via weighting factors, each with their own uncertainty maps and spatial structure, may be characterized by quite a different level of spatial autocorrelation, and it is not easy to predict the spatial structure of the autocorrelation in the final uncertainty map based on the input maps. Naively, we would expect the spatial autocorrelation structure to be more fragmented in the vulnerability map than in any of the source data coverages, but it would be useful to test this assumption. We do not know of any direct work dealing with the propagation of spatial autocorrelation during overlay, except via Monte Carlo simulations. We did not attempt to characterize this aspect of the data in any way.

Finally, it is worth reiterating the consequences of managing the uncertainty in the context of government policy-making. At the present time, aside from a recognition that error and uncertainty need to be managed, there is some reluctance to do so because of the increased analysis required and fears that the results of such analysis will show poor reliability and hence invalidate early efforts at environmental control. Our analysis indicates, however, that studying uncertainty leads to increased understanding of the roles played by different types of data, without necessarily invalidating the predictions of the model itself. Furthermore, it becomes possible to evaluate the relative costs and benefits of using different types of data. Although a complex problem, the management of uncertainty is beginning to show clear benefits to policy-makers.

ACKNOWLEDGMENTS

This research has been funded by the Canadian Natural Sciences and Engineering Research Council (NSERC) and Québec's Association des industries forestières (AIFQ) through the establishment of an Industrial Chair in Geomatics Applied to Forestry, and also from a provincial FCAR team grant awarded to the authors for work on spatial uncertainty.

REFERENCES

Aller, L., T. Bennett, J.H. Lehr, R.J. Petty, and K. Hackett. *DRASTIC: A Standardized System for Evaluating Ground Water Pollution Potential Using Hydrogeologic Settings,* National Water Well Association, EPA/600/2-85/018, U.S. Environmental Protection Agency, Ada, OK, 1987.

Aubert, E. Quantification de l'Incertitude Spatiale en Photo-Interprétation Forestière à l'Aide d'un SIG pour le Suivi Spatio-Temporel des Peuplements, *M.Sc. Thesis*, Université Laval, 1995.

Aubert, E., G. Edwards, and K.E. Lowell. A Raster Method for Determining Boundary Change in the Presence of Uncertainty, submitted to *International Journal of GIS*, 1998.

Aubert, E., G. Edwards, and K.E. Lowell. Quantification des Erreurs de Frontière en Photo-Interprétation Forestière pour le Suivi Spatio-Temporel des Peuplements, *Proceedings of the Canadian Conference on GIS*, Ottawa, 4–10 June, 1994, pp. 195–205.

Bevington, P.R. *Data Reduction and Error Analysis for the Physical Sciences,* McGraw-Hill Company, New York, 1969.

Edwards, G. Aggregation and Disaggregation of Fuzzy Polygons for Spatio-Temporal Modelling, Proceedings: *International Workshop on Advanced Geographic Data Modelling*, Delft, Netherlands, September 1994, pp. 141–154.

Edwards, G. and K.E. Lowell. Modeling Boundary Uncertainty in Photointerpretation, *Photogrammetric Eng. Remote Sensing,* 62(4), pp. 377–391, 1996.

Goodchild, M.F., S. Guoquing, and Y. Shiren. Development and Test of an Error Model for Categorical Data, *Int. J. GIS,* 6(2), pp. 87–104, 1992.

Lunetta, R.S., R.G. Congalton, L.K. Fenstermaker, J.R. Jensen, K.C. McGwire, and L.R. Tinney. Remote Sensing and Geographic Information System Data Integration: Error Sources and Research Issues, *Photogrammetric Eng. Remote Sensing,* 57, pp. 677–687, 1991.

Merchant, J.W. GIS-Based Groundwater Pollution Hazard Assessment: A Critical Review of the DRASTIC Model, *Photogrammetric Eng. Remote Sensing,* 60(9), pp. 1117–1127, 1994.

Monfet, J. *Evaluation du Coefficient de Ruissellement à l'aide de la Méthode SCS Modifiée,* Ministère des richesses naturelles du Québec, Québec, rapport HP-51, 1979.

Pouliot, J., K.P.B. Thomson, G. Edwards, and M. Rheault. Exploitation des Images ERS-1 pour la Cartographie de la Vulnérabilité de la Nappe Souterraine. *Actes de la Conférence sur l'Extraction de Paramètres Biogéophysiques à partir de Données ROS pour les Applications Terrestres.* Toulouse, 10–13 October, 1996.

Index